Quaternary of
South-West England

THE GEOLOGICAL CONSERVATION REVIEW SERIES

The comparatively small land area of Great Britain contains an unrivalled sequence of rocks, mineral and fossil deposits, and a variety of landforms which encompass much of the Earth's long history. Well-documented ancient volcanic episodes, famous fossil sites, and sedimentary rock sections, used internationally as comparative standards, have given these islands an importance out of all proportion to their size. The long sequences of strata and their organic and inorganic contents have been studied by generations of leading geologists, giving Britain a unique status in the development of the science. Many of the divisions of geological time used throughout the world are named after British sites or areas; for instance the Cambrian, Ordovician and Devonian systems, the Ludlow Series and the Kimmeridgian and Portlandian stages.

The Geological Conservation Review (GCR) was initiated by the Nature Conservancy Council in 1977 to assess and document the most scientifically-important parts of this rich heritage. The GCR reviews the current state of knowledge of key earth-science sites in Britain and provides a firm basis upon which site conservation can be founded in years to come. Each GCR volume describes and assesses networks of sites of national or international importance in the context of a portion of the geological column, or a geological, palaeontological, or mineralogical topic. The full series of 42 volumes will be published by the year 2000.

Within each individual volume, every GCR locality is described in detail in a self-contained account, consisting of highlights (a précis of the special interest of the site), an introduction (with a concise history of previous work), a description, an interpretation (assessing the fundamentals of the site's scientific interest and importance), and a conclusion (written in simpler terms for the non-specialist). Each site report is a justification of a particular scientific interest at a locality, of its importance in a British or international setting, and ultimately of its worthiness for conservation.

The aim of the Geological Conservation Review series is to provide a public record of the features of interest in sites being considered for notification as Sites of Special Scientific Interest (SSSIs). It is written to the highest scientific standards but in such a way that the assessment and conservation value of the sites is clear. It is a public statement of the value placed upon our geological and geomorphological heritage by the earth-science community that has participated in its production, and it will be used by the Joint Nature Conservation Committee, the Countryside Council for Wales, English Nature and Scottish Natural Heritage in carrying out their conservation functions. The three country agencies are also active in helping to establish sites of local and regional importance. Regionally Important Geological/Geomorphological Sites (RIGS) augment the SSSI coverage, with local groups identifying and conserving sites which have educational, historical, research or aesthetic value, enhancing the wider earth-science conservation perspective.

All the sites in this volume have been proposed for notification as SSSIs; the final decision to notify or re-notify sites lies with the governing councils of the appropriate country conservation agency.

Information about the GCR publication programme may be obtained from:

GCR Unit,
Joint Nature Conservation Committee,
Monkstone House,
City Road,
Peterborough, PE1 1JY.

Titles in the series

Published by Chapman & Hall, 2–6 Boundary Row, London SE1 8HN, UK

First edition 1998

Typeset in 10/12pt Garamond ITC by Columns Design Ltd, Reading, Berkshire
Printed in Great Britain by T.J. International Ltd, Padstow, Cornwall

ISBN 0 412 78930 2

A catalogue record for this book is available from the British Library

Library of Congress Catalog Card Number: 96–85905

Sold and distributed in North, Central and South America
by Kluwer Academic Publishers.
101 Philip Drive, Norwell, MA 02061, USA

In all other countries sold and distributed
by Kluwer Academic Publishers Group.
P.O. Box 322, 3300 AH Dordrecht, The Netherlands.

∞ Printed on acid-free text paper, manufactured in accordance with ANSI/NISO Z39.48-1992
(Permanence of Paper).

Quaternary of South-West England

S. Campbell
Countryside Council for Wales, Bangor

C.O. Hunt
Huddersfield University

J.D. Scourse
School of Ocean Sciences, Bangor

D.H. Keen
Coventry University

and

N. Stephens
Emsworth, Hampshire.

GCR Editors: **C.P. Green and B.J. Williams**

Contents

Contents

Contents

Contributors

S. Campbell Countryside Council for Wales, Plas Penrhos, Ffordd Penrhos, Bangor, Gwynedd LL57 2LQ.

S. Collcutt Oxford Archaeological Associates Limited, Lawrence House, 2 Polstead Road, Oxford OX2 6TN.

R. Cottle English Nature, Norman Tower House, 1–2 Crown Street, Bury St Edmunds, Suffolk IP33 1QX.

D.G. Croot Department of Geographical Sciences, University of Plymouth, Drake Circus, Plymouth, Devon PL4 8AA.

A.P. Currant Department of Palaeontology, The Natural History Museum, Cromwell Road, London SW7 5BD.

N.D.W. Davey GCR Publications Unit, Joint Nature Conservation Committee, Monkstone House, Peterborough PE1 1JY.

A.J. Gerrard Department of Geography, Birmingham University, P.O. Box 363, Edgbaston, Birmingham B15 2TT.

A. Gilbert Department of Geographical Sciences, University of Plymouth, Drake Circus, Plymouth, Devon PL4 8AA.

J.E. Gordon Scottish Natural Heritage, 2 Anderson Place, Edinburgh EH6 5NP.

C.P. Green Department of Geography, Royal Holloway and Bedford New College, University of London, Egham, Surrey TW20 0EX.

C.O. Hunt Department of Geographical and Environmental Sciences, University of Huddersfield, Queensgate, Huddersfield HD1 3DH.

D.H. Keen Centre for Quaternary Science, School of Environmental and Natural Sciences, Coventry University, Priory Street, Coventry CV1 5FB.

J.D. Scourse School of Ocean Sciences, University College of North Wales, Menai Bridge, Bangor, Gwynedd LL59 5EY.

R.A. Shakesby Department of Geography, University College Swansea, Singleton Park, Swansea, West Glamorgan SA2 8PP.

N. Stephens 8 Christopher Way, Emsworth, Hampshire PO10 7QZ.

A.J. Stuart Castle Museum, Norwich NR1 3JU.

B.J. Williams Earth Resources Centre, University of Exeter, North Park Road, Exeter EX4 4QE.

Acknowledgements

The short-listing and selection of the GCR sites contained in this volume began with a widespread consultation exercise, in the early 1980s, co-ordinated by John Gordon, Bill Wimbledon and Chris Hunt. Between 1987 and 1990, preparatory work for compiling the volume, involving field visits and site descriptions, was begun by Stewart Campbell. The writing of this volume was initiated by the Nature Conservancy Council in 1990, and has been seen to completion by the Joint Nature Conservation Committee on behalf of the three country agencies, English Nature, Scottish Natural Heritage and the Countryside Council for Wales. Each site account bears the name of its author(s). Draft text was produced by the named contributors from 1990 to 1994. All the draft site descriptions, introductory passages and figures were edited and unified into the designated format of this publication series by Stewart Campbell between 1994 and 1997. Within this volume, all published source material is duly referenced. In addition, the authors of the volume have contributed their own personal knowledge of sites, and numerous extra notes, concepts and descriptions have been incorporated from unpublished thoughts and discussions: several sites are described in detail for the first time here.

The selection of the 63 GCR sites described in this volume involved the assessment of several hundred potential localities. In addition to the named contributors, many members of the earth-science community assisted with information or advice during site selection and documentation. Without their assistance, the volume could not have been produced. The help of the following colleagues is therefore gratefully acknowledged: J. Alan, A. Bolt, D.Q. Bowen, D. Brunsden, C. Caseldine, J.A. Catt, R.A. Cullingford, D.C. Davies, K.H. Davies, D.D. Gilbertson, N. Glasser, A.B. Hawkins, A. Heyworth, S.A.V. Hill, C.E. Hughes, H.C.L. James, C. Kidson, M. Macklin, D. Maguire, D. Mottershead, N. Perkins, H. Prudden, J. Rooke, J. Rose, P. Sims, A. Straw, A.J. Sutcliffe, N. Thew, R.S. Waters, R.C. Whatley and R. Wolton.

Several sites required excavation prior to assessment, and the members of the Creswell Natural History and Archaeology Society and I.P. Brooks, G. Coles, D. Francis and J. Rooke kindly helped with these site investigations.

Stewart Campbell is particularly grateful to David Keen and Brian Williams for reviewing the entire text in detail and on several occasions, and also to the following for reviewing or revising substantial parts of the text: Simon Collcutt, John Gerrard, John Gordon, Chris Green, Chris Hunt, Les James, James Scourse, Nick Stephens and Anthony Sutcliffe. Professors David Bowen and Nick Stephens kindly provided much published and unpublished data, and have given constant support for the project: John Gordon and Bill Wimbledon provided guidance and encouragement throughout its long gestation.

Acknowledgements

It is a pleasure to acknowledge the help of the GCR publication production team: Neil Ellis (Publications Manager), Valerie Wyld (former Sub-editor) and Nicholas Davey (former Editorial Assistant), Carolyn Davies, Tania Davies, Mark Diggle, Caroline Mee and Margaret Wood also provided invaluable bibliographic, scientific and administrative support.

The various topographic and geological maps, which make up many of the figures in the volume, have been compiled from numerous, individually acknowledged, sources. Many rely, at least in part, upon the high-quality maps produced by the British Geological Survey and the Ordnance Survey, whose underpinning contribution is gratefully acknowledged. All the diagrams were drawn by Dave Williams of Lovell Johns, St Asaph, Clwyd.

Access to the countryside

This volume is not intended for use as a field guide. The description or mention of any site should not be taken as an indication that access to a site is open or that a right of way exists. Most sites described are in private ownership, and their inclusion herein is solely for the purpose of justifying their conservation. Their description or appearance on a map in this work should in no way be construed as an invitation to visit. Prior consent for visits should always be obtained from the landowner and/or occupier.

Information on conservation matters, including site ownership, relating to Sites of Special Scientific Interest (SSSIs) or National Nature Reserves (NNRs) in particular counties or districts may be obtained from the relevant country conservation agency headquarters listed below:

Countryside Council for Wales,
Plas Penrhos,
Ffordd Penrhos,
Bangor,
Gwynedd LL57 2LQ.

English Nature,
Northminster House,
Peterborough PE1 1UA.

Scottish Natural Heritage,
12 Hope Terrace,
Edinburgh EH9 2AS.

Preface

STRUCTURE OF THE VOLUME AND TERMINOLOGY USED

This book contains scientific descriptions of 63 localities (Figure A) of at least national importance for Quaternary geology, geomorphology and environmental change in South-West England. These sites were selected by the Geological Conservation Review and are accordingly designated 'GCR' sites. Chapter 1 provides an introduction to the Quaternary. Chapter 2 synthesizes the geomorphological development and Quaternary history of the region, and outlines the principles involved in site selection.

The individual GCR site descriptions form the core of the book. In the following chapters, sites are arranged and described in broad geographic areas and by research topic. This is necessitated by the widely disparate nature of the field evidence in South-West England: sites demonstrating the full range of Quaternary and geomorphological features are not evenly and conveniently dispersed throughout the region, and some areas have significant gaps. Neither do the individual chapters contain sites that necessarily equate with particular site selection networks. Rather, the chosen chapter headings provide the least repetitive means of describing the sites and background material. Where possible, a chronological approach, from oldest to youngest, has been used to describe sites within a given chapter. Again, this approach is not always possible, and a group of sites may show variations on landform or stratigraphic evidence broadly within one major time interval or chronostratigraphic stage; inevitably there are many overlaps.

Each chapter introduction provides an overview of the region or topic, highlighting the particular aspects of Quaternary or geomorphological history which are of special significance. The individual site reports each contain a synthesis of currently available documentation and interpretations of the site evidence. A key part of each site account is the 'Interpretation' section, which explains the site's importance in a regional, national or international context, and justifies its conservation. Where sites were chosen as part of a closely related network, cross-reference is made to the related sites to provide a fuller understanding of their respective attributes and the justification for their selection. Where sites are of particular historical significance, the history of study at the site is presented in detail.

There is currently no universally agreed system of terminology for the subdivision of Quaternary deposits in Britain. Mitchell *et al.* (1973b) proposed a correlation scheme based on standard stages. Since that date, however, not only has there been a great increase in knowledge of the Quaternary succession, so that the 1973 system is now incomplete, but also certain of the stage names proposed at the time have been questioned as to their suitability or even the existence of the sediments to which they refer.

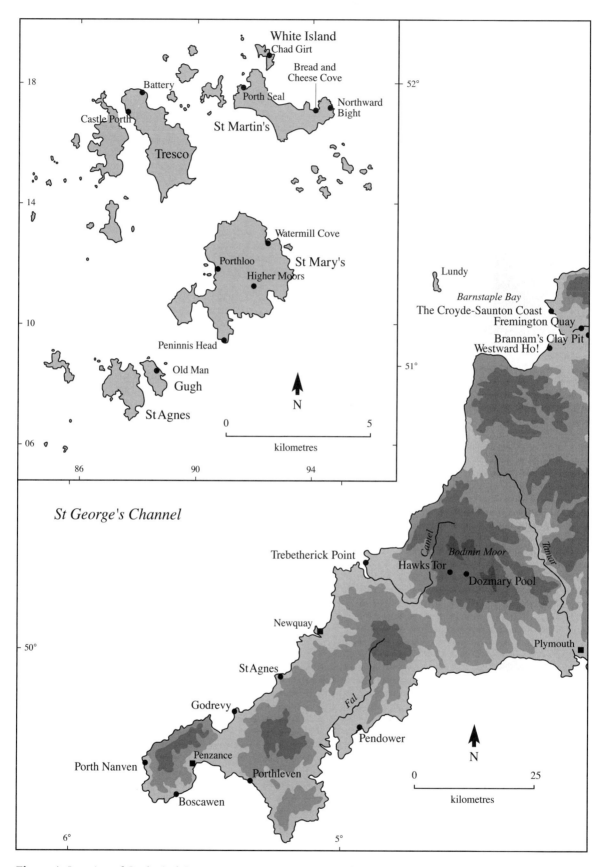

Figure A Location of Geological Conservation Review (GCR) sites described in this volume.

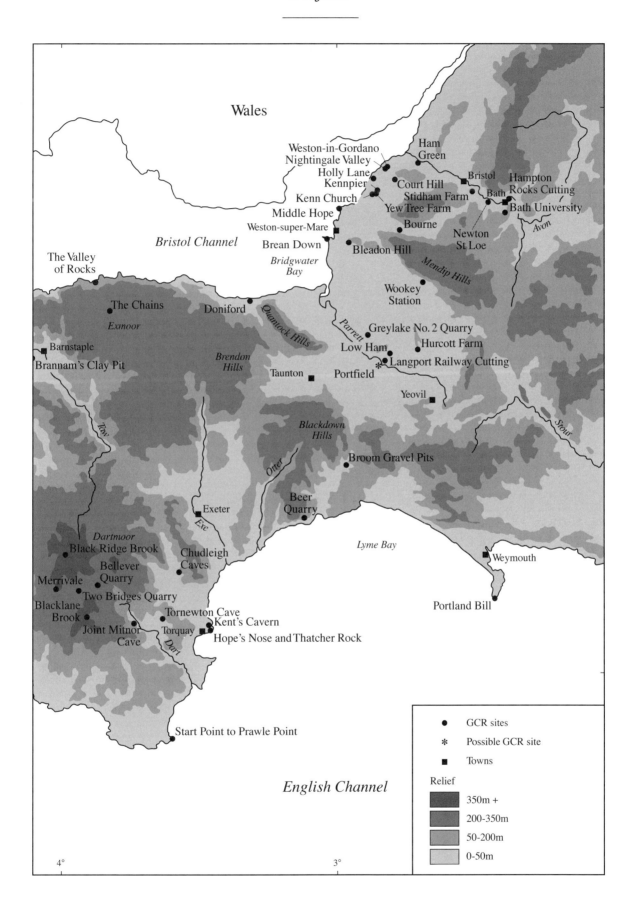

Oxygen Isotope Stage	Age (ka BP)	Region				
		Isles of Scilly	Cornwall (mainland)	Devon	Dorset	Somerset & Avon
1	10	**Higher Moors** peat	**Hawks Tor** peat **Dozmary Pool** peat **Godrevy** sand	**Merrivale** earth hummocks **Blacklane Brook** peat **Black Ridge Brook** peat **Kent's Cavern** Granular Stalagmite **Westward Ho!** submerged forest & peat beds		**The Chains** peat **Greylake No. 2 Quarry** silts **Wookey Station** colluvial and aeolian deposits **Kennpier** estuarine silts **Yew Tree Farm** estuarine silts
2	24	**Porthloo** solifluction deposits **Watermill Cove** solifluction deposits **Old Man** solifluction & aeolian ('sandloess') deposits **Peninnis Head** tors **Porth Seal** solifluction deposits **Bread & Cheese Cove** Scilly Till, outwash gravel & solifluction deposits **Chad Girt** solifluction deposits **Northward Bight** solifluction deposits **Battery** meltwater & solifluction deposits **Castle Porth** solifluction deposits	**Hawks Tor** slope & organic deposits **Pendower** aeolian & solifluction deposits **Boscawen** solifluction deposits **Porth Nanven** solifluction deposits **Godrevy** solifluction deposits **Trebetherick Point** solifluction deposits & 'boulder gravel'	**Merrivale** tors, slope deposits & patterned ground **Bellever Quarry** bedded growan & solifluction deposits **Two Bridges Quarry** solifluction deposits **Kent's Cavern** Stony Cave Earth **Tornewton Cave** Glutton Stratum **Chudleigh Caves** Pixie's Hole cold fauna and deposits **Hope's Nose & Thatcher Rock** head **Start Point to Prawle Point** solifluction deposits **Brannam's Clay Pit** Fremington Clay? **Croyde-Saunton Coast** head **Westward Ho!** head **The Valley of Rocks** tors & solifluction deposits	**Portland Bill** cryoturbation structures	**Doniford** head, gravels & loams **Brean Down** upper breccias & blown sand **Wookey Station** fan gravels **Holly Lane** breccias and loams
3	59	**Watermill Cove** organic deposits **Porth Seal** organic deposits **Bread & Cheese Cove** organic deposits	**Boscawen** organic deposits	**Kent's Cavern** Loamy Cave Earth **Tornewton Cave** Elk Stratum		**Brean Down** 'Bone Bed'
4	71	**Porth Seal** solifluction deposits	**Pendower** head? **Boscawen** solifluction deposits? **Trebetherick Point** 'sandrock'	**Tornewton Cave** *Rangifer* and *Bison* fauna **Croyde-Saunton Coast** 'sandrock'	**Portland Bill** head	**Brean Down** lower breccias & silty sands **Holly Lane** breccias and loams?
5a-d	116					**Low Ham** estuarine deposits
5e	128	**Porthloo** raised beach deposits **Watermill Cove** raised beach deposits **Porth Seal** raised beach deposits **Chad Girt** raised beach deposits **Northward Bight** raised beach deposits	**Pendower** raised beach deposits & dune sand? **Porth Nanven** raised beach deposits? **Godrevy** raised beach deposits & 'sandrock'? **Trebetherick Point** raised beach deposits & blown sand	**Kent's Cavern** Crystalline Stalagmite **Tornewton Cave** Bear Stratum & Hyaena Stratum **Chudleigh Caves** Pixie's Hole speleothem **Joint Mitnor** Cave Earth **Hope's Nose & Thatcher Rock** Thatcher Rock raised beach deposits & Hope's Nose dune sand **The Valley of Rocks** Lee Bay raised beach deposits **Croyde-Saunton Coast** raised beach deposits **Westward Ho!** raised beach deposits	**Portland Bill** East Beach **Portland Bill** weathering of Loam	**Langport Railway Cutting** calcrete **Greylake No. 2 Quarry** palaeosol and marine deposits **Middle Hope** raised beach deposits **Bourne** soil **Hampton Rocks Cutting** calcrete
6	186		**Pendower** head? **Godrevy** head?	**Tornewton Cave** minor debris flow **Croyde-Saunton Coast** head	**Portland Bill** Loam **Broom Gravel Pits** terrace gravels?	**Portfield** sand **Bourne** fan gravels **Newton St Loe** fluvial gravels **Hampton Rocks Cutting** fluvial gravels
7	245		**Pendower** raised beach deposits & dune sand? **Porth Nanven** raised beach deposits? **Godrevy** raised beach deposits?	**Tornewton Cave** Otter Stratum **Hope's Nose & Thatcher Rock** Hope's Nose raised beach deposits **Croyde-Saunton Coast** raised beach deposits	**Portland Bill** West Beach **Broom Gravel Pits** temperate floodplain silts & clays?	**Greylake No. 2 Quarry** estuarine deposits **Portfield** silts **Kenn Church** sands and gravels **Weston-in-Gordano** sands and silts
8	303			**Hope's Nose & Thatcher Rock** head		**Langport Railway Cutting** fluvial gravels **Stidham Farm** fluvial gravels
9	339			**Kent's Cavern** Breccia?		**Hurcott Farm** fluvial gravels
10	?					**Ham Green** fluvial gravels?
11	423			**Kent's Cavern** Breccia?		
12	478			**Brannam's Clay Pit** Fremington Clay? **Croyde-Saunton Coast** erratics **The Valley of Rocks** ?? **Fremington Quay** 'stony clay' = glacigenic deposits?		**Ham Green** fluvial gravels?
13-21			**Porthleven** Giant's Rock (Stage 16?)	**Kent's Cavern** *Homotherium* & *Ursus deningeri* remains **Brannam's Clay Pit** Fremington Clay (Stage 16?)		**Court Hill** glacigenic sediments (Stage 16??) **Bath University** glacigenic sediments (Stage 16??) **Nightingale Valley** glacigenic sediments (Stage 16??) **Bleadon Hill** sands and gravels **Kennpier** estuarine deposits (Stage 15??) **Kennpier** diamictons (Stage 16??) **Yew Tree Farm** channel-fill (Stage 15??) **Yew Tree Farm** glacigenic gravels (Stage 16??)
Pre-Pleistocene		**Peninnis Head** tors	**St Agnes Beacon** Miocene & mid-Oligogene sands & clays; Reskajeage Surface	**Beer Quarry** solutional pipes & 'clay-with-flints' **Merrivale** tors **Bellever Quarry** altered granite or growan **Two Bridges Quarry** altered granite		

Figure B A stratigraphical correlation of the Geological Conservation Review sites described in this volume. Sites appear more than once where they have multiple interests, or interests of different ages. Many of the ascriptions are highly provisional, and reference should be made to the individual site reports in this volume for a fuller discussion of the possible ages of the site evidence. Particular uncertainties are denoted by question marks.

Preface

The most recent attempt to correlate Quaternary deposits across Britain (Bowen, in prep.) has been based on lithostratigraphy and a time-frame founded on oxygen isotope stages. This scheme has been applied to South-West England (Campbell *et al.*, in prep.) and, wherever possible in this volume, site evidence is referred to the oxygen isotope framework. The basis of such a chronology is the oxygen isotope signal recognized in deep-sea sediments. This signal has been shown to be a function of the Earth's orbital parameters (Hays *et al.*, 1976), and astronomical data have been used to 'tune' the geological timescale (cf. Imbrie *et al.*, 1984; Prell *et al.*, 1986; Ruddiman *et al.*, 1986, 1989; Martinson *et al.*, 1987). For the period back to about 620 ka, the timescale is that developed by Imbrie *et al.* (1984), which has been substantiated by later work (Prell *et al.*, 1986; Shackleton *et al.*, 1990). For the earlier part of the Quaternary, the revised timescale of Shackleton *et al.* (1990) is adopted.

The correlation of the selected GCR sites for South-West England (Figure B), must however be regarded as highly provisional. The fact is that most site evidence remains undated, and even those sites which have yielded 'absolute' and 'relative' dates are often open to widely disparate interpretations. This volume also describes some 'pre-Quaternary' sites which lend an important insight into the long-term evolution of the British landscape: placing their evidence into a precise timescale is even more tenuous.

Where radiocarbon 'dates' (age estimates) are cited, they are quoted in radiocarbon years before present (AD 1950), with laboratory reference number and associated standard error if available. It should be noted that the radiocarbon timescale diverges from the calendrical one, and although calibration is available back to 9 ka in detail (cf. Pilcher, 1991) and to 30 ka in outline (Bard *et al.*, 1990), the interpretation of radiocarbon measurements, particularly during parts of the Late Devensian, is additionally complicated (cf. Ammann and Lotter, 1989; Zbinden *et al.*, 1989).

Where possible, modern taxonomic nomenclature has been used and the Geological Society's 'Instruction to Authors' guidelines for taxonomic quotation followed. Nomenclature for marine Mollusca follows Seaward (1982); for freshwater and brackish Mollusca it follows Kerney (1976a); and for land Mollusca it follows Kerney and Cameron (1979).

Chapter 1

Introduction to the Quaternary

S. Campbell and J.E. Gordon

Climate change in the Quaternary

INTRODUCTION

This chapter is a general account of the Quaternary. Together with the conclusions presented at the end of each site account, it is intended for the less specialist reader.

The Quaternary is the most recent major subdivision of the geological record, spanning the late Cenozoic. Traditionally, it is divided into two intervals of epoch status – the Pleistocene and Holocene. The Holocene occupies only the last 10 000 years of geological time and is the warm interval or interglacial in which we now live. Consequently, it is often regarded as part of the Pleistocene rather than a separate epoch. In a strict geological sense, the base of the Pleistocene (and also, by default, that of the Quaternary) is defined in Italy at the type locality of Vrica, where it is dated to about 1.64 Ma (million years) (Aguirre and Pasini, 1985). However, it is now known from many parts of the world that glacial advances occurred well before the start of the Pleistocene *sensu stricto*, and recent evidence from ocean-floor sediments shows a major climatic deterioration in the northern hemisphere at around 2.4 Ma (Shackleton *et al.*, 1984; Ruddiman and Raymo, 1988). Indeed, it is now well established that the current warm period, the Holocene, is simply the latest interglacial in a long series of profound climatic fluctuations which has characterized the last 2.4 Ma.

The term Quaternary is used here in a somewhat loose sense, following its traditional usage for the last part of Cenozoic time. Where greater precision is required, the terms Pleistocene and Holocene are used.

THE CHARACTER OF THE QUATERNARY

The Quaternary is often assumed to be synonymous with the 'Ice Age', an association which can be traced back to Forbes (1846) who equated Pleistocene with the 'Glacial Epoch'. In reality, however, there have been many ice ages (glacials) separated by warmer episodes (interglacials), when the Earth's climate was similar to that of today, if not warmer. Indeed, the deep-sea sedimentary record shows that up to 50 'warm' and 'cold' climatic oscillations have occurred within the last 2.4 Ma (Ruddiman and Raymo, 1988; Ruddiman *et al.*, 1989) (Figure 1.1). Equally, the glacial and interglacial periods cannot be characterized simply as 'cold' or 'warm', respectively; the ice ages were not unbroken in their frigidity since the exceptionally cold phases (stadials) were punctuated by warmer periods (interstadials), in some cases lasting for several thousand years (Gordon and Sutherland, 1993). The fundamental characteristic of the Quaternary is therefore one of change through time and space in geomorphological processes, floras, faunas and environmental conditions, all modulated by the changing climate. The record of such changes is preserved in a variety of forms of interest to the Quaternary scientist, including landforms, sediment sequences and organic remains.

CLIMATE CHANGE IN THE QUATERNARY

The abrupt onset of the late Cenozoic ice ages is, as yet, unexplained. However, the succession of ice ages (glacials) and interglacials has occurred at known frequencies (Campbell and Bowen, 1989), and changes in insolation (the receipt of solar radiation at the Earth's surface and throughout its atmosphere) associated with orbital rhythms are now established as the principal external driving forces of the Earth's climatic system.

Terrestrial evidence for long-term climate change is incomplete and fragmentary. This is because the continental environment is largely one of net erosion; evidence for earlier events has been, and is being, continually destroyed. In contrast, long, more or less continuous sequences of sediments have built up on the deep ocean floors. Studies of cores from these presumed uninterrupted sedimentary sequences have provided a finely detailed stratigraphic and climatic framework unrivalled in its completeness (Shackleton and Opdyke, 1973; Shackleton *et al.*, 1990). Cycles of changing ice volumes, and hence climatic variability as shown by the deep-sea record, can be matched with the pattern of the Earth's orbital rhythms – namely its cycles of eccentricity (with a periodicity of *c.* 100 ka (thousand years)), tilt (41 ka) and precession (23 ka and 19 ka) (Hays *et al.*, 1976; Imbrie and Imbrie, 1979; Imbrie *et al.*, 1984; Berger, 1988).

Although orbital changes have played a crucial role in triggering the succession of Quaternary ice ages (and interglacials), it is still not clear how climate change actually occurs and how the relatively small variations in radiation occasioned by the orbital changes are amplified by the Earth's climate

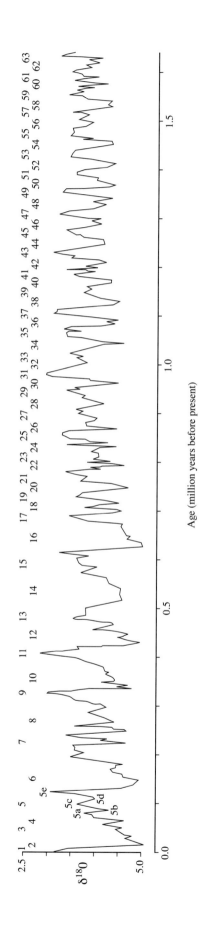

Figure 1.1 The oxygen isotope record, as represented in a borehole (Site 607) in the Mid-Atlantic at latitude c. 41°N. Numbered stages are shown at the top; even-numbered ones are relatively cold (more ice) and odd-numbered ones relatively warm (less ice). Note that the amplitude and wavelength of the curve increases at c. 0.7 million years ago (the δ18O scale is a ratio obtained by comparing the proportion of 18O to 16O in samples to that in a mean sea-water standard). (From Bridgland, 1994, compiled from data published by Ruddiman *et al.*, 1989).

system. However, it is likely that the orbital fluctuations trigger a complex set of interactions and feedbacks involving the atmosphere, oceans, biosphere, cryosphere (the realm of snow and ice) and global tectonics (which in part determine the configuration of the world's oceans and landmasses). Although no convincing unifying theory has yet been put forward, several potential key links have been identified. It appears that the role of atmospheric carbon dioxide is fundamentally important. It has been shown that changes in the atmospheric content of the gas follow orbital periodicities (e.g. Barnola *et al.*, 1987), but precede changes in ice volume (Shackleton *et al.*, 1983; Shackleton and Pisias, 1985). Although the interchange of carbon dioxide between the oceans and atmosphere may amplify orbital effects (cf. Saltzman and Maasch, 1990), modelling experiments have shown that reductions in atmospheric carbon dioxide alone are insufficient to produce the magnitude of global cooling necessary for ice-sheet growth (Broccoli and Manabe, 1987; Broecker and Denton, 1989).

The changing configuration of the Earth's landmasses and oceans has also been seen as a key factor; for example, uplift in the American southwest and in the Tibetan Plateau may have affected the pattern of atmospheric circulation in the northern hemisphere, causing waves in the upper atmosphere, bringing colder air to lower middle latitudes and producing the cooling necessary for the onset of Quaternary glaciation (Ruddiman and Raymo, 1988; Ruddiman and Kutzbach, 1990). Likewise, variations in the configuration of land and sea and in oceanic salinity have produced major changes in global oceanic circulation and thus in climate (Broecker and Denton, 1989, 1990). The oceans, in particular, may play a crucial role, and much interest has focused on the pattern of circulation in the North Atlantic which appears to undergo sudden changes between glacial and interglacial modes (Broecker and Denton, 1990), although the 'switching' mechanism is uncertain. Other factors which do not involve a clear linear relationship with climate change have also been identified in the growth and decay of ice sheets. Perhaps the most notable is the inherent instability of marine-based ice sheets (e.g. Hughes, 1987; van der Veen, 1987; Jones and Keigwin, 1988) and, in particular, the effects of topography on ice-sheet growth (e.g. Payne and Sugden, 1990) and decay (e.g. Eyles and McCabe, 1989). It is thus clear that climate change is not simply a direct linear result of astronomical 'forcing'.

THE DEEP-SEA RECORD

One of the most significant developments in Quaternary science during the last 40 years or so, has been the recovery of sediment cores from the floors of the world's oceans and the resolution of the climatic and other environmental records which they contain (Bowen, 1978; Imbrie and Imbrie, 1979). Since these incomparably detailed sedimentary sequences are recognized in oceans throughout the world, they provide a stratigraphic/climatic framework of global significance and applicability.

Ocean sediments have provided three major lines of evidence for climate change in the Quaternary. First, oxygen isotope analysis of the carbonate in the shells of marine microfossils (e.g. foraminifera) preserved in the sediments provides a measure of isotopic variability in the $^{18}O/^{16}O$ ratio through time in the global ocean (Emiliani, 1955, 1957; Shackleton, 1969; Shackleton and Opdyke, 1973). During cold episodes, water molecules of the light isotope (^{16}O) are evaporated from the sea more readily and 'stored' as ice, leaving the remaining sea water relatively enriched with the heavier isotope (^{18}O). The changing pattern of isotope composition in the ocean thus provides a signal of ice-sheet growth and decay, and a record of such changes has been reconstructed through time (Figure 1.1): this also forms the basis for oxygen isotope stratigraphy (see below). Although this signal provides an indication of changing ice volumes, it must be stressed that this is not a direct measure of global climate or vegetational development (Bridgland, 1994).

Second, variations in past sea surface temperatures (SST) may be determined from assemblages of marine microfossils and biomarkers (molecular stratigraphy), and trace element ratios from calcareous microfossils found in the sea-bed sediments (Brassell *et al.*, 1986; De Dekker and Forester, 1988). Third, variations in the percentages of calcium carbonate (often higher during warm episodes), and the occurrence of ice-rafted or wind-blown detritus (higher during cold periods) can also provide measures of climatic variability (Ruddiman *et al.*, 1986).

The isotopic and dust records in the polar ice sheets have also provided detailed proxy climate records, extending back some 250 ka (e.g. Johnsen *et al.*, 1972; Jouzel *et al.*, 1987; Oeschger and Langway, 1989; Jouzel *et al.*, 1993). In particular, new cores from the Greenland ice sheet have revealed exceptional, high-resolution records of climatic fluctuations during the Devensian (Johnsen *et al.*, 1992; Dansgaard *et al.*, 1993; Taylor *et al.*, 1993). These fluctuations are notable for their frequency and rapidity; for example 24 interstadial phases interrupted the last glacial cycle, and abrupt rises in temperatures of 5–8°C occurred within a few decades. At the end of the last glaciation, a warming of 7°C occurred within a period of 20–25 years (Dansgaard *et al.*, 1989). Comparison of recent ocean-core data with the Greenland ice-core records indicates a good correlation between the different sets of information and demonstrates a close coupling between the atmosphere, oceans and ice sheets in the North Atlantic region (Bond *et al.*, 1993; Bond and Lotti, 1995). The results from this work have given new impetus to the search for links between the offshore marine records, the Greenland ice-core records and the responses of terrestrial systems and biota to the inferred fluctuations in the climate (e.g. Lowe *et al.*, 1994; Lowe *et al.*, 1995; McCabe, 1996; Whittington *et al.*, 1996). Many of these recent advances and their significance are described in more detail in the papers presented by Gordon (1997).

SUBDIVIDING THE QUATERNARY

The oxygen isotope chemistry of the deep-sea sediment pile now provides the main basis for subdividing the Quaternary, with a number of successive oxygen isotope stages recognized globally (Figure 1.1). These stages, running counter to normal geological practice, are numbered backwards in time and down through the geological column. Warm periods with correspondingly low volumes of ice are given odd numbers; the present interglacial, the Holocene, is numbered as Stage 1. Times of high ice volume (glacials) are given even numbers, the last main cold phase in Britain, the Late Devensian, being numbered as Stage 2. Stages are also divided into sub-stages, for example, Stage 5 into sub-stages 5a–5e, often reflecting stadial or interstadial events.

The position in the deep-sea sediment cores of a major reversal in the Earth's magnetic field, the Matuyama–Brunhes Reversal at 780 ka, provides a yardstick with which to calibrate the oxygen isotope record. The boundaries of the different isotope stages have also been adjusted and refined with respect to known orbital patterns (Imbrie *et al.*, 1984; Martinson *et al.*, 1987; Shackleton *et al.*, 1990).

THE HISTORY OF THE ICE AGES

The earliest evidence for significant climatic deterioration in late Cenozoic times comes from planktonic fossils found in North Atlantic ocean sediments. A cycle of northern hemisphere glaciation may have been initiated about 3.1 Ma with the closure of the Straits of Panama and the creation of the Isthmus of Panama by crustal upheaval (Keller and Barron, 1982). This ended the latitudinal movement of surface waters between the Atlantic and Pacific oceans, and replaced it with a meridional flow in the North Atlantic. In this way, warmer waters were able to reach higher latitudes, thus creating the potential for increased precipitation and ice-sheet growth on adjacent continents. As sea temperatures fell, polar ice sheets grew and launched ice floes into the surrounding seas (Ruddiman and Wright, 1987; Ruddiman and Raymo, 1988). Shortly afterwards, glaciers developed in Iceland (Einarsson and Albertson, 1988), although it was not until about 2.4 Ma that large ice sheets spread to middle latitudes (Shackleton *et al.*, 1984). These ice sheets, between about 2.4 and 0.9 Ma, seem to have fluctuated in harmony with the 41-ka cycle of obliquity (variations in the tilt of the Earth's axis). Subsequently, northern hemisphere glaciation intensified and ice sheets grew to maximum volumes about twice their previous size, and fluctuated with a 100-ka rhythm, namely that of orbital eccentricity. After about 0.45 Ma, the 100-ka rhythm became dominant, although other fluctuations in ice volume were superimposed on the long-term pattern. These smaller-scale fluctuations occurred at frequencies of 41 and 23 ka, corresponding to the cycles of orbital precession (Campbell and Bowen, 1989).

Why there should be a shift in ice-age rhythms at about 0.9 Ma is unknown, although some have attributed this to continued uplift of the Tibetan Plateau which would have caused reorganization in global atmospheric circulation (Ruddiman and Raymo, 1988; Ruddiman *et al.*, 1989). However, the change may be related to variations in the behaviour of the ice sheets themselves. It is known that after this time, marine-based ice sheets developed on continental shelf areas such as Hudson's Bay and the Baltic and Irish seas. They grew rapidly and reached considerable thicknesses, prompted perhaps by cooler air brought to lower middle latitudes as a result of topographic uplift elsewhere in the world. Equally, their collapse and disappearance were also probably dramatically swift.

In Britain, the area covered by ice varied considerably during different glaciations. During the last (Late Devensian) glaciation, ice extended as far south as the north Midlands, impinging on the north coast of East Anglia and covering most of South Wales. During earlier glaciations ice sheets were more extensive, but probably never reached farther south in south-central and south-east England than the present Thames Valley (Bridgland, 1994). In the South-West, there is a longstanding debate over whether pre-Devensian ice masses reached the northern shores of Devon and Cornwall and even the Isles of Scilly (Chapters 6, 7 and 8). The sedimentary record of the Quaternary in Britain is far from complete; the most comprehensive record comes from East Anglia and the Midlands where a series of glacials and interglacials has been recognized. The correlation of Quaternary deposits across Britain has long been based on a series of reference localities and stratotypes (e.g. Mitchell *et al.*, 1973b), but the problem remains, however, to relate the fragmentary terrestrial record to the oxygen isotope timescale, particularly before the last interglacial (Ipswichian Stage) (Gordon and Sutherland, 1993). Nevertheless, in recent years progress has been made in relating the British terrestrial evidence to the comprehensive deep-sea record, particularly in Wales (Bowen and Sykes, 1988; Campbell and Bowen, 1989; Bowen, 1991), the Midlands and the Thames Valley (Bridgland, 1994) and East Anglia (Bowen *et al.*, 1989). Significant advances in this respect have also come from South-West England, the subject of this volume.

BRITISH QUATERNARY ENVIRONMENTS

The major shifts of climate which characterize the Quaternary were accompanied by equally profound changes in environmental conditions which left a strong imprint on the landforms, fossils and sediments of Britain. During the cold or glacial stages, substantial areas were subjected to the effects of glacial erosion and deposition and a wide range of landforms and deposits was produced. The erosional and depositional legacy of the Pleistocene glaciers has been documented in many publications; for Wales and Scotland the evidence has been reviewed comprehensively in recent Geological Conservation Review volumes (Campbell and Bowen, 1989; Gordon and Sutherland, 1993, respectively). One of the main controversies in the Quaternary history of South-

West England has been the extent to which the Pleistocene ice sheets affected the region, and this aspect is explored fully in subsequent chapters.

As ice sheets melted, vast quantities of meltwater were liberated, giving rise to characteristic suites of landforms and deposits. Even in South-West England, an area relatively untouched by the ice sheets, a variety of controversial landforms may bear witness to meltwater erosion at glacial margins (Chapters 7 and 10).

Repeated climate change also subjected the flora and fauna of Britain to stress: fundamental changes in the distribution of plants and animals took place. Beyond the margins of the ice sheets and during the cold climatic phases of the Quaternary, periglacial conditions prevailed. Such environments were characterized by frost-assisted processes and by a range of frost- and ground ice-generated landforms and deposits. Mass wasting (downslope movement of soil on both large and small scales) and increased wind action were prevalent, also producing a range of characteristic features. In the fossil record, the flora and fauna of these cold periods is, not surprisingly, restricted in diversity and dominated by cold-tolerant species; large areas were dominated by tundra vegetation. Although periglacial processes are still active today in some of Britain's uplands, their Pleistocene legacy is nowhere better recorded than in South-West England, where many of the most spectacular landforms and most intricate sequences of deposits were formed (Chapters 4, 6, 7, 8 and 9).

Conversely, the warmer or interglacial periods of the Quaternary are characterized by the absence of glacial, periglacial and glaciofluvial features, and there were times when chemical weathering, soil formation and the accumulation of organic sediments took place. Variations in the quantity and type of pollen grains preserved in organic deposits, such as peats and lake muds, have been used to define systems of pollen zones or pollen biozones. These zones are characterized by particular vegetational assemblages which can be used to chart sequences of vegetational, climatic and environmental change. Traditionally, these have been used as the principal basis for distinguishing between various interglacial phases in the land-based Quaternary record and for the definition of chronostratigraphic stages (West, 1963, 1968; Mitchell *et al.*, 1973b). Unfortunately, although several distinctive interglacial episodes in the British Pleistocene can be distinguished, very little evolution of the flora actually occurred, thus hindering biostratigraphic correlation. However,

Figure 1.2 Portland Bill, Dorset, was chosen as a GCR site for its complex sequence of terrestrial and marine sediments. The site is exceptional in demonstrating raised beach deposits which can be assigned to two separate interglacials. These have been correlated with Stage 7 (*c*. 200 ka BP) and Stage 5e (*c*. 125 ka BP) of the deep-sea oxygen isotope record. The photograph shows the Stage 7 raised beach deposits, overlain by loam and head, on the west side of Portland Bill. (Photo: D.H. Keen.)

interglacials can be differentiated broadly on the basis of pollen assemblage zone biostratigraphy, with individual parts of interglacial cycles (sub-stages) being recognized; for example, pre-temperate, early temperate, late temperate and post-temperate sub-stages (Turner and West, 1968). The pollen-based 'climato-stratigraphic' model which has formed the basis of most British Pleistocene studies, is now evidently too simple to represent the full complexity of the climatic record as revealed by the deep-sea sequences (Bridgland, 1994). Also, many former plant communities lack direct modern counterparts, and the geological dictum that 'the present is the key to the past' is not always applicable. Interglacial environments in the British Isles were generally characterized by a climax vegetation of mixed deciduous oak forest. The last time Britain experienced conditions similar to today was about

125 ka, when the interglacial (part of the Ipswichian Stage) lasted about 10 ka.

Unlike the flora, some elements of the Quaternary fauna have evolved. Therefore, certain glacial and interglacial periods can be characterized broadly by distinctive fossil assemblages, particularly those of large mammals. During the last interglacial, for example, creatures such as the hippopotamus, lion and elephant were indigenous to Britain. Likewise, fossils of both terrestrial and marine molluscs and Coleoptera (beetles) can be sensitive indicators of changing climatic conditions by analogy with their present-day environmental tolerances and geographical ranges.

The succession of glacials and interglacials and the growth and decay of ice sheets have been accompanied by equally profound changes in the coastal zone. World sea level has varied in time with the amount of water locked up in the ice sheets, and during glacial stages, world or eustatic sea levels have been lowered. The converse is true during warmer interglacial phases. The level of the land has also varied, sinking under the weight of advancing ice sheets and rising up or rebounding when they melted (isostasy). This complex interplay of changing land and sea levels has left a widespread legacy in Britain, manifested by the many beaches, shore platforms and marine sediments which now lie above the present sea level. This is particularly true in South-West England where prominent shore platforms and raised beach deposits have provided the focus for much research and interest in the Quaternary history of the region (Chapters 6, 7 and 8; Figure 1.2). Equally, a range of submerged shoreline features, drowned forests and valleys provide important evidence for sea levels which were relatively lower in the past. The complex story of changing land and sea levels is superbly illustrated by the sediments and fossils of the Somerset lowlands, a key area for sea-level studies in Britain (Chapter 9).

Significant changes in the courses of rivers and their channel patterns have also occurred in the Quaternary. These are related to changes in discharge, sediment supply and sea level. Some rivers have reworked and built up large quantities of glacially derived sediments along their floodplains. Subsequent down-cutting has sometimes resulted in 'staircases' of terraces both in rock and superficial materials. In some valleys, terraces have been traced for considerable distances and been assigned specific names and ages with respect to their contained fossils and stratigraphic position; in many cases they can be ascribed with some certainty to particular interglacial or glacial phases or, more recently, to the oxygen isotope timescale (Bridgland, 1994). Although terrace stratigraphy has not featured prominently in research on the Quaternary history of South-West England, evidence of this nature abounds: the Axe Valley terraces are significant as the most prolific source of Palaeolithic hand-axes in the South-West, and reveal a unique record of Quaternary events (Chapter 9), while the terraces of the Bath Avon furnish some of the most detailed Quaternary sedimentary evidence found in the Bristol and Bath areas (Chapter 10). Spectacular abandoned channels, such as the Valley of Rocks in north Devon, may have been formed by the combined effects of river capture and coastal retreat, reflecting major landscape changes during the Quaternary; alternatively, glacial meltwater may have played a role in their formation (Chapter 7).

THE CHALLENGE FOR QUATERNARY SCIENCE

Quaternary science has made a fundamental contribution to understanding landscape evolution and past environmental change. This understanding has much broader relevance in areas such as climate dynamics, sea-level changes, the history of flora and fauna and engineering site investigations. Interest in the Quaternary has therefore burgeoned in recent years, and many texts explain the fundamentals of the science and environmental changes in Britain (e.g. West, 1977a; Bowen, 1978; Lowe and Walker, 1984; Bradley, 1985; Dawson, 1992; Jones and Keen, 1993). Change through time is a fundamental aspect of Quaternary studies. Often, successive climatic and environmental changes are recorded by a variety of indicators in layers or sequences of sediments. For example, raised beach sediments may be overlain by periglacial slope deposits and in turn by glacial sediments such as till, thus recording changes from times of relative high sea level and temperate conditions through a periglacial regime to fully glacial conditions. Quaternary environmental changes and processes have not been uniform in their operation throughout Britain. This, coupled with a phenomenal complexity of geological composition and structure, has given rise to a remarkable regional diversity of landforms and Quaternary deposits (Gordon and Sutherland, 1993). The challenge for the Quaternary scientist is to establish the precise nature of the deposits and the sequence of inferred environmental and climatic events at a given locality, and then to classify the results and correlate the site evidence with other sequences and areas (Bowen, 1978). In doing so, the history of geomorphological events and processes can be pieced together, changes in the development of the flora and fauna charted and new insights gained into the processes which drive climate and other environmental changes.

A crucial challenge at present is to relate the fragmentary land-based evidence to the deep-sea oxygen isotope record (Shackleton and Opdyke, 1973; Sibrava *et al.*, 1986). The latter provides the standard stratigraphic scale with which sequences are now routinely correlated (e.g. Bowen *et al.*, 1989; Campbell and Bowen, 1989; Ehlers *et al.*, 1991; Gordon and Sutherland, 1993; Bridgland, 1994). A variety of improved dating techniques, both numerical and relative, extending back beyond the range of the radiocarbon method (e.g. amino-acid, U-series and ^{36}Cl methods), is assisting greatly in this respect (Kukla, 1977; Bowen, 1978; Bowen *et al.*, in prep.). Because of the complexity of the evidence and the range of techniques now available, Quaternary science is based increasingly on a multidisciplinary approach. Although the heart of the subject still relies heavily on geological, stratigraphical methods, the combined efforts of geologists, geographers, geomorphologists, botanists, zoologists and archaeologists are necessary to achieve the maximum resolution of the available evidence.

Chapter 2

The geomorphological evolution and Quaternary history of South-West England: a rationale for the selection and conservation of sites

THE PRINCIPLES AND METHODOLOGY OF THE GEOLOGICAL CONSERVATION REVIEW
S. Campbell and J.E. Gordon

Introduction

The prime aim of the Geological Conservation Review (GCR) was to select sites for conservation which are of at least national, that is British, importance to the sciences of geology and geomorphology; more than 3000 sites have been selected. The full rationale of the GCR and the detailed criteria and guidelines used in site selection are given elsewhere (Crowther and Wimbledon, 1988; Allen *et al.*, 1989; Gordon and Campbell, 1992; Gordon and Sutherland, 1993; Gordon, 1994; Wimbledon *et al.*, 1995; Ellis *et al.*, 1996): this volume presents the detailed scientific justification for the selection of sites representing the Quaternary interests of South-West England. A broad categorization of geological and geomorphological subject matter (e.g. the major subdivisions of the geological timescale) is a prerequisite to site selection. Of the *c.* 100 site selection categories used by the GCR, 24 are concerned with geomorphology and the Quaternary, reflecting the particularly widespread nature and diversity of the evidence. This volume describes sites selected in two 'blocks' of the GCR – the 'Quaternary of South-West England' and the 'Quaternary of Somerset'.

This chapter reviews the most significant topics and themes for which South-West England furnishes important field evidence. In doing so it provides an overview of the geomorphological evolution and environmental history of the area during the Quaternary, which forms an important background to subsequent chapters. In presenting these topics and themes broadly chronologically, the critical elements required in a site conservation network are clearly demonstrated. Further detailed justification for the inclusion of sites within this volume is given in the individual subject chapters and particularly within the 'Interpretation' sections of individual site accounts.

Site selection guidelines and site networks

This volume describes 63 sites which merit conservation because of their significance to the geomorphological evolution and Quaternary history of South-West England. Sites for coastal and fluvial geomorphology, in the sense of modern landforms and processes, and large-scale mass movement features are reviewed elsewhere in the relevant thematic volumes of the GCR Series. For the site selection blocks considered here, site selection was underpinned by the premise that the particular block should be represented by the minimum number of sites. Only those sites absolutely necessary to represent the most important aspects of the geomorphological and Quaternary history of South-West England were therefore selected; unnecessary duplication of interest was thus minimized (cf. Gordon and Sutherland, 1993). To compile the site network, extensive consultations were carried out with appropriate Quaternary scientists, geomorphologists and other specialists, and several hundred sites were assessed before the final listing was produced. The principal authors of this volume were involved extensively in this process but were not the only consultees.

The landscape of Britain displays a rich diversity of Quaternary features and evidence of environmental change, often with distinct regional associations, related for example to a combination of geology, evolution of river systems, mountain glaciation or patterns of sea-level change. South-West England, with its unglaciated granite landscapes, periglacial formations and raised beach deposits, is quite distinctive in its characteristics from, say, the Thames Valley, East Anglia or Scotland. Each of these regions offers a distinctive contribution to the overall picture of Quaternary landscapes and environments in Britain. A prime aim of site selection was to reflect this diversity and to select networks of sites representing the major regional variations in landscape evolution and the history of environmental change during the Quaternary in Britain; hence the regional approach adopted for site selection as in the *Quaternary of Wales* (Campbell and Bowen, 1989), the *Quaternary of Scotland* (Gordon and Sutherland, 1993), the *Quaternary of the Thames* (Bridgland, 1994) and the *Quaternary of South-West England* (this volume). These 'regions' provided a practical basis for site selection; they not only demonstrate distinctive Quaternary features and sequences, and therefore research themes, but also form manageable units. A corollary is that many Quaternary scientists specialize in the history of a particular region or topic within a region, and thus a pool of expertise with detailed knowledge could be established.

The themes and issues that form the focus of the

Quaternary history and research in a given region primarily reflect the nature of the field evidence available. For example, South-West England, by virtue of its position mostly beyond the margins of the Pleistocene ice sheets, preserves an exceptional record of the long-term evolution of the British landscape. Likewise, Scotland is particularly noted for its assemblage of glacial landforms and raised shorelines formed during the Devensian late-glacial. For the Quaternary scientist concerned with conservation, the wealth of detail present in such assemblages of Quaternary landforms and deposits presents problems as well as opportunities, namely in deciding which sites merit conservation. Within the general regional framework, the approach adopted was to identify *networks* of sites that represent the main landscape features, distinctive aspects of Quaternary history and the principal research themes. Such features and themes were recognized at two levels: (a) those relating to the specific characteristics of the area in question (e.g. granite landforms and periglacial features in South-West England); and (b) those relating to national interests or distributions (e.g. pollen biostratigraphy and sea-level changes during the Holocene) for which regional representative sites were required. It should be noted that this categorization relates to the occurrence of the interests and does not imply differences in the importance of sites in the different categories. Thus sites selected for a regionally occurring interest are nevertheless of *national* importance, for example the tors of Dartmoor or the raised beaches of Devon. Indeed, all the sites selected for the GCR Series are considered to be at least of national importance. For South-West England, the site networks considered are those representing:

- long-term landscape evolution
- granite landforms
- Pleistocene cave sequences
- Pleistocene sea-level changes
- periglacial landforms and deposits
- key sequences of deposits for interpreting the distinctive Quaternary history of the region
- Holocene vegetation history

Potential sites representing the key elements of these themes were then identified from extensive consultations and compared to ensure selection of the 'best' sites and to minimise duplication of similar features within each regional block. Site selection decisions were made on the basis of the following guidelines, fuller details of which can be found elsewhere (e.g. Crowther and Wimbledon, 1988; Allen *et al.*, 1989; Gordon and Campbell, 1992; Gordon and Sutherland, 1993; Gordon, 1994; Wimbledon *et al.*, 1995; Ellis *et al.*, 1996): 1. international importance; 2. classic examples; 3. representativeness; 4. uniqueness; 5. being part of a site network; 6. providing an understanding of present environments; 7. historical importance; and 8. research potential and educational value. Sites will fall into one or more of these categories; all things being equal, sites with multiple interests are preferred.

Some sites and areas are recognized as internationally important; they may have formal status as stratotypes, type sites/areas or World Heritage sites, or they may have informal, but widely held, international recognition. None of the sites in this volume has gained formal international status, although the 'Burtle Beds' of Somerset, reflected here by a sub-network of interrelated sites (Chapter 9), and the lengthy Pleistocene sequences such as those at Kent's Cavern and the Torbryan Caves, are demonstrably of international significance; the Palaeolithic site at Broom aspires to this level of importance on archaeological grounds alone.

Likewise, some sites are recognized as classic examples of particular features, and as such are frequently cited in standard textbooks. Examples in the Quaternary of South-West England include the Dartmoor tors (see Merrivale) and the Giant's Rock at Porthleven in Cornwall.

It can be argued that all sites are unique in one way or another, and it was not an aim of the GCR to include features or sites simply because of their 'uniqueness': rather, a unique feature, sequence or landform was only included if it contributed something special or of great significance to an understanding of the Quaternary of South-West England or Britain as a whole. For example, such unique features could be the only known part of the geological record, thus filling an important stratigraphic gap, or landscape feature of their type; they may have no counterpart in the South-West, Britain or indeed internationally. In South-West England, the site at St Agnes Beacon falls into this category, providing the only tangible evidence for the age of a widespread erosion surface (Walsh *et al.*, 1987; Jowsey *et al.*, 1992). Other examples are the unique faunal remains of Kent's Cavern, the landform assemblage of the Valley of Rocks, the remarkable Palaeolithic finds of the Axe Valley and the controversial Fremington Clay (Chapter 7).

Within most GCR site networks, by far the most common type of site selected is usually that representative of a particular rock unit, fossil-bearing bed or landform. It must be stressed that not any representative, characteristic or typical site will suffice, and a series of weightings was applied to the general guidelines to help to distinguish which is the best or most suitable site in each category: for example, preference was given to sites with the most extensive or best preserved record; with the most detailed geochronological evidence; with a particularly long history of study; or with an additional attribute or related geological interest. Thus, a site may be selected as showing the most complete regional representation of phenomena which are quite widespread. Examples include many of the Quaternary stratigraphic sites in this volume, such as those showing the best development of raised beach deposits coupled with the most intricate and informative overlying periglacial sequences (e.g. Pendower, Porth Nanven and the Croyde–Saunton Coast) or the most intensively studied of the 'Burtle Beds' sites with the most detailed faunal and geochronological evidence (Chapter 9).

In some cases, sites were selected for their contribution to a network or sub-network of related sites, for example, the Quaternary sites of the Somerset lowland. This applied especially where there is a geographical component in the scientific interest. Thus, a sub-network of sites may include different aspects of one type of phenomenon which shows significant regional variations in its characteristics, for example, in relation to factors such as underlying geology, climate or relief. Such sub-networks may comprise unique, representative, classic or other types of sites, or a combination thereof. Within a sub-network, there may be 'core' sites, perhaps those showing the most extensive and best researched sequences, while 'accessory' sites may demonstrate significant variations on the main theme. A good example of a smaller sub-network is that covering aspects of granite landform evolution in the South-West. Here, Merrivale can be regarded as the core site, demonstrating an unparalleled range of granite and related landforms – including tors, blockfields, stone runs, altiplanation terraces and earth hummocks – while Two Bridges and Bellever quarries provide critical stratigraphic evidence concerning related weathering and slope processes, which underpins theories on how the Merrivale landform assemblage and granite landscapes in general originated.

Some Quaternary interests have a widespread distribution of evidence that forms part of a national continuum, a prime example being that for vegetation history following the climatic amelioration at the end of the Devensian Stage. Nevertheless, the timing and patterns of vegetation change, for example the spread of trees during the Holocene, shows significant regional variations. Since no single site in any part of the country can encapsulate such an aspect of Quaternary history, a network of reference sites was compiled as far as possible for type regions on a wider national basis (cf. Berglund *et al.*, 1996) and incorporated within the GCR regional blocks.

Certain sites have been selected because they throw light on the development of the present ecological landscape. This is particularly true of the Holocene pollen sites in this volume, such as the Chains (Exmoor), Blacklane Brook and Black Ridge Brook (Dartmoor) which show the progressive modification and management of the landscape by human activity; also the submerged forest and associated beds at Westward Ho!, which provide evidence of a pattern of sea-level changes on a very recent timescale.

A further justification for conserving sites is their contribution to the development of geology in general, and Quaternary science in particular. In Wales, and especially Scotland, certain sites provided critical evidence which firmly established the Glacial Theory in Great Britain, for example, Agassiz Rock (Gordon and Sutherland, 1993) and Moel Tryfan and Cwm Idwal (Campbell and Bowen, 1989). In this context, Kent's Cavern in South-West England was fundamental in demonstrating the connection between early humans and extinct ice-age mammals.

Another reason for including certain sites is their research and educational value, often arising from the controversial interpretations placed upon the evidence they provide. There is some overlap here with the previous guideline, and deciding the point at which a given site-based interpretation or field-tested theory becomes a major advance in geological thought is a moot point and one that only an historical perspective can clarify. Nonetheless, some sites illustrate particularly well the development of scientific thinking on the subject of, for example, landscape history and indeed the debates about processes and chronology which characterize Quaternary studies. In this respect, Two Bridges Quarry on Dartmoor is one of the most significant geomorphological sites in South-West England: ever since D.L. Linton's classic

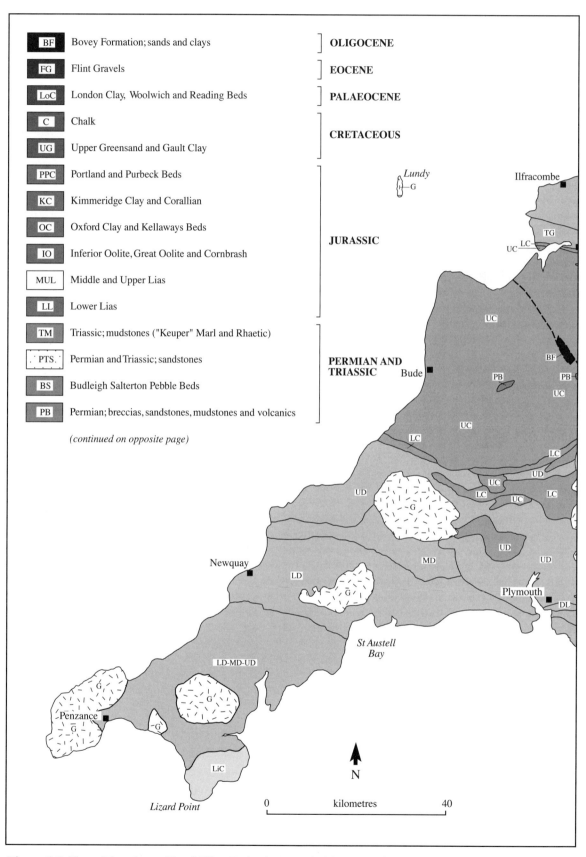

Figure 2.1 The solid geology of South-West England. (Compiled from British Geological Survey sources.)

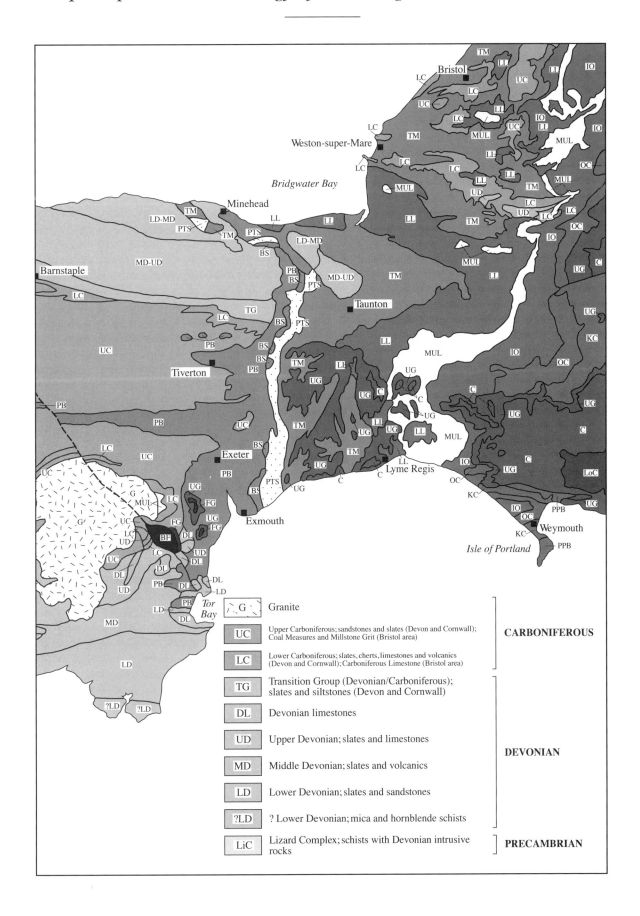

study, *The problem of tors* (Linton, 1955), it has been the focus of repeated field studies and interpretations which have underpinned different theories of granite landscape evolution (Campbell, 1991).

The potential of sites for future research is also another consideration. Although new sites become available, it is important that existing sites are maintained for the application of new research techniques and for testing the latest theories. Thus, the long-term research potential of many sites has been seen as a key factor in their selection; also, some sites will provide the 'standard' against which the evidence from new sites can be compared.

The sites described in this volume can be regarded as the 'building blocks' for reconstructing the Quaternary history and geomorphological evolution of South-West England, as recorded in constituent landforms and sediments. On an entirely practical level, all selected sites must be conservable, meaning in essence: (a) that development planning consents do not exist or else amendments can be negotiated; and (b) that sites are physically viable, for example, in terms of the long-term stability of exposures and their location with respect to the water-table (Gordon and Campbell, 1992).

It is clear from the foregoing that many factors have been involved in selecting the sites proposed for conservation and described in this volume. Sites rarely fall neatly into one category or another; normally they have assets and characteristics which satisfy a range of the guidelines and preferential weightings used. A full appreciation of the reasons for the selection of individual sites cannot be gained from these few pages. The full justification and arguments behind the selection of particular sites are only explained satisfactorily by the full site accounts given in subsequent chapters. This, after all, is the *raison d'être* of the GCR Series of publications.

The following synthesis of geomorphological evolution and Quaternary history in South-West England, addresses the major topics and themes which have concerned scientists over many years, and the field evidence on which many theories and interpretations have been based; in doing so, it develops the principal themes required in a network of conservation sites.

THE GEOMORPHOLOGICAL AND QUATERNARY EVOLUTION OF SOUTH-WEST ENGLAND: A SYNTHESIS
S. Campbell

The pre-Quaternary inheritance

Introduction

The pre-Quaternary geology of South-West England is detailed in a number of texts (e.g. Edmonds *et al.*, 1975, 1979, 1985; Durrance and Laming, 1982; Green, 1985; Goode and Taylor, 1988) and only a broad outline, sufficient as geomorphological background, is given here: outcrop details are summarized in Figure 2.1. Although entitled the *Quaternary of South-West England*, this volume also describes sites which furnish important evidence for reconstructing the longer-term geomorphological history of Britain, and these essentially 'Tertiary Period' sites encapsulate a range of geological and climatic controls which were fundamental in establishing the broad outline of the landscape of South-West England in pre-Quaternary times.

The pre-Cenozoic basement

The oldest rocks in South-West England occur in the Lizard Peninsula, where igneous and metamorphic types of pre-Devonian and possibly Precambrian age are exposed. From a geomorphological point of view, the most prominent features of South-West England's relief are the granite bosses of late Carboniferous to Permian age. Although now exposed on the surface in five separate major outcrops, gravity surveys show them to be linked at depth in one major granite batholith, which extends from Dartmoor westward to the Isles of Scilly and beyond to the south-west (Figure 2.2). The granites intrude Devonian and Carboniferous sedimentary 'country' rocks which are arranged in a major east to west-trending syncline formed during Armorican (Hercynian) crustal movements (Figure 2.1).

In contrast, Permian and Triassic rocks, principally sandstones and marls, dominate the Avon and Somerset lowlands considered here. They are punctuated, particularly in the Weston-super-Mare, Bristol and Cheddar districts by inliers of Carboniferous Limestone Series rocks (Whittaker and Green, 1983). The approximate eastern limit of the area under consideration here, roughly

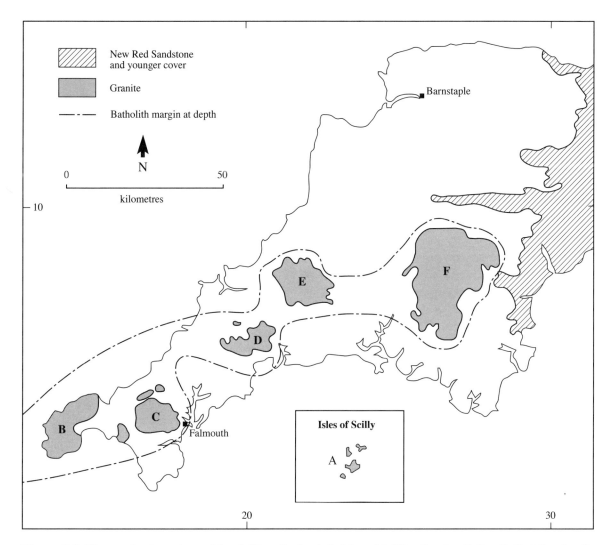

Figure 2.2 The granite intrusions of South-West England: A, Isles of Scillies Granite; B, Land's End Granite; C, Carnmellis Granite; D, St Austell Granite; E, Bodmin Moor Granite; F, Dartmoor Granite. (Adapted from Floyd *et al.*, 1993.)

between Bath in the north and Portland in the south, is dominated by outcrops of Jurassic and Cretaceous strata. Over much of South-West England, however, Upper Cretaceous strata are absent: marginal facies of this age occur west and north-west of the Isles of Scilly and in the English Channel, but although the Dartmoor granite was probably transgressed by the Cenomanian Sea, the Chalk cover was thin and has since been removed. There is no evidence that the transgression extended to the granite masses farther west (Hancock, 1969).

The main outline of South-West England, as a landmass separating the Celtic Sea to the north and the English Channel to the south, evolved probably after the granites were emplaced, through Permian and Mesozoic times and into the Cenozoic (King, 1954; Hancock, 1975; Stephens, 1980). In the mid-Tertiary, earth movements reactivated a number of Permo-Carboniferous structures and imposed a set of NNW to SSE-trending dextral wrench faults, and perhaps also caused a general north to south tilting of the landmass. These faults are significant from a geomorphological point of view, controlling the NNW–SSE alignment of the western edge of Exmoor with the Bovey Tracey Basin, the trend of the Petrockstow Basin and perhaps the alignment of the troughs and ridges of east St Austell Moor (Dearman, 1963; Shearman, 1967; Everard, 1977; Stephens, 1980). Major displacements of the Lizard–Dodman–Start igneous and metamorphic complexes may also have occurred at this time.

British Geological Survey maps and Memoirs

The South-West England Unit of the British Geological Survey was formed in 1961, and the subsequent output of 1:10 000-scale geological maps and accompanying Memoirs has been invaluable to geologists and geomorphologists alike. The programme started with the production of the Okehampton (338) Sheet and Memoir (Edmonds *et al.*, 1968), and continued westwards on to Boscastle and Holsworthy (322/323; McKeown *et al.*, 1973), these sheets being dominated by Devonian and Carboniferous rocks. A Memoir was also produced on the geology of the coast between Tintagel and Bude (Freshney *et al.*, 1972). The mapping then spread northwards into predominantly Carboniferous rocks, resulting in the production of the Chulmleigh (309; Freshney *et al.*, 1979a) and Bude and Bradworthy (307/308; Freshney *et al.*, 1979b) maps and Memoirs. Moving eastwards, the Bideford (292; Edmonds *et al.*, 1979) and Ilfracombe/Barnstaple (277/293; Edmonds *et al.*, 1985) sheets mostly depict Devonian rocks.

In Somerset, the Taunton (295; Edmonds and Williams, 1985) and Weston-super-Mare (279 and parts of 263 and 295; Whittaker and Green, 1983) maps and Memoirs followed, as did Shaftesbury (313; map 1995). These sheets illustrate mostly Mesozoic strata. As the Unit expanded, work in western Cornwall led to the Penzance (351–358; Goode and Taylor, 1988) and Falmouth (342; Leveridge *et al.*, 1990) maps and Memoirs.

Maps and Memoirs have also been produced by the University of Exeter under contract to BGS. These comprise Newton Abbot (339; Selwood *et al.*, 1984), Tavistock (337; map only) and Trevose/Camelford (335–336; Selwood *et al.*, 1997). Memoirs describing the complex geology of the Bristol (Kellaway and Welch, 1993) and the Wells and Cheddar districts (280; Green and Welch, 1965) are also available.

Pre-Quaternary landscape evolution

The oldest sedimentary evidence for Tertiary conditions and landscape evolution comes from south and east Devon, where the Chalk and other strata are succeeded by a complex series of plateau deposits comprising residual flint gravels (e.g. *in situ* clay-with-flints and the Aller and Buller's Hill gravels) and associated pedogenic and diagenetic horizons (Edwards, 1973; Hamblin, 1973b, 1974a,

1974b; Isaac, 1979, 1981, 1983a, 1983b; Green, 1985). These deposits provide crucial evidence for a continuum of processes from the end of the Cretaceous into the Eocene: they show evidence for a series of superposed weathering profiles which reflect a protracted pedological, diagenetic and geomorphological history. This began with the emergence of the post-Chalk land surface at the end of the Cretaceous, and continued with several phases of soil development in tropical and subtropical conditions (evidenced by a variety of silcrete and laterite horizons), and ended with redistribution of these products by fluvial agencies (e.g. the Aller and Buller's Hill gravels) (Chapter 3; Figure 3.1). A Palaeocene age for the controversial clay-with-flints seems highly likely because their redistributed facies and weathering products are overlain by Upper Eocene to Oligocene sediments of the Bovey Formation (Wilkinson *et al.*, 1980; Edwards and Freshney, 1982; Isaac, 1983b). The latter contain kaolinitic clays derived from the weathering of granite (Dartmoor) and Palaeozoic sediments, as well as alteration products from the granite (Bristow, 1969; Green, 1985). This suggests that the Dartmoor granite was intensely kaolinized and/or deeply weathered (Chapter 4) prior to Upper Eocene times, most probably during the Upper Cretaceous and Palaeocene (Green, 1985). Post-Eocene modification to the upland landscape of South-West England would seem to have been slight in comparison.

However, over much of the South-West Peninsula, particularly in west Cornwall, a vast break in the geological succession exists, on the one hand, between the Palaeozoic and igneous basement rocks and the small remnants of rocks of Tertiary age, such as those at St Agnes Beacon (mid-Oligocene and Miocene), Polcrebo Downs and Crousa Down (unknown) and St Erth (late Pliocene) (Chapter 3). With the exception of the Palaeocene deposits of east Devon and substantial Oligocene sediments in the Bovey Tracey and Petrockstow basins, these small outliers constitute the most significant sedimentary evidence for establishing the nature and timing of land-shaping events. The remaining evidence is entirely erosional in nature (e.g. planation surfaces) and lends little precision to the interpretation of events (Kidson, 1977).

Traditionally, geomorphologists have viewed the early to mid-Tertiary as a time when the landmass of South-West England was subjected to alternating phases of marine inundation and subaerial exposure and weathering: these conditions have been used to account for a multiplicity of perceived ero-sion surfaces (Chapter 3; St Agnes Beacon) (e.g. Balchin, 1937, 1946, 1952, 1964). Recent evidence from St Agnes Beacon in west Cornwall, where Miocene and mid-Oligocene sands and clays over-lie a prominent erosion surface (*c.* 75–131 m OD), however, suggests that surfaces above this general level can be no younger than Miocene. This lends support to the view that much of the South-West Peninsula has existed, more or less in its present form, since perhaps as early as the Eocene. Landscape evolution subsequent to this is likely to have been slow, the sole depositional evidence for a subsequent marine incursion onto the South-West Peninsula being a minor transgression of late Pliocene age at St Erth (Mitchell *et al.*, 1973a; Chapter 3).

In recent years, there has been an increasing ten-dency to invoke only one complex polygenetic erosion surface (e.g. Coque-Delhuille, 1982; Battiau-Queney, 1984), the constituent landforms having been shaped in tropical and subtropical environments (Upper Cretaceous and Palaeocene) and uplifted since the late Miocene (Green, 1985; Bowen, 1994b).

Quaternary events prior to the Devensian Stage

Introduction

Although in current classification, the St Erth marine beds are placed in the late Pliocene, it can be argued that they more properly represent the early part of the cycle of profound climatic fluctua-tions which characterizes the Quaternary (Chapter 3). This series of fluctuations, hallmarked by major falls of sea level during cold and sometimes glacial conditions, and correspondingly higher sea levels during more temperate phases, is now compre-hensively demonstrated from the deep-sea sedimentary evidence (e.g. Shackleton and Opdyke, 1973). Increasingly, the land-based evi-dence for these profound climatic and environmental shifts is being correlated with the developing deep-sea oxygen isotope framework through a variety of geochronological and other techniques.

Early glaciation

In South-West England, the earliest evidence extends the known record to around 0.7 Ma, but even this is insecurely dated; no known evidence survives from the earlier part of the Quaternary.

The oldest known glacial deposits occur in the Bristol district (Hawkins and Kellaway, 1971; Gilbertson and Hawkins, 1978a, 1978b; Andrews *et al.*, 1984; Bowen, 1991, 1994b), where they under-lie fluvial and estuarine sediments at Yew Tree Farm and Kennpier (Chapter 10). The latter con-tain *Corbicula fluminalis* (Müller), dated by amino-acid geochronology to the Cromerian Complex: formerly they were attributed to Oxygen Isotope Stage 13 (Bowen and Sykes, 1988; Bowen *et al.*, 1989), but current thinking places them in Stage 15 (Bowen, 1994b). The underlying glacial deposits therefore pre-date these stages, and since the oxygen isotope record indicates a major cli-matic deterioration and one of the largest ice-volume episodes during Stage 16 (*c.* 600 ka), the latter ascription would seem possible (Campbell and Bowen, 1989; Bowen, 1994b). (It is worth noting, however, that the freshwater fauna from Kenn differs significantly from faunas at Cromerian Complex sites farther east, such as Sugworth, Oxfordshire (Gilbertson, 1980) and Little Oakley, Essex (Preece, 1990; Bridgland, 1994).) It is also possible that the glacial deposits of the Bristol area are similar in age to the scattered outcrops of 'Older Drift' glacial sediments found in South and south-west Wales, having been deposited by the same coeval Welsh and Irish Sea ice sheet which moved eastwards, effecting depo-sition on both sides of the Bristol Channel (Campbell and Bowen, 1989).

Whether this Stage 16 glaciation was responsible for the giant erratics found around the coasts of Devon and Cornwall is not known. Their often extremely large size and isolated, low-level distrib-ution have traditionally led to the view that they are the product of ice-rafting, not a major glacial invasion *per se*. This possibility is given much cre-dence in theories which propose the likelihood of substantial glacio-isostatic depression beyond the margins of former ice sheets (e.g. Eyles and McCabe, 1989, 1991; Bowen, 1994b), and the con-sequent tidewater or glaciomarine character of these ice sheets. Recently, the intriguing possibil-ity that some of these erratics could have originated from as far away as Greenland or been emplaced on ice floes from a disintegrating, pre-last glacial maximum Laurentide ice sheet, have been raised. Bowen (1994b) has also raised the possibility that enigmatic deposits on Crousa Down and glacial gravels on Lundy and in the Isles of Scilly were introduced at this time – although the latter had subsequently been redistributed by periglacial processes (but see Chapter 8).

The extent to which this earliest-known glacial event modified the landscape of the South-West is not fully known, although the survival of Tertiary sediments and other landscape palimpsests, such as erosion surfaces, and the localized distribution of erratic material, which is confined to the Somerset lowland and the Bristol and Bath areas (Hunt *et al.*, 1984; Chapters 9 and 10), probably mitigate against the Peninsula having been completely overrun by ice during this or any other stage of the Pleistocene, as does the lack of landforms characteristic of glacial erosion. Indeed, the geomorphological position of the glacial deposits in the Bristol district and the 'Older Drift' of South and south-west Wales – in the case of the latter, confined to occurrences in the entrance to Milford Haven (West Angle Bay) and to highly dissected coastal plateau deposits – suggests that the geomorphology of the coastal fringe was then similar to that of the present (Campbell and Bowen, 1989; Bowen, 1994b).

Cromerian Complex events

Although events and conditions of the Late Pleistocene in South-West England, particularly those of the Devensian Stage and later, are generally well documented and understood, evidence for the Early and Middle Pleistocene is much less clear. However, four main phases of temperate, probably interglacial, climate (not including the Holocene) can be adduced from evidence in the region, particularly raised beach sediments (see below). Cromerian Complex events, probably equivalent to Oxygen Isotope Stages 13–21 (Bridgland, 1994; Bowen, 1994b), are recorded by the estuarine deposits at Kennpier and perhaps by 'temperate' mammal remains at Kent's Cavern near Torquay (the 'sabre-tooth' cat fauna). Although a fossil mammal fauna at Westbury-sub-Mendip (Westbury 1 fauna) also probably dates from this broad period (Heal, 1970; Bishop, 1974, 1982; Stringer *et al.*, 1996), no other deposits of this age are known. High relative sea levels during this period, however, may have fashioned, in part, a variety of prominent shore platforms (e.g. Mitchell, 1960; Stephens, 1966b; Stephens and Synge, 1966; Kidson, 1971). An absence of associated fossiliferous marine deposits nonetheless precludes correlation via amino-acid geochronology with the oxygen isotope timescale. Although evidence from fossils and artefacts is far from clear, it seems highly

likely that humans were well established in the region by late Cromerian Complex times (see Kent's Cavern; Cullingford, 1982). Recent work carried out on the later Westbury faunas (Westbury 2 and 3 faunas) shows them to be post-Cromerian *sensu stricto* (i.e. West Runton) and pre-Hoxnian (Stringer *et al.*, 1996).

Anglian and Saalian events

The events of the profoundly cold Anglian Stage which followed (Oxygen Isotope Stage 12; *c.* 450 ka), are not clear from landform or sedimentary evidence in South-West England. Although extensive glaciation occurred in eastern England and the Midlands at this time, and may have left its mark in South Wales (the Paviland Moraine, Gower), no specific features in the South-West are clearly attributable to the Anglian ice sheet. Possible glacial deposits on the Isles of Scilly, Lundy Island and in the Barnstaple area (the Fremington Clay) and at other locations along the north Devon and north Cornwall coasts, have attracted considerable research interest and have been ascribed to an Irish Sea ice sheet of disputed but possibly Anglian, Saalian (Wolstonian; see below) or Devensian age (see also Chapters 6, 7 and 8): Stephens (1966b) also attributed a variety of dry and misfit valleys in north Devon to meltwater erosion at the margins of a Saalian ice sheet (Chapter 7). Indeed, until the last 15 years or so, there was broad consensus that the maximum limit of Pleistocene glaciation in the South-West could be marked by a line running from the Isles of Scilly broadly north-west, fringing the coast of north Cornwall and north Devon (Figure 2.3): some workers maintained that this glacial 'maximum' was attained in the Anglian (Mitchell, 1960; West, 1968, 1977a), but many others have ascribed a 'Wolstonian' age (*sensu* Mitchell *et al.*, 1973b) to this major glacial event (e.g. Stephens, 1966b, 1970a; Kidson, 1971, 1977; Edmonds, 1972). However, in the last ten years or so, the Wolstonian Stage of the British Pleistocene, as defined by Mitchell *et al.* (1973b), has been intensely scrutinized: many deposits formerly held to be Wolstonian are now believed to be Anglian in age (Perrin *et al.*, 1979: Sumbler, 1983a, 1983b; Rose, 1987, 1988, 1989, 1991). There is also increasing evidence to suggest that more than a single climatic cycle separates the Hoxnian and Ipswichian (warm) stages (e.g. Bridgland, 1994). For convenience, therefore, the continental term 'Saalian', which recognizes this climatic complexity, is used

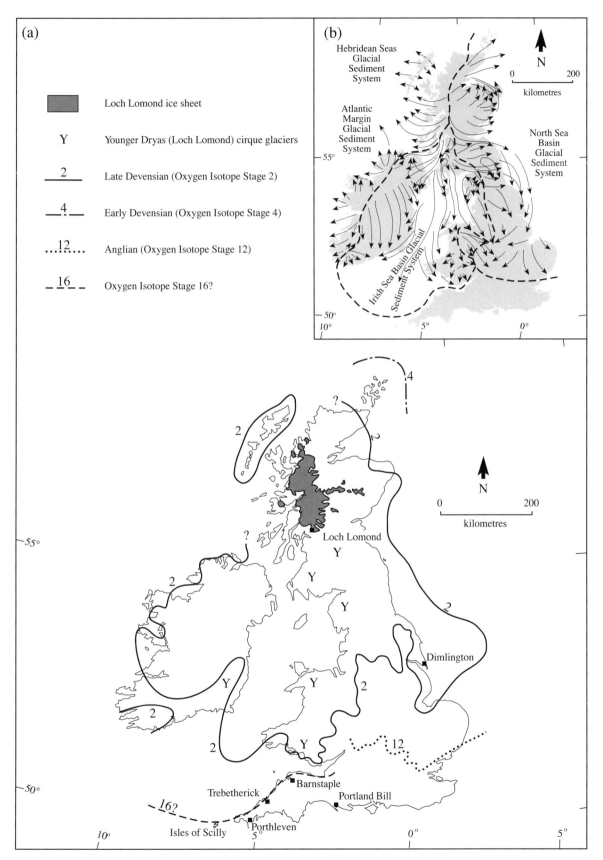

Figure 2.3 (a) Reconstructed Pleistocene maximum ice limits after Bowen (1994a) and Gray and Coxon (1991). (b) British glacial sediment systems. After Charlesworth (1957), and Bowen (1991). (But also see Figure 8.4.)

Figure 2.4 Brannam's Clay Pit in the 1960s, showing extensive working faces in the Fremington Clay. Although long held to be a 'Wolstonian' glacial deposit, the age and origin of the Clay remain highly controversial (Chapter 7). (Photo: N. Stephens.)

in this volume to represent the time-stratigraphic intervals between the Anglian and Ipswichian stages.

This redefinition raises the important question of whether deposits in the South-West, formerly regarded as Wolstonian, can now simply be redefined as Anglian in age? The answer seems to be no. Moreover, the 'glacial' sequences in the region have themselves been the subject of considerable reinterpretation, with the possibility emerging that some, such as those in the Isles of Scilly and at Fremington, are as recent as Late Devensian in age and perhaps even glaciomarine in origin (Scourse, 1985a, 1987; Eyles and McCabe, 1989; Scourse *et al.*, 1991; Bowen, 1994b) (see Devensian Stage; this chapter). For much of its length, therefore, the Pleistocene 'maximum' ice limit of previous workers is now in even more doubt.

The problematic 'Boulder Gravel' at Trebetherick on the north Cornish coast, long held by some workers to be glacigenic and Saalian in age (Clarke, 1969; Stephens, 1970a), is the subject of reinterpretation in this volume (Chapter 6): a glacial origin can no longer be sustained for this unit. The most recent attempts to date the Fremington Clay, using optically stimulated luminescence (OSL) techniques, have provided ambiguous results, although an age at least as old as Anglian seems likely (Croot *et al.*, 1996).

The contentious proposition that Anglian Stage ice (*c.* 450 ka) transported the 'bluestones' of Preseli to Salisbury Plain and ultimately Stonehenge (Kellaway, 1971; Thorpe *et al.*, 1991), has now been shown to be untenable: a ^{36}Cl age determination on an igneous rock from the Stonehenge collection, in the Salisbury Museum, shows that probably the rock was still buried at source during the Anglian (*c.* 400 ka), and did not become exposed to the atmosphere by denudation until the Late Devensian (Green, 1993, 1997; Bowen, 1994b; Bowen *et al.*, in prep.). This evidence adds to the cogent arguments put forward against transport of the bluestones by ice (e.g. Green, 1973, 1997; Kidson and Bowen, 1976; Darrah, 1993; Scourse, 1997), lending further weight to the possibility that the stones were quarried from Preseli by prehistoric people, after the Late Devensian, and taken to the site of Stonehenge. An Anglian age cannot, however, be ruled out for the giant erratics scattered along the coasts of South-West England (e.g.

Figure 2.5 Exposures in the Axe Valley terrace gravels at Kilmington Gravel Pit during the late 1960s. (Photo: N. Stephens.)

Stephens, 1961a, 1966b; Edmonds, 1972; Edmonds *et al.*, 1979).

What is clear, however, is that during the Pleistocene cold stages between the Cromerian Complex and Devensian times (Oxygen Isotope Stages 12, 10, 8 and 6), much of the landscape of South-West England was subject to considerable modification through periglacial processes. Bowen (1994b) observes that periglacial denudation through some 50 ice-age cycles since 2.4 Ma was probably the main agent of geomorphological development. Although the uplands of the South-West, especially the elevated granite terrains of Dartmoor and Bodmin Moor, probably suffered significant cryonival modification during Pleistocene cold stages, the effects of intense periglacial activity during the Devensian Stage have tended to mask or remove evidence of earlier phases. Thick 'head' deposits are one of the most characteristic features of the region (Stephens, 1970a; Mottershead, 1971, 1977a, 1977b; Scourse, 1985a, 1987), but their chronostratigraphic subdivision is often tenuous, and there are few sites where there is unequivocal evidence for pre-Devensian periglacial activity. However, Saalian slope deposits are known from beneath the Swallow Cliff (Middle Hope) raised

beach deposits (Briggs *et al.*, 1991). These contain fossil molluscs which indicate an unstable grassland environment. It is highly likely that other periglacial deposits and landforms from pre-Devensian cold stages do survive locally in South-West England, although their differentiation from Devensian products is, in most instances, highly problematic.

Even the many caves, where deposits might have accumulated and been protected, bear little clear evidence for pre-Devensian cold conditions. One of the most notable exceptions, however, is Tornewton Cave (Chapter 5), where a debris flow, containing a 'cold' fauna of possible Stage 6 age, is present. Linking cave deposits with sedimentary sequences outside them has proved difficult. There is some suggestion, however, that an ancient terrace of the River Dart, lying at *c.* 83 m OD, pre-dates Ipswichian Stage sediments lying in Joint Mitnor Cave below (Chapter 5): a Saalian age for the terrace has thus been tentatively suggested (Cullingford, 1982).

The famous terraces of the River Axe provide unique sedimentary and fossil evidence for this controversial timespan (see Broom Gravel Pits, Chapter 9; Figure 2.5), and have yielded numerous

Acheulian hand-axes. Here, a series of cold-climate, braided stream gravels and sands are interrupted by pollen-bearing floodplain deposits indicative of more temperate conditions. The terrace gravels probably accumulated during cold-climate conditions in the Saalian Stage; the still-water floodplain deposits bear witness to a more temperate intra-Saalian event (Oxygen Isotope Stage 7, 9 or 11?) for which there is currently no firm ascription (Stephens, 1970a; Green, 1974b; Scourse, 1984; Shakesby and Stephens, 1984). There is also evidence for this timespan in the terrace deposits of south-east Somerset, at Langport Railway Cutting, Portfield and Hurcott Farm (Chapter 9) and in the Avon Valley (Chapter 10).

It has been speculated that permafrost conditions during the Saalian were responsible for massive contortions in the Oligocene beds of the Bovey Basin (Dineley, 1963; Gouldstone, 1975; Jenkins and Vincent, 1981). It is possible, however, that the features were formed by basin subsidence or loading (Straw, 1974), and even if a periglacial origin is accepted, a Devensian age cannot be ruled out. A large alluvial 'cone' at the mouth of the Erme Gorge, on the southern edge of Dartmoor, has also been attributed to fluvial and mass movement processes in a Saalian (Wolstonian) periglacial environment (Gilbertson and Sims, 1974).

Raised beach sediments and amino-acid geochronology

The most significant developments in determining pre-Devensian events within the region have come from the application of amino-acid dating techniques to the widespread raised beach deposits and, most recently, to fluvial deposits. The ages of the raised beach sediments have long been disputed, with both Hoxnian (Mitchell, 1960, 1972; Stephens, 1966b, 1970a, 1974, 1977) and Ipswichian (Zeuner, 1959; Edmonds, 1972; Kidson, 1974, 1977; Kidson and Wood, 1974) ages having been suggested. In general, the coastal Pleistocene sequences of the region appear relatively simple with the shore platform(s) at many localities being succeeded by raised beach deposits and a variety of head, blown sand and colluvial horizons. However, determining the precise age(s) of the raised beach deposits in these sequences, absolutely vital to the interpretation and chronological subdivision of the overlying beds, has led to numerous disputes.

In the late 1970s/early 1980s, amino-acid and U-series studies offered, for the first time, a sound

basis for dating and correlating British raised beach deposits. This work, much of which has been carried out on coastal sequences in South Wales and South-West England, has been fundamental in assigning British terrestrial sequences to the oxygen isotope framework. Early amino-acid work by Andrews *et al.* (1979), on fossil protein in gastropods and bivalves from raised beaches in south-west Britain, suggested that the latter had been formed during at least two, possibly three, high stands of Pleistocene sea level: further refinement of the technique and wider surveys upheld the hypothesis of two separate interglacial beaches (Davies, 1983). In this respect, the evidence from Portland Bill has been vital, with the raised beaches there now having been ascribed with confidence to Oxygen Isotope Stage 7 (Figure B in Preface; Bowen *et al.*, 1985; Bowen and Sykes, 1988; Campbell and Bowen, 1989) and Stage 5e (Davies, 1983; Davies and Keen, 1985; Campbell *et al.*, in prep.).

The work of D.Q. Bowen and colleagues in both South Wales and South-West England has been fundamental in providing statistical corroboration for the numbers and ages of these high sea-level events (Bowen *et al.*, 1985; Bowen and Sykes, 1988). Aminostratigraphy, standardized to *Littorina littorea*, now shows strong evidence for three high sea-level stands in South-West England. Two of these stands are correlated with Oxygen Isotope Stage 7 (*c.* 200 ka) and Stage 5e (*c.* 125 ka): an earlier high sea-level event, identified from an older reworked fauna, is ascribed to Oxygen Isotope Stage 9 (*c.* 320 ka), although *in situ* raised beach sediments attributable to this oldest event are unknown either in South Wales or South-West England (cf. Campbell and Bowen, 1989).

In situ occurrences of the Stage 7 raised beach deposits occur at Portland Bill and several other sites, and with the reworked Stage 9 fauna provide critical evidence for two intra-Saalian temperate episodes and high sea-level stands, and introduce the intriguing possibility that parts of the overlying terrestrial, largely periglacial, sequences may date from a variety of Saalian cold-climate stadial events (cf. Oxygen Isotope Stages 6, 8 and 10). The raised beach deposits at Swallow Cliff have yielded amino-acid ratios consistent with an intra-Saalian or Ipswichian age. It is still clear, however, that probably the bulk of terrestrial materials found in the coastal sections, attests to cold-climate processes in the Devensian Stage.

Virtually nothing is known about sea levels around South-West England during the colder, pre-

Ipswichian, stages of the Pleistocene, although it is generally assumed that they were low: a variety of submarine valleys and cliffs may have been cut, at least partly, during such low sea-level stands (Kidson, 1977).

Although not currently a GCR site, the sea-caves of Berry Head, Torbay, have yielded the densest concentration of Uranium-series dates on speleothem for any site in the region. This recent and detailed work (Proctor and Smart, 1991; Baker and Proctor, 1996) gives valuable evidence for dating the temperate climatic events which allowed speleothems to develop and also adds to the evidence for dating the high sea levels accompanying interglacials.

The caves on Berry Head were mainly exposed by quarrying of the Devonian limestone during the 19th century. Baker and Proctor (1996) list twelve named caves in the rocks of the Head, and provide a plot of their height against area (their fig. 58) which shows that the caves cluster strongly around 3 m, 8.5 m and 25 m above mean sea level. Proctor and Smart (1991) and Baker and Proctor (1996) conclude that these cave heights represent former sea levels, and thus the sediments and speleothems in the caves provide a datable sequence of Middle Pleistocene temperate and marine events. The dates obtained from the caves were mostly from Corbridge Cave, heights below 10 m above mean sea level, but some dates were also determined from The Hole-in-the-Wall No. 1 and Hogberry Caves at around 25 m OD.

The dates obtained also centre on three timespans which match the three height concentrations. The lower levels have yielded age determinations of between 155 and 116 ka BP. The only temperate stage to fall within that time slot is Oxygen Isotope Stage 5e, whose deposits at Thatcher Rock on the north side of Torbay were dated by Mottershead *et al.* (1987), primarily by amino-acid geochronology, to the same temperate phase. The levels around 8.5 m gave dates concentrated around 210 ka BP (range 184–244 ka BP), within Oxygen Isotope Stage 7, and match the suggested estimates of Mottershead *et al.* (1987) for the age of the Hope's Nose raised beach which occurs at a slightly higher level on an open site on the north side of Torbay. The higher levels, at 25 m, which have no expression in raised beach deposits outside caves in the area described in this volume, have given dates with a mean of 328 ka BP (range 287–412 ka BP). These cannot be younger than Stage 9.

These dates provide good evidence for the ages of the major late Middle Pleistocene high sea-level events in South-West England, and also provide calibration of amino-acid chronologies derived from raised beach sites outside the caves (see Bowen, 1994b). Continuing work by Baker and Proctor (1996) is beginning to use annual laminae within the speleothems to determine depositional palaeotemperatures. The undoubted importance of these caves clearly necessitates future consideration of their conservation status.

The Ipswichian Stage

Raised beach deposits attributable to the Ipswichian Stage (Oxygen Isotope Stage 5e) are very common in the coastal sequences of the South-West. They bear witness to temperate conditions with sea levels close to those of the present day, perhaps a few metres higher (Bowen, 1994b). The most widespread interglacial marine deposits in the region are the Burtle Beds of the Somerset lowland (Bulleid and Jackson, 1937, 1941; Kidson *et al.*, 1974; Kidson, 1977; Hughes, 1980; Chapter 9). Until recently, it has been assumed that these deposits bear witness to a major marine transgression in the Ipswichian, and while this holds true in some areas, recent aminostratigraphic studies have revealed a range of ages for marine sediments in this area (Hunt *et al.*, 1984; Chapter 9). The same is likely to be the case for interglacial marine sediments in the Vale of Gordano, Bristol (ApSimon and Donovan, 1956; Chapter 10).

The most modern amino-acid measurements also suggest that there may even have been two separate sea-level stands during Oxygen Isotope Stage 5e itself (cf. Sherman *et al.*, 1993), and that these could correspond to orbitally tuned Oxygen Isotope events 5.53 (*c.* 125 ka) and 5.51 (*c.* 122 ka) (Martinson *et al.*, 1987; Bowen, 1994b). It cannot therefore be assumed that Oxygen Isotope Stage 5e was a period of climatic and sea-level uniformity, and raised beach deposits of both stands are likely to be present in the region although they have yet to be differentiated with any certainty. This lends support to Sutcliffe's (1976, 1981) contention that two 'temperate' faunas of different but Ipswichian ages are present in some cave sequences (see below).

Away from the coast, the evidence for temperate conditions both in the Ipswichian and Saalian stages is more isolated, again being largely restricted to 'protected' locations such as caves. Whereas rich 'cold' mammal faunas from the Devensian Stage are common in the Devon and

Mendip caves, evidence for temperate events and conditions is again more sparse. Probably the richest faunal assemblage attributed to the Ipswichian Stage comes from Joint Mitnor Cave in Devon (Sutcliffe, 1960) where the remains of at least 18 species including *Hippopotamus*, hyaena and *Palaeoloxodon antiquus* provide the hallmark of the stage and clearly bear witness to temperate conditions. On the assumption that a widely recorded British interglacial fauna with *Hippopotamus* is everywhere of Ipswichian age (e.g. Sutcliffe, 1981), then fossiliferous deposits which include *Hippopotamus* at Durdham Down near Bristol (Donovan, 1954; Stephens, 1970a), Tornewton Cave near Buckfastleigh (Sutcliffe and Zeuner, 1962) and Eastern Torrs Quarry and Milton Hill caves, Devon (Sutcliffe, 1960) are also likely to date from this time: an Ipswichian age for the remarkable accumulation of *Hippopotamus* material, the bone remains of at least 17 individuals, found during excavations for the Honiton by-pass (Turner, 1975) is also likely. Further evidence for temperate conditions during the Ipswichian Stage came from a brick pit in Barnstaple, where remains of straight-tusked elephant were discovered in 1844 (Arber, 1977), in an area mapped subsequently by Edmonds (1972) as an Ipswichian terrace.

Other evidence of Ipswichian temperate conditions comes from the Hutton Bone Cave (Weston district), where a single bone of *Hippopotamus amphibius* was found, and possibly Bleadon Bone Cave where a 'temperate' to 'cold' fauna including *Palaeoloxodon antiquus* is also recorded (Donovan, 1954, 1964; Hawkins and Tratman, 1977; Whittaker and Green, 1983). Molars of *Dicerorhinus kirchbergensis* recorded from a fissure in a quarry at Milton, Weston-super-Mare, also indicate a warm climate: an Ipswichian age is possible, but an earlier episode cannot be ruled out (Whittaker and Green, 1983). The lower cave earth at Rhinoceros Hole (Wookey Hole) contained bones and teeth of straight-tusked elephant, extinct rhinoceros (*Stephanorhinus (Dicerorhinus) hemitoechus*) and *Hippopotamus amphibius* for which an Ipswichian age is likely (Stuart, 1982a, 1982b).

During the Ipswichian Stage (Oxygen Isotope Stage 5e), it is likely that substantial weathering and soil formation took place. Clear evidence for this is generally lacking from the region, although a variety of weathering profiles found in presumed deposits of Saalian age (such as the Fremington Clay and a variety of head deposits), have been attributed to subaerial chemical weathering in the Ipswichian (Stephens, 1970a, 1977). Any products of Ipswichian weathering of the granite terrains of the South-West have either been redistributed by periglacial processes in the Devensian Stage (see below), or are thus far indistinguishable from the products of earlier weathering episodes, such as those which occurred during previous interglacials and particularly during Upper Cretaceous and Palaeocene times.

Good evidence for Ipswichian rubification and calcrete development, however, is known from the pre-Ipswichian terrace gravels at Hampton Rocks Cutting and at Langport Railway Cutting, and a well-marked rubified palaeosol of probable Ipswichian age is present at Bourne (Chapters 9 and 10).

The Devensian Stage

Early and Middle Devensian sub-stages

Much landform and sedimentary evidence in the South-West can be attributed with confidence to cold-climate processes during the colder parts of the Devensian Stage (Oxygen Isotope Stages 5d, 5b, 4 and 2), particularly during the Late Devensian Dimlington sub-stage (occurring within Oxygen Isotope Stage 2): evidence for temperate episodes and conditions within the stage, however, is extremely rare. Some such evidence is available for the southern Somerset lowland, where isolated deposits are interbedded with cold-stage fluvial gravels (Hunt, 1987).

Although recent work has demonstrated the possibility of an Early Devensian glaciation in North Wales (Addison and Edge, 1992) and eastern Scotland (Bowen, 1991; Gordon and Sutherland, 1993), there is no evidence that ice sheets reached the South-West at this time. Indeed, convincing evidence of Early and Middle Devensian sub-stages is generally lacking, although it is highly probable that Oxygen Isotope Stage 5e raised beach deposits at many localities are succeeded by head deposits which include Early, Mid- and Late Devensian facies (e.g. Scourse, 1985a, 1987, 1991, 1996b). The complex climatic fluctuations which characterize the transition from Ipswichian to Devensian times, so clearly demonstrated by stratigraphical, fossil and geochronological evidence from Bacon Hole Cave on the northern side of the Bristol Channel (Stringer *et al.*, 1986), and in Jersey to the south (Keen *et al.*, 1996), are rarely recorded in South-West England. Nonetheless, falling Oxygen Isotope

Figure 2.6 The massive 'terrace' of raised beach, wind-blown sand ('sandrock') and periglacial head flanking Saunton Down in north Devon. (Photo: N. Stephens.)

Stage 5e sea levels may be demonstrated by the transition from marine to aeolian sand along the Croyde–Saunton coastline (Gilbert, 1996; Figure 2.6). The non-periglacial character of immediately post-Stage 5e times may also be also demonstrated by cliff-fall materials in the head sequence at Brean Down, and, almost certainly, at many other coastal locations: the Pleistocene sequences of the South-West are closely comparable to those of South Wales, especially Gower, in this respect (Campbell and Bowen, 1989).

The deterioration of the climate in the Early Devensian, to fully periglacial conditions, is marked by head formation at a number of localities. At Holly Lane, Clevedon, a series of alternating breccias and aeolian sands may represent cold moist phases (breccia) separated by cold arid periods (sand) within a periglacial environment (Gilbertson and Hawkins, 1974). Recent fossil evidence from the site, however, indicates progressively deteriorating conditions in a cold steppe environment (Chapter 10).

It is quite likely that with the onset of periglacial conditions in the Devensian, the first deposits to be moved downslope and rearranged by solifluction would be older unconsolidated Quaternary sedi-

ments: there is clear evidence at some sites for parts of raised beach sequences to have been redistributed by solifluction (e.g. Westward Ho! in north Devon and Porth Seal in the Isles of Scilly), and certainly any remnant river gravels, glacial or glaciofluvial sediments from the Saalian would have been highly susceptible to remobilization during an early periglacial phase or phases. The Trebetherick Boulder Gravel in north Cornwall and glacial gravels on Lundy and the Isles of Scilly may have been partially redistributed in this way during the Devensian (Scourse, 1991, 1996b). Again, a close analogue comes from Gower where pre-Devensian glacial materials (Western Slade Diamicton), colluvial beds and limestone scree were solifluected down the coastal slopes under periglacial conditions in Early and Middle Devensian times (Bowen, 1970, 1971, 1973a, 1973b, 1974; Henry, 1984a, 1984b; Campbell and Bowen, 1989).

The generally deteriorating climate of Early Devensian times was, as elsewhere in Britain, punctuated by warmer phases. One such period is demonstrated at Low Ham in the Somerset lowland, where sediments and fossils provide unique evidence in the South-West for high relative sea levels during an Early Devensian interstadial: a correlation with Oxygen Isotope Stage 5a is tentatively suggested, providing a close analogue with the evidence from Bacon Hole Cave, Gower (Stringer *et al.*, 1986). Other evidence in the South-West for warmer episodes within the Devensian Stage prior to the Devensian late-glacial is rare: the ascription by ApSimon *et al.* (1961) of mollusc- and vertebrate-bearing beds at Brean Down to the Allerød has, however, been questioned (Currant, unpublished data), and an earlier interstadial event may be present (Chapter 9). With the exception of a Stage 3 faunal assemblage at Tornewton Cave (Chapter 5), there is little clear evidence in the South-West for conditions which prevailed in the Middle Devensian.

Glaciation

There is little doubt that periglacial denudation was the chief land-forming process throughout the Devensian Stage in the South-West. With the exception of the Fremington Clay and mixed lithology drift (Scilly Till and associated meltwater sediments) on the Isles of Scilly, no glacigenic sediments of Late Devensian age are known: even these recent ascriptions are controversial (Chapters 7 and 8). The Fremington Clay or 'till', near Barnstaple, has been a source of debate since its discovery in 1852 (Maw, 1864), and most workers have argued that it was deposited by Irish Sea ice during the Saalian Stage (e.g. Mitchell, 1960; Stephens, 1966b, 1970a, 1977; Kidson, 1977): Saalian glacio-lacustrine conditions have also been invoked to account for its unusually fine texture and 'dispersed' erratics (Edmonds, 1972). Recent interpretations of Pleistocene sequences throughout the Irish Sea Basin, however, have led to the suggestion that the Fremington Clay may have been deposited as a distal mud 'drape', well beyond the southern margin of a Late Devensian, marine-based (floating) Irish Sea glacier (Eyles and McCabe, 1989, 1991; Campbell and Bowen, 1989; Bowen, 1991, 1994b). The possibility that glacial materials in the Isles of Scilly were deposited either directly from the Late Devensian Irish Sea ice sheet (Scourse, 1985a, 1987) or a glaciomarine sediment plume from it (Eyles and McCabe, 1989, 1991) is given consideration in Chapter 8, and rests on the validity of radiocarbon dates (Bowen, 1994b), thermoluminescence (TL) dates, the correct diagnosis of sedimentary provenance and type and whether the deposits lie *in situ* or have been reworked (Scourse, 1985a, 1987; Scourse *et al.*, 1991). However, the maximum southward extent of the combined Welsh and Irish Sea Late Devensian ice sheet, particularly across South Wales, is well charted (Bowen, 1970, 1971, 1974, 1977, 1982; Campbell, 1984) and the interpretation of materials in the Isles of Scilly as *in situ* Late Devensian glacial sediments necessitates radical revisions of previously held views regarding maximum offshore ice limits (Garrard and Dobson, 1974; Garrard, 1977; Doré, 1976; Scourse *et al.*, 1991). Scourse *et al.* (1990) have argued that the glacial advance which affected the Isles of Scilly and the Celtic Sea was a short-lived thin-ice surge beyond a more established terminus in St George's Channel.

If glaciomarine conditions persisted in the Irish Sea Basin, as suggested by some workers, then the distal margins of the ice sheet would have been controlled by sea level (not climate). Ice-rafting under such circumstances could account for some erratics found around the shores of the South-West. With the exception of the Tregarthen Gravel of the Isles of Scilly (Scourse, 1991; Chapter 8), unequivocal deposits associated with deglaciation of the Late Devensian ice sheet, that is glaciofluvial rather than glaciomarine, are unknown.

Devensian sea levels

There is little unequivocal evidence in and around South-West England for establishing the pattern of Devensian sea levels. It has normally been assumed that during cold or stadial phases, sea levels were generally well below those of the present: eustatic levels between 100 m and 130 m below OD have generally been suggested for the Late Devensian (Oxygen Isotope Stage 2), with significant volumes of water becoming effectively 'locked-up' in ice sheets (Aharon, 1983; Lowe and Walker, 1984). There is, however, some evidence for a transition from grounded to floating ice at -135 m OD in the Celtic Sea associated with an extra-glacial shoreline at between $c. -60$ and -90 m OD (Scourse *et al.*, 1990). The warmer, interstadial, phases almost certainly saw higher sea levels: Donovan (1962) assigned the Swallow Cliff (Middle Hope) platform and beach to high relative sea levels during the Upton Warren Interstadial and suggested that the main terrace of the River Severn and a variety of erosional features around the Bristol Channel were related to the same base level. Although Wood (in Callow and Hassall, 1969) reported radiocarbon dates from the Middle Hope raised beach deposit to support this ascription, most authors have assigned the Swallow Cliff raised beach to the Ipswichian (Gilbertson, 1974;

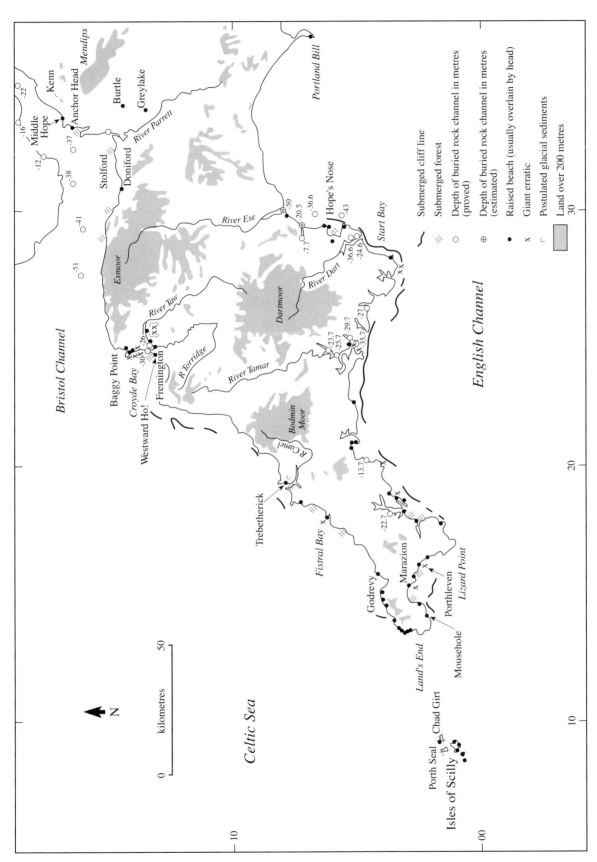

Figure 2.7 Quaternary coastal landforms and deposits around South-West England. (Adapted from Kidson, 1977.)

Gilbertson and Hawkins, 1977) or an even earlier event (Davies, 1983; Chapter 9). In fact, the only reliable evidence for high Devensian sea levels comes from Low Ham (see above; Chapter 9), where faunal and stratigraphic evidence indicates a high sea-level stand during Oxygen Isotope Stage 5a (Hunt, 1987; Chapter 9), possibly equivalent to the Upton Warren Interstadial (cf. Campbell and Bowen, 1989; Addison and Edge, 1992).

In contrast, a variety of submarine features, including ridges, benches, cliffs and buried rock channels, has been described around the shores of South-West England and ascribed to low Devensian sea levels (e.g. Codrington, 1898; Clarke, 1970; Kidson and Heyworth, 1973; Durrance, 1974; Kidson, 1977; Goode and Taylor, 1988; Figure 2.7). Much of the interpretation is speculative, particularly regarding the age of the features, and it seems highly likely that the buried rock channels, for example, were cut during low sea levels not only in the Devensian but also in earlier cold, low sea-level stages of the Pleistocene and perhaps even before (cf. Stride, 1962; Anderson, 1968; Donovan and Stride, 1975; Kidson, 1977). Only in a very few cases have the offshore infills of the channels been examined, and their Holocene (Pollen Zone IV upwards) contents do little to resolve the issue (Clarke, 1970; Kidson, 1977).

Although establishing the lowest levels to which the Devensian sea fell around the coasts of South-West England has proved impossible, there seems to be some consensus that it paused at − 43 m OD for perhaps a significant time (Kidson, 1977). Evidence for this includes a submerged cliffline with buried channels grading to it in Start Bay (Hails, 1975a, 1975b; Kelland, 1975); similar features off Plymouth (Cooper, 1948); a buried channel of the Exe also grading to a buried cliffline (Clarke, 1970); and a variety of river long-profiles in south Devon (River Erme) and north Devon (the rivers Taw and Torridge) which have also been related to a base level of *c.* − 43 to − 45 m OD (McFarlane, 1955; Kidson, 1977). Acceptance of this particular Devensian low sea-level stand, however, has not been universal. Donovan and Stride (1975) interpreted submerged shoreline features found at the general level of − 44 m OD as a degraded coastline of 'late Tertiary' age, while recent work in Barnstaple Bay has shown that the buried channel of the Taw–Torridge Estuary in fact grades to a level of *c.* − 24 m OD, and not to *c.* − 46 m as suggested by McFarlane (1955). Further complexity has been introduced by the work of

Durrance (1969, 1971, 1974) who identified a series of buried channels and terraces beneath the present floors of the Exe and Teign valleys: these, he argued, were related to two distinct low sea-level stands in Early and Late Devensian times. There is emerging evidence for Devensian late-glacial sea levels at lower elevations than previously thought; for example, at − 120 m OD in the central Celtic Sea at *c.* 11 ka BP (Scourse and Austin, 1994). In reality, the pattern of sea-level change in the Devensian is poorly understood, and the ascription of particular landforms to specific stands of the sea is highly speculative. Indeed, recent suggestions that glaciomarine conditions prevailed in the Irish Sea Basin and beyond during the Late Devensian (e.g. Eyles and McCabe, 1989, 1991) has thrown wide open the question of Late Devensian relative sea levels in this area: while a large body of data indicates low eustatic sea levels during the Late Devensian, regional isostatic depression of the landmass may have led to widespread high relative sea levels and to glaciomarine conditions (Eyles and McCabe, 1989, 1991; Campbell and Bowen, 1989; Bowen, 1994b). These issues remain to be resolved.

Periglaciation

With the possible minor exceptions noted above, it is abundantly clear that the vast majority of South-West England remained glacier-free during the Late Devensian. Cold-climate, frost-assisted processes were dominant: copious quantities of head blanket not only the coastal cliffs, but many inland slopes (Waters, 1965; Mottershead, 1971, 1977a, 1977b; Kidson, 1977); blockfields and 'clitter' surround the spectacular granite tors of Dartmoor and Bodmin Moor and demonstrate the efficacy of freeze-thaw activity and solifluction processes; and a range of intricate landforms including patterned ground, altiplanation terraces and cryoturbation features bear further witness to the importance of periglacial processes in shaping the landscape throughout the Devensian (and before). Also of considerable importance are widespread fluvial terrace gravels (Hunt, 1987) and alluvial fan deposits, for instance in Mendip (Macklin and Hunt, 1988). The widespread development of such features gives the region a special geomorphological character and importance, reflected in the relatively large number of GCR sites which have been selected for their periglacial sediments and landforms (e.g. Merrivale; Chapter 4).

Head and loess

The most widespread evidence for Devensian periglacial activity occurs in the form of slope deposits, for which De la Beche (1839) aptly coined the term 'head'. Throughout the coastlands of the South-West, head has arrived in its cliff-top locations from the slopes and coastal plateaux above and behind as a result of freeze-thaw and solifluction processes (Mottershead, 1971, 1972, 1977a, 1977b; Kidson, 1977). The surface of such deposits invariably adopts a terrace or apron form, and detailed work by Mottershead (1971, 1972, 1977a, 1977b) has shown that the constituent particles show a clear preferred orientation related to the local slopes (cf. Gower and the Isles of Scilly; Harris, 1973; Henry, 1984a, 1984b; Scourse, 1987). For the most part, these materials are composed of local 'country' rocks and erratics are scarce. Frequently, colluvial layers interdigitate with the head deposits and show that sheet-wash processes over partially vegetated surfaces also occurred.

There has been a widespread tendency to divide the head deposits in the coastal sequences in the South-West into lower (or main) and upper heads (Barrow, 1906; Stephens, 1966b, 1970a; Kidson, 1977). Workers who have attributed a Hoxnian age to the underlying raised beach deposits, traditionally have assigned the lower head to the Saalian and the upper to the Devensian, invoking evidence for temperate (Ipswichian) weathering in between (e.g. Stephens, 1966b, 1970a). Although such dating of the head may remain valid at localities where an Oxygen Isotope Stage 7 (intra-Saalian; see Figure B in Preface) raised beach deposit has been confirmed, head sequences found above Stage 5e raised beach sediments, perhaps most cases, must belong to the Devensian: facies variation within the head in the latter context could be accounted for by cold conditions in separate stadials of the Devensian, but could also reflect variations in processes and sediment supply. In the Isles of Scilly, some sites show organic sediments interbedded with head. The former deposits have yielded pollen assemblages indicating open-grassland vegetation and suggesting deposition in a periglacial environment. Organic sediments from these sites, which include Carn Morval and Watermill Cove (St Mary's), Porth Seal and Bread and Cheese Cove (St Martin's) and Porth Askin (St Agnes), have yielded radiocarbon dates indicating deposition between 35 and 21 ka BP, that is late Middle to early Late Devensian (Scourse, 1991). Not only do these sites provide the earliest vegetational record for the Isles

of Scilly, but they cover a period very poorly represented in terms of organic sediments in the rest of the British Isles: as such they are of national significance (Chapter 8). Normally, however, precise chronological subdivision is impossible (e.g. Gilbertson and Mottershead, 1975; Scourse, 1987, 1991) and a Late Devensian age is frequently assumed for the bulk of periglacial materials in many sections (cf. Gower; Campbell and Bowen, 1989). It is important to note that at locations in the Channel Islands and Normandy, Keen (1993) and Keen *et al.* (1996) have demonstrated ages for the head deposits extending at least as far back as Stage 8: even where there are no raised beach deposits, it is clear that some head is pre-Devensian in age and has a vertical unconformity (of Stage 5e age) which has been re-exposed by the Holocene sea. The same is clearly possible for other sections in South-West England.

Many coastal and inland head sequences are capped by finer-grained silty materials. These testify to a variety of colluvial and aeolian processes alternating with and post-dating solifluction of the head. In some cases, clear downslope bedding is visible, and sheet-washing in a sparsely vegetated environment, probably during latest Devensian times, seems to have been responsible. At many locations, however, the capping deposits are silty, almost stoneless, and show no sign of bedding. Usually, these deposits have been interpreted as loess, derived from underlying head and glacial deposits in Late Devensian times. Barrow (1906) was the first to recognize the significance of these deposits in the Isles of Scilly, and compared them with deposits along the Brittany coast. Thick interbedded aeolian sands are known from the Avon coastlands, notably at Brean Down (ApSimon *et al.*, 1961) and Holly Lane (Gilbertson and Hawkins, 1974), but also at a variety of other locations (Gilbertson and Hawkins, 1983).

More recently, the distribution, nature and source of such materials have been more thoroughly investigated; for the Mendips (Findlay, 1965), for Devon (Harrod *et al.*, 1973) and especially for mainland Cornwall and the Isles of Scilly (Coombe *et al.*, 1956; James, 1968, 1975a, 1975b; Catt and Staines, 1982; Roberts, 1985; Scourse, 1985a, 1987; Catt, 1986). A Late Devensian age is normally assumed and has, in a few cases, been confirmed by thermoluminescence dating (Wintle, 1981). The source(s) of these materials is more problematic, and while a small proportion may be locally derived, Catt and Staines (1982) have shown that the bulk of material in the Cornish loess was

Figure 2.8 Haytor Rocks, Dartmoor. Although tors such as these have evolved over an extremely protracted timescale, their final form, and that of the slopes around them, was fashioned by periglacial processes in the Devensian (Chapter 4). (Photo: S. Campbell.)

blown from glacial outwash deposits in the southern part of the Irish Sea Basin (cf. Gower; Case, 1977, 1983, 1984). The glacigenic sediments of the Celtic Sea and Isles of Scilly are mineralogically similar to the Cornish loess and a genetic association has been proposed (Scourse *et al.*, 1990). The implications of the presence of subaerially exposed, offshore glacial sediments during the Late Devensian, with respect to proposals for glaciomarine conditions at this time, have yet to be explored fully.

Many sections in head and loess display a marked discontinuity some 0.5 m below the surface, where upper layers with friable consistency and crumb or blocky structure give way to lower, extremely well-compacted or indurated layers with a platy structure (Clayden, 1964; Stephens, 1980). These differences have been interpreted as the result of permafrost development, with the upper horizon representing the active layer where seasonal thawing occurred, and the lower layer attaining its structure under permanently frozen ground condi-

tions (Stephens, 1980). The upper parts of many coastal sequences also exhibit clay-filled cracks, some with margins of light-grey clay or re-precipitated iron (e.g. Prah Sands (Cornwall), Fremington Quay (north Devon) and the Camel Estuary (Scourse, 1987)). The precise environmental conditions under which these cracks formed are unknown, although desiccation and frost-cracking are possible mechanisms (cf. Gower; Bowen, 1970; Campbell, 1984). Such features are often intimately related to cryoturbation structures such as involutions and frequently cross-cut them: collectively they bear witness to a variety of post-solifluction periglacial processes in latest Devensian times (see Devensian late-glacial; below).

This pattern of activity is also repeated inland, and the stripping of frost-shattered and weathered materials by solifluction from around the Dartmoor and other tors is well documented (Chapter 4; Figure 2.8), and was fundamental in establishing present slope configurations. The division of inland head sequences, such as those on Dartmoor, into lower (fine) and upper (coarse) heads, reflecting an inversion of normal granite weathering profiles (Waters, 1964, 1974; Mottershead, 1976), has been shown to be unrealistic: many sites show the reverse relationship or even greater facies complexity (Green and Eden, 1973; Gerrard, 1983, 1989a).

However, despite the many well-documented periglacial sedimentary sequences in the South-West, the extent of permafrost during the various cold phases of the Devensian Stage (and before) is still disputed (Williams, 1969, 1975; Straw, 1974; Ballantyne and Harris, 1994).

Although the development of periglacial landforms such as altiplanation terraces (see Merrivale; Chapter 4) is unlikely to have been accomplished during a single Devensian stadial, patterned ground such as stone stripes and polygons, and the widespread clitter on many hillslopes in the vicinity of tors, probably reflect a late stage of periglacial ornamentation, certainly of Late Devensian, possibly even partly Younger Dryas age (Gerrard, 1983). It is quite likely, however, that much of the material, especially that making up the more substantial landforms such as solifluction lobes, is older and has merely been rearranged by later periglacial activity.

Fluvial deposits and landforms

Fluvial deposits and landforms of Devensian age are widespread in the South-West, although they have received relatively little recent attention from researchers. Morphostratigraphic work, for instance on the Dart (Brunsden *et al.*, 1964) and Exe (Kidson, 1962), has rarely been followed up by investigation of the deposits underlying the terrace surfaces.

In the Somerset lowland, east Dorset and in the valleys around Dartmoor, Exmoor and other uplands, cold-stage fluvial sedimentation was widespread. Typically, most of the valley-floor was occupied by flashy ephemeral gravel- and sand-bedded streams, leading to the accumulation of often substantial thicknesses of planar- and trough cross-bedded sand and gravel. Periods of aggradation alternated with periods of incision: Hunt (1987) reported four substantial sand and gravel aggradations of Devensian age from the Somerset lowland (Chapter 9). The fluvial sediments usually pass laterally into, and interdigitate with, valley-side head deposits. At Doniford, for instance, coastal exposures over 2 km in length are cut through 5 m of head interdigitating with trough cross-bedded

gravel and sand (Chapter 7). Interstadial fluvial sedimentation was from small meandering streams, and the deposits from these are seldom preserved.

Around upland areas like Mendip, large alluvial fans can in part be attributed to Devensian fluvial activity (Findlay, 1965, 1977; Pounder and Macklin, 1985; Macklin, 1986; Macklin and Hunt, 1988). In the Bourne fan, Devensian aeolian and fluvial deposits were laid down over an interglacial palaeosol. At Wookey Station, fossil evidence suggests that the fan gravels accumulated in an Arctic desert landscape with little vegetation cover. The deposition of fan gravel was synchronous with a phase of aeolian sedimentation (Chapter 9).

Cave deposits

Fossil and archaeological evidence from the Devensian, particularly the Late Devensian, is profuse in the South-West, and lends an important insight to changing climatic and environmental conditions during the stage. Most of this evidence is preserved in cave sequences, where the bulk of material is Devensian in age and characterized by accumulations of cold-climate thermoclastic scree often containing a characteristically 'cold' fauna, typically including woolly mammoth, woolly rhinoceros, reindeer, horse, brown bear, spotted hyaena and the narrow-skulled vole (Sutcliffe, 1969, 1974; Cullingford, 1982), as well as Upper Palaeolithic artefacts (Campbell, 1977). Excellent summaries are provided by Donovan (1954, 1964) and Hawkins and Tratman (1977) for sites in the Mendip, Bristol and Bath areas and by Sutcliffe (1969, 1974) for those in Devon.

Notable sites include Kent's Cavern, where a rich Devensian fauna including woolly rhinoceros, woolly mammoth and reindeer is found in association with Upper Palaeolithic artefacts; and Tornewton Cave (Torbryan) where the Reindeer Stratum with its distinctive small mammal assemblage has been taken as evidence for cold conditions in the Devensian Stage (see Tornewton Cave; Chapter 5).

Although the totality of artefact, sedimentary and fossil evidence from the cave sites of the South-West is enormous, its contribution to a precise reconstruction of Devensian Stage events and environments has not been fully realized, and the interpretation of 'derived' faunas and pollen sequences is, as elsewhere, fraught with difficulties. The major process evidenced by the cave sequences appears to have been the formation of thermoclastic scree, by the collapse of materials from cave walls and roofs; in some places, mass flow events appear to have been responsible for the accumulations and were instrumental in reworking older deposits (e.g. Tornewton Cave). In some caves, evidence from pollen, land snails and artefacts shows a progression of climatic conditions through the coldest part of the Late Devensian into the Devensian late-glacial (see below).

The Devensian late-glacial and Holocene

Introduction

Devensian late-glacial and particularly Holocene environments have been the focus of much attention in South-West England. Although sites demonstrating evidence from these periods were selected as a discrete national network of the GCR, they are included within subsequent regional accounts for convenience (Chapters 4, 7 and 8): to avoid unnecessary repetition, a general synthesis of Devensian late-glacial and Holocene events, and an introduction to these sites, is given here.

The Devensian late-glacial

Latest Devensian Stage events which occurred between the deglaciation of the main Late Devensian ice sheet and the beginning of the Holocene (= Flandrian or 'post-glacial') fall within Devensian late-glacial time. There have been major problems regarding definition of this period, and a variety of climatic, geological, geochronological and chronostratigraphic terms has been used, interchangeably and unhelpfully, to classify evidence from this time interval. Although the start of the Devensian late-glacial is diachronous, a timespan for the period of *c.* 14.5 to 10 ka BP is generally involved.

Originally, a threefold sequence of cold–temperate–cold environments was adduced from the Late Weichselian (= Late Devensian) in south Scandinavia from plant macrofossil and lithostratigraphic evidence. Three chronostratigraphic stages – the Older Dryas (cold), Allerød (warm) and Younger Dryas (cold) were erected and have been recognized in sequences elsewhere by distinctive, climate-dependent, pollen assemblage biozones. An additional temperate event, the Bölling Interstadial, within the Older Dryas has also been recognized. Late-glacial pollen diagrams in Britain have customarily been zoned on this basis (Mitchell *et al.*, 1973a).

The Scandinavian sequence of events, however, is not precisely comparable with the British evidence, and there was considerable regional diversity in the development of vegetation in Britain after the Late Devensian glacial maximum: ice-free conditions were not attained everywhere at the same time and the diachronous nature of events has, in part, led to the different terminology used to describe the changing climate and conditions of the period. The terms Older and Younger Dryas and Allerød have generally been applied to evidence described in South-West England (e.g. Conolly *et al.*, 1950; Brown, 1977; Pennington, 1977; Caseldine, 1980). The latter is broadly equivalent to the Windermere Interstadial in northern England (e.g. Pennington, 1977), and the Younger Dryas is more or less synonymous with the Loch Lomond Stadial. The widely used term 'Lateglacial Interstadial', derived from studies of the Scottish Devensian late-glacial (e.g. Gordon and Sutherland, 1993), accommodates the Older Dryas, Bölling and Allerød events. Devensian late-glacial events fall within Stage 2 of the oxygen isotope framework.

A simpler interpretation of the Devensian late-glacial is to regard it as a period of climatic warming during and after deglaciation of the main Late Devensian ice sheet. This warming was interrupted latterly by a deterioration in climate leading to the establishment of glaciers in upland Britain between *c.* 11 000 and 10 000 radiocarbon years ago, that is during the Younger Dryas (Watts, 1980; Lowe and Walker, 1984; Campbell and Bowen, 1989).

Detailed records of the Devensian late-glacial are rare in the South-West, reflecting a lack of suitable depositional basins. The fullest and most reliable record comes from Hawks Tor on Bodmin Moor, where the classic tripartite stratigraphy is well demonstrated: pollen, plant macrofossil and lithological evidence shows clearly two periods of Arctic conditions, characterized by a sparsely vegetated, treeless landscape and active solifluction and sheet washing (equivalent to the Older and Younger Dryas), separated by a warmer interlude in which birch woodland developed in sheltered locations (Allerød). This was the first terrestrial site in Britain where the Devensian late-glacial was recognized. It is highly likely that this broad pattern of environmental change was repeated elsewhere in the South-West, and significant landscape modification, particularly during the Younger Dryas, is almost certain.

However, unlike areas of Britain covered by the Late Devensian ice sheet, there is no profound lithological change from glacial to periglacial sediments in the South-West (perhaps with the exception of parts of the Isles of Scilly; Chapter 8), and periglacial conditions are likely to have continued, albeit diminished in intensity, until the climatic warming of the Allerød: there is no evidence in the region of temperate conditions or deposits dating from the full glacial phase of the Late Devensian (*c.* 21 to 13 ka BP), and none for a temperate oscillation (cf. the Bölling) prior to the Allerød. Solifluction, sheet-washing and the formation of patterned ground probably continued into the Older Dryas and was renewed, particularly in upland locations such as Dartmoor and Bodmin Moor, during the Younger Dryas. It is highly likely that the final periglacial 'ornamentation' of the landscape occurred in the Younger Dryas, although differentiating the sedimentary products and landforms of this phase from those of earlier Devensian times is impossible in the vast majority of locations. Certainly, large areas would have resembled tundra if not polar desert, and there may have been discontinuous permafrost with snowbeds persisting in suitable locations (Williams, 1975).

Evidence for Devensian late-glacial conditions is probably widely recorded in the cave sequences of the South-West, although it has rarely been demonstrated convincingly. Campbell (1977) has shown that the formation of thermoclastic scree in a number of Mendip caves continued unabated from the coldest part of the Late Devensian right through into the late-glacial (the equivalent of Older Dryas time): pollen evidence from Wookey Hole Hyaena Den, Sun Hole and Gough's Cave shows clearly the climatic improvement of the Allerød (marked by a peak in birch pollen) and also the return to colder conditions and renewed scree formation in the Younger Dryas. The presence of humans is widely indicated in these sequences by the presence of later Upper Palaeolithic (Creswellian) artefacts (Campbell, 1977). The reliability of pollen evidence from caves, however, is generally low and some sequences, for example that at Badger Hole, show no differentiation of the Older Dryas, Allerød and Younger Dryas events: instead, a general improvement of conditions from the peak of the last glacial to the Holocene is indicated.

The only reliable dated evidence for the beginning of warm conditions in the Allerød comes from Hawks Tor, where organic sedimentation began at *c.* 13 ka BP (Brown, 1977). Here, open grassland gave way to juniper scrub by about 12 ka BP and was followed by the spread of birch between *c.* 11.5 and 11 ka BP. Although birch woodland appears to have been restricted to carrs in sheltered valleys in the uplands of Bodmin Moor, it seems highly probable that the lowlands of the South-West were more densely covered, although there is no clear evidence for the actual extent of interstadial woodland (Caseldine, 1980). Coleoptera from Hawks Tor show that the period of maximum interstadial warmth may have been somewhat earlier than the thermal maximum adduced from pollen evidence. It is also probable that the transition from Arctic to interstadial conditions was vastly more rapid than suggested from the pollen data (Caseldine, 1980): this is in keeping with pollen and insect evidence from other areas of the British Isles (e.g. Coope and Brophy, 1972; Coope, 1977).

Other evidence in the South-West for conditions in the Allerød is generally lacking. Although a palaeosol at Brean Down may be attributable to this phase, widespread evidence for soil development is not recorded elsewhere: the products are likely to have been destroyed and redistributed during the more rigorous climate of the succeeding Younger Dryas. Neither is there firm evidence for establishing the pattern of Devensian late-glacial sea levels; all such evidence in the South-West now lies under the sea or is deeply buried. Eustatic sea levels at *c.* 14 ka BP, however, are likely to have been between *c.* 60 m and 90 m below those of the present (Fairbridge, 1961; Jelgersma, 1979; Fairbanks, 1989; Scourse and Austin, 1994; Lambeck, 1995; Bard *et al.*, 1996).

The cold Younger Dryas event, which saw renewed glacier development in other parts of upland Britain, is most clearly recorded in the South-West by palynological evidence at Hawks Tor (Conolly *et al.*, 1950; Brown, 1977; Caseldine, 1980). A return to periglacial conditions, an unstable soil cover, the disappearance of woody elements from the flora and the existence of widespread snowbeds, are all hallmarks of the stage (Caseldine, 1980). The onset of colder conditions at Hawks Tor has been established at *c.* 11 ka BP, but a lack of other suitable sites precludes the precision of environmental reconstruction achieved elsewhere in the British Isles, particularly with respect to charting the movements of the Polar Front. It is likely that the processes of frost-heaving (patterned ground) and aeolian transport of silt and sand (loess and coversand) were all renewed during the Younger Dryas, but few, if any, deposits or landforms can be attributed to the phase with certainty. A return of large areas to a tundra vegetation and discontinuous permafrost is likely. Eustatic sea level towards the end of the period is likely to have been in the region of − 43 m OD (Hawkins, 1971).

The Holocene

Introduction

By international agreement, the Holocene (Flandrian or 'post-glacial') commences at 10 ka BP. Evidence for Holocene conditions in the South-West is widely recorded, both inland and offshore: numerous peat and 'forest' beds, now lying partly beneath the sea, bear witness to formerly more extensive wooded lowland areas and allow the progressive rise of the Holocene sea to its present position to be charted in detail (Figure 2.4). Organic sequences such as those in the Somerset Levels and in upland areas such as Dartmoor and Bodmin Moor, have, in parallel, enabled a reconstruction of Holocene terrestrial environments and vegetational history.

Pollen analysis and radiocarbon dating have provided the principal means of establishing the rate at which the rising Holocene sea flooded the land surface, and for correlating the offshore organic sequences with those inland.

Sea levels

Around the coastline of South-West England, as in Wales, the Holocene transgression can be reconstructed from a series of alternating peat, 'forest' and marine clay beds, the most obvious manifestations being the widely recorded 'submerged forest' beds which are normally covered by sand or mud and only exposed at low tides and times of low beach levels (e.g. Hawkins, 1971, 1973; Kidson and Heyworth, 1973; Kidson, 1977; Figure 2.4). These 'submarine forests' drew much early attention (e.g. Borlase, 1757, 1758; Boase, 1828; Carne, 1846; Henwood, 1858; Ussher, 1879a). In describing tree remains in the intertidal areas of Mount's Bay, Borlase (1757) correctly concluded that St Michael's Mount had once been surrounded by woodland, but argued that the trees had subsequently been drowned as a result of 'subsidence of sea-shores'.

Often, these forest beds contain large trunks and stumps of oak and pine trees in growth position: frequently, the tree remains are rooted in thin soils which developed on bedrock or solifluction deposits, and can be traced inland for some distance (Kidson, 1977; Durrance and Laming, 1982). The forest beds and associated peats occur down to and below the levels of the lowest tides. Clarke (1970) recorded a peat bed at − 24 m OD off Teignmouth in south Devon, while Austen (1851) reported tree stumps rooted in terrestrial sediments and overlain by marine deposits at − 20 m OD off Pentuan in Cornwall: boreholes in the Somerset Levels have shown peat to rest on bedrock at depths in excess of − 21 m OD (Kidson and Heyworth, 1973; Kidson, 1977).

While important evidence for changing Holocene sea levels has been documented widely around the coastline of South-West England – for example, at Westward Ho! in north Devon (Rogers, 1946; Churchill and Wymer, 1965; Balaam *et al.*, 1987) and in the English Channel (Boase, 1828; Reid and Flett, 1907; Clarke, 1970; Pascoe, 1970; Welin *et al.*, 1973; Hails, 1975a, 1975b; Goode and Taylor, 1988) – the pattern of change is undoubtedly best demonstrated in the Somerset Levels and Bridgwater Bay (Godwin, 1943, 1948; Godwin *et al.*, 1958; Kidson and Heyworth, 1973, 1976; Kidson, 1977). The following reconstruction relies heavily on evidence from these areas.

It is likely that much of the continental shelf, including all the Bristol Channel area and parts of the Western Approaches, was dry land during the Late Devensian: at the beginning of the Holocene, sea levels may have been as low as − 43 m OD (Hawkins, 1969, 1971). The submerged and buried terrestrial deposits, including the peat and forest beds, cover a vertical range of at least 47 m (Kidson, 1977) and record stages in the drowning of the land surface by the transgressing Holocene sea.

Sea-level rise began very early in the Holocene and, as Kidson (1977, p.286) observed, ' … took place initially so rapidly that by 6000 BP only about 6 metres of rise remained to be achieved at a decelerating pace' up to the present day. As sea level rose, many deeply entrenched valleys, rock channels and shoreline features, cut at times of lower base level, were drowned progressively along with their vegetation (e.g. Goode and Taylor, 1988). The sub-Holocene surface of the Somerset Levels and Bridgwater Bay is particularly well established, and here it has been possible to reconstruct the migration of the shoreline inland at 1 ka intervals, from 9 to 4 ka BP (Kidson and Heyworth, 1973; Kidson, 1977). With the rise in sea level, superficial deposits including glacial and periglacial detritus were reworked, and much sand-, pebble- and cobble-sized material was transferred from offshore areas to the nearshore zone and on to modern beaches (e.g. Hails, 1975a, 1975b; Kelland, 1975; Kidson, 1977). With the exception of a major basal peat bed, the deposits of the early Holocene are largely thick, inorganic marine clays, silts and sands – reflecting the fast initial rate of sea-level rise: other peat beds from this time are rare and generally thin (Kidson, 1977).

By about 6 ka BP, the rate of rise had slowed and sea levels were sufficiently high to affect groundwater tables, cause waterlogging and accelerate peat growth. A succession of alternating peats and clays, particularly in the Somerset Levels, reflects the complex interplay between marine and terrestrial sedimentation and peat growth. At times, the rate of terrestrial sedimentation was more rapid than sea-level rise, and vegetation colonized a near-horizontal land surface and peat formation resumed; at others, the reverse was true (Kidson, 1977). In some areas of the Somerset Levels, peat growth consistently outstripped the rate of sea-level rise, resulting in a thick peat or 'raised bog' layer which now forms the land surface, and which effectively excluded the sea from large areas for as much as 5000 years. In other areas, the intercalations of peat and clay record a more complex story: where fen and carr peats occur, the relationship between sea level, groundwater and peat growth is particularly clear (Kidson, 1977). Indeed, in these areas, Godwin (1960) identified a series of flood episodes which he suggested had prompted the building of wooden trackways across the bogs in prehistoric times: it is likely that these flooding episodes can be related to contemporaneous high sea levels (Kidson and Heyworth, 1973; Kidson, 1977).

With the exception of areas of raised bog and dune sand, it is clear that all Holocene sediments which infill the Somerset Levels were laid down at or very close to sea level. The regional sea-level curve established by Kidson and Heyworth (1973) shows a smooth, almost exponential, rise. There is no evidence during later Holocene times for minor transgressions or regressions of the sea, either in the Somerset Levels or more generally in South-West England. Indeed, Kidson and Heyworth (1973) and Kidson (1977) have argued that the complex intercalations of peat and clay which date from the last 5000 years or so, accumulated in response to minor variations in the relative rates of sedimentation and sea-level rise. Even the thicker marine clays which fringe the seaward side of the Somerset Levels, and which were interpreted by Godwin (1943, 1956) as indicating a significant transgression in Romano-British times, are more likely to reflect the cumulative effects of occasional high sea levels caused by exceptional storms, storm surges and long-period tides (Kidson and Heyworth, 1973; Kidson, 1977).

The diminishing rate of sea-level rise in later Holocene times is perhaps most graphically illustrated by various parts of the Isles of Scilly where hut circles, stone cists and field walls are sometimes exposed on the foreshore at low tide: this archaeological evidence shows that most of the islands (except Agnes, Annet and Gugh) were still connected as a single landmass until roughly between the 7th and 13th centuries AD, when the central low-lying area was finally flooded by the sea (Fowler and Thomas, 1979; Bell *et al.* 1984; Thomas, 1985). Interestingly, the rate of relative sea-level rise around the Isles of Scilly today appears to be greater than in neighbouring areas, at 4 mm per annum (Scourse, 1996b).

In terms of the overall scale of environmental and landscape change during the early Holocene, changes in the coastal zone since mid-Holocene times have been relatively minor (e.g. Steers, 1946; Arber, 1949; Robson, 1950; Everard *et al.*, 1964; Todd, 1987). The principal geomorphological effects have been the re-occupation and undercutting of ancient clifflines (e.g. Arber, 1949, 1974; Savigear, 1960, 1962), the development of coastal barriers, spits and storm beaches (e.g. Spratt, 1856; Martin, 1872, 1876, 1893; Ussher, 1904; Worth, 1904, 1909, 1923; Rogers, 1908; Toy, 1934; Robson, 1944; Kidson, 1950; Robinson, 1955, 1961), the siltation of estuaries, in part caused by the obstruction of estuary mouths by spits and barriers (Symons, 1877; Everard, 1960a; Witherick, 1963; Everard *et al.*, 1964) and the reactivation of coastal landslips (Arber, 1940, 1946). Perhaps the most significant coastal development during late Holocene times, however, has been the widespread accumulation of dune sand, for example, the Braunton and Northam Burrows in north Devon (Rogers, 1908; Kidson *et al.*, 1989; Chisholm, 1996) and the Cornish 'towans' such as those at Penhale and Gwithian (Crawford, 1921; Harding, 1950; Burrows, 1971).

The chronology of these recent sand accumulations is in general poorly understood, despite a series of molluscan studies (e.g. Spencer, 1974, 1975; Evans, 1979) and a wealth of archaeological evidence (e.g. Crawford, 1921; Harding, 1950; Bruce Mitford, 1956; Megaw *et al.*, 1961; Megaw, 1976; Whimster, 1977). From an archaeological point of view, the dunes effectively seal extensive areas of former land surface (Bell *et al.*, 1984) and appear to post-date significant forest clearance in the region (see below). A clear pattern of Holocene dune formation, however, has not been established, although archaeological and documentary evidence shows that movement of sand was particularly active during both Roman and Mediaeval times (cf. Jones *et al.*, 1990).

Finally, there is no evidence in the South-West for isostatic rebound during the Holocene, and the sea-level curve derived for Bridgwater Bay (Kidson and Heyworth, 1973; Kidson, 1977) matches well with data derived more recently from west Wales (Heyworth *et al.*, 1985), indicating no differential movement between the two areas. Sea levels during the Holocene never attained levels higher than those of the present day (Kidson and Heyworth, 1973; Kidson, 1977).

Terrestrial environments and anthropogenic effects

The rapidly rising sea levels of the early Holocene were accompanied by significant climatic warming, the return of North Atlantic Drift waters to the coast and renewed vegetational and soil development. Within 500 years, the tundra conditions of the Younger Dryas had been replaced by a climate at least as equable as that of the preceding Allerød. Evidence based on indications of marine temperatures from elsewhere in Britain suggests that the bulk of this thermal improvement may have taken place in as little as 40 years (Peacock and Harkness, 1990). On land, one of the first effects was the immigration of thermophilous insects, although this is scantily recorded in the South-West (Caseldine, 1980). While the temperature rise indicated by the pollen is more gradual, the same general improvement in conditions is clear. The Younger Dryas vegetation, dominated by taxa indicative of disturbed ground and by those of open habitats, was quickly succeeded in the earliest Holocene by open grassland and shrub and scrub vegetation with juniper and crowberry.

By about 9 ka BP, it is likely that substantial areas of South-West England had been colonized by birch woodland. Trees, however, were less quick to colonize more exposed and elevated areas such as Bodmin Moor (see Dozmary Pool) and Dartmoor (see Blacklane Brook). By *c.* 9 ka BP, these areas were still probably open moorland, with birch copses and hazel and oak scrub being restricted to sheltered valleys and hillsides (Brown, 1977;

Caseldine, 1980). Despite the lack of trees in these areas, there is evidence that Mesolithic communities were already exploiting the moorland (Jacobi, 1979). This compares with Scotland, for example, where the earliest evidence of human activity in the post-glacial dates to a little before 5 ka BP (Gordon and Sutherland, 1993).

The spread of forest vegetation had profound effects on fluvial planform and depositional style, with a change from gravel aggradation to single-channel incising streams around 9 ka BP (Macklin and Hunt, 1988).

From about 9 to 5 ka BP, climatic conditions continued to improve, with woodland expanding to its maximum in the late Boreal/early Atlantic (the 'climatic optimum'). In the lowlands and more sheltered locations, the forest thickened and diversified in composition, with hazel and oak becoming dominant. In many upland areas, open woodland developed, although it is likely that, even at the time of maximum forest cover, the most exposed areas of Cornwall remained under grassland or heather moor (Caseldine, 1980). Even where a tree cover was established in exposed upland areas, species diversity is likely to have been low: oak faced little competition, pine never became established and elm was restricted by a lack of suitable soil parent materials. Brown earth soils were probably developed over wide areas (Caseldine, 1980).

Estimating the treeline of the early to mid-Holocene in the upland areas of South-West England has proved contentious: there are few blanket bogs of this age in which tree remains are preserved. Throughout the Holocene, tree pollen found in the peats of Bodmin Moor never exceeds *c.* 50–70% of the total, reflecting a persistent suppression of woodland cover by exposure to strong westerly airflows (Brown, 1977; Caseldine, 1980). While some areas of the moors remained open and treeless, the environs of Dozmary Pool were colonized by open oak forest, and tree cover was certainly extensive in even more elevated areas such as Dartmoor (Simmons, 1964a; Maguire and Caseldine, 1985). Tree remains in peat at Blacklane Brook on Dartmoor show that trees extended to at least 457 m OD, prompting the suggestion that all but the most exposed and elevated summits of the moor were tree-covered (Simmons, 1964a, 1969). The best estimates place the mid-Holocene forest maximum treeline on Dartmoor at altitudes between *c.* 497 and 547 m OD: this provides, with the exception of Bodmin Moor, one of the lowest treeline estimates in Britain for this time

(Caseldine, 1980; Maguire and Caseldine, 1985).

Despite the uncertainties about the exact position of the treeline in upland areas during the mid-Holocene forest maximum, it is highly likely that the remainder of the South-West was thickly wooded. The distribution of tree remains in sub-merged forest and peat beds all around the coast of South-West England (Kidson, 1977; Figure 2.4), bears witness to the greater extent of woodland prior to its submergence, by about 6 ka BP, by the rising Holocene sea. Offshore peat beds with tree remains provide convincing evidence that large areas, now drowned, were once wooded. Pollen and plant macrofossil evidence from Higher Moors on St Mary's, shows that woodland even extended to the Isles of Scilly during the mid-Holocene. Such evidence has significant implications for studies of tree migration and dispersal given that the islands are situated 45 km south-west of the mainland, against the direction of the prevailing winds.

Following establishment of the 'climatic optimum' vegetation, during the later Boreal and early Atlantic periods, vegetational development in the South-West has been influenced by two main factors – climate and human activity; as with other parts of Britain, divorcing the effects of one from the other has proved difficult. The most notable recurrent pollen change of the post-glacial in Britain, namely the 'elm decline' at around 5 ka BP (Smith and Pilcher, 1973), is revealed at some sites but not at others. Because of consistently low levels of elm pollen in the upland areas, such as Bodmin Moor, the decline is not apparent, and even in areas where it is recorded, such as in the Somerset Levels, its timing and causes are complicated (see below) (Pennington, 1977; Beckett and Hibbert, 1978, 1979; Caseldine, 1980).

Changes in pollen assemblages (e.g. Blacklane Brook) and the presence of charcoal and artefacts (Dozmary Pool), show that by about 7.5–7 ka BP Mesolithic humans were already clearing the forest and modifying the landscape of Dartmoor and Bodmin Moor (Simmons, 1969, 1975; Jacobi, 1979; Caseldine, 1980). Although traditionally Mesolithic people have been regarded as 'hunters, fishers and gatherers who did not affect their environment much beyond their immediate surroundings' (Godwin, 1956), it has become increasingly clear that in the South-West, Mesolithic forest clearance in marginal areas, although relatively minor in extent, may have led to irreversible environmental changes, including the podsolization of soils on high ground after clearance, and perhaps even, indirectly, the inception of blanket bog growth

(Simmons, 1969, 1975; Merryfield and Moore, 1974).

It is equally clear from evidence in the region that, by about 6.5 ka BP, the climate had become wetter, resulting in the development of raised and blanket bogs and the rapid spread of alder (Scaife, 1984).

Many sites in the South-West, including those on Dartmoor, Bodmin Moor and the Isles of Scilly, demonstrate the progressive clearance of wood-land by humans during the Neolithic and later. While the Mesolithic episodes were probably minor and restricted to the mainland, those of the Neolithic were more extensive and long-lasting; evidence for post-forest clearance regeneration, however, is evident at some sites. By the Bronze Age, significant areas of woodland had been cleared, allowing light-demanding hazel to become dominant in some areas (Dimbleby, 1963; Bayley, 1975; Caseldine, 1980). The progressive modification of the woodland stand to allow cultivation from Neolithic times onward, is indicated by the pollen of cereals, weeds and ruderals. Certainly, the general view is that the removal of woodland in many areas had become irreversible by the Bronze Age, and that by Iron Age times a relatively open landscape, similar to today's, had been created (Christie, 1978; Caseldine, 1980).

Nowhere in the South-West, however, is the influence of humans on the environment better illustrated than in the Somerset Levels. Here, the discovery of the 'lake villages' at Glastonbury and Meare stimulated much early interest (Bulleid and Gray, 1911, 1917, 1948; Gray and Bulleid, 1953; Gray, 1966). A series of palaeobotanical studies by Godwin and his colleagues established both the vegetational history of the peatlands and set the burgeoning archaeological discoveries in an environmental context (Godwin, 1941, 1943, 1948, 1955a, 1955b, 1960, 1967; Clapham and Godwin, 1948; Godwin and Willis, 1959; Dewar and Godwin, 1963). The basic sequence of vegetational development has since been elaborated in studies which have used modern palynological techniques and radiocarbon dating (e.g. Beckett, 1977, 1978a, 1978b, 1979a, 1979b, 1979c; Beckett and Hibbert, 1976, 1978, 1979), plant macroscopic remains (e.g. Beckett, 1978a, 1978b, 1979a, 1979b; Coles and Orme, 1978; Caseldine, 1980(a); Orme *et al.*, 1981), beetle remains (e.g. Girling, 1976, 1977a, 1977b, 1978, 1979a, 1979b, 1980; Coles and Orme, 1976, 1978) and wood decay and tree ring evidence (e.g. Girling, 1976, 1979b, 1980; Morgan, 1976a, 1976b, 1977, 1978, 1980, 1982; Carruthers,

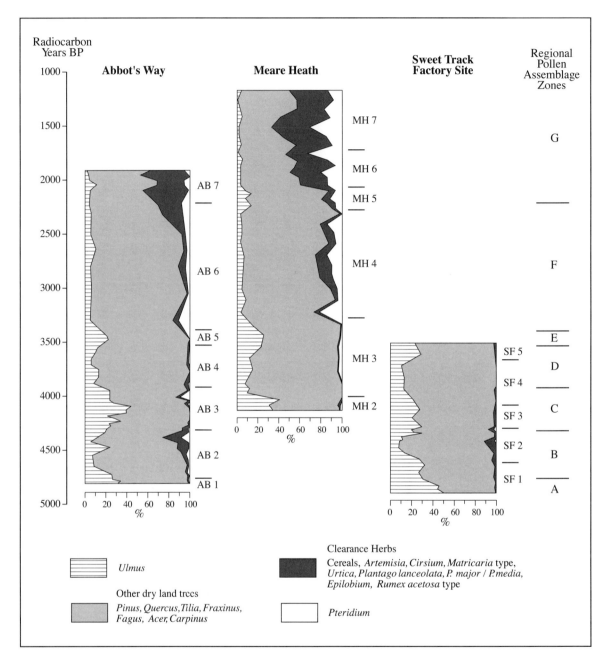

Figure 2.9 Summary Holocene pollen diagrams from the Abbot's Way, Meare Heath and Sweet Track Factory sites in the Somerset Levels (selected taxa only). The vertical scale is based on dates in uncorrected radiocarbon years BP (before present 1950). (Adapted from Beckett and Hibbert, 1979.)

1979; Bevercombe, 1980) to chart vegetational development in the area and anthropogenic effects on the environment: over 40 sites in the Levels have furnished palaeoenvironmental information, summaries of which are provided by Bell *et al.* (1984) and Beckett and Hibbert (1978, 1979).

Suffice to say here, that the Holocene deposits of the Somerset Levels consist, towards the coast,

of intertidal and brackish clays, silts and sands with intercalated peats (Godwin, 1941; Bell *et al.*, 1984). Inland, thick peat, often divisible into a highly humified lower layer and a poorly humified upper layer, overlies estuarine clay and is itself overlain in places by clay (Godwin, 1941). Preserved within the peat are prehistoric wooden trackways, which evidently linked the surrounding higher drier land – the 'islands' of bedrock and Burtle Beds (Bell *et al.*, 1984).

Peat formation in the Somerset Levels began at about 5.7 ka BP, as reedswamp colonized a waterlogged surface of estuarine clay (Kidson, 1977; Bell *et al.*, 1984). By *c.* 4 850 BP, trees had started to colonize the reedswamp, resulting in a 'fenwood' (Bell *et al.*, 1984). At this time, the area above the Levels appears to have been densely wooded: alder and birch fringed the waterlogged areas, while elm, oak and lime dominated the higher, better-drained land (Beckett and Hibbert, 1979: Regional Pollen Zone A; Figure 2.9). This woodland is comparable to the 'climax forest' found throughout much of southern Britain in Atlantic times. There is no evidence in the pollen record to suggest that Mesolithic people had a significant effect on the vegetation cover at this time.

At about 4.7 ka BP, a major change in the vegetational composition of the drier land, both within and surrounding the peatlands, is suggested by the pollen record (Regional Pollen Zone B; Figure 2.9). A major decline of elm pollen, as well as that of other trees, indicates the first major phase of forest clearance to support a largely pastoral economy. It was around this time, at the Atlantic/Sub-Boreal transition, that some of the earliest Neolithic trackways, such as the Sweet Track and Abbot's Way, were constructed. Also around this time, there was a major change in the peatlands themselves, with fenwood vegetation giving way to the development of raised *Sphagnum* bog. This change, which began in the vicinity of Abbot's Way at *c.* 4 650 BP, reaching other areas such as Shapwick Heath at a much later date (*c.* 4 050 BP), reflects increased waterlogging (see Holocene sea levels; this chapter) (Beckett and Hibbert, 1979; Bell *et al.*, 1984). The clearance phase itself lasted until *c.* 4.3 ka BP, when increasing amounts of elm pollen and decreasing values of herb pollen denote a likely diminution of human activity in the area, lasting perhaps until about 4 ka BP (Regional Pollen Zone C; Figure 2.9; Beckett and Hibbert, 1979).

Between about 4 and 3 ka BP (Regional Pollen Zones D and E; Figure 2.9), a minor phase of forest clearance followed by modest regeneration is indicated (Beckett and Hibbert, 1979; Bell *et al.*, 1984). Regeneration, however, was shortlived: a further fall in elm values and a rise in the pollen of *Plantago lanceolata*, *Rumex* and other herbs (Regional Pollen Zone F; Figure 2.9) suggest increased human activity, substantial forest clearance and re-establishment of a pastoral economy in and around the Levels, from between *c.* 3.3 to 2.2 ka BP. These changes coincide with continued raised bog development and the construction of later Neolithic and Bronze Age trackways (Clapham and Godwin, 1948; Dewar and Godwin, 1963; Coles *et al.*, 1975). Many of the tracks appear to have been built in response to flooding of the raised bog, a first phase of which lasted from about 2.7 to 2.3 ka BP (Clapham and Godwin, 1948; Beckett and Hibbert, 1979; Bell *et al.*, 1984). Towards the end of this period, the Iron Age village at Meare was constructed on higher land at the bog's margins.

From about 2 200 to 1 350 BP, the uppermost layers of peat in the Levels were formed. Their pollen content, dominated by herbs (grasses, docks and sorrels), denotes massive forest clearance during the Iron Age and into Romano–British times (Beckett and Hibbert, 1979). By these times, wide areas had been cleared for agriculture, and the presence of cereal pollen grains and those of their associated weeds, implies that both arable and pastoral farming were being practised throughout the higher drier land of the region, such as at Meare 'island'. It seems likely that during this period, birch growth was renewed on bog surfaces, while scrub woodland may have invaded once-cleared areas, already abandoned as a result of soil exhaustion (Beckett and Hibbert, 1979). Peat formation appears largely to have ceased by this time, although a precise reconstruction of events and conditions has been hampered by peat cutting.

One of the chief effects of the establishment of soil and vegetation covers in South-West England during the Holocene, together with its milder climate, was a marked diminution of geomorphological activity, especially in comparison with the preceding Younger Dryas. However, large volumes of unconsolidated periglacial debris have been remodelled by rivers, as witnessed by Holocene river terraces (Macklin, 1985, 1986; Macklin and Hunt, 1988), and significant landslides and mudflows have taken place (e.g. Arber, 1940, 1973; Brunsden and Jones, 1976; Grainger *et al.*, 1985, 1996; Brunsden, 1996; Grainger and Kalaugher, 1996; Kalaugher *et al.*, 1996). There is no clear evidence in the region for the climatic deterioration known as the 'Little Ice Age', and, with the exception of controversial earth hummocks on Dartmoor (Gerrard, 1983, 1988; Bennett *et al.*, 1996), virtually none for contemporary periglacial processes. Although the transport of sand and silt by wind and slope processes may have continued from the Younger Dryas into the earliest Holocene, the mid- and later Holocene appear to have been mostly times of relative geomorphological inactivity. Human impact on fluvial processes and sedimentation patterns has, however, been significant, with mining phases on Mendip causing the destabilization of fluvial regimes and the aggradation of significant alluvial units (Macklin *et al.*, 1985; Macklin, 1986). However, in recent centuries, as elsewhere in Britain, sand movement was reactivated (see above), although dune accumulation was not synchronous everywhere.

Chapter 3

Pre-Quaternary and long-term landscape evolution

THE PRE-QUATERNARY INHERITANCE
C. P. Green and S. Campbell

Introduction

The main outline of South-West England's landscape owes its origin to a combination of geological and tectonic controls and geomorphological processes in pre-Quaternary time. Together with other areas of southern Britain which escaped the major erosive and depositional effects of Pleistocene ice sheets, South-West England became a focus for numerous studies concerned with establishing the nature, distribution, ages and origins of various perceived erosion surfaces and related drainage networks (e.g. Balchin, 1937, 1946, 1952, 1964; Trueman, 1938; Wooldridge and Linton, 1939; Wooldridge, 1950; Bradshaw, 1961; Weller, 1959, 1960, 1961). These morphological studies mark an important stage in the development of geomorphological thought and technique. However, over much of South-West England there are relatively few on-land pre-Pleistocene deposits, and the great majority of relict features in the landscape are erosional and therefore almost impossible to date precisely. Overwhelmingly, the morphological evidence has proved profoundly unsatisfying, and until recently has lacked any serious confirmation from deposits (Kidson, 1977). However, two sites in particular provide significant evidence for establishing the nature and timing of major pre-Quaternary landshaping events and processes in the South-West, and are the subject of this chapter. St Agnes Beacon provides unique evidence for establishing the relative age(s) of erosion surfaces in the region (Walsh *et al.*, 1987; Jowsey *et al.*, 1992), while Beer Quarry shows a spectacular example of the controversial clay-with-flints, and has a major bearing on the pattern and nature of Palaeogene weathering processes. A brief synopsis of the early work on erosion surfaces and drainage networks, and a more detailed account of pre-Quaternary weathering residues, sediments and landform development are given here, both as an introduction to the subject and to the selected GCR sites. The long-term evolution of the characteristic granite terrains of the South-West merits separate consideration in Chapter 4.

Erosion surfaces and drainage networks: a brief history of research

Erosion surfaces have been recognized at a variety of heights throughout South-West England, up to and including the summits of Exmoor, Dartmoor and Bodmin Moor. The first examples were probably recognized by Reid (1890), who described a narrow shelf around the south and west coasts of Cornwall. The location of fossiliferous marine deposits in a valley cut in this shelf at St Erth, and the presence of what appeared to be a degraded cliffline backing the shelf above *c.* 430 ft (131 m), led Reid (1890) and Reid and Flett (1907) to suggest a marine origin and early Pliocene age for the surface (cf. Milner, 1922; Wooldridge, 1950). A similar coastal plateau was recognized around Torquay by Jukes-Browne (1907), while Barrow (1908) drew attention to possible marine erosion surfaces (Miocene) on Bodmin Moor at 750 and 1000 ft (229 and 305 m, respectively). The first proponents of a subaerial origin for surfaces in the South-West were Davis (1909) and Sawicki (1912).

The 1930s and 1940s saw a proliferation of erosion surface studies in the South-West: Gullick (1936), Balchin (1937, 1946) and Pounds (1939) described further surfaces in Cornwall; Green (1936) extrapolated Barrow's (1908) Bodmin surfaces to east Devon; while Macar (1936) and Hollingworth (1939) provided more general accounts and attempted wider correlations. In 1941, Green described the high platforms of east Devon, distinguishing at least six erosion surfaces between 440 and 920 ft (134 and 280 m): these were believed to range in age from Miocene to Pliocene and to be marine in origin (Green, 1941).

In 1952, Balchin provided details of planation surfaces in the Exmoor region, describing a 'staircase' of eight surfaces ranging in height up to *c.* 1225 ft (373 m) and being separated by bluffs or 'worn-down clifflines'. At this time, the assumption was that the lower six surfaces were marine in origin. The possibility was acknowledged, however, that the summit level and the surface below it could have been formed subaerially in 'sub-Cretaceous' and 'early Tertiary' times, respectively (Balchin, 1952). An acceptance that even the marine-formed surfaces had undergone substantial subaerial modification was by now becoming implicit in the literature (e.g. Balchin, 1952; Stephens, 1952).

The 1960s saw substantial interest in the erosion

surfaces and drainage networks of the South-West, with an increasing emphasis on measurement and statistical correlation. Detailed studies were undertaken on Bodmin Moor and in east Cornwall by Weller (1959, 1960, 1961); in north Devon (Arber, 1960); in west Cornwall (Everard, 1960b); in the Lizard Peninsula (Fryer, 1960); and in north-west Devon (Bradshaw, 1961). This work on erosion surfaces was to become intimately related to models concerning the development of drainage networks (e.g. Waters, 1951, 1953, 1960c; Kidson, 1962; Brunsden, 1963; Brunsden *et al.*, 1964). Fryer's (1960) study is notable in that it rejected a marine origin for all surfaces, save the 430 ft (131 m) level (cf. Reid, 1890), and invoked a single extensive surface, formed subaerially as a peneplain from Triassic times onward: this echoes views put forward by Jones (1951) concerning the evolution of the Welsh landscape. This shift in thinking is to some extent mirrored in subsequent studies. Orme (1961, 1964), for example, recognized four high-level planation surfaces around southern Dartmoor (Chapter 4): the upper two belonged to the early and mid-Tertiary respectively, the lower two to the late Tertiary. All were believed to have been formed subaerially. According to Orme (1964), the late Tertiary landscape was then drowned in early Pleistocene times to a height of *c*. 700 ft (213 m): a 'staircase' of forms below this level marked stillstands of the falling Pleistocene sea. Although Wooldridge (1950) noted similar surfaces up to heights of 1200 ft (366 m) (Calabrian), he ascribed the lower levels to retreat stages of the Pliocene sea.

In contrast, Kidson (1962) found no evidence in the South-West to support sea levels or marine planation above the general level of 210 m. The latter level was regarded as the most prominent surface in the region, representing the limit of an early Pleistocene marine transgression (Kidson, 1962, 1977; Brunsden, 1963; Brunsden *et al.*, 1964). Even the latter ascription has not gone unchallenged: Simpson (1964) dismissed this feature as an exhumed shoreline of Upper Cretaceous age (see below). Such datings would seem all the more unconvincing in view of the evidence from St Agnes Beacon, which shows that any geomorphologically significant transgressions of the sea above *c*. 75 m OD in post-Miocene times are highly unlikely (see St Agnes Beacon; Walsh *et al.*, 1987; Jowsey *et al.*, 1992).

Neither have attempts to link drainage development to the erosion surfaces and to general schemes of denudation chronology proved particularly rewarding. While many of the earliest studies on drainage systems were largely incidental to enquiries on the age and origin of the region's erosion surfaces (e.g. Clayden, 1906; Dewey, 1916), several notable attempts at a more holistic approach have been made. These include the work of Waters (1951, 1953, 1960c) in south-west Devon, and Wooldridge (1954), Brunsden (1963) and Brunsden *et al.* (1964). In general, an hypothesis of a falling, albeit oscillatory, base level from late Tertiary through Pleistocene times, producing both erosion surfaces and river rejuvenations, has been upheld (Balchin, 1964, 1981): Brunsden (1963) claimed that up to 17 separate stages in this development could be recognized in the catchment of the River Dart.

Pre-Quaternary weathering residues, sediments and landform development

Introduction

A more precise reconstruction of pre-Quaternary landform development, however, has been based on the nature and distribution of pre-Quaternary weathering residues and sediments and upon structural information. Evidence of weathering and soil formation is particularly useful since it may demonstrate the presence and environmental significance of palaeosurfaces. Datable, unconsolidated deposits found at given locations within the landscape, for example, the Miocene and Oligocene sediments of St Agnes Beacon, provide not only evidence of processes and environments at the time of deposition, but also important constraints on the extent and magnitude of later events and processes: in the context of St Agnes Beacon, it is highly unlikely that the sands and clays there could have survived a significant post-Miocene marine transgression or an incursion of Pleistocene ice. The importance of the age of the relict deposits remaining both as conformable and unconformable outcrops in the landscape is immediately evident. In the subsequent section, a brief outline of the chief surviving sedimentary evidence is given, followed by a possible model for pre-Quaternary landform development: the latter relies heavily on the synthesis given by Green (1985).

Pre-Quaternary sedimentary evidence in South-West England

In situ and reworked clay-with-flints

The oldest outcrops of post-Cretaceous/pre-Quaternary sediments in the South-West are found in east Devon, south Somerset and west Dorset. Here, relatively large areas of clay-with-flints and -chert cap the Chalk and Upper Greensand. Where *in situ*, these deposits have been classified as the Combpyne Soil (e.g. Isaac, 1979, 1981, 1983a, 1983b) and the Tower Wood Gravel (Hamblin, 1973a, 1973b) (Figure 3.1). Residual flint gravels also occur in solution pipes formed in Devonian limestone near Kingsteignton (Figure 3.3; Brunsden *et al.*, 1976). A Palaeocene age for these residual deposits has been established from several lines of evidence (see Beer Quarry), but not least because locally they underlie Upper Eocene–Lower Oligocene beds of the Bovey Formation (Edwards, 1976; Isaac, 1981). In many areas, these residual deposits were reworked by Palaeocene and Eocene fluvial processes, giving rise to a series of deposits which include the Peak Hill, Mutters Moor, Buller's Hill and Aller gravels (Edwards, 1973; Hamblin, 1973a, 1973b; Brunsden *et al.*, 1976; Isaac, 1981). Although evidence for dissolution of the Chalk and the redistribution of weathering residues is fragmentary and dispersed, Beer Quarry has been chosen by the GCR to demonstrate both an excellent example of *in situ* clay-with-flints and a spectacular series of solution-formed, clay-filled chalk pipes.

The Bovey Tracey and Petrockstow basins

Significant outcrops of post-Cretaceous/pre-Quaternary sediments are found in the Bovey Tracey and Petrockstow basins (Figure 3.1). These freshwater Eocene gravels and Oligocene sand, clay and lignite deposits, lie in fault-guided basins which were tectonically active at various times in the Tertiary: in places, these sediments overlie residual gravels of proposed Palaeocene age. They contain weathering products which have a bearing on the development of adjacent landmasses, especially the granite terrain of Dartmoor.

St Agnes Beacon

Mid-Oligocene and Miocene sands and clays occupy an area of some 1.6 km^2 around St Agnes Beacon in west Cornwall. Although the age and origin of these sediments have long been disputed, recent work has established the presence of two distinct Mid-Oligocene and Miocene outliers, the sediments in which were formed by a variety of lacustrine, aeolian and colluvial processes (Walsh *et al.*, 1987; Jowsey *et al.*, 1992). The site, described in detail in this chapter, is of great significance for constraining the age of landforms and erosional surfaces in the region.

St Erth

The St Erth Beds consist of sands and clays located in a valley at a height of *c.* 30 m OD in west Cornwall (Figure 3.1). The clays contain a rich marine fauna with strong Mediterranean affinities and many extinct species. Reid (1890) argued that the fauna indicated water depths of around 90 m, and postulated that the sediments had been laid down in a sea which was also responsible for cutting an extensive platform at about 430 ft (131 m) OD (Reid, 1890; Reid and Flett, 1907; see above). Reid correlated the St Erth Beds with the Lenham Beds of south-east England and ascribed both to the Pliocene.

Reid made no reference to earlier suggestions that the clay at St Erth was a 'boulder clay' containing marine shells. However, this possibility was revived by Mitchell (1965) who argued that the fauna had been deposited during the Cromerian and had been reworked by an ice sheet of Anglian age. Subsequently, Mitchell *et al.* (1973a) re-examined molluscs, foraminifera and ostracods from the clay and concluded, like Reid, that the deposits were marine in origin and Pliocene in age: a water depth of only 10 m, however, was suggested, giving a projected sea level some 45 m above present OD (Kidson, 1977).

Although a glacial origin for any part of the St Erth sequence is no longer considered likely, considerable controversy still surrounds the chronostratigraphic classification of the beds. In this context, the site is particularly significant in providing evidence for arguments over the position of the Pliocene/Pleistocene boundary which, in the United Kingdom succession, has a particularly controversial history. Currently, the boundary is based on a stratotype section at Vrica in southern Italy, which establishes an age of *c.* 1.6 Ma BP, more or less coincident with the end of the Olduvai magnetic event (Aguirre and Pasini, 1985). In the North Sea region, however, there is much evidence to suggest that the boundary should be older: exact definitions vary, but the base of the Praetiglian Stage (2.3 Ma) (Zagwijn, 1989; Gibbard *et al.*, 1991) or the transition between the Matuyama and Gauss magnetic polarity chronozones at 2.45 Ma

BP (Harland *et al.*, 1982) have been suggested (Balson, 1995).

Recent work on planktonic foraminifera indicates an age for part of the Red Crag between 3.2 and 2.4 Ma BP, that is Late Pliocene by either definition (Funnell, 1987, 1988). Foraminifera from the St Erth Beds indicate an age of *c.* 2.1–1.9 Ma BP (Jenkins *et al.*, 1986) – Pliocene according to the international definition of the boundary, but Pleistocene by most current United Kingdom practice. The St Erth site has been selected as part of the Pliocene GCR site network and will be described in the *Tertiary of Great Britain* stratigraphic volume of the GCR Series. The St Erth deposits, however, have been described widely in Quaternary and geomorphological literature pertaining to the region, and further consideration of their relevance is given below.

Crousa Common and Polcrebo Downs

There are other localities in South-West England where high-level deposits of post-Cretaceous/pre-Quaternary age are found. The most notable, perhaps, are found on Crousa Common and Polcrebo Downs in west Cornwall. The deposits on Crousa Common were described as early as 1843 by Budge, and consist of yellow clay with copious quantities of water-worn quartz pebbles (Stephens, 1980; Campbell, 1984), and occur at a height of about 110 m OD. Unlike the Polcrebo deposits (*c.* 152 m OD), they contain fossils (spores and pollen) but their age and origin are also unknown (Scourse, 1996b). Correlations, however, have been made between the Crousa/Polcrebo deposits and those at St Erth; a Pliocene age has often been alluded to (Hill and MacAlister, 1906; Reid and Scrivenor, 1906; Milner, 1922; Hendricks, 1923). Both marine and fluvial agencies have been suggested to account for the deposits, and although recent Scanning Electron Microscopy (SEM) work upholds a waterlain origin for those at Crousa Common (Campbell, 1984), precise environmental and age inferences cannot yet be made. Bowen (1994b) speculated that the deposits on Crousa Common could be glacial in origin, formed during an Early Pleistocene (Oxygen Isotope Stage 16) ice advance.

Similar gravels above 82 m OD in the Bristol district have been described by Palmer (1931). Mitchell (1960) correlated gravels at Hele near Barnstaple (56 m OD) with the St Erth Beds (Pliocene) and argued that neither ice nor sea level had attained this height since their emplacement. Kidson and Wood (1974), however, have argued that the Hele gravels are not marine but glacio-

fluvial in origin, and have ascribed them to the glacial sequence which includes the Fremington Clay (see Chapter 7).

A controversial sandy flint gravel has also been described at Orleigh Court near Bideford in north Devon. This small outcrop, some 1.2 km long by 0.4 km wide, was regarded by Ussher (1879b) as re-sorted Cretaceous material of Tertiary age; by Boswell (1923) as Eocene; and by Rogers and Simpson (1937) as a derived deposit of at least post-Eocene age. Although the age and origin of these deposits are poorly understood, it is possible that they can be correlated with residual flint gravels of proposed Palaeocene age in the Haldon Hills (Tower Wood Gravel) and Beer/Sidmouth area (Combpyne Soil) (Edwards and Freshney, 1982; Green, 1985).

Pre-Quaternary landform development in South-West England: a synthesis

The pre-Quaternary origin of significant elements in the relief of the British Isles is now widely accepted. Historically, there have been two main schools of thought regarding the shaping of the pre-Quaternary landscape, and debate has centred on the key question of whether the bulk of denudation took place in the Palaeogene (e.g. Pinchemel, 1954) or in the Neogene (e.g. Wooldridge and Linton, 1955). On the Palaeozoic rocks of South-West England, there is clearly the added possibility that some landforms could be of Mesozoic age (Guilcher, 1949; Linton, 1951; Green, 1985). Recent work throughout southern England has certainly raised considerable doubts concerning the reality and age-range of the various geomorphological 'staircases' which have been proposed (see above; Jones, 1980; Green, 1985), and sedimentary and landform evidence from South-West England, in particular, has been fundamental in shifting opinion towards the Pinchemel school of thought. Although it is beyond the scope of the present work, which is overwhelmingly concerned with reviewing the Quaternary evidence, to examine all aspects of work and field evidence concerning pre-Quaternary landscape evolution, the following synthesis highlights the most important aspects with respect to the landforms and deposits of the South-West.

Pre-Tertiary geomorphological development

There is significant evidence that Permian–Triassic erosion effected the primary shaping of the present relief of the Palaeozoic rocks of western Britain:

massive denudation of the rocks of South-West England occurred at this time. The granites of Dartmoor (Chapter 4) range in age from Late Carboniferous to Early Permian (Hawkes, 1982) and are believed to have been intruded at depths of between five and nine kilometres below the surface. Despite this depth of intrusion, debris from the Dartmoor igneous complex occurs in Early Permian sediments, indicating substantial erosion and exposure of parts of the pluton within a surprisingly narrow time interval (Laming, 1982; Green, 1985). Material from the Dartmoor igneous complex is also present within Triassic conglomerates, and both Permian and Triassic sedimentary rocks locally provide excellent and familiar evidence of geomorphological and climatic conditions at the time of their deposition: breccias, aeolian and fluvial sands, conglomerates and mudstones indicate deposition under semi-arid conditions around the margins of a dissected upland (Green, 1985). Certainly, by the end of Triassic times, the rocks of South-West England appear to have been reduced to a surface of very low relief: progressive denudation is reflected in the passage upward in the thick (> 1000 m) Permian succession in Devon from basal breccias through sandstones to mudstones (Green, 1985).

Only relatively minor modifications to the Permian–Triassic landscape of South-West England are believed to have occurred during Jurassic and Lower Cretaceous times. Hart (1982) argued that South-West England remained a land area for much of the Jurassic: most of the Jurassic rocks of the Wessex Basin are marine shelf/shallow water sediments, and any terriginous inputs, such as clay minerals, appear to have originated from a landmass of low relief (Cosgrove, 1975). Certainly, no surfaces of Jurassic age can be recognized on the Palaeozoic rocks of the South-West, and evidence of Jurassic terrestrial weathering in southern England is confined to the Lulworth Beds of the uppermost Portlandian in south Dorset (Green, 1985).

The culmination of the Upper Cretaceous marine transgressions (Hancock, 1969) was a turning point of far-reaching significance in the long-term development of landforms in southern Britain (Green, 1985). The most elevated parts of the South-West Peninsula may have escaped submergence (Hancock, 1969), but over the rest of southern Britain a continuous cover of chalk was deposited. Towards the end of the Lower Cretaceous, the intensity of erosion appears to have increased in South-West England, presumably as a result of uplift: arenaceous sediments in the Lower and Upper

Greensands reflect this intensification. In many areas of the South-West, the only evidence of a former chalk cover is the preservation of flint in residual deposits (clay-with-flints) or in later sediments derived from them (Isaac, 1981). Whatever the character of the relief across which the Cenomanian transgression extended, details of the subchalk surface must have been shaped by marine agencies. Although unconformities between the Upper Cretaceous rocks and older Mesozoic sediments are effectively planar, the planation effected by the transgression on the harder rocks of the Peninsula was less complete: some benches on the flanks of Dartmoor, however, have been explained as the product of the transgressive Upper Cretaceous seas (Simpson, 1964). The surviving evidence, however, fails to show whether the submergence of Cornubia was complete (Green, 1985).

Notwithstanding, it is highly unlikely that terrestrial surfaces and associated features of pre-chalk age in South-West England survived without some modification. However, the gross morphology of the region could well pre-date deposition of the Chalk and be the product of earlier Cretaceous or Mesozoic denudation (Green, 1985). It is also possible for rocks deeply weathered in pre-chalk times to have survived, and a pre-chalk age for the kaolinized granites of the South-West has been proposed (cf. Millot, 1970; Lidmar-Bergström, 1982; Esteoule-Choux, 1983; Chapter 4).

Chalk sedimentation in South-West England appears to have ended in mid-Campanian times (Figure 3.1). Since this is also likely to reflect the termination of marine conditions, removal of chalk by subaerial agencies could have begun as early as the Upper Cretaceous (late Campanian to Maastrichtian) (Green, 1985). Maastrichtian (Upper Cretaceous) and Danian (Lower Palaeocene) sediments are absent in southern England, but offshore they consist of pure limestones, consistent with denudation of chalk from contemporary land areas such as parts of South-West England.

Palaeogene geomorphological development
The uplands of east Devon, south Somerset and west Dorset furnish some of the most compelling evidence in Britain for Palaeocene denudation of the Chalk cover under tropical climatic conditions (Green, 1985; see Beer Quarry). This weathering reduced the Chalk over the Palaeozoic basement to a mantle of weathering residues, essentially the widespread clay-with-flints, the Tower Wood Gravel found on the summits of the Haldon Hills (Hamblin, 1973a) and the Combpyne Soil of the

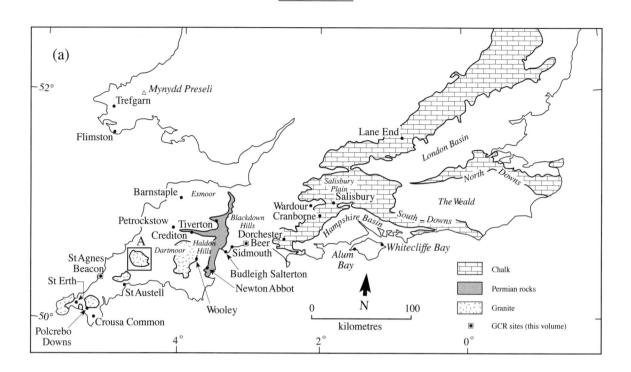

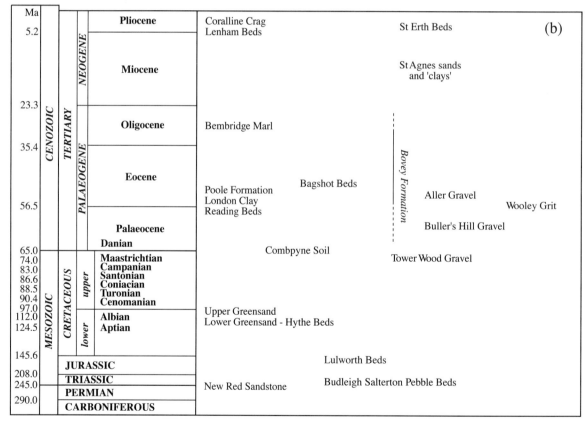

Figure 3.1 (a) Southern and South-West England, showing localities referred to in the text, and selected geological outcrops. (b) Significant deposits and events in the geomorphological evolution of southern Britain. (Adapted from Green, 1985, with timescale based on Harland *et al.*, 1982.)

Sidmouth and Beer area (Isaac, 1981, 1983b). Isaac (1983b) places the formation of the Combpyne Soil and Tower Wood Gravel in the Danian (Lower Palaeocene), and regards this as a time of intense lateritic weathering, when any former chalk cover was reduced to a mantle of weathering residues: in a few places, he identified *in situ* lateritic weathering characteristics such as pallid and mottled zones overlying red earth horizons. A Danian age is inferred by analogy with lateritic weathering of that age in the Interbasaltic Formation of Northern Ireland (Isaac, 1983b), although there appears to be no good reason why weathering of the Chalk could not have begun earlier, in the Upper Cretaceous (Green, 1985).

Late Palaeocene and Eocene erosion then occurred under subtropical climatic conditions. Evidence for this comes from the Tertiary gravel deposits of south and east Devon. Resting on the Tower Wood Gravel of the Haldon Hills, and directly on the Upper Greensand in the Kingsteignton pipes (Brunsden *et al.*, 1976) and around the margin of the Bovey Basin, are flint-rich gravels, the Buller's Hill Gravel of Hamblin (1973b) (see Figure 3.5). These gravels, and their lateral equivalents, the Peak Hill Gravels, are thought to have been reworked by fluvial processes from the residual Tower Wood Gravels and Combpyne Soil, respectively. In the Kingsteignton pipes, the Buller's Hill Gravels are overlain by the Aller Gravels (Edwards, 1973; Brunsden *et al.*, 1976): these gravels appear to be overlain by the main part of the Bovey Formation (Figure 3.5; Green, 1985).

Small amounts of Palaeozoic debris present in the Buller's Hill Gravel may indicate a renewed exposure of the basement at this time, and there are clear indications that the Late Palaeocene and Early Eocene saw repeated reworking of a thin veneer of sediments and weathering residues over substantial areas of the South-West. Isaac (1983b) has shown that kaolinite in deposits of the Bovey Formation, including the Buller's Hill Gravel, is less well ordered than in the *in situ* Tower Wood Gravel (residual). This has been taken to indicate a weathering phase separate from and shorter (or of lesser intensity) than the phase responsible for the Tower Wood Gravel and Combpyne Soil. Similar indications of *in situ* deep weathering profiles have been described beneath Oligocene Bovey Formation sediments in the Petrockstow Basin (Bristow, 1969), confirming the existence of a deeply weathered terrain in South-West England during the Palaeogene (Green, 1985). Silicified

deposits (often termed Sarsens) are widespread in southern Britain, and most workers have proposed a Palaeogene age for their formation (see Beer Quarry; Clark *et al.*, 1967; Isaac, 1979, 1981, 1983a, 1983b).

Deposition of the bulk of the Bovey Formation appears to have occurred during the Eocene and part or all of the Oligocene: throughout this period, sediment was derived from erosion of deep but relatively immature weathering profiles developed on both granite and Upper Palaeozoic metasediments under subtropical or warm temperate climatic conditions (Green, 1985). It is likely that the erosional morphology of the summit relief on these rocks was acquired during this interval. An extensive erosional surface in the region between Salisbury Plain in the east and Dartmoor in the west appears to have developed, becoming refined by Early Eocene (London Clay) times by marine and fluvial agencies (Green, 1985). In south-east England, the Chalk inherited its present summit morphology before the end of the Palaeocene, prior to quite deep burial beneath later Tertiary sediments. Areas of Palaeozoic rocks in the west were drained by rivers running into the Early Eocene sea of southern England: there is no evidence that this sea, even at its maximum extent in London Clay times, extended farther west than the basin of the upper Otter (Green, 1974b, 1985). Faulting contributed significantly to relief development at this time, and also later in the Tertiary (Green, 1985).

Throughout southern Britain, there are consistent indications that denudation of the Chalk was largely effected under tropical or subtropical conditions before the end of the Palaeocene. In the Late Palaeocene and Early Eocene, the Chalk around the western fringe of the Hampshire Basin, and the rocks exposed by the removal of the Chalk both to the west and north of the surviving chalk outcrop in southern England, formed an erosional province in which a surface of low relief was widely developed (Cope, 1994). During the same interval, the area occupied by the Chalk outcrop in south-east England became an essentially depositional province and was buried beneath the Thanet Sands, the Reading Beds and a substantial thickness of Eocene sediments, derived largely from the northern part of the aforementioned erosional province (Morton, 1982). Towards the end of the Palaeogene (mid-Eocene onwards), the area of marine sedimentation in southern Britain became progressively smaller (Murray and Wright, 1974), although there is no evidence for the substantial production of terrigenous sediment. In fact, in the offshore record of Palaeogene sedimentation in the English Channel (Curry *et al.*, 1970), carbonate rocks predominate throughout and form the whole recorded sequence for the Middle and Late Eocene. The volume of Oligocene sediments is small, comprising in addition to the onshore outcrops, only one offshore record – of freshwater limestone. This scarcity of terrigenous sediment when most of southern England formed a land area strongly suggests a Late Palaeogene terrain of low relief close to base level (Green, 1985).

Neogene geomorphological development

The record of Neogene landform development in southern Britain is extremely difficult to interpret. This arises both from the deficiencies of the sedimentary record and the possibility that the Neogene consisted of a prolonged morphostatic phase – a time in which geomorphological activity was limited through one mechanism or another. Certainly, Neogene deposits in Britain are limited and scattered (e.g. the Coralline Crag and the St Erth Beds), and until recently, Miocene sediments were only known offshore around southern Britain (Curry *et al.*, 1970; Evans and Hughes, 1984): St Agnes Beacon, west Cornwall, however, is one of five known sites in the British Isles where fossiliferous non-marine sediments (in the case of St Agnes Beacon, sands and 'clays') of possible Miocene age have been preserved (Walsh *et al.*, 1987; Jowsey *et al.*, 1992; Walsh *et al.*, 1996), while the St Erth Beds provide the only sedimentary evidence for the incursion of the Pliocene sea in south-west Britain.

However, the recognition of a supposed Neogene erosional surface on the Chalk of south-east England (the Miocene–Pliocene peneplain) was central to the work of Wooldridge and Linton (1955) and their belief that the bulk of post-Cretaceous denudation had occurred in the Neogene. The reality of this peneplain has often been challenged (Pinchemel, 1954; Clark *et al.*, 1967; Green, 1985), and most indications are that the bulk of chalk denudation occurred much earlier in the Tertiary (see above; Jones, 1980; Green, 1985).

Certainly in the South-West, the disposition of Cretaceous and Early Tertiary sediments and weathering residues in relation to the broad pattern of relief on the Palaeozoic rocks, seems inconsistent with the production of a significant part of that relief by Neogene erosion: indeed, relief elements which can be referred confidently to the Neogene are virtually absent throughout southern Britain, and in the South-West, the late Palaeogene surface

survives without convincing evidence of either Neogene erosion or significant solutional lowering. However, in other areas of southern Britain, such as parts of the Weald, erosional surfaces do appear to have developed, and have removed evidence of older Palaeogene surfaces and deposits (Jones, 1980). Such erosional areas now form the summit relief and clearly pre-date the Quaternary terrace sequences found in many river valleys: a Neogene age is therefore indicated. The summit relief of southern Britain is evidently not everywhere of the same age: substantial areas of the South-West appear to have escaped significant Neogene modification, while the latter's effects elsewhere are more clearly demonstrable. The lack of Neogene landform development over large areas of the South-West is best explained by a generally low level of relief in relation to base level, while differential development of relief is most readily explained by localized structural activity (Green, 1985).

The evidence from St Agnes Beacon demonstrates very clearly that, excepting unexplained local structural activity, there can have been no significant regional marine incursions above the general level of *c.* 75 m OD since the Miocene: the largely unconsolidated Miocene sands there are unlikely to have survived such an incursion. Evidence for a subsequent marine transgression on to the Peninsula in Pliocene times is witnessed by the deposits at St Erth (and possibly by those at Crousa Common and Polcrebo Downs), but the wider geomorphological effects of this incursion are unknown. Although, through amino-acid geochronology, the raised beach deposits of southern Britain have started to provide a framework for determining the changing pattern of Pleistocene land and sea levels (e.g. Bowen, 1994b), the overwhelmingly erosional nature of the evidence lends vast uncertainty to determining the nature and timing of geomorphological conditions and events in later Tertiary and early Pleistocene times. The low-level erosion surfaces, shore platforms and raised beach deposits with a major bearing on these issues are the subject of Chapters 6, 7 and 8.

BEER QUARRY
S. Campbell

Highlights

One of the best exposures of clay-filled chalk 'pipes' in Britain, Beer Quarry provides important evidence for interpreting solutional processes oper-

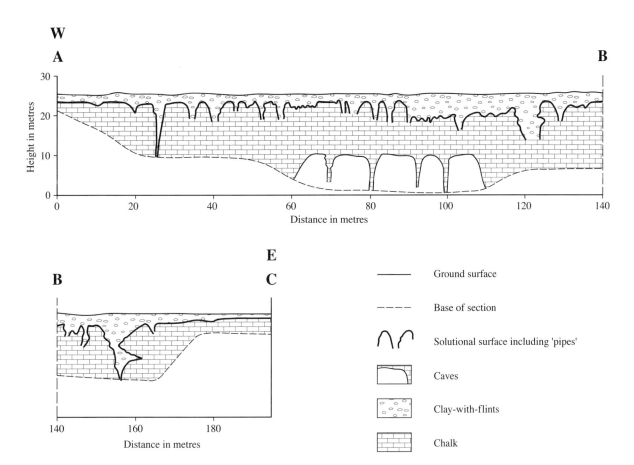

Figure 3.2 Clay-filled chalk 'pipes' exposed along the northern working face of Beer Quarry in 1990.

ating in chalk landscapes, and shows a classic example of the controversial 'clay-with-flints'. Recent evidence suggests that the pipes may have formed largely as the result of solutional processes which accompanied lateritic (tropical) weathering in the Palaeocene.

Introduction

Beer Quarry exposes important evidence for long-term, including pre-Pleistocene, landscape evolution in South-West England. A superb series of 'pipes' here was probably formed by chalk solution in the Tertiary: an infill of 'clay-with-flints' may be the product of dissolution of Upper Cretaceous rocks. Described in detail for the first time here, the features at Beer Quarry have a major bearing on reconstructions of the Tertiary palaeoenvironment of east Devon: these have formerly been adduced from weathering profiles and lithostratigraphic evidence found elsewhere in the plateau deposits of the region (e.g. Woodward and Ussher, 1911;

Green, 1974a; Isaac, 1979, 1981, 1983a, 1983b). Little agreement exists as to the origins of solutional cavities and associated infill materials in general: Beer Quarry provides important evidence for both, and adds to the growing belief that these features are probably polygenetic and have formed at different times.

Description

Beer Quarry (SY 215895), one mile west of Beer in Devon, is noted for the famous Beer Freestone, worked here for centuries, and for an extensive network of caves cut into the Chalk during its extraction. The main area of caves, excavated as early as Roman times (Perkins, 1971), lies to the south of Quarry Lane, and is now a visitors' centre. The current workings, including the GCR site, lie to the north of Quarry Lane and consist of a large disused face between *c*. SY 213896 and 215896. Present freestone workings are restricted to a small area at the extreme eastern end of the face, and

Figure 3.3 A large 'pipe' structure towards the western end of the north quarry face, showing the abrupt transition from chalk to the infill material (clay-with-flints). Even in monochrome, the profound darkening of the infill towards the pipe's margins can be seen clearly. (Photo: S. Campbell.)

other beds in the Middle Chalk are now worked for agricultural lime in the south-eastern part of the quarry. The quarry floor is occupied by plant and buildings and, in its deepest part, excavated caverns run north for about a quarter of a mile into a large chamber supported by chalk pillars (De la Beche, 1826; Ager and Smith, 1965). These adits underlie the main disused face (Figure 3.2) and were used as ammunition stores during the Second World War.

Jukes-Browne (1903) recorded a succession of Middle Chalk up to 24 m thick overlying Cenomanian limestone at Beer Quarry. The Beer Freestone, some 4 m thick, is the basal part of the Middle Chalk and is itself overlain by a series of more highly jointed, flinty, nodular and brecciated chalk beds (Jukes-Browne, 1903). These less coherent beds are penetrated from the top by solution cavities, here conveniently referred to as 'pipes'. Details of the pipe structures found in the northern quarry face are shown in Figure 3.2. These sediment-filled pipes are well exposed along the entire 200 m-long section (Figure 3.2; section A–C), and merge upwards into an extremely poorly sorted, flinty gravel which forms a continuous layer, some

2 m thick, at the top of the sections. Individual pipes are commonly 2–3 m deep by 2–3 m wide at the top; most, although not all, taper with depth. The pattern of piping, however, is extremely irregular. Many small pipes (< 1 m wide by 1 m deep) occur, in addition to several much larger examples (at 25, 120 and 155 m along the section; Figure 3.2). Features at 25 m and 155 m extend beneath the base of the exposed face and therefore exceed 15 m in depth. Some pipes are narrow, others much wider (e.g. those at *c.* 120 m and 155 m); between *c.* 90 and 115 m along the section, the Chalk has been replaced/dissolved over a broad area now occupied by flinty gravel at least 3.5 m in depth.

Similar pipes also occur elsewhere in the quarry, but are generally less frequent and less well developed. An extremely large, clay- and gravel-filled pipe occurs, however, at the easternmost end of the quarry. One example (Figure 3.2; section B–C) shows the infill material to occur beneath an *in situ* 'shelf' of unaltered chalk. In several pipes, the clay-with-flints is completely surrounded by unaltered chalk.

Most of the pipes show a rapid transition from

Figure 3.4 Detail of a typical pipe margin, showing the abrupt transition from chalk to clay-with-flints. (Photo: S. Campbell.)

chalk to infill (Figures 3.3 and 3.4). The latter consists almost entirely of whole and broken flints in a black, chocolate-brown or red clay-rich matrix: clay-with-flints is an apt description for these mostly structureless, unbedded and undifferentiated deposits. Material infilling the pipes is visually strongly similar to the capping layer: there is some suggestion that flints are more densely packed towards the base of pipes where, as a result, the deposit is almost clast-supported. Variations in the colour of the infill material do, however, occur. Generally it is reddest towards the top of the sections and dark chocolate-brown or black towards the base. Along pipe margins, the infill is frequently colour-zoned with a blackened band, almost 0.5 m wide, occurring next to the Chalk (Figures 3.3 and 3.4). Elsewhere, the different colours are abruptly juxtaposed without apparent pattern, perhaps having been disrupted by frost.

Interpretation

Details of the petrography and sedimentology of the deposits filling the pipe structures at Beer are not available, although the pipe structures were recorded by Ager and Smith (1965) and Perkins (1971). However, much information regarding the nature and origin of the infill can be gleaned from published studies of adjacent plateau deposits in east Devon (e.g. Woodward and Ussher, 1911; Waters, 1960d, 1960e; Hamblin, 1968, 1973a,

1973b, 1974a, 1974b; Edwards, 1973; Brunsden *et al.*, 1976; Isaac, 1979, 1981, 1983a, 1983b). The formation of the solutional cavities or 'pipes' may also be related to work on comparable karstic structures and associated infills found elsewhere in southern England and northern France (e.g. Osborne White, 1903; Kirkaldy, 1950; Avery *et al.*, 1959; Hodgson *et al.*, 1967; Mathieu, 1971; Thorez *et al.*, 1971; Pepper, 1973; Walsh *et al.*, 1973; Chartres and Whalley, 1975; Brunsden *et al.*, 1976).

The Tertiary plateau deposits of east Devon were first mapped and described in detail by Woodward and Ussher (1911), in the Geological Survey Memoir for the Sidmouth area (Sheet 326/340). Large areas of clay-with-flints and -chert were shown to cap the Chalk and Upper Greensand of the east Devon tableland. These deposits were divided into two groups (both Eocene): 1. clays and gravels with quartz, quartzite and other materials in addition to flint and chert; 2. clay with flint and chert – at least, in part, true clay-with-flints (Woodward and Ussher, 1911). Material infilling the pipes at Beer would appear, albeit visually, to belong to the latter category.

More recently, Waters (1960d, 1960e), Edwards (1973) and Hamblin (1968, 1973a, 1973b) have studied the plateau deposits: an origin involving the dissolution of the Chalk beneath a cover of Tertiary deposits has generally been invoked. The broad two-fold division of deposits (Woodward and Ussher, 1911) has been upheld and, in the Haldon Hills, a lower residual unit and an upper fluvial unit have been identified (Hamblin, 1968, 1973a, 1973b).

Similarly, Isaac (1979) originally divided Tertiary plateau sediments in the Sidmouth area into two lithostratigraphic formations: 1. the Peak Hill Gravels (composed of unabraded flints in a matrix of kaolinite, and formed by the removal of calcium carbonate from the Chalk under lateritic weathering conditions); and 2. the Mutters Moor Gravels (derived from the Peak Hill Gravels by aeolian and fluvial processes). These formations are separated by an unconformity and, locally, silcretes are present.

Subsequently, Isaac revised this scheme and termed the *in situ* stony clays the Combpyne Soil, and their reworked equivalents the Peak Hill Gravels (Isaac, 1981). The former, he argued, originated as a Palaeocene lateritic weathering profile (with pallid and mottled zones), formed as the Chalk underwent dissolution beneath Tertiary overburden. This led to the development of kaolinitic residual flint gravels up to 10 m in thickness over much of the east Devon tableland: well-differentiated lateritic weathering profiles were preserved in irregular deep pockets in the Chalk, and it seems likely that the material infilling the pipes at Beer can be assigned, at least in part, to the Combpyne Soil lithostratigraphic formation.

Tertiary history in south and east Devon: a synthesis

Isaac (1981, 1983b) has argued that the Tertiary plateau deposits of east Devon are characterized by a complex of superposed weathering profiles which reflect a protracted pedological, diagenetic and geomorphological history, beginning with the emergence of the post-chalk land surface at the end of the Cretaceous. In essence, the plateau deposits represent a series of Tertiary palaeosols in part reworked during the Tertiary and disturbed by subsequent frost-action in the Pleistocene.

The oldest, and principal, Tertiary stratigraphic unit recognized in the Beer/Sidmouth area is the Combpyne Soil, developed as the Chalk underwent dissolution in a tropical (Palaeocene) climate on a relatively stable land surface (Isaac, 1981, 1983b). The depth of chalk dissolved and removed is unknown, but likely to have been substantial. Residual deposits (initially the Combpyne Soil) are mainly kaolinitic, non-indurated, lateritic weathering products. They include, however, a range of silcretes and siliceous indurated deposits reflecting protracted soil formation and diagenesis: the silcretes may indicate several separate periods of desiccation in an arid environment (Kerr, 1955; Isaac, 1979, 1981, 1983a, 1983b). A cycle, beginning with decalcification of the Chalk (and Greensand), followed by kaolinization accompanied by dissolution of residual quartz, and ending with localized silicification at various depths in the weathering profile, can thus be identified (Isaac, 1983b).

Subsequent reworking of the Combpyne Soil altered the structure of the original profile and led to the formation of the Peak Hill Gravels (Isaac, 1979, 1981, 1983b). West and north of Sidmouth, the Combpyne Soil was eroded differentially: where it was removed completely, Pleistocene gravels rest directly on deeply weathered Upper Greensand (Isaac, 1981). The Peak Hill Gravels (themselves reworked Combpyne Soil) have been correlated by Isaac (1979) with plateau deposits on the Haldon Hills to the east of the Bovey Basin. The residual flint gravel there, the Tower Wood Gravel, rests on decalcified Upper Greensand and is overlain by fluvial gravels (Hamblin, 1973a, 1973b). It has a similar clay mineralogy and lithological content to the Peak Hill Gravels, and is thus regarded

as a westward continuation of the Sidmouth/Beer kaolinitic residual deposits (Isaac, 1979, 1981).

A Palaeocene age for the residual Tertiary sediments is based on several lines of evidence (Isaac, 1983b). First, the residual gravels underlie Upper Eocene to Lower Oligocene beds of the Bovey Formation (Edwards, 1976; Isaac, 1981), and are cut by faults thought to be of late Middle Eocene or early Upper Oligocene age (Isaac, 1981). Second, these lateritic weathering products have been correlated with the Interbasaltic Formation of Northern Ireland (Eyles, 1952), where clay-with-flints rests on chalk and is overlain by basalt. Radiometric dates (Evans *et al.*, 1973) show the Irish residual gravels to span 65–66 Ma BP, and thus a lowest Danian to Maastrichtian age has been proposed (Figure 3.1). A Palaeocene age for the Tower Wood Gravels of the Haldon Hills, with which the Peak Hill Gravels are correlated (Isaac, 1981, 1983b), was proposed independently by Hamblin (1973a, 1973b). A Palaeocene age thus seems appropriate for the Combpyne Soil, Peak Hill Gravels and associated silcretes (Isaac, 1979).

Following deposition of the Peak Hill Gravels and before the Mutters Moor Gravels were emplaced, a second deep weathering profile was established in the Sidmouth area – the Seven Stones Soil (Isaac, 1981). This zone of weathering, again characterized by mottled and pallid zones, occurs at some localities beneath Mutters Moor Gravels, and elsewhere affects the surfaces of the Peak Hill Gravels, Combpyne Soil and even the Upper Greensand where the former sediments are absent (Isaac, 1981) (Figure 3.5).

In addition to the residual flint gravels and associated pedogenic and diagenetic horizons, the plateau deposits of east Devon include the Buller's Hill Gravel (Hamblin, 1973a, 1973b) and the Aller Gravel (Edwards, 1973), both thought to be fluvial in origin (Figure 3.5). The former is believed to have originated from erosion and subsequent redeposition of the Tower Wood Gravel and, in addition, contains material from deeply weathered Upper Palaeozoic sediments (Isaac, 1983b). The Aller Gravel consists of up to 20 m of rounded quartz and flint gravels and sands with subordinate bodies of white to pale-grey silt and sandy clay (Isaac, 1983b). Although Hamblin (1974a) and Edwards (1973) have differed as to the origin of the Buller's Hill and Aller gravels, Isaac (1983b) is of the opinion that they are 'spatially and temporally closely related'. However, differences in lithological content could indicate that the two deposits are separated from one another perhaps by a fairly substantial time interval – in fact the time

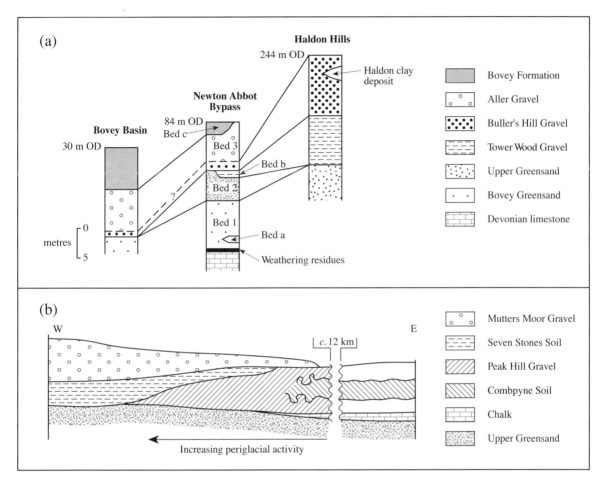

Figure 3.5 (a) Stratigraphic correlations of successions in the Bovey Basin, Newton Abbot bypass and Haldon Hills. (Adapted from Edwards, 1973, Hamblin, 1973b and Brunsden *et al.*, 1976.) (b) Schematic representation (not to scale) of field relations of major lithostratigraphic units in the Sidmouth area. (Adapted from Isaac, 1981.)

necessary to remove the Chalk cover from most, if not all, of South-West England. The Buller's Hill Gravel could therefore be Palaeocene, the Aller Gravel possibly mid-Eocene. Important evidence for the stratigraphic relationships of these deposits comes from solution cavities in Devonian limestone exposed along the margins of the Newton Abbot bypass (Brunsden *et al.*, 1976). Here, Aller Gravel overlies Buller's Hill Gravel which in turn overlies Tower Wood Gravel resting on Upper Greensand (Brunsden *et al.*, 1976; Isaac, 1983b; Figure 3.5). Since all four units are overlain by beds of the Bovey Formation (Upper Eocene to Oligocene), it is likely that the fluvial flint-rich gravels represent a phase of erosion and destruction of the flint-rich Palaeocene weathering profiles (residual sediments) before Bovey Formation sedimentation began: this must have occurred before Upper Eocene times (Isaac, 1983b).

Isaac (1983b) has argued that intra-Eocene tec-tonism played a key role in the destruction of the residual mantle formed by the deep weathering (Palaeocene) of Cretaceous sediments, and culmi-nated in the initial downwarping of deep tectonic basins along wrench-fault zones: the Bovey Basin, for example, saw subsequent redeposition of the weathered Cretaceous mantle together with weath-ering products from the adjacent Dartmoor granite (Bristow, 1968) as well as more recently weathered (during the Eocene) Upper Palaeozoic meta-sedi-ments which became exposed as the downwarping progressed.

In the Sidmouth area (Figure 3.5), a further flint-rich gravel, the Mutters Moor Gravel, has been identified (Isaac, 1981, 1983b). Unlike the Tower Wood and Buller's Hill gravels, which are offset by faulting associated with downwarping of the Bovey Basin (Hamblin, 1973a, 1973b), the Mutters Moor Gravels were not so affected (Isaac, 1981). These flinty gravels, with profuse angular chalk-flints in a sandy clay matrix, contain many silcrete fragments. They appear to have been eroded from Palaeocene weathering products (= Combpyne Soil, Peak Hill Gravels and Seven Stones Soil), and to have been redeposited during the Pleistocene (Isaac, 1981). Isaac has argued that part of the level plateau at *c.* 180 m OD in the Sidmouth area (northern Mutters Moor), may be an exhumed palaeo-deflation surface. Intensely mechanically weathered Peak Hill Gravels and Seven Stones Soil kaolinite, found in the Mutters Moor Gravels, is consistent with the material having been reworked by aeolian and fluvial agencies in the Pleistocene, when cold desert conditions at times prevailed. In such an environment, cryoturbation, solifluction and other periglacial processes were instrumental in reworking the older weathering profiles (Isaac, 1981).

The origin and significance of the infilled pipes at Beer must therefore be viewed against the extremely complex history of Tertiary weathering and the subsequent destruction of the weathered mantle both during later Tertiary and Pleistocene times. Without detailed laboratory analyses of the pipe-sediments, it is impossible to reach firm conclusions as to their origin. However, as a working hypothesis, it seems likely that the pipe structures themselves began to form at the end of the Cretaceous as a stable land surface emerged, and subsequent tropical weathering (Palaeocene) developed a thick mantle (flint-rich residual gravels) from beneath a covering of clay-rich Tertiary sediment. From Isaac's descriptions, it is not entirely clear whether the clay-with-flints contained in the pipes at Beer would represent the earliest *in situ* Palaeocene lateritic weathering products, the Combpyne Soil, or material reworked from these beds (Peak Hill Gravels). Certainly, two main fabrics can be discerned within the Beer pipe infill; the first, which occurs near the base of the pipes, shows a dense packing of tightly interlocking flints with little matrix. The second occurs towards the top of the pipes and in the continuous gravel capping. Here, the flints are less profuse, being supported in a red clay matrix. It seems reasonable to assume that the developing pipe structures, with steep sides and restricted lateral extent, would have prevented erosion and downslope reworking of the residual gravels, under almost any but the most severe conditions. The lower, densely packed flint material in the pipes may therefore have formed as lines of tabular and nodular flints were gradually let down, passively, from above as the Chalk was removed by solution (cf. Isaac, 1979). Small 'rafts' of chalk found towards the bottom of some pipes

may have escaped complete solution, giving further credence to this hypothesis: a correlation of these deposits with the Combpyne Soil would thus seem appropriate. The overlying, less densely packed flint material, in the upper part of the pipes and in the capping layer, may, on the other hand, have undergone some reworking, and may thus be equivalent to the Peak Hill Gravels (Isaac, 1979, 1981, 1983b). The uppermost layers of flint gravel at Beer were, alternatively, probably subjected to reworking and, at the very least, severe disruption by Pleistocene frost-action. That some part of the infill should more properly be ascribed to the Mutters Moor Gravel lithostratigraphic formation cannot therefore be ruled out. Indeed, comparable solution cavities at Dunscombe (2 km east of Sidmouth) were recorded by Isaac (1983b). Here, they are filled with black montmorillonitic clays, and residual flint gravels protrude down into limestone and decalcified sediments beneath (Isaac, 1983b).

Solution of the Chalk at Beer has clearly been on a large scale. Small-scale features have been described from the area, for example, in the Devonian limestone (Ussher, 1913), although larger structures have been described from the Carboniferous Limestone and the Chalk elsewhere in southern Britain (Kirkaldy, 1950; Walsh *et al.*, 1972, 1973). Indeed, similar solutional cavities in excess of 100 m diameter have been described from Dorset (Arkell, 1947), Wiltshire (Jukes-Browne, 1905) and Kent (Worssam, 1963). Large-scale solution pipes were more recently described from Devonian limestone exposed during construction of the Newton Abbot bypass (Brunsden *et al.*, 1976), and comparable features in the Chalk near South Mimms, Hertfordshire, have long attracted attention (Wooldridge and Kirkaldy, 1937; Kirkaldy, 1950; Thorez *et al.*, 1971); there, the largest pipe noted was 10 m in diameter at the base of the pit.

Thorez *et al.* (1971) described the solution pipes at Castle Lime Works, South Mimms, as being filled with sands and pebbly deposits believed to be part of the Thanet Beds (Palaeogene). They argued that the Chalk had dissolved *in situ* beneath a cover of the Thanet Sands, by percolating water, the effects of which were heightened by the anticlinal structure of the Chalk and by a resultingly very low water-table (cf. Walsh *et al.*, 1973; see below) (Wooldridge and Kirkaldy, 1937). Initially, the Chalk was dissolved beneath the Bullhead Bed at the base of the Thanet Sands: at first, this bed supported the sand, but with continued dissolution of the Chalk and enlargement of the cavities, eventually collapsed with the overlying sands into the pipes. Deposits overlying the Tertiary infill are believed to be colluvial and solifluction deposits, which smoothed out irregularities in the local ground surface during the Pleistocene (Thorez *et al.*, 1971).

A dark brown, porous clay lining to the cavities has a more complex origin. Although closely related to dissolution of the Chalk, this is not a pure chalk residue: much of it is oriented and laminated (Thorez *et al.*, 1971), and it is interpreted as an illuvial deposit, washed in from the Tertiary sediments above. Small chalk fragments remain incompletely dissolved towards the base of some cavities (cf. Beer). Here, the Chalk dissolved along joints, eventually becoming sealed on all sides by the illuviated clay. Continued solution of the isolated chalk 'rafts' was effected by ions passing through the relatively porous clay: once completely dissolved, further deposition of clay in the space provided by the dissolved chalk was impossible, and voids were left as the remaining chalk was dissolved (Thorez *et al.*, 1971). The large pore space and low bulk density of the clay have been taken as indicating its relatively recent formation. In older clay-with-flints (e.g. at Beer?), the voids in the clay have frequently been reduced by continual collapse, alternate swelling and shrinking and the continued deposition of illuvial clay (Thorez *et al.*, 1971). Thus, although the same basic geological controls (namely a stable chalk surface overlain by Tertiary sediment) exists at both South Mimms and Beer, albeit in an entirely residual state at the latter, solution of the Chalk may have taken place at radically different times. This is also a conclusion drawn by Chartres and Whalley (1975) who studied dissolution features in chalk at Basingstoke. They argued that a brown clay lining to the cavities there had formed relatively recently, indicating that truly residual clays can form at any time by dissolution of underlying chalk. A similar explanation was put forward by Worssam (1981) to account for clay linings in loess-filled gulls (cambered fissures) at Allington in Kent. Here, large limestone blocks are surrounded by silt and are capped with clay linings: the linings must have been deposited after the blocks became incorporated into the silt, or they would show significant signs of tilting and/or disruption (Worssam, 1981).

In contrast, the large solutional features described in the Devonian limestone near Newton Abbot (Brunsden *et al.*, 1976) could show that the solutional processes there operated over a much more protracted timescale (Figure 3.5). These pipes contain sediments representing the Upper Greensand and the Chalk as well as the Eocene Buller's Hill and Aller gravels, and the beds of the Upper Eocene to Oligocene Bovey Formation (Brunsden *et al.*, 1976; Figure 3.5). It has been suggested that these pipes could therefore be the result of subsidence related either to: 1. prior hydrothermal alteration of the limestone; 2. a cover of permeable Cretaceous and Tertiary sediments; or to a combination of both factors (Brunsden *et al.*, 1976).

Walsh *et al.* (1973) attempted to synthesize evidence from a wide range of sites throughout Britain exhibiting solutional cavities and associated infill materials. They concluded that a common age and origin was unlikely: pipes had clearly developed more than once, and post-Eocene, post-Pliocene and Pleistocene interglacial phases of piping could all be recognized. On balance, most pipe structures were considered to have formed by selective solution effected from beneath by artesian groundwater. The process was considered to be self-propagating: the more chalk eaten away from below, the more progressive the collapse of solution residues and overlying fill-sediments (Walsh *et al.*, 1973). Gravitational collapse of infill sediments was only likely to cease when the local water-table dropped to a position where fissure systems were effectively bridged or shut off from the artesian source. Thus, in areas where water-tables and hydrological conditions had changed radically and repeatedly, complex histories and sequences of pipe development were likely. Walsh *et al.* also recommended that the varied, and often confusing, terminology used to describe 'pipes', 'pockets', 'fissures' and their associated infill, be rationalized to include three principal terms: 1. solution subsidence (the process); 2. solution subsidence deposit (the materials involved); and 3. solutional subsidence mass (the cavity or host body).

The evidence from Beer Quarry, in conjunction with detailed work carried out throughout the Beer and Sidmouth areas (Isaac, 1979, 1981, 1983a, 1983b), is of considerable relevance to understanding the origin of the controversial clay-with-flints and solutional processes operating in limestone terrains. Overwhelming evidence from this area suggests that solutional subsidence deposits, namely flinty residual gravels (= clay-with-flints) were formed as chalk underwent dissolution (solution subsidence) beneath Tertiary deposits. The pattern of structures (largely vertical or slightly oblique pipes) probably reflects major vertical joints in the Chalk subsequently exploited by solution. Well differentiated lateritic weathering profiles, developed in local sediments comparable to the infill material, together with locally developed silcretes, point to weathering, with complex patterns of diagenesis and pedogenesis, in a tropical environment: distinct phases of humid and arid conditions are likely to have prevailed at different times (Isaac, 1981, 1983b).

There is strong local stratigraphic evidence that most of this weathering occurred during the Palaeocene (see above). The concentration of illuvial clay towards the pipe margins at Beer (where there is a distinctly darker or blackened zone) has not yet been proven: in any case, there is evidence elsewhere to suggest that eluviation of clay minerals from overlying sediments could have occurred almost at any time from the earliest Tertiary to the present day. However, the identification of lateritic weathering products in residual flinty gravels found between the Chalk or Upper Greensand and Bovey Formation sediments (Upper Eocene to Oligocene) in east Devon, tightly constrains the age of the weathering event(s) and also, therefore, the principal phase of solutional activity, to the Palaeocene.

This degree of precision contrasts markedly with evidence from elsewhere in Britain and France where the age of the clay-with-flints has been much disputed (Pepper, 1973): most British workers have favoured a Pleistocene age, whereas workers in France have generally accepted ages within the Palaeogene for similar deposits. Neither is there agreement as to the age relationship of the clay-with-flints and its associated silcretes, in the form of 'sarsens' elsewhere in southern England (Isaac, 1979; Summerfield and Goudie, 1980). Thus although: 1. radiometric dates and the stratigraphic relations of the clay-with-flints in Northern Ireland (e.g. Wright, 1924; Fowler and Robbie, 1961; Evans *et al.*, 1973); 2. the evidence from east Devon (e.g. Isaac, 1979, 1981, 1983a, 1983b; Hamblin, 1973a, 1973b); together with 3. climatic evidence (Dury, 1971); and 4. a knowledge of the conditions necessary for silcrete formation (e.g. Watkins, 1967; Watts, 1978) would all support a Palaeocene age for the clay-with-flints at Beer, this can by no means be accepted for all other occurrences of similar deposits. Chartres and Whalley (1975), for example, provided convincing evidence to show that solution of the Chalk at Basingstoke had taken place almost entirely within the Quaternary. Using evidence from clay mineralogy, particle-size analyses and the morphology of chalk rubble within the infill sediments, they showed that the Chalk was affected by frost-action before the infill material (clay-with-flints) was

formed, and that solution of the Chalk has since continued to produce an irregular weathering front. The clear implication is that there, no significant solution of the Chalk, in the Tertiary, preceded this sequence of events, or if it did, that evidence was removed first. Such a view has been upheld by Pepper (1973) although it is at variance with French (e.g. Dewolf, 1970; Mathieu, 1971) and some English workers (e.g. Loveday, 1962).

The interpretation of solutional cavities and associated infills is therefore fraught with difficulties: rarely is any firm dating possible. Controversy even exists regarding the origin of the clay-rich zones found in some of the pipes, namely whether they are autochthonous (that is derived directly by decalcification of the Chalk with only superficial pedogenesis) or allocthonous (that is derived from material other than the Chalk, for example, washed in from a permeable cover) (Chartres and Whalley, 1975). Whether the kaolinite found in the residual gravels of east Devon was formed by *in situ* weathering (e.g. Green, 1974a) or derived from the granite mass of Dartmoor (e.g. Hamblin, 1973a, 1973b) where it may have formed by hydrothermal processes, is significant: a supergene origin, at least in part, for the Dartmoor growan is given some credence by the substantial evidence for weathering found in the Tertiary and Cretaceous rocks of east Devon (Bristow, 1968) (see Chapter 4).

Conclusion

The lack of detailed work so far carried out at Beer is surprising because the site demonstrates probably the finest examples of solutional pipes found in the Chalk anywhere in southern England: its conservation status can be justified on this basis alone. Although a preliminary description and interpretation of the site have been given here, precise details of the nature and origin of the pipe infill must await more comprehensive laboratory and field examination. Nonetheless, the infill material strongly resembles that studied elsewhere in the region and ascribed, on the basis of pedological and diagenetic characteristics and stratigraphic relationships, to Palaeocene tropical weathering. If, as seems likely, a correlation can be proved between the residual Palaeocene clay-with-flints of the east Devon tableland and the infill sediments at Beer, the latter will be one of very few examples of solutional activity in Britain which can be related to a precise stratigraphic timescale and to a well-established sequence of landscape evolution. The east

Devon tableland deposits, typified by those at Beer, and the residual deposits (growan) of the Dartmoor area (Chapter 4) together record vital evidence for the climatic and denudation history of South-West England during the Palaeogene. Collectively, they demonstrate a long and complex history of pedogenesis and diagenesis which is also typical of other deeply weathered regions of the world (Isaac, 1983b).

Until more detailed evidence is available from a wide range of sites, it is perhaps prudent to regard solutional cavities in general, and indeed the clay-with-flints as polygenetic (e.g. Hodgson *et al.*, 1967; Mathieu, 1971; Walsh *et al.*, 1973; Chartres and Whalley, 1975). It is certain that the debate concerning the origin of both will continue. In this context, Beer Quarry together with other GCR sites at South Mimms, Allington Quarry, Spot Lane Quarry and Bath University, provide a range of contrasting evidence for processes operative in limestone landscape development: these sites will be central to resolving the outstanding controversies.

ST AGNES BEACON
S. Campbell and R. A. Shakesby

Highlights

Rare, non-marine Miocene deposits occur here, and provide unique evidence for the long-term evolution of South-West England's landscape. The survival of the St Agnes sands and 'clays' indicates that this part of the South-West cannot have been overrun by glacier ice nor inundated by the sea since the Miocene.

Introduction

St Agnes Beacon is an important site for the interpretation of Tertiary stratigraphy. The site is also of considerable geomorphological interest because the deposits and the underlying bedrock surface have significant implications for regional landscape evolution – especially for the age and mode of formation of the sub-deposit surface, for its relationship to other 'erosion' surfaces in South-West England, and for establishing the extent and intensity of Pleistocene glaciation in the region. On the basis of these merits, the site has attracted wide interest from Quaternary scientists; hence its inclusion in this volume. The site was apparently first

referred to by Borlase (1758) and subsequently by Pryce (1778), Boase (1832), Hawkins (1832), De la Beche (1839), Belt (1876), Davies and Kitto (1878), Ussher (1879a) and Whitley (1882). It was also studied by Reid (1890), Reid and Scrivenor (1906), Milner (1922) and Boswell (1923). More recently, the site has been discussed by Mitchell (1965), Atkinson *et al.* (1974, 1975), Hall (1974), Atkinson (1975, 1980), Edmonds *et al.* (1975), Wilson (1975), Campbell (1984), Coque-Delhuille (1987) and Goode and Taylor (1988). A detailed account of the stratigraphy, sedimentology and palynology of the deposits and their significance was given by Walsh *et al.* (1987). Further investigation of part of the deposits was published by Jowsey *et al.* (1992). The site has also been referred to in general texts by Austen (1851), Whitley (1866), Davison (1930), Gullick (1936); Robson (1944) and Macfadyen (1970).

Description

Although formerly regarded as a single outlier, the deposits at St Agnes are now believed to comprise two distinct outliers; the St Agnes Outlier and the Beacon Cottage Farm Outlier (Walsh *et al.*, 1987; Jowsey *et al.*, 1992; Figures 3.6 and 3.7). The sediments in the St Agnes Formation consist mainly of sands and clays and are arranged in an arc around the north and east slopes of St Agnes Beacon. This outcrop covers 1.6 km^2 and has a residual volume of some 5×10^6 m^3 (Walsh *et al.*, 1987), reaching a maximum exposure thickness of *c.* 10 m. Active workings only occur in New Downs Pits (Doble's Sandpits) (*c.* SW 706509). The following stratigraphy was proposed for the St Agnes Formation by Walsh *et al.* (1987):

3.	Upper Sands	– Beacon Member
2.	Middle 'Clays'	– New Downs Member
1.	Lower Sands	– Doble Member

The St Agnes Formation is underlain by Devonian slates ('killas') or by the St Agnes Granite. There is evidence that this sub-deposit floor is extensive; in New Downs Pits it is sub-horizontal and covers an area of some 1000 m^2. Its junction with the upper slopes of the Beacon appears to take the form of a steep stepped cliff, with an overall gradient of 45° or more (Walsh *et al.*, 1987). Davies and Kitto (1878) raised the possibility that the abrupt break of slope on the east side of the Beacon was a buried sea cliff. In New Downs Pits, the sub-deposit

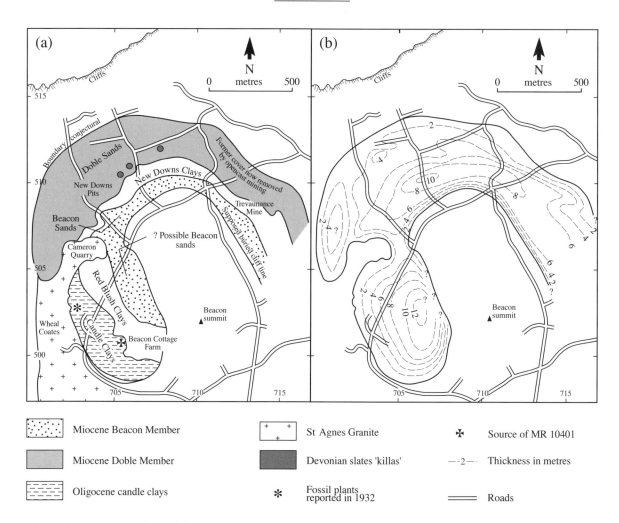

Miocene Beacon Member

Miocene Doble Member

Oligocene candle clays

St Agnes Granite

Devonian slates 'killas'

✳ Fossil plants
reported in 1932

✱ Source of MR 10401

—-2—- Thickness in metres

═══ Roads

Figure 3.6 (a) The geology of the St Agnes and Beacon Cottage Farm outliers as interpreted by Walsh *et al.* (1987). The area between Cameron Quarry and the Beacon was regarded as problematic and has been re-mapped by Jowsey *et al.* (1992) (Figure 3.7); (b) Isopachs of combined Tertiary and Quaternary sediment. (Adapted from Walsh *et al.*, 1987.)

basement is stained red, lilac and orange to a depth of at least 1 m; derived clasts of stained killas in the overlying beds show clearly that the killas was weathered prior to deposition of the Doble Member (Walsh *et al.*, 1987).

The Doble Member is heavily cemented for a depth of up to *c.* 0.5 m at the base of the bed. The iron content sometimes exceeds 10% by mass and forms tubular structures up to 2 m long and 0.3 m wide – sometimes filled with uncemented yellow sand (Hosking and Pisarski, 1964; Atkinson *et al.*, 1974; Walsh *et al.*, 1987). Reid and Scrivenor (1906) recorded a bed of pebbles towards the base of the St Agnes deposit, although this is no longer evident.

The Doble Member (Lower Sands – bed 1) is around 5–6 m thick and consists largely of yellow or buff, well-sorted and fine-grained silty, quartz-rich sand. The bed becomes paler towards its junction with the overlying bed 2, with which it has a gradational contact (Walsh *et al.*, 1987). Epsilon-type planar cross-beds are evident, varying considerably in size; palaeocurrent directions indicate a source from the north-west (Walsh *et al.*, 1987). This sand bed is interrupted by a 10 cm-thick band, 1.8 m above the base of the bed, comprising rounded pebbles of vein quartz and sandstone and large angular cobbles of stained killas.

The succeeding New Downs Member (bed 2) is a pale-grey deposit up to *c.* 3.5 m thick. It has frequently been described as a clay, but in fact comprises mostly silt and sand with some clay. Isolated vein quartz pebbles occur towards the top of the bed, which is sharply truncated. Sediments

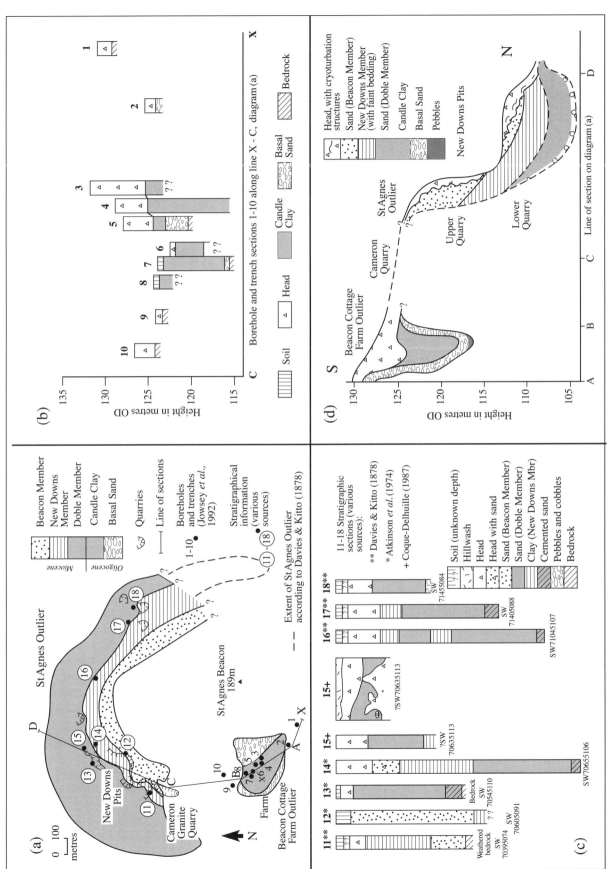

Figure 3.7 (a) A revision of the St Agnes and Beacon Cottage Farm outliers by Jowsey et al. (1992). (b) Borehole and trench sections along line X–C (diagram a), adapted from Jowsey et al. (1992). (c) Stratigraphic sections along line D–C (diagram a), compiled from various sources. (d) Schematic reconstruction of the Beacon Cottage Farm and St Agnes outliers, based on Jowsey et al. (1992).

Figure 3.8 Members of the Quaternary Research Association examine the sequence at St Agnes during the Annual Field Trip to west Cornwall in 1980. The sands are overlain unconformably by periglacial head. (Photo: S. Campbell.)

from this bed have yielded a Miocene microflora. It is a poorly sorted deposit, and exhibits only faint stratification, with a consistent dip of 3–8° to the north (Walsh *et al.*, 1987). Davies and Kitto (1878) noted that the 'clays' were subdivided into two beds in one exposure north of St Agnes Beacon (Site 16; Figure 3.7).

The overlying Beacon Member (bed 3) is represented by *c.* 3 m of cross-bedded, very well-sorted, fine- to medium-grained yellow and orange sand; the sediments are closely comparable to the basal member (bed 1). Some of the sand is weakly iron-cemented, and irregular lenses of green silty sand are also present. Stratification in this bed is usually less marked than in bed 1 (cf. Coque-Delhuille, 1987), especially near the junction with the overlying head, where the sediments are greatly distorted. Where the bedding is relatively undisturbed, foresets are dominantly inclined to the SSE or south (Walsh *et al.*, 1987).

The capping Pleistocene head (up to 1.5 m thick) is an unsorted deposit comprising dominantly angular and subangular clasts of killas and the St Agnes Granite set in sand (Figure 3.8). It displays cryoturbation structures, with sands of the

New Downs Member thrust up into the head in places (Coque-Delhuille, 1987).

The Beacon Cottage Farm Outlier (Figure 3.7) has no current exposures: the stratigraphy and interpretation of these beds therefore rests heavily on historical records (e.g. the field notes of H. Dewey and the work of Davies and Kitto (1878)), 61 boreholes and trenches (Jowsey *et al.*, 1992) and the palynological evidence, which shows clearly a flora of mid-Oligocene age. The relationship of this outlier to the St Agnes Outlier has been clarified by Jowsey *et al.* (1992), who have demonstrated that the two formations are discrete and that there is no overlap by the Beacon Member on the deposits of the Beacon Farm Outlier as had been thought by Walsh *et al.* (1987) (Figures 3.6 and 3.7). The boreholes and trenches excavated by Jowsey *et al.* (1992) revealed that the Beacon Cottage Farm Outlier is up to 8.9 m thick and comprises two members: Basal Sand which is often pebbly and Candle Clay, which in fact comprises sandy silts. These deposits are overlain by up to *c.* 6.4 m of head. The Candle Clay has yielded a mid-Oligocene flora (BGS Sample MR 10401; Mitchell, 1965; Walsh *et al.*, 1987).

Interpretation

William Borlase in 1758 (Macfadyen, 1970) regarded the St Agnes beds as marine, having been formed by an event that could be 'no other than the universal deluge'. The sediments were also referred to by Pryce (1778), Boase (1832) and Hawkins (1832), the latter considering them to be 'alluvial'. Davies and Kitto (1878), however, returned to the proposition that the sands had formed in a marine environment, with the clay having been deposited in 'a sheltered embayment of the sea'. An apparently 'shingle-worn' cliff some 4–8 m high and a postulated sea-stack, caves and hollows – exposed by mining in 1875 to the east of the Beacon – led Davies and Kitto and, later, Reid and Scrivenor (1906) to regard the sands overlying and banked against these features as marine. Such an explanation for the beds was supported by particle-size analyses and by the interpretation that individual sand grains showed signs of marine abrasion (Milner, 1922). Such an origin was considered likely by Boswell (1923), who also noted that a dune (aeolian) origin was feasible. Recently, Coque-Delhuille (1987) favoured a marine origin. In marked contrast, Atkinson *et al.* (1975) and Goode and Taylor (1988) favoured a fluviatile origin. Such an origin for the 'clay', with its consistent northerly 3–8° dip of bedding structures, would require post-depositional tectonic action for which there is no evidence. Detailed analyses of quartz grain surface textures (Campbell, 1984; Walsh *et al.*, 1987) show an aeolian origin for the sands and a colluvial (i.e. slope-wash) origin for the 'clay'. The latter contains a mixture of grains of aeolian origin (presumably derived from the underlying sand) and of source rock origin (rock weathering products transported downslope). Similarly, mainly aeolian and colluvial origins were suggested respectively for the Basal Sands and Candle Clay of the Beacon Cottage Farm Outlier by Jowsey *et al.* (1992).

The beds were first classified as Tertiary by De la Beche (1839). More specifically, Reid (1890) assigned the deposits to 'Older Pliocene' times: a summary of the early findings was given by Reid and Scrivenor (1906).

Mitchell (1965) referred briefly to pollen extracted from a lignite sample (originally collected by H. Dewey in 1932; BGS Sample MR 10401) from a poorly defined location at the base of the beds at Beacon Cottage Farm: the results indicated an Oligocene age for the beds – thus discounting Reid's view that the platform underlying the sand beds (the 130 m platform of Cornwall) was Pliocene in age, and that the St Agnes Beacon had been an island in the sea at this time. Further palynological examination of Dewey's sample (Atkinson *et al.*, 1975) led to a more precise ascription of the beds to the Middle–Upper Oligocene.

New dating evidence for the St Agnes beds was provided by Walsh *et al.* (1987). Their work shed additional light on the ages of the beds which have important implications for the long-term landscape evolution of west-central Cornwall. A re-analysis of the microflora in the BGS Sample (MR 10401) from sediments in the Beacon Cottage Farm Outlier confirmed a mid-Oligocene age for the flora (cf. Atkinson *et al.*, 1975) and suggested low-energy deposition for the 'clay' in a lacustrine environment (Walsh *et al.*, 1987), although later sedimentological analysis indicates a colluvial origin, as for similar deposits in the St Agnes Formation (Jowsey *et al.*, 1992). These sediments were tentatively correlated with equivalent beds of the Bovey Formation of Devon (that is of Palaeogene age).

By contrast, lignitic material from sediments in the New Downs Pits (St Agnes Formation) yielded an impoverished, although distinctive, pollen assemblage of probable Miocene age (Walsh *et al.*, 1987). These different dates, together with the lack of superposition, mean that the sand beds around St Agnes Beacon can no longer be regarded as a single formation – and the beds have been re-classified accordingly (see site description).

Walsh *et al.* (1987) argued that the prominent planation surface beneath the beds at St Agnes, which also occurs throughout much of Cornwall between *c.* 75–131 m, be termed the Reskajeage Surface. They concluded that sea level never reached this height in mid- and late Tertiary times, thus eliminating the possibility that the feature formed through marine activity. Rather, they suggested that the surface originated as a tropical or subtropical etch plain which was formerly covered by a saprolite of varying thickness with upstanding inselbergs (such as St Agnes Beacon), and had formed over a protracted period up to late Miocene times. The St Agnes sediments are thus considered to be small remnants of tropical or subtropical subaerial weathering products which underwent redistribution by wind action and slope-wash processes in mid-Oligocene and Miocene times (Walsh *et al.*, 1987; Jowsey *et al.*, 1992).

St Agnes Beacon is therefore not only of outstanding interest for Tertiary stratigraphy, but demonstrates important evidence for long-term landscape evolution in South-West England. The sediments of the St Agnes Outlier, with their microflora of Miocene age, are one of only five on-land Miocene deposits known in the British Isles (Walsh *et al.*, 1996): this has major implications for interpreting the Cornish landscape. Ascription of the Beacon Cottage Farm Outlier to an earlier date than the St Agnes Formation, despite similar sedimentological characteristics, hinges on a museum sample for which there is no good locational or stratigraphic control. Indeed, Coque-Delhuille (1987) regards the provenance of this sample as too uncertain, and on these grounds rejected an Oligocene age for the Beacon Cottage Farm Outlier. If, as suggested, however, the sand and 'clay' beds around St Agnes Beacon do comprise two distinct outliers of mid-Oligocene and Miocene ages (Walsh *et al.*, 1987; Jowsey *et al.*, 1992), then the relationship of the Reskajeage Surface to these sediments is vital to understanding the age and mode of formation of this important macro-element in the Cornish landscape. Although it is clear that the surface underlies the Miocene St Agnes Formation, until recently the relationship of the surface to the older (mid-Oligocene) Beacon Cottage Farm Outlier has been less certain. Walsh *et al.* favoured the view that part of the St Agnes Outlier overlay that of the Beacon Cottage Farm (Figure 3.6), the surface of the latter possibly representing a stained and deeply weathered soil of mid-Tertiary age underlying the sub-Miocene unconformity. This view, however, is no longer tenable as there appears to be no overlap of the two formations (Jowsey *et al.*, 1992). If, as Walsh *et al.* have suggested, the Reskajeage Surface had been cut in Miocene times, and the overlying Miocene sediments (wind-blown and colluviated) have never been subject to marine inundation, then the same must be true of the mid-Oligocene Beacon Cottage Farm Outlier which lies at a similar altitude (Walsh *et al.*, 1987). The surface cannot therefore have been formed by marine agencies in the interval between the mid-Oligocene and Miocene. This tends to demolish the often-expressed view that this surface is of marine origin; especially implausible is the suggestion of marine planation cutting across wide areas of hard Devonian metasediments and forming steep slopes in granite at Carn Brea, yet at the same time not also removing the weak Tertiary sands and 'clays' exposed on the northern slopes of the St Agnes Beacon (Walsh *et al.*, 1987). The simplest hypothesis then is to regard the Reskajeage Surface as a subaerial surface; by implication, any marine transgression or fashioning of the west-central Cornish landscape by marine agencies has been confined to levels below the Reskajeage Surface, and/or to pre-mid-Oligocene times.

This interpretation supports very strongly the view that the Cornubian Peninsula was more or less in its present form as early as the Eocene (Freshney *et al.*, 1982; Walsh *et al.*, 1987), and has subsequently only undergone what may be regarded as minor geomorphological alterations. Post-Eocene landscape evolution must therefore have been extremely slow; the only evidence for marine incursion on to the peninsula is a minor transgression at St Erth (Mitchell, 1965; Mitchell *et al.*, 1973a) during the Late Pliocene, and even this left no obvious bevel in the landscape (Walsh *et al.*, 1987).

This interpretation of the west Cornish landscape finds a close analogue with work carried out in Wales by Battiau-Queney (1984, 1987). She argued that in Wales there is only one, polygenetic planation surface, the original constituent landforms having been shaped in a tropical or subtropical environment with associated weathering products (e.g. Trefgarn Rocks – see Campbell and Bowen, 1989). Large-scale altitudinal variations were attributed by her to late Tertiary warping along relatively few major structural axes, the smaller-scale landforms, such as the tors at Trefgarn and Preseli (St Davids), being regarded as true 'inselbergs' (Battiau-Queney, 1984). Walsh *et al.* (1987) speculated that the Reskajeage Surface is also therefore present in Wales, citing tightly folded and planed Upper Palaeozoics in south Dyfed overlain by postulated Oligocene clays at Flimston (Murchison, 1839). This view has been strengthened by the recent discovery of fossiliferous Miocene deposits on Anglesey, which lends further support to the concept of a widespread 'Reskajeage/Menaian Surface', and to the possibility that large areas of Britain were formerly smothered by extensive sheets of Miocene sediment (Walsh *et al.*, 1996).

The Tertiary deposits near St Agnes Beacon therefore provide critical evidence to suggest that the macro-elements of the landscape in Cornwall (and perhaps farther afield) have changed comparatively little since the early Tertiary. In post-Tertiary times, geomorphological change has been essentially limited to comparatively small-scale modifications such as coastal denudation, valley incision and the redistribution of Tertiary weathering products (and less weathered rock) mainly by periglacial activity during Pleistocene cold phases.

The dating of the St Agnes beds as Tertiary, their highly eroded nature and location on a prominent headland on the north Cornish coast, make the proposition that Cornwall was ever inundated or even impinged on by a southward-moving ice sheet (even as early as the Anglian Stage) unlikely. The giant erratics of the Cornish coast (see Porthleven; Chapter 6), however, pose an intriguing problem as regards their mode of emplacement, as does the evidence for glacier ice having reached the nearby Isles of Scilly (e.g. Mitchell and Orme, 1967; Scourse, 1985a, 1987, 1991).

Conclusion

The St Agnes beds are not only of interest because of their Tertiary stratigraphy and implications for tectonic activity and high-level marine action, but also because of the important implications for land-scape development during the Quaternary. First, the survival of these unconsolidated deposits on a promontory of the north Cornish coast is important evidence against glacier ice ever reaching Cornwall during the Pleistocene. Second, it also implies that repeated periglacial activity in the Pleistocene has been comparatively ineffective in redistributing unconsolidated sands and sandy silts on exposed slopes, even though freeze-thaw activity led to the production of angular debris from the bedrock slopes of St Agnes Beacon: thicknesses of up to *c.* 6.4 m of solifluction debris (head) were laid down and cryoturbation of upper sediment layers occurred. At the macro-scale, however, the implication is that the Cornish landscape has undergone only minor alteration to its late Tertiary form, notably with coastal modifications, valley incision and more limited alteration to general relief through periglacial action than some workers have advocated.

Chapter 4

Granite landscapes

INTRODUCTION
S. Campbell

This chapter examines the granite terrains of South-West England, and contains descriptions of two groups of sites: 1. those with a direct bearing on the genesis of major granite landforms; and 2. those which have allowed a detailed reconstruction of Devensian late-glacial and Holocene environmental changes in such terrains (Figure 4.1). The granite landscapes of the South-West, particularly those of Dartmoor, are some of the best known and most intensively studied in Britain: the principal GCR sites selected to represent the geomorphology of non-glaciated granite terrains are located here. In contrast, GCR sites demonstrating key features of glaciated granite landscapes have been selected in north-east Scotland and in the Cairngorms (Gordon and Sutherland, 1993). The distinctiveness of the non-glaciated granite landscapes of the South-West, the controversies over their evolution and the importance of the selected GCR sites, merit the detailed introduction given below. In addition, an introduction to the geomorphology of Dartmoor and a brief history of relevant local research is given as a preface to the selected GCR sites.

A synthesis of Devensian late-glacial and Holocene environmental history for South-West England is provided in Chapter 2: only a brief introduction to the sites pertaining to this interval, and occurring within granite terrains, is given here.

Two further GCR sites on the Isles of Scilly – Peninnis Head (granite landforms) and Higher Moors (Holocene vegetational history) – also have a bearing on the evolution of granite landscapes. For convenience, they are considered in a regional account of the geomorphological development and Quaternary history of the Isles of Scilly (Chapter 8).

GRANITE LANDFORMS AND WEATHERING PRODUCTS
S. Campbell, A.J. Gerrard and C.P. Green

The granite terrains

Accounts of the characteristic tors and associated landforms and deposits, developed largely in and from granite, have occupied a substantial part of the geomorphological literature on South-West England. The granite intrusions of the South-West Peninsula (Figure 4.1) form the basis of a distinctive landscape: the selected GCR sites at Merrivale,

Two Bridges Quarry and Bellever Quarry on Dartmoor provide some of the finest examples of granite landforms (e.g. tors and clitter) and associated weathering products (e.g. decomposed granite or growan, and slope deposits) anywhere in Britain. The scale and superb development of these features led to a number of pioneering geomorphological studies in the region (e.g. Linton, 1955; Te Punga, 1956; Palmer and Neilson, 1962; Waters, 1964).

The origin of the rugged landscape of upland South-West England has been speculated upon for many years. Many early workers suspected that glacier ice had played at least some part in the evolution of the landscape (e.g. Nathorst, 1873; Belt, 1876; Somervail, 1897; Worth, 1898; Pillar, 1917; Pickard, 1943). The postulated evidence for glacial activity on Dartmoor, however, has never been substantiated, and the landforms have usually been explained in other ways (Gerrard, 1983). Although unglaciated, it was subject to periglacial activity during the cold phases of the Pleistocene, and many of the landforms are relicts from such activity. Manley (1951) has argued that the permanent snow-line, at the period of maximum glaciation, was about 30 m above the highest Dartmoor summits.

As the only unglaciated upland region in Britain, its importance in elucidating late Tertiary and Pleistocene landform evolution is considerable (Brunsden *et al.*, 1964). The landscape of Dartmoor, and other parts of South-West England, has thus seen a continuous geomorphological development since at least early Tertiary times without the imprint of glaciation. Dartmoor in particular has long been seen as a key to understanding both Tertiary and Quaternary landscape evolution (Wooldridge, 1950, 1954; Gerrard, 1983). Intense chemical weathering of the granite under warm climatic conditions probably occurred during parts of the Tertiary. It was also probably subjected to weathering under warm conditions in the 'interglacial' stages of the Pleistocene, as well as to intense cryonival (periglacial) processes during a number of the cold Pleistocene stages. Some workers have argued that the landforms and deposits seen in the Dartmoor area today evolved above all else in response to periglacial conditions (e.g. Waters, 1964, 1974). While there is a firm foundation for arguing that many of the characteristic landforms such as tors and clitter, screes, valley-side buttresses, terraces of 'rubble drift' and benched hillslopes were formed, or at least substantially modified, by periglacial processes, other landforms, particularly some of the tors, are likely

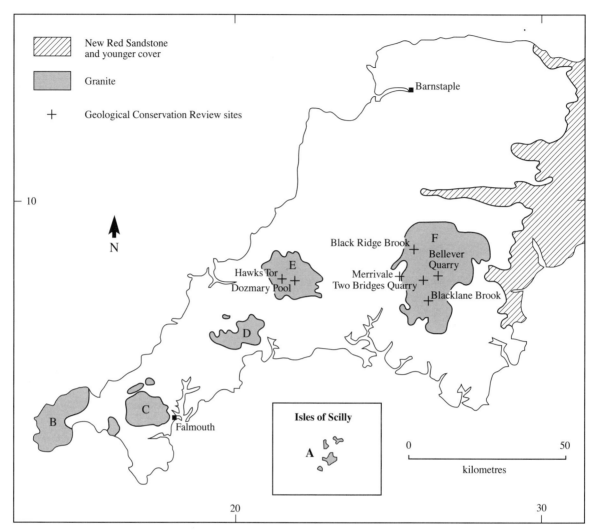

Figure 4.1 Location of GCR sites in relation to: A, Isles of Scilly Granite; B, Land's End Granite; C, Carnmellis Granite; D, St Austell Granite; E, Bodmin Moor Granite; and F, Dartmoor Granite. (Adapted from Floyd *et al.*, 1993.)

to have a much more protracted and complicated history.

The various mechanisms, processes and possible timescales involved in the evolution of the distinctive granite landscape of the South-West are considered fully within the individual site accounts (Merrivale, Two Bridges Quarry and Bellever Quarry): these sites alone have attracted considerable scientific interest and illustrate many of the theories on granite landscape evolution. Of fundamental importance to understanding this evolution, however, is the nature and origin of the granite and its weathering products.

The granite

The granite intrusions of the South-West have been described in considerable detail (e.g. Reid *et al.*, 1912; Brammall and Harwood, 1923, 1932; Brammall, 1926a, 1926b; Worth, 1930; Exley, 1959, 1965; Stone and Austin, 1961; Exley and Stone, 1964; Durrance and Laming, 1982; Floyd *et al.*, 1993), and are widely held to be linked at depth in one huge batholith, a continuous body of rock some 250 km long by 50 km wide (Durrance and Laming, 1982; Floyd *et al.*, 1993; Figure 4.1). The intrusions have caused the surrounding Devonian and Carboniferous 'country' rocks to become folded, faulted and, to varying degrees, metamorphosed in an 'aureole' between 1 km and 3 km wide (Durrance and Laming, 1982). The granites of the South-West were intruded as the result of a much larger-scale tectonic episode – the Variscan or Hercynian Orogeny, occasioned by substantial plate movements.

The first map showing an outline of the Dartmoor granite was published by De la Beche (1835), and the origin of the mass attracted considerable early interest (e.g. De la Beche, 1839; Henwood, 1843; Ormerod, 1869; Ussher, 1888; Hunt, 1894). In general, an igneous origin was suggested, and Ussher argued that the intrusion was probably laccolithic in form. Hunt (1894) disagreed with this origin, preferring a metasomatic explanation, namely that the mass had arisen through *in situ* alteration of sedimentary material by 'silicic alkalic' fluids.

From the earliest work it was accepted that the pluton consisted of two main rock types, a coarse-grained granite and a fine-grained variety. Further subdivisions were proposed by Reid *et al.* (1912), and the detailed work of Brammall and Harwood in the 1920s and 1930s added significantly to the knowledge of the various mineralogical and chemical properties of the rock. Brammall and Harwood (1923, 1932) recognized that the granite was a composite intrusion which had involved successive sheet-like injections. Initially, they considered there had been four major intrusive phases, but later this was reduced to three, which had resulted in the formation of: (a) the 'giant granite', characterized by abundant large feldspar phenocrysts; (b) the 'blue granite', with fewer large phenocrysts; and (c) a variety of minor intrusions, commonly aplitic and finer-grained (Durrance and Laming, 1982). The giant granite forms most of the tors, and as a result it is sometimes referred to as the 'tor granite'. The principal conclusions of Brammall and Harwood's work were that the pluton was igneous, composite, probably laccolithic (cf. Ussher, 1888), and that the magmas had been derived from the melting of sedimentary rocks at depth, although they had undergone extensive changes as the result of assimilating material at higher levels.

Numerous studies have since been published on the granites of the South-West. These have dealt in detail with many diverse topics including: the nature of the granites; their field relations with the surrounding country rock and with regional structures; aspects of petrogenesis, including metasomatism and recrystallization; and alteration of the granites including tourmalinization, greisening and kaolinization. Excellent reviews are provided by Exley and Stone (1964) and Durrance and Laming (1982) among others. It is useful here, however, to consider some broad structural aspects of the Dartmoor granite, since these have a direct bearing on landscape evolution.

The outcrop of the Dartmoor granite is irregular in shape. According to Bott *et al.* (1958), the magma may have risen in the south and spread northwards as a laccolithic 'tongue'. Another possibility is that it rose through relatively resistant Devonian rocks in the south-central region, spreading out on reaching the Carboniferous–Devonian interface both to the north and, to a lesser extent, southwards (Durrance and Laming, 1982). It has often been suggested by geologists that the upper domed surface of the intrusion represents the original roof of the pluton (Durrance and Laming, 1982). Contacts between the granite and the country rocks are generally sharp, but rarely exposed.

Joints and faults

Jointing in the granite has been seen as a major factor influencing the development of granite landforms such as tors (e.g. Waters, 1954, 1957; Gerrard, 1974, 1978, 1982; Durrance and Laming, 1982). Indeed, the shapes of most tor outcrops are closely controlled by major joint planes which fall into two main categories: high angle or vertical plane joints; and subhorizontal joint planes usually termed floor or sheet joints (Gerrard, 1974, 1982; Durrance and Laming, 1982). A third set of less well-developed joints, inclined broadly at angles between 20 and 80°, is also present. Although these three sets probably have different origins, all show one common feature, namely that they occur in greater frequency (that is in greater numbers per unit volume of rock) towards the top of exposures (Durrance and Laming, 1982). Gerrard (1982) has argued that the relationship between landforms and jointing is also complicated because some joints are of primary origin, whereas others are secondary, and that it is essential to be able to distinguish between the two main types.

It is normally accepted that the jointing in these rocks results, at least partly, from stored stress: the upward increase in the number of open joints is therefore a reflection of pressure release caused by the erosion of overlying rock (e.g. Gilbert, 1904; Jahns, 1943; Kieslinger, 1958; Bradley, 1963; Brunner and Scheidegger, 1973). Such a mechanism is appropriate for explaining the floor or sheet joints of the Dartmoor granite, although some may be primary sheet structures formed during emplacement and cooling (e.g. Oxaal, 1916; Meunier, 1961; Gerrard, 1982). These joint planes are approximately horizontal where seen on ridges and hill tops, and are inclined on the flanks of hills towards neighbouring valleys at angles of up to 20–25° (Durrance and Laming, 1982). Consequently, the Dartmoor granite is characterized by broadly curved sheet or floor joints which closely mirror the surface contours of the landscape (Gregory, 1969; Gerrard, 1974, 1982).

Although the unloading process theory is difficult to test in the field and should not be assumed to be the universal cause of such joints (Addison, 1981; Gerrard, 1982), the sheet joints or so-called 'pseudo-bedding planes' are widely seen in the tor outcrops of Dartmoor, and have been regarded as having a major bearing on their genesis (Gerrard, 1974, 1978, 1982, 1983, 1989b; see below).

On the other hand, the high angle and vertical joints strike in all directions, although with marked maxima running in broadly N-S, E-W, NW–SE and NE-SW directions: according to Durrance and Laming (1982), they show no geometrical relationship with the boundary of the pluton, and indeed there is considerable doubt as to their mode of origin: a variety of mechanisms involving both tensional and compressive forces has been suggested (Crosby, 1893; Becker, 1905; Hodgson, 1961; Roberts, 1961; Blyth, 1962). Whatever their origin, the spacing and frequency of these vertical and near-vertical joints has clearly influenced the detailed form of granite landforms such as tors (Linton, 1955; Palmer and Neilson, 1962; Gerrard, 1974, 1982, 1989b).

The third group of joints (less well developed) merges at one extreme with the high-angle and vertical joints, and at the other with the subhorizontal floor or sheet joints (Durrance and Laming, 1982). Although their strike directions are not yet well documented, they may have been caused by an imbalance of rock densities within the Dartmoor pluton (Bott *et al.*, 1970; Durrance and Laming, 1982). Further understanding of the joints has been provided by a fractal analysis (Gerrard, 1994a).

Some joints are grooved (slickensided) showing a limited degree of movement (Durrance and Laming, 1982; Gerrard, 1982). Larger-scale faulting, however, has produced more significant movements from a geomorphological point of view: towards the centre of the pluton, rivers have become incised into N-S and NW–SE courses controlled by pronounced fracturing and weakening. Estimating the lateral and vertical displacements in this central area is difficult, and clear evidence of measurable faulting is restricted to the granite boundary (Durrance and Laming, 1982). This wrench-faulting, affecting both the granite and its envelope of surrounding rocks, is believed to be of Tertiary age, although it may also have rejuvenated older structures (Dearman, 1963, 1964; Shearman, 1967; Durrance and Laming, 1982).

Early work on granite landforms

The granite landforms of the South-West, and particularly the tors of Dartmoor, have long attracted the attention of writers. De la Beche (1839, 1853) propounded that tors had formed by differential weathering: an early phase of formation underground was envisaged, followed by erosion of the more decomposed parts. MacCulloch (1848) also provided a useful early account, supplemented by Ormerod in 1858; superb illustrations of many of the tors were provided both by Ormerod (1858) and Jones (1859). The latter account gives some perceptive views on the formation of 'tors, cheesewrings and logging stones' – and clearly relates their formation to frost and associated subaerial weathering guided along both vertical and horizontal joint planes, followed by subsequent removal of the weathered detritus. On the other hand, Mackintosh (1867, 1868b) argued that tors were relict sea stacks, and Woodward (1876) ascribed them to the action of 'wind-driven sand'. There was also much speculation on the importance of the decomposed granite. Indeed, Reid *et al.* (1910) argued that the decomposed or 'kaolinized' granite on Bodmin Moor strongly influenced the distribution of all major landforms. In effect, the kaolinized areas appeared to coincide with valleys and marshy depressions, whereas upstanding hills were composed of intact and unaltered granite. Peat has subsequently accumulated in many of these 'kaolin-floored' depressions (see Hawks Tor).

Other early writers were much taken with the apparently layered structure of the Dartmoor granite and its manifestation in the tors (De la Beche, 1839; Mackintosh, 1868a; Ormerod, 1869; McMahon, 1893; Albers, 1930). Ormerod (1869) observed that the dips of many of these curved or 'pseudo' beds mirrored the form of the local hills, an association later viewed to be of great importance in models of landform genesis (Gerrard, 1974). Both Ormerod and Mackintosh also observed sections where weathered and unweathered granite were juxtaposed and argued that selective 'spheroidal' weathering had played an important part in the formation of the tors, much as Brayley (1830) had earlier argued using evidence from Cornwall. Sub-surface weathering was also invoked to account for the tors by Dawkins (in Sandeman, 1901; see below).

Bate (1871) described the 'clitter of the tors of Dartmoor', and noted that 'Around the base of most of the huge tors that give a mountainous character to Dartmoor cluster large masses of granite rocks in wonderful confusion' (Bate, 1871, p. 517). He argued that the clitter had been derived from the surface of the tors by frost-action. Where the masses of angular rubble occurred at some distance downslope from the tors, he suggested that the upper hillslopes had been 'glazed' by the perennial accumulation of thin layers of ice. This had facilitated the downslope movement over the ice of even quite large granite boulders to lower levels, where they accumulated in piles in much the same manner as a protalus rampart. A similar mechanism, but involving snow patch rather than ice accumulation, was also later suggested by the Geological Survey (Reid *et al.*, 1912).

Alternatively, Belt (1876) suggested that large granite blocks sprinkled all over the surface of Dartmoor had been glacially transported, and then deposited from floating ice in an immense proglacial freshwater lake, which he believed had covered much of northern Europe. This echoed a widespread early view, noted previously, that Dartmoor had been glaciated, and indeed Campbell (1865) thought that the tors themselves had been shaped by 'floating ice'. Whitley (1885) argued that the clitter surrounding the tors in both Cornwall and Devon was flood-formed, 'large masses of solid granite having been severed from their native beds and swept down the slopes of local hills' by a catastrophic 'post-glacial flood'.

Albers (1930), also much taken with the clitter accumulations of Dartmoor, argued that the material occurred on both level ground and in distinct piles towards the base of slopes. He suggested that it had been pushed downhill by 'some force', and invoked Hoegbom's (1913–1914) mechanism of solifluction to account for the accumulations. The clitter was thus believed to be the result of downhill sludging of sediment and granite blocks over a permafrost layer, comparable to that found in tundra regions today. In arguing for a freeze-thaw/solifluction origin for the clitter, Albers also thought it reasonable to suggest that the tors themselves had been formed, probably substantially, by frost-action during the Pleistocene – 'Thus the evidence would point to the tors and clitters of Dartmoor being due to the splitting action of frost, and to movement occasioned by solifluction' (Albers, 1930; p. 378). Albers' study forms an important landmark in the development of thought pertaining to landscape evolution on Dartmoor, particularly in its use of a modern-day analogue (Spitsbergen) to explain relict (periglacial) landforms. The same arguments were later to form the basis of one of the main theories of tor formation put forward by Palmer and Neilson (1962) (see below).

Models of tor formation

Since the work of Albers (1930), a steady stream of papers and textbooks has attested to continued interest in the genesis of Dartmoor landforms, especially tors (Linton, 1955; Waters, 1957, 1964, 1966a, 1966b; Palmer and Neilson, 1962; Linton and Waters, 1966; Brunsden, 1968; Perkins, 1972; Gerrard, 1978, 1982; Twidale, 1982): various 'models' of tor formation have been propounded, the most notable being those of Linton (1955) and Palmer and Neilson (1962), which, to a large extent, have formed the basis for most subsequent theories and discussion. However, the suggestion that retreat of scarps across bedrock to leave tors and pediments (King, 1958) is also worthy of consideration. King argued that his theory was only applicable to skyline tors and that tors in other positions might have formed differently.

Linton's (1955) paper is a classic, although in many ways it enunciates ideas less explicitly formulated by earlier workers (e.g. De la Beche, 1839, 1853). The significance of this paper has been assessed recently by Gerrard (1994b). Linton proposed a two-stage model for the formation of tors. First, deep chemical weathering under warm humid conditions (Linton favoured the Neogene, but the bulk of recent evidence would suggest the Palaeogene) produced a thick regolith, with corestones (ellipsoidal masses of granite separated from the bedrock by regolith) occurring where joint planes were most widely spaced (Figure 4.2). He argued that vertical joints and pseudo-bedding planes were fundamental in guiding this rotting, which itself had been effected by percolating acid groundwater. Second, the products of weathering (the regolith) were removed by mass-wasting processes, leaving the 'sound' granite and

(a) Theoretical arrangement of initial joint sets in granite

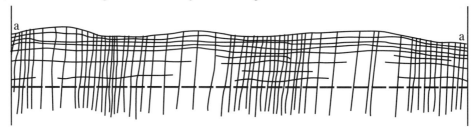

(b) Tertiary weathering

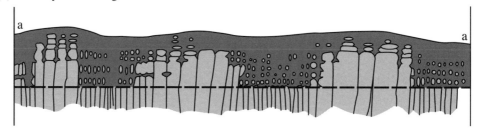

(c) After Pleistocene stripping of Tertiary weathering products

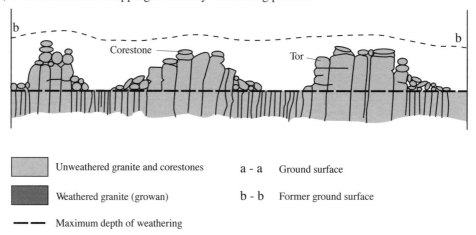

Figure 4.2 Linton's (1955) classic two-stage model of tor formation.

corestones as upstanding tors (Figure 4.2). Linton proposed that tors had probably been exhumed under periglacial conditions during the Pleistocene when solifluction and meltwater would have been efficient agents in removing the regolith. During this period, periglacial activity may also have modified the tors. Linton thus defined a tor as ' ... a residual mass of bedrock produced below the surface level by a phase of profound rock rotting effected by groundwater and guided by joint sys-tems, followed by a phase of mechanical stripping of the incoherent products of chemical action.' (Linton, 1955; p. 476). Such processes were believed to have operated over large areas and pro-tracted timescales, producing the distinctive tors in the Pennines, at Trefgarn and Preseli in Wales, and at the Stiperstones, Shropshire, as well as on Dartmoor (Linton, 1955). The two-stage process on a larger scale has been embodied in Budel's double surface of planation. But, as developed by Thomas

(1978) and suggested by Lewis (1955), in the discussion of Linton's article, the process is likely to be more continuous.

Following work on the 'gritstone' tors of the Pennines (Palmer and Radley, 1961) and a brief note on the origin of those on Dartmoor (Palmer and Neilson, 1960), Palmer and Neilson (1962) put forward a single-stage periglacial mechanism to account for the Dartmoor tors: their model has since formed the principal alternative to the two-stage process of Linton (1955). They argued that the tors and associated clitter were formed by frost-action and solifluction which occurred throughout the Pleistocene. They argued that the distribution of the tors and clitter was related to the origin and depth of the incoherent or decayed granite. Unlike Linton, who argued that the granite decomposed by sub-surface chemical weathering during the Tertiary, Palmer and Neilson suggested that the incoherent granite had, in most cases, been kaolinized by pneumatolytic processes. Some might also have been produced by physical processes, namely frost-shattering along crystal and cleavage boundaries (see below). Thus the two sets of processes (pneumatolysis and physical disintegration) were believed to be quite distinct and unrelated. In support of this argument, Palmer and Neilson cited the occurrence of decayed granite on ridges where tors were absent, and the fact that the weathered material was never found around the tors themselves. The rounded nature of some constituent 'blocks' in the tors was explained by post-glacial, atmospheric chemical weathering. The tors were thus regarded as 'upward projections of solid granite left behind when the surrounding bedrock was broken up by frost-action and removed by solifluction' (Palmer and Neilson, 1962; p. 337). They could thus be termed 'palaeo-arctic tors'.

Many workers have since suggested that all three main sets of processes – pre-Tertiary pneumatolytic and hydrothermal alteration of the granite, Tertiary (and later) sub-surface chemical weathering, and physical weathering and disintegration (periglacial) – have probably been involved in the formation of the tors and associated features (e.g. Brunsden, 1964, 1968; Gerrard, 1978). Nonetheless, the two main models of tor formation have since provided the basis for much discussion, and considerable efforts have been made to determine the crucial origin(s) of the weathered granite or 'growan' (see below; Two Bridges and Bellever quarries).

A further concept regarding tor formation, however, is worthy of note. As previously observed, many tors demonstrate prominent floor or sheet joints (pseudo-bedding) running parallel to the surface of the tor or local hill. Likewise, many tors consist of several rock masses arranged in an avenue, often with the central part absent. Brunsden (1968) has argued that, if reconstructed, these separate rock masses with their missing centres, would form a large, single dome structure, with the linked floor joints demarcating once-continuous sheets of granite. Consequently, it has been suggested that the original form of the tors was neither a pile of corestones (Linton, 1955) nor a soil-covered hill (Palmer and Neilson, 1962), but rather a granite dome analogous to the granite inselbergs of Africa (Brunsden, 1968).

These arguments have been extended by Twidale (1982) who has suggested that domed, block-strewn or castellated structures worldwide are generically related: the dome structure is the starting point of an evolutionary sequence (encompassing landforms such as nubbins (analogous to tors), castle koppies and castellated inselbergs) in which the major differences are based simply on the radii of the original domes, as well as the relative depth of sub-surface formation. St Agnes Beacon in Cornwall has also been interpreted as an inselberg exhumed from a deep Tertiary saprolite (Walsh *et al.*, 1987; Jowsey *et al.*, 1992).

Kaolinization or Tertiary chemical weathering?

The presence of decomposed or altered granite throughout the South-West is little disputed. However, there has been very little firm agreement as to its origin, and opinion has alternated between deep chemical weathering and hydrothermal and physical alteration processes. Further, there has been a proliferation of technical terms which has perhaps further added to the confusion. The nature and origin of these 'altered' granitic materials, however, are of prime importance to understanding the evolution of granite landforms: many workers (e.g. Waters, 1954, 1957; Linton, 1955) have attempted to relate Dartmoor landforms (including the tors, erosion surfaces and drainage nets) to the presence or absence of the altered granite and to the differential action of chemical and mechanical weathering (Brunsden, 1964).

The main arguments have arisen as to whether the altered granites are due to post-emplacement weathering (mainly chemical but also physical processes) or to effects penecontemporaneous with the intrusion of the granite (e.g. hydrothermal processes). The latter, of which kaolinization is perhaps the most important, involves chemical alteration of the granite by heated and superheated waters. In this respect, it may be distinguished from pneumatolytic processes, the latter more strictly being the action of hot gases associated with the igneous activity. Although the distinction is perhaps arbitrary, it is useful since distinctive minerals and rock types are produced. The term 'chemical weathering' is used here to mean alteration of the granite by supergene processes, that is normal subaerial weathering effected largely by the circulation of weakly acid groundwater. 'Physical' weathering is taken to mean alteration of the granite by mechanical processes, primarily frost-action.

It has long been held that the kaolin deposits of the South-West are distinct from any altered material that may have been formed as a result of chemical weathering (Exley and Stone, 1964). The hypothesis that kaolinization was the result of weathering has had few early supporters, an exception being Hickling (1908), and a good case has been put forward for a hydrothermal origin (e.g. Collins, 1878, 1887, 1909; Reid *et al.*, 1910; Exley, 1959, 1964; Exley and Stone, 1964). However, the controversy was re-opened by the suggestion that the china clay (kaolin) deposits have a weathering or supergene origin (Sheppard, 1977). Thus clear differentiation in the field between hydrothermal and supergene weathering products is still tenuous.

For clarity, the arguments over the nature and origin of the altered materials can be divided into three main 'schools'. First, there are those who have argued that there is evidence for the widespread development of a substantial, chemically weathered regolith or saprolite, probably during warmer, more humid, conditions in the Tertiary (Waters, 1954, 1957; Linton, 1955; Linton and Waters, 1966). This view is central to the Linton theory of tor formation (see above), and has wide repercussions for general landscape evolution including the development of erosion surfaces, tor and basin topography and drainage nets (Waters, 1954, 1957, 1960c, 1960d, 1960e, 1964). A second school holds that the altered granite is largely hydrothermal in origin (e.g. Reid *et al.*, 1910; Palmer and Neilson, 1962; Exley and Stone, 1964), and it can therefore accommodate landscape development without invoking a thick, chemically weathered regolith; other mechanisms, principally physical weathering (frost-action and solifluction) are deemed to have played a substantial, if not dominant, role in landform genesis (Te Punga, 1957; Palmer and Neilson, 1962). Finally, many workers have suggested that a combination of all three main sets of processes – hydrothermal alteration, chemical and physical weathering – has been involved in the evolution of granite landforms and the Dartmoor landscape (e.g. Brunsden, 1964, 1968; Eden and Green, 1971; Doornkamp, 1974; Gerrard, 1983). This school deals with the conflicting evidence by invoking a sequence of events involving both hydrothermal and supergene processes, with hydrothermal activity 'softening-up' the granite and rendering it extremely susceptible to later supergene alteration (e.g. Bristow, 1977, 1988; Sheppard, 1977; Durrance and Laming, 1982). Most recently, Floyd *et al.* (1993) have provided a detailed evolution and alteration scheme for the St Austell Granite.

Many early workers favoured the first school of thought, and argued that the granite had been rotted differentially by chemical action (e.g. De la Beche, 1839, 1853; Reid *et al.*, 1910, 1912; Ussher, 1912; Worth, 1930), and these arguments were to reach their fullest expression with Linton (1955). Others have argued that at least some of the altered or incoherent granite is hydrothermal or pneumatolytic in origin (e.g. Reid *et al.*, 1912; Worth, 1930; Guilcher, 1950; Dines, 1956). Resolution of these problems clearly requires precise parameters against which the field and laboratory evidence can be assessed. Although many recent studies have provided classifications and diagnostic characteristics for recognizing the various alteration products caused by weathering or hydrothermal processes (e.g. Brunsden, 1964; Dearman and Baynes, 1978; Irfan and Dearman, 1978a), there is still no firm agreement. If anything, opinion in general appears to have swayed farther away from the seemingly well-established view that much of the kaolin and altered granite is purely of hydrothermal origin.

Perhaps the strongest arguments put forward for hydrothermal alteration of the granite are: 1. the great depth and form of some deposits – in particular the fact that unweathered rock overlies kaolinized granite which itself can reach great depths. Some china clay pits work areas of altered granite well over 500 m in diameter. Proven depths of such material in the St Austell Granite are in excess of 250 m (Exley, 1959, 1976; Brunsden, 1964; Bristow, 1977; Durrance and Laming, 1982); 2. the often close association of the altered material with greisen-bordered quartz-tourmaline veins; the crystallinity index of the kaolinite increases towards such major quartz-tourmaline veins (e.g. Brunsden, 1964; Durrance and Laming, 1982; Gerrard, 1983). Good reviews of the characteristics to be expected in hydrothermally/pneumatolytically altered granites are given by Brunsden (1964), Exley and Stone (1964), Durrance and Laming (1982) and Floyd *et al.* (1993).

However, there is a large and growing body of data which suggests that the contribution of hydrothermal processes to altering the granites of the South-West may have been overestimated. Many workers have long suspected that the effects of hydrothermal and chemical weathering processes often occur together (e.g. Reid *et al.*, 1910, 1911, 1912). Brunsden (1964) argued that the evidence on Dartmoor showed that the decomposed granite had been formed by a combination of chemical weathering, frost-pulverization and pneumatolysis. Evidence for all three processes was deemed to occur within single profiles (Two Bridges Quarry), although distinguishing between them was not easy, since the lines of weakness in the granite (the joints and fissures) had provided a focus for all the processes (Brunsden, 1964): Brunsden, however, proposed a classification to distinguish between the processes. He argued that if a section showed evidence of mineral breakdown (e.g. a biotite weathering front, eluviation of clay minerals, progressive stages of physical disintegration, spheroidal weathering and grus formation, a zoning of the weathering horizon, an increase of solid rock, and corresponding decrease of decayed rock with depth), then chemical weathering had been the cause of decomposition. On the other hand, physical weathering (frost-action) would be characterized by a comminution of particles (without a loss of mass), by leaching of minerals in solution and by eluviation of clays. Frost-wedges, involutions and head deposits were considered to aid identification, but in reality the two former features are virtually absent on the granites of South-West England, and the latter feature is so widespread that none of them has diagnostic value in this context. Finally, he suggested that hydrothermal/pneumatolytic alteration could be recognized by tourmalinization and ore deposits – although this could be confirmed only if the sections of altered material increased and widened with depth, if a cover of solid, unaltered granite is present or if there is an upward increase in alteration products (Brunsden, 1964).

Although some details of Brunsden's work have been questioned (Eden and Green, 1971; Green and Eden, 1973), subsequent studies have shown clearly that chemical decomposition has played a role in the development of altered profiles. A variety of detailed classifications of weathering grades,

based largely on engineering properties, has been proposed (Fookes *et al.*, 1971; Dearman and Fookes, 1972; Dearman and Fattohi, 1974; Dearman *et al.*, 1976, 1978; Baynes and Dearman, 1978; Dearman and Baynes, 1978; Irfan and Dearman, 1978, 1979a, 1979b). Dearman and Baynes (1978) devised a system for differentiating the alteration products based on a combination of field mapping and engineering grades. It was argued that by mapping the distribution of grades of 'equal intensity of effect', the effects of hydrothermal alteration, chemical weathering and frost-shattering could be distinguished. It was admitted, however, that ascertaining the precise extent to which each of the three potential processes had affected the rock was still difficult to determine (Dearman and Baynes, 1978). Laboratory techniques (including the use of Scanning Electron Microscopy) have also been used to improve the recognition of different weathering grades in sound and altered granite (Irfan and Dearman, 1978, 1979a, 1979b). These grades have been based on recognizing changes in the microfabric of the granite, and they have been used to show that the initial ingress of weathering agencies occurs along primary cracks, pores and open-cleavages (Baynes and Dearman, 1978).

Eden and Green (1971) applied textural and mineralogical investigations to samples of the altered or decomposed granite from sites throughout Dartmoor, and distinguished between the products of pneumatolytic/hydrothermal alteration and chemical weathering. They argued that the altered granite or 'growan' was less decomposed than true kaolin deposits elsewhere, for example in the St Austell Granite (Exley, 1959; Exley and Stone, 1964), for which they felt a hydrothermal origin had been securely established. Their results showed that the growan had originated as a weathering product: in comparison with the kaolinized material it contained much less silt and clay, more intact feldspar crystals, and quartz and mica constituents showing little alteration. They suggested that occurrences of pneumatolytically altered granite on Dartmoor were rare (cf. Two Bridges Quarry). Even the presence of tourmaline in the sections at Two Bridges Quarry, which Palmer and Neilson (1962) had associated with 'kaolinized granite', was rejected as an indication of hydrothermal alteration: tourmaline is found widely in solid, unaltered granite and is likely to have been formed prior to kaolinization (Exley, 1959; Eden and Green, 1971; Floyd *et al.*, 1993). In conclusion, Eden and Green suggested that the occurrences of growan on Dartmoor were only 'moderately decomposed'. This, they argued, indicated that the material had probably not been formed in a hot humid environment (Waters, 1954, 1957; Linton, 1955; Linton and Waters, 1966), although in warmer conditions than today, perhaps in a meso-humid, subtropical climate (see Two Bridges and Bellever quarries). Although this led Eden and Green (1971) to accept Linton's two-stage hypothesis of tor formation, they suggested that chemical weathering had been less effective and widespread on Dartmoor than previously thought: the tors had been exhumed from a sandy, not clayey, weathering zone, principally located in or near the main river valleys, and thus any deep weathering had been extremely localized.

Eden and Green (1971) therefore argued that the growan bore little resemblance to the kaolin deposits, stressing the lack of clay, high feldspar content and lack of feldspar alteration as prime evidence. X-ray diffraction studies also revealed the presence of gibbsite, which Green and Eden (1971) suggested was further evidence of chemical weathering. This mineral has frequently been noted in weathered granite in the humid tropics, and its occurrence in France has been used in support of a former hot and humid climate (Maurel, 1968): subtropical (Bakker, 1967) and temperate (Dejou *et al.*, 1968) weathering regimes have also been invoked to explain its presence elsewhere. Gerrard (1994d) has asked a number of questions. Does its presence in the Dartmoor weathered granite imply that humid tropical conditions formerly existed or is our understanding of the factors favouring gibbsite formation at fault? Are there special circumstances which have led to its production? In comparisons with tropical areas, Green and Eden concluded that the gibbsite in the Dartmoor growan occurred as an initial product of weathering, showing that any weathering here was at an early stage; its presence did not necessarily imply a humid tropical environment. The production of gibbsite is an example of where it is difficult to relate a specific clay mineral to specific climatic characteristics. It seems to be related to the stage of the weathering process and the particular leaching conditions. Gerrard (1994d) concludes that gibbsite, in appreciable amounts, probably indicates lateritic-type weathering, but small amounts can be produced under a variety of conditions. Such a view is supported by recent work in northeast Scotland where two main granite weathering products have been differentiated (Hall, 1983; Hall *et al.*, 1989). These comprise 'clayey grus' (the pro-

posed product of intense weathering, possibly under subtropical conditions between Miocene and mid-Pliocene times) and 'grus' (the product of less intense weathering perhaps during warm interglacial conditions in the Pleistocene). In the context of these Scottish granite weathering products, Mellor and Wilson (1989) have concluded that gibbsite is a feature which pre-dates the last glaciation, but its precise time of formation is uncertain; it could have formed under humid, warm-temperate to subtropical conditions during the Tertiary and/or during Pleistocene interglacials (Hall, 1983). However, its status as an indicator of warm environments is uncertain because the mineral is also believed to form at the initial stages of rock breakdown (Hall *et al.*, 1989).

Doornkamp (1974) studied micromorphological characteristics of weathering products from Dartmoor (head deposits, bedded growan and *in situ* growan) using Scanning Electron Microscopy (SEM). He concluded that most material showed evidence of mechanical weathering. Evidence for granite which had been altered chemically was found, however, at Two Bridges Quarry. Doornkamp suggested that these results supported Eden and Green's view that chemical weathering had occurred, but had been much more selective than previously thought.

Gerrard (1983) has argued that the most critical evidence for chemical weathering comes from oxygen and hydrogen isotope studies (e.g. Sheppard *et al.*, 1969; Savin and Epstein, 1970; Sheppard, 1977) which have enabled differentiation between hydrothermally formed kaolin and that formed by chemical weathering (supergene) processes. On this basis, Sheppard (1977) has argued that some of the kaolin deposits of South-West England owe their origin to weathering, and similar conclusions have been drawn from SEM studies (Keller, 1976), and by Ollier (1983). Because weathering is a widespread phenomenon and occurrences of hydrothermal alteration rare (cf. Konta, 1969), Ollier argued that the former should always be assumed unless the latter could be rigorously proven. Thus, even many economic deposits once attributed to hydrothermal activity could now be related to weathering (e.g. Amstutz and Bernard, 1973; Ollier, 1977, 1983). In particular, the presence in altered material of chlorite and some cracked quartz grains (e.g. Moss, 1966; Bisdom, 1967; Baynes and Dearman, 1978) ' ... falls far short of proof, and even short of a reasonable suggestion' (Ollier, 1983; p. 58).

Similar changes of view have been happening in

work on other granite areas. Bird and Chivas (1988), using oxygen isotopes in Australian weathering profiles, were able to distinguish profiles formed in the late Mesozoic and early Tertiary from profiles formed in post-mid-Tertiary times. The deeply altered Bega Granite of south-east Australia was once attributed to hydrothermal alteration, but isotope studies have shown it to be weathered. However, alteration of the Conway Granite, in New Hampshire, has now been shown to be due to hydrothermal alteration and not weathering. Therefore there is still much work to be done. For South-West England granites, Durrance *et al.* (1982) have suggested that, although alteration resulted from reaction with meteoric water, the system was driven by geothermal heat.

Further, although more indirect, evidence for Tertiary chemical weathering comes from elsewhere in the region. Bristow (1968) has shown the presence of a weathered mantle beneath Late Oligocene sediments in the Petrockstow Basin. Chemical and mineralogical analyses have shown that these weathered deposits formed under humid subtropical or warm-temperate conditions (Bristow, 1968). Similar types of weathering have been described elsewhere in the South-West (Fookes *et al.*, 1971; Dearman and Fookes, 1972; Dearman and Fattohi, 1974; Dearman *et al.*, 1976). Isaac (1979, 1981, 1983a, 1983b) has also provided evidence for Tertiary weathering profiles in the plateau deposits of east Devon: the distribution of laterites and silcretes reflects a complex pedological, diagenetic and geomorphological history (Chapter 3). If the evidence presented above is correct, then there would appear to be an ample basis for arguing that chemical weathering did occur on Dartmoor during the Tertiary (Gerrard, 1983), irrespective of any previous hydrothermal effects.

Notwithstanding the growing evidence for chemical weathering as an important process in the alteration of the Dartmoor and other granites, and in the formation of the 'growan', other workers have propounded that physical weathering also produces similar material. Te Punga (1957) originally argued that the weathered Bodmin Moor Granite (see Hawks Tor) had been formed by Pleistocene frost-shattering. Comparable material in the Massif Central has been ascribed to processes operating in a cool-temperate environment (Collier, 1961). Incoherent granite in the Sierra Nevada has also been attributed to frost-riving (Prokopovich, 1965), but Wahrhaftig (1965) has shown that the chemical alteration of biotite to chlorite in this material produces a 14Å-size clay

residue which causes expansion and mechanical shattering of the rock. It has thus been suggested that similar processes may have produced the Dartmoor growan which, it is claimed, resembles 'sandy weathering products' elsewhere in Europe (Jahn, 1962; Bakker, 1967; cf. Eden and Green, 1971). However, such a 'grussification' process has not been accepted by all workers.

These deliberations will continue, because the nature and origin of the 'altered' material are central to theories of granite landscape evolution. The view is taken here that the evidence most probably reflects the operation of hydrothermal, chemical and physical weathering processes, sometimes all combined, over an extremely protracted timescale: the evolution/alteration sequence proposed by Floyd *et al.* (1993) would seem to offer an appropriate working model for geomorphologists. Whatever the relative contribution of each process, there is no doubt that the nature and distribution of the altered granite itself have strongly influenced the operation of periglacial processes and the development of characteristic landforms during the Pleistocene (Gerrard, 1983).

Periglaciation, slope development and landform assemblages

Most recent workers, while commenting on the controversial origin of the altered granite, have simply accepted its occurrence as a fact. At the simplest level, the weathered granite is treated as a soft material extremely susceptible to erosion, particularly by periglacial mass wasting processes. The material's importance in influencing the development of granite landforms is therefore fully acknowledged, but the main emphasis since the 1950s has been the role of periglacial conditions and processes. Certain landforms and deposits can be shown to have originated from these processes, and there is little doubt that distinctive elements of the granite landscape were produced during various cold phases of the Pleistocene.

Although the effects of frost-action on the granites of the South-West were appreciated long ago, Albers (1930) provided the definitive link between modern-day periglacial processes (principally frost-shattering and solifluction) and fossil landforms (see above). Preliminary descriptions of a wide range of fossil periglacial features within the region were subsequently given (Dines *et al.*, 1940), and the role of frost-action was considerably heightened by the work of Guilcher (1949, 1950) and Te

Punga (1956, 1957) who described periglacial land-forms and deposits throughout southern England, including Dartmoor. The evidence they described included fossil ice-wedges, involutions, stone poly-gons, blockfields, loess and, on Dartmoor itself, altiplanation terraces and earth hummocks (see Merrivale). Te Punga (1957) concluded that much of the landscape had been severely denuded during different periglacial episodes and had been subject to 'vigorous down-wearing by mass wasting', mainly solifluction. He argued that

'It seems probable that the effects of successive periglacial episodes were cumulative, each later episode emphasizing the landforms produced in earlier episodes; it is unlikely that interperiglacial erosion, seeing that it was restricted essentially to linear processes, was competent to obscure or obliterate earlier developed periglacial landscape form' (Te Punga, 1957; p. 410).

The fact that the present relict periglacial land-scape is so obvious would add weight to this argument. Significant to later ideas was Te Punga's belief that vast quantities of material had been transported during periglacial conditions to pro-duce a subdued landscape characterized by convex upper slopes and concave lower slopes: following Tricart (1951), he argued that the convex upper slope had been a zone of wastage, while the lower slope had been a zone of deposition. Periglacial fea-tures were widely preserved because present-day processes were relatively ineffective, due to bind-ing vegetation, and because the duration of post-periglacial time had been relatively short (Te Punga, 1956).

These concepts were reinforced by Waters during the 1960s (e.g. Waters, 1960a, 1960b, 1961, 1962, 1964, 1965, 1966b, 1971), who attempted to link individual deposits in the region (head and solifluction deposits) to specific periglacial episodes. Although periglacial landforms and deposits were widespread in Britain, he considered that the wholesale redistribution of pre-Pleistocene, deeply weathered regolith by periglacial, mainly solifluction, processes, made Dartmoor ' ... proba-bly the purest relict periglacial landscape in Britain' (Waters, 1960a; p. 174). He argued that two main sets of these processes had been operative in the South-West during the Pleistocene: 1. gelifraction or the weathering of coherent rock (mainly freeze–thaw activity); and 2. 'geliturbation' – the disturbance and removal of material principally by solifluction. The effects of these processes were manifested in landscape features such as tors, the modification of slopes, patterned ground and

solifluction debris or head (Waters, 1964).

Returning to the ideas of Tricart (1951) and Te Punga (1956), Waters suggested that two separate phases of periglacial activity could be discerned in various head layers throughout Dartmoor. The most complete sections showed evidence for two cryogenic episodes (Figure 4.3), each of which was marked by the downslope transfer ('geliturbation') of different debris types. Thus, during the first cold episode, successive layers of the existing weather-ing profile (growan) were removed from upper slopes and deposited in reverse order lower down, forming layers of 'bedded growan' and the main head (Figure 4.3). Not judged to have been signifi-cantly affected by subsequent weathering (either interglacial or interstadial), material on the upper slopes again became exposed to periglacial processes. This time, more coherent blocks of bedrock, derived from tors and other surface expo-sures, were detached by frost-action and removed by solifluction to lower levels, where a second, 'blocky' or upper head accumulated on the older deposits (e.g. Waters, 1964, 1965; Linton and Waters, 1966). This inversion of the weathering profile was believed to be widespread on Dartmoor (Waters, 1964, 1965) and the presence of two sep-arate head deposits, formed during different periglacial phases, was widely accepted at the time (e.g. Brunsden, 1968; Gregory, 1969).

Waters therefore paints a picture of the evolving Dartmoor landscape where a pre-existing weath-ered regolith (the growan), of variable thickness, is transferred from higher to lower levels, creating a smoothed topography punctuated only by tors and buttresses of the most massive and resistant mate-rials. Many upstanding masses of sound granite were completely destroyed by frost-action, and clit-ter accumulated where the rate of frost destruction exceeded the rate of removal: some of this material was rearranged into patterned ground consisting of stripes and nets (Waters, 1964). The removal of material from the higher levels by solifluction caused aggradation of head on lower slopes and the development of large valley-floor terraces. Where suitable lithological conditions prevailed (see Cox Tor, Merrivale), benched hillslopes or alti-planation terraces were formed. Likewise, Brunsden (1968) concluded that three main types of periglacial landform and deposit were present on Dartmoor: 1. frost-shattered rock outcrops and boulder-strewn slopes; 2. frost regoliths of head and solifluction debris; and 3. small-scale landforms cut into upland slopes.

Building on the work of Te Punga and Waters,

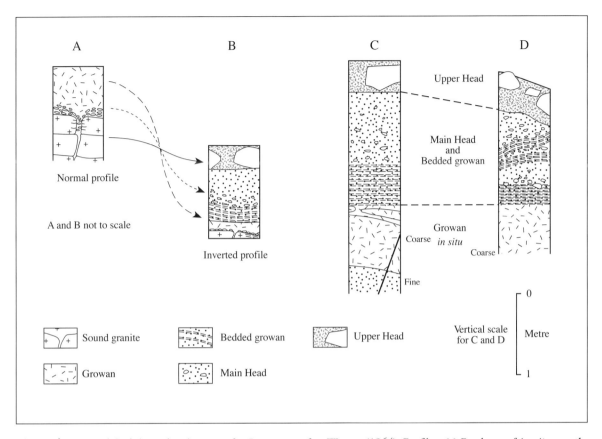

Figure 4.3 A model of slope development for Dartmoor, after Waters (1964). Profiles: (a) Products of *in situ* weathering on a granite substrate; (b) Inversion of normal weathering profile following two separate periods of periglacial mass wasting; (c) and (d) Measured sections at Shilstone Pit (SX 659902), Dartmoor. Many slope configurations, however, do not conform to this model (see text).

more recent research in the area has confirmed the widespread role of periglacial processes on landform development, and stressed the importance of structural control (particularly jointing patterns in the granite) (e.g. Green and Eden, 1973; Cullingford, 1982; Gerrard, 1983, 1989b). Detailed analysis of slope deposits within the region, however, shows Waters' two-stage periglacial inversion model to be an oversimplification. Green and Eden (1973) studied the sources and distribution of material in the slope deposits of Dartmoor. They showed that all parts of the slope contributed material to the deposits and that a simple two-fold division of slopes into 'source' and 'accumulation' areas was therefore untenable. They demonstrated that even on lower slopes, movement of slope deposits had been accompanied by erosion of the underlying granite (see Merrivale), and that this basal material had been incorporated into the transported layer: widespread inversion of the weathering profile was not apparent. These principles also applied to the 'bedded growan' commonly found on Dartmoor. Green and Eden suggested that this material was locally derived having been displaced downslope by solifluction deposits moving over weathered granite (see Two Bridges and Bellever quarries). An origin as surface-wash sediment (cf. Waters, 1971) was therefore ruled out. Green and Eden's study also provided new information on the relationship of the clitter to local head deposits. Clitter and its rock sources were encountered in a variety of different slope positions. The clitter was not therefore simply derived from ridges, summits and tors as had been suggested previously (Waters, 1964).

Green and Eden concluded that because the clitter, head deposits and bedded growan were present in a wide variety of locations on Dartmoor slopes, they could not be the product of progressive stripping of a normal weathering profile – that is, from the upper parts of valley-side slopes to the lower ones. Support for this contention comes from recent work carried out on head deposits throughout Dartmoor (Gerrard, 1989a). Two main types of head, a fine and a coarse variety, do exist, but the relationships between the two are complex. Frequently, both types intermingle: gullies have been cut and infilled at different times, indicating complicated sequences of slope modification rather than a simple reworking of weathered granite. Large blocks of granite occur throughout the head and not just at its surface. Bedded growan is common, particularly in mid-slope situations. The bedrock surface is extremely variable and frost-shattered with solid stacks of rock reaching almost to the surface. In places, head rests directly on striated bedrock (Gerrard, 1989a). More detailed relationships between head and growan have been reported (Gerrard, 1990, 1994c).

The granite landforms of the South-West were reinvestigated by Gerrard who provided detailed accounts of periglacial landforms in the Cox Tor-Staple Tors area (Gerrard 1983, 1988; see Merrivale), the influence of rock type and structure on granite landforms on Dartmoor and eastern Bodmin Moor (Gerrard, 1974, 1978) and the origin of Dartmoor slope deposits (see above; Gerrard, 1989a). Considerable emphasis has been placed on the effects of granite jointing on the distribution of landforms, particularly tors (see above – joints and faults; see below – erosion surfaces and drainage development). On the basis of the density and pattern of jointing, Gerrard devised a classification of tors into: 1. summit tors; 2. valley-side and spur tors; and 3. small tors cropping out on the flanks of low convex hills. He has argued that areas with closely spaced joints become the focus of initial weathering and erosion. This leads to 'compartmentalization' of the landscape into positive and negative areas (cf. Waters, 1957). Erosion, guided by joint density, has long been a matter of speculation, but Knill (1972) has shown that joints in some valley-floor areas are separated by only c. 0.5 m. Gerrard suggests that joints in the areas of upstanding relief (ridges and domes) would initially have been in a state of compression, but as erosion occurred along lines of weakness (stream erosion along zones with dense joints), the joints in the domes themselves would open and allow weathering (Gerrard, 1982). Such a mechanism provides the basis for a model of tor formation, and its emphasis on the spacing of joints makes it similar to Linton's (1955) in this respect. It differs because the continued removal of weathered material, especially from the valley areas, is seen as releasing

further stresses stored in the domes, thus progressively exposing them to further weathering (Gerrard, 1974, 1982; see Merrivale). It is therefore the dense jointing which is picked out by the weathering agencies and stream courses, and the tors found crowning many of the 'unloading' or 'dilatation' domes that owe their formation and distribution to further incision by streams around the domes (Gerrard, 1974).

In addition, Gerrard (1982) argued that these processes occur at different rates throughout the landscape, and that there is therefore a variety of relationships between the accumulation and removal of weathered products. He envisaged three different situations: 1. where the ground surface was relatively stable and a deepening of the weathered regolith would occur; 2. where the surface was unstable and regolith was gradually removed; and 3. where a steady-state situation occurred and the renewal and removal of material occurred at similar rates. It was considered possible that each state might exist concurrently at different sites: the Linton explanation of weathering followed by removal was therefore too simple to explain the major elements of a granite landscape (Gerrard, 1982).

On this basis, Gerrard devised a classification of tors on Dartmoor and eastern Bodmin Moor (see above). Detailed measurements on 65 tors show major variations with respect to size and intensity of jointing, and the slope angles at their base. Both summit and valley-side tors possess relatively closely spaced vertical joints, whereas those of the emergent tors are much more widely spaced (Gerrard, 1978).

In combining the evidence of structural control with that for periglacial processes, Gerrard (1983) produced a composite diagram to explain the main geomorphological elements seen in the Dartmoor landscape (Figure 4.4). Although individual measured slopes rarely fit the model exactly, those in the vicinity of Great Staple Tors (see Merrivale) show a close correspondence.

A very similar classification of tors was produced by Ehlen (1994). She was able to demonstrate significant differences between these groups. Summit tors generally possessed high relative relief (mean 126 m), the rock was megacrystic and possessed the widest joint spacing. For primary vertical joints, the spacing was usually > 300 cm; for primary horizontal joints the mean spacing was 73 cm and for secondary horizontal joints the mean was 13 cm. Summit tors are usually controlled by vertical joints or by vertical joints and horizontal joints combined.

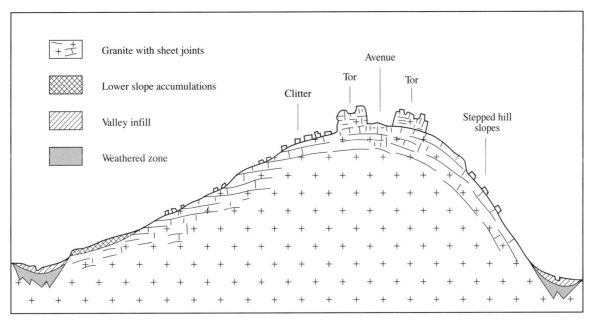

Figure 4.4 A schematic composite representation of the main geomorphological features of Dartmoor. (After Gerrard, 1983.)

Feldspar is usually abundant in the rock (> 30% potassium feldspar; > 18% plagioclase). Spur tors generally possess narrower vertical joint spacing and horizontal joint spacing is intermediate. The rocks are fine-grained (< 1 mm) and feebly megacrystic or equigranular. Potassium feldspar abundance is low. Valley-side tors have narrow joint spacing, and horizontal joints control tor shape. The rocks are finer grained (< 2 mm), feebly megacrystic and quartz abundance is low.

In terms of spatial distribution within Dartmoor, multivariate analysis produced five tor groups. Tors in the first group occur mainly south of a line from Great Mis Tor to Bell Tor. They are characterized by medium to high numbers of megacrysts, medium- to coarse-grained feldspar, narrow to intermediate vertical joint spacing and low to intermediate quartz abundances. Most of the tors are summit tors. Tors of the second group are scattered across Dartmoor, many of them lamellar in form. They are characterized by fine- to medium-grained feldspar, widely spaced vertical joints, low secondary joint spacing ratios and low to intermediate quartz abundances. Only two tors are present in the third group and occur to the north-west and east. The rock possesses no megacrysts and vertical joint spacing is narrow. Most of the tors in the fourth group occur in the east and possess medium to high numbers of megacrysts, intermediate vertical joint spacing, low quartz

abundances, moderate to highly abundant plagioclase and occur in the form of summit tors. The fifth group is the largest and the tors are present throughout Dartmoor, although there is a tendency for them to occur near the granite boundary. The rock has few megacrysts, narrow to intermediate vertical joint spacing, low to intermediate plagioclase abundances and forms summit and valley-side tors. Tourmaline veins are typically present. In general throughout the granite of Dartmoor, relationships exist between grain size, rock texture, jointing and landforms (Ehlen, 1989, 1991, 1992).

Towards a composite model of landscape evolution

Theories on granite landscape evolution have ranged widely, and a variety of mechanisms has been proposed to account for the tors and associated landforms of the South-West. The significance of periglacial processes in shaping this landscape seems to be the only major area of agreement: less has been reached regarding the precise origins of the decomposed or altered granite (growan), and the classic models of tor formation are perhaps too simple to explain wide variations in slope and tor morphology. The possibility must also exist that tors are an example of equifinality (White, 1945;

Selby, 1977; Gerrard, 1984). Also, tors in the same landscape may have been formed by different processes. It is likely that many of the valley-side tors on Dartmoor are the result of periglacial processes, but that large summit tors have a composite origin from both chemical and frost action (Gerrard, 1994b). Many summit tors, especially Great Staple Tor, seem to possess rounded upper portions and significantly more angular basal portions. As French (1976; p. 233) has noted, ' ... it is conceivable that both exhumation and modification of two-cycle tors and the formation of one-cycle tors could have occurred at the same time in different parts of Dartmoor depending upon the localisation of the deep weathering process' and ' ... tors of different forms may exist adjacent to each other and develop under the same climatic conditions but by two different processes' (French, 1976; p. 234).

Recent work, however, goes some way to providing an integrated approach to the study of these landforms. The complexity of depositional sequences now demonstrated is at variance with former reconstructions where perhaps only one or two main phases of periglacial modification were envisaged. Although the basic configuration of the 'dome and basin' topography may have been inherited from the Tertiary (and earlier), the smaller (and some meso-scale) details of the slopes and landforms reflect clearly the cumulative operation of periglacial and other processes during the Pleistocene. Since it is widely agreed that Dartmoor was never glaciated, it is reasonable to assume that substantial landscape changes occurred during the multiple periglacial phases now known to have affected the region (Bowen, 1994b): as a result, depositional evidence is likely to be complicated, and the effects of the periglacial modification cumulative. Little, however, is known about the age(s) of the various slope deposits in granitic inland areas, and their relationship to the better-dated coastal 'head' sequences has yet to be firmly established (cf. Mottershead, 1971; Stephens *in* Linton and Waters, 1966). In some areas, the legacy of Pleistocene periglacial activity may be substantial. The efficacy of such processes, however, is not universal as attested by the survival of relatively fragile sands and 'clays' at St Agnes, west Cornwall (Chapter 3), and the selective survival of given landscape features is, as yet, unexplained. There is good reason to believe, however, that the morphological detail of much of the present landscape is the result of periglacial activity during the various cold episodes which have characterized the Devensian Stage alone.

The selected GCR sites on Dartmoor (at Merrivale, Two Bridges and Bellever quarries) and in the Isles of Scilly (Peninnis Head) demonstrate between them a huge variety of granite landforms and associated weathering products: collectively, they illustrate many of the theories which have been propounded to account for the formation of tors and the altered granite or growan, and demonstrate, impressively, the range of periglacial processes which were operative in this area during the Pleistocene. This small network of sites will remain central to future reconstructions of granite landscape evolution in Britain.

Dartmoor: the physical background

The Dartmoor pluton gives rise to an elevated region of widespread moorland dotted with tors. The granite areas cover *c.* 250 square miles, and extend some 22 miles from north to south and 18 miles from east to west (Worth, 1930). The highest ground occurs in the north-central parts of the granite where most of Dartmoor's radially draining rivers begin their courses. Here, the principal summits range between *c.* 1600 to 2000 ft (488–610 m) above sea level; in the southern area they range between 1200 and 1600 ft (366–488 m) (Worth, 1930).

Worth's (1930) early work on the physical geography of Dartmoor is worthy of special note. He divided the area into different terrains based on relief and elevation, and provided comprehensive accounts of the landforms, peatlands and present vegetation. In his superb illustrations of the many famous Dartmoor tors (e.g. Littaford, Chat, Blackingstone, Bowerman's Nose, Staple, Great Mis, Cox, Thornworthy, East Mill and Oke tors, among others) lay the key to his belief that the present form of the region was largely inherited from the upper surface of the granite when it cooled in contact with the overlying sedimentary rocks (Worth, 1930, 1967). His principal line of evidence for this assertion was the striking coincidence of the 'pseudo-bedding planes' (sheet or floor joints) with the slopes of local hillsides and summits.

Erosion surfaces and drainage development

The difference in elevations between the north and south parts of Dartmoor, noted by Worth, was seen as evidence by Waters (1957, 1960c, 1960d, 1960e; and *in* Brunsden *et al.*, 1964) for a series of erosion surfaces. These were related to different base levels and, on Dartmoor, several major erosive episodes had modified what were considered to be remnants of a higher and older pre-existing peneplain which had formed the basis of the generally southward-sloping Dartmoor plain. An analysis of specific (relative) relief in the 541 km squares which cover the upland granite area of Dartmoor, at a scale of 1:25 000, clearly demonstrates the plateau-like nature of the area (Gerrard, 1993).

The uppermost and oldest surface (the remnant peneplain) was considered to occur at elevations between *c.* 1900 and 1500 ft (580–457 m), and was represented by three main residual land masses. A middle surface, occurring on both the granite and adjacent country rocks at elevations between *c.* 1300 and 1050 ft (396–320 m), was represented by piedmont benches of varying width, various valley-side benches, valley-floor segments and basins (Waters, 1960c). This tor-crowned surface with its elongated basin-like depressions was considered to show considerable dissection, and had been much affected by differential erosion and weathering. A further, much more pronounced and extensive surface (the lower surface), was separated from the middle surface by a group of facets or relatively steep slopes. Lying between *c.* 950 and 750 ft (290–229 m) it was, according to Waters, represented on Dartmoor only by river terraces and valley-floor segments, although it was extensively developed elsewhere (the 'Bodmin Moor Platform' of Green (1941)).

The final surface (690 to 550 ft (210–168 m) OD), widely developed in south-west Devon, was also present on Dartmoor, and was shown by bevelled spurs to the north of the moor and, even less reliably, by an accordance of summit heights to the south (Waters, 1960c). Waters argued that the highest upstanding 'residuals' (those above *c.* 1500 ft (457 m)) had survived as the most resistant elements of an extensive chemical etch plain. This subaerially formed and much dissected peneplain, bearing tors and rotted granite (growan), was believed to have been created over a protracted period through Miocene and even into Pliocene times. It was considered to form the basis of the gently sloping Dartmoor plain into which the later, successively lower, surfaces had been cut (Waters 1957, 1960c, 1960d; Brunsden *et al.*, 1964). The latter were also considered to be of late Tertiary age, having formed by a variety of subaerial and marine processes. Waters argued that only relatively minor modification to this basic landscape occurred during the Pleistocene, and although mass wasting 'exposed summit tors, moulded slopes and plastered valley floors with rubble-drift', no further base-levelled surfaces were created on the upland (Waters *in* Brunsden *et al.*, 1964).

The origin of these and comparable 'erosion surfaces' elsewhere in South-West England is discussed widely in the earlier geomorphological literature (e.g. Jukes-Browne, 1907; Barrow, 1908; Davis, 1909; Wooldridge and Linton, 1939; Green, 1941, 1949; Wooldridge, 1950, 1954; Balchin, 1952, 1964, 1981), and forms a protracted and important element in the development of geomorphological thought. These aspects are more fully considered in Chapter 3, and suffice to say here that the widespread occurrence of these many different 'planation' surfaces, either marine- or subaerially formed, is now disputed (e.g. Coque-Delhuille, 1982, 1987; Battiau-Queney, 1984, 1987).

Nonetheless, the belief in these erosion surfaces has formed the basis for many interpretations of the Dartmoor landscape, including models of drainage development as well as attempts to link the characteristic granite landforms into lengthy models of landscape evolution and denudation chronology (Waters, 1957; Brunsden *et al.*, 1964). Waters (1957), for example, related the pattern of Dartmoor rivers to the form of the upper erosion surface, characterized by a 'basin and tor' topography. He argued that the region's rivers were quite incapable of producing the basins in which they now lie, and that basins were therefore in existence before the drainage net. Waters suggested that differential chemical weathering of the granite, strongly influenced by variations in joint spacing, had resulted in the creation of basin forms where the weathered granite was most deeply developed (cf. Linton, 1955; see above; models of tor formation). Brunsden (*in* Brunsden *et al.*, 1964) suggested that the earliest drainage pattern on Dartmoor probably ran eastwards, and indeed that it had been a major agent in producing the summit plain or the highest erosion surface. Subsequent uplift and southward tilting of this surface may have led directly to the next phase of planation which created the middle erosion surface (*c.* 1300–1050 ft (396–320 m)), and to the initiation of dominantly north to south drainage lines.

Many studies of the rivers of Dartmoor and adjacent areas have been made, based on reconstructions of valley long-profiles and terrace gradients (e.g. Green, 1949; Waters, 1957; Kidson, 1962; Brunsden, 1963; Brunsden *et al.*, 1964). However, treating drainage evolution in the wider context of denudation chronology, and relating knick-points and various gradient curves to particular Tertiary and Pleistocene base levels, has involved a number of assumptions now believed to be false or, at least, highly dubious (Cullingford, 1982). A detailed morphometric analysis of all third-order drainage basins on Dartmoor has been conducted by Gerrard (1989b). This analysis demonstrated the essential uniformity of basin characteristics. Groupings of basins, obtained from a hierarchical cluster analysis, seem to be related to size and relative relief. However, groups combine at an early stage in the clustering process, indicating the integration and stability of the drainage net. The small variation in drainage density and the high correlation between the total stream length and area also suggest that the drainage networks are relatively stable. The grouping of the basins has added to the interpretation of long-term evolution based on remnants of erosion surfaces, river terraces and river long-profiles.

In more recent studies, greater emphasis has been placed on the role of jointing in the Dartmoor granite in influencing the pattern of streams (Blyth, 1962; Palmer and Neilson, 1962; Gregory, 1969; Gerrard, 1974, 1978). Gregory (1969) argued that the arrangement of valleys and interfluves shows a generally rectilinear pattern, reflecting very strongly the influence of jointing. This argument has been carried further by Gerrard (1974) who argued that the influence of jointing on the evolution of the Dartmoor landscape had been dominant. The horizontal joints or pseudo-bedding planes, which closely follow the contours of the land surface, evolved, Gerrard argued, as the land surface was denuded, thus reducing primary confining pressures in the granite. This resulted in a series of 'unloading' or dilatation domes, picked out clearly by the evolving drainage net. The dominant vertical joints, trending both north to south and east to west, had been the focus of subsequent weathering and were therefore instrumental in determining the local pattern of drainage (cf. Blyth, 1962; Palmer and Neilson, 1962; Gregory, 1969). Indeed, Gerrard (1974) argued that the tors crowning many such domes, had evolved as stream incision and erosion removed overburden, thereby releasing further compressive stresses in the granite. (Hawkes (1982) suggests that the granite was originally intruded beneath a cover of at least 1–3 km thickness.) This unloading in turn opened up additional joints to weathering processes, and continued a progressive cycle of landform development. Gerrard argued that the tors, located on the summits of these unloading

domes, therefore owed their distribution and formation, at least in part, to stream incision, although their form was also related to sub-surface chemical weathering and cryonival processes (Gerrard, 1974). In his synthesis, the horizontal and vertically developed joints were of the utmost significance, first in guiding the developing drainage net and thus delimiting the evolving unloading domes and, secondly, in governing the form and distribution of the tors (Gerrard, 1974).

The differential erosion of the Dartmoor granite massif was also stressed by Coque-Delhuille (1982). She argued that the original petrography of the granite had been relatively unimportant in subsequent landscape evolution and argued, like Gerrard, that the role of structure, particularly jointing, had been a prime determinant in the ensuing and selective pattern of erosion. This erosion occurred on what she regarded as only two erosion surfaces: the 'Dartmoor surface', derived from a post-Hercynian/pre-Permian surface, and a lower polygenetic 'Devon–Cornwall surface' which had evolved since Cretaceous times (Chapter 3). Some parts of the Dartmoor landscape have been more sensitive to change than others (Gerrard, 1991). Interfluves and plateaux have been affected by changes in weathering regime but have been essentially unaffected by hillslope changes initiated along river courses. The inner plateaux appear to have changed little in general form throughout the Quaternary and owe their extent to geological factors. Valley-side slopes have been more sensitive and soil and slope materials indicate the scale of landscape change.

MERRIVALE
S. Campbell

Highlights

Merrivale is one of the classic British localities for tors and associated periglacial landforms. It exhibits many of the most significant features of tor morphology, and demonstrates their relationships to bedrock lithology and structure. It demonstrates some of the most widely accepted evidence in Britain for cryoplanation.

Introduction

Merrivale GCR site encompasses the Cox Tor and Staple Tors area of Dartmoor, and provides one of

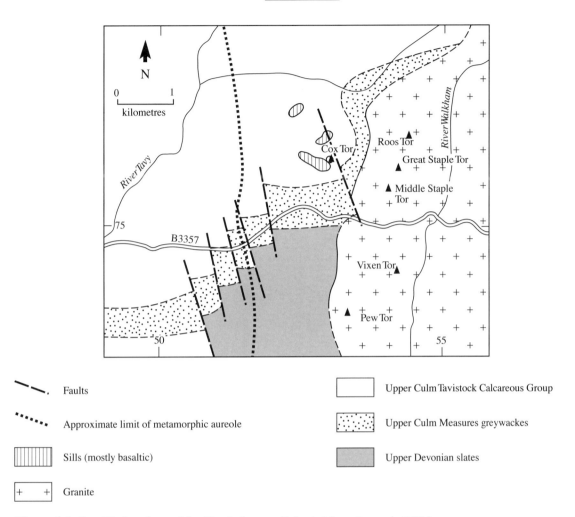

Figure 4.5 Simplified geology of the Merrivale area. (Adapted from Gerrard, 1983.)

Legend:

— ‑ ‑ . Faults

• • • • • • • Approximate limit of metamorphic aureole

[||||||] Sills (mostly basaltic)

[+   +] Granite

[   ] Upper Culm Tavistock Calcareous Group

[:::::] Upper Culm Measures greywackes

[▓▓] Upper Devonian slates

the most spectacular assemblages of granite land-forms anywhere in Britain. The site appeared in Linton's (1955) classic paper *The problem of tors*, while altiplanation terraces around Cox Tor were among the first examples to be described in the British landscape (Te Punga, 1956, 1957; Waters, 1962). Merrivale is also notable for one of the most extensive occurrences in southern Britain of fossil periglacial earth hummocks (Te Punga, 1956; Gerrard, 1983; Bennett *et al.*, 1996). The site has also featured in many other important geomorphological studies (e.g. Worth, 1930, 1967; Palmer and Neilson, 1962; Brunsden, 1964, 1968; Waters, 1964; Stephens, 1970a; Green and Eden, 1971; Green and Gerrard, 1977; Cullingford, 1982) . More recent accounts of the landforms were provided by Gerrard (1983, 1988, 1989a), Ballantyne and Harris (1994) and Harrison *et al.* (1996). The granite tors also possess excellent examples of rock (weathering) basins.

Description

The Merrivale area, encompassing Cox Tor (SX 530761), Staple Tors (SX 542760) and Roos Tor (SX 544765), is located on the western fringe of Dartmoor overlooking the Tamar Valley and adjacent lowlands which separate Dartmoor from Bodmin Moor. The principal landforms exhibited by the site are tors, blockfields and stone stripes (clitter), altiplanation terraces and earth hummocks. The site also shows sections through superficial deposits (head and weathered regolith). The structure and composition of the granite, which underlies part of the site, is well shown in Merrivale Quarry (SX 546753).

A. Geology

The local geology is shown in Figure 4.5 and controls, significantly, the distribution of landforms

Figure 4.6 Aerial photograph (scale *c.* 1:10 000) showing: (a) Cox Tor; (b) Roos Tor; (c) Great Staple Tor; (d) Middle Staple Tor; (e) Little Staple Tor; (f) Merrivale Quarry. Distinct 'boulder runs' of clitter are particularly evident around the Staple Tors. (Cambridge University Collection: copyright reserved.)

(Figures 4.6 and 4.7). The Staple Tors are developed in granite and are separated from Cox Tor by a col cut in metamorphic rocks of the aureole (mainly Devonian slates) (Figure 4.5; Reid *et al.*, 1912; Dearman and Butcher, 1959; Gerrard, 1983). Cox Tor is itself formed of diabase (metadolerite) and is surrounded by a belt of calcareous hornfels, bordered by non-calcareous Culm Measures of Carboniferous age (Figure 4.5). The diabase of Cox Tor is more densely jointed than the granite of the Staple Tors and Roos Tor. The latter consists predominantly of medium- to coarse-grained granite but with common veins, inclusions of tourmaline and dykes of finer-grained granite. A large resistant

vein of microgranite (aplite) runs through Roos Tor (Dearman and Butcher, 1959; Gerrard, 1983).

The coarse-grained grey granite, which comprises most of the tors in the Staple Tors complex, is well exposed in fresh faces in the adjacent Merrivale Quarry (Dearman and Baynes, 1978). Near the surface, however, joint surfaces show pitting caused by the decomposition of plagioclase feldspar. Widely spread (3–7 m), orthogonal planar joints occur with granulated selvages 50–100 mm wide, composed of fresh, mechanically disintegrated interlocking granite gravel. Some joints are stained red. The origin of the red staining is believed to be hydrothermal, although near the ground surface the joints have also been affected by chemical weathering. Some joints were formed by stress release (Dearman and Baynes, 1978).

B. Landforms

Tors

The principal tors of this site are Roos Tor, Great Staple Tor, Middle Staple Tor and Cox Tor (Figure 4.6). These tors occupy summits at elevations above 430 m OD: the smaller and lower-lying feature of Little Staple Tor (*c.* 380 m OD) occurs on an interfluve. Great Staple Tor is of the 'avenue' type with a missing central portion (Green and Gerrard, 1977). It consists mostly of massive stacks of joint-bounded blocks. Sheet joints (pseudo-bedding planes) in the granite of this tor dip downslope at an angle slightly less than that of the local ground surface (Green and Gerrard, 1977).

Blockfields, boulder runs and boulder stripes (clitter)

Clitter is present on most of the hillslopes surrounding the principal tors at Merrivale (Te Punga, 1956; Gregory, 1969; Green and Gerrard, 1977; Gerrard, 1988; Bennett *et al.*, 1996), particularly on the western slopes of the granite which forms the Staple Tors complex (Figure 4.6). Many of the north to south-trending valleys exhibit marked cross-valley asymmetry, with west-facing slopes possessing gentler angles. Clitter is often more prominent on these western slopes. Although formerly thought to be randomly distributed, the blocks and boulders (clitter) mantling the local slopes show distinct organization (Te Punga, 1957; Gerrard, 1988; Figure 4.6). First, blockfields are common, particularly towards the base of the main slopes leading from the tors. This detritus is composed of boulders which exhibit considerable variation in long-axis orientation and which are often inclined at steep angles (Gerrard, 1988).

Second, the clitter is arranged, particularly on the western slopes, into stripes, runs and other patterns (Te Punga, 1957; Green and Gerrard, 1977; Gerrard, 1983, 1988; Figure 4.6). A vast variety of these forms exists around the Staple Tors, including narrow stone stripes, boulder runs and garlands. Narrow stripes (up to 3 m wide) start and finish in mid-slope positions. In places, they coalesce to form wider runs; elsewhere they diverge or coalesce apparently at random (Gerrard, 1983). However, according to Gerrard (1983), there is an order to the arrangement of the individual blocks. Boulders in the centre of runs often stand on-end

or with their long-axes pointing downslope. In some stripes, smaller boulders rest against or override larger ones. Some small stripes show a central hollow (Gerrard, 1983). The stripes are present on slopes as gentle as 6° (Green and Gerrard, 1977).

It is significant that many runs and garlands lead directly from the base of tors. On the other hand, stripes often start mid-slope at a great distance from the nearest outcrop. There is little differentiation in the size of material found in the various boulder structures (Gerrard, 1983).

Altiplanation terraces

The flanks of Cox Tor consist of well-defined rock-cut terraces or 'benched hillslopes'. These are particularly well developed on the north and south slopes where the features are most frequent but small; on the eastern slopes they are more regular and extensive and to the west it is not clear whether the breaks of slope are terrace forms or are the result of clitter accumulations (Te Punga, 1956, 1957; Waters, 1962; Green and Gerrard, 1977; Gerrard, 1983, 1988) (Figure 4.7). These terraces were first described by Te Punga (1956, 1957), and were re-mapped in detail by Gerrard (1983). The highest are cut in the diabase of Cox Tor itself, the lower ones in hornfels and Culm Measures. The inclination of the 'treads' varies from 3–9°; the 'risers' from 11–20° (Green and Gerrard, 1977; Gerrard, 1983, 1988). Towards Cox Tor, the risers become small vertical rock cliffs cut in the diabase. Tread widths range from 13–65 m and the features can be as much as 800 m long (Figure 4.7). Local exposures confirm that the terraces are cut in rock: only a very thin veneer of debris, rarely thicker than 1 m, rests upon the treads (Te Punga, 1956; Gerrard, 1983).

Earth hummocks

The Cox Tor area contains one of the most extensive fields of earth hummocks in Britain. Thousands of small sub-hemispherical mounds, up to 2 m in diameter and 0.25 m high, occur on the terraced and adjacent slopes around Cox Tor (Te Punga, 1956; Green and Gerrard, 1977; Gerrard, 1983, 1988; Bennett *et al.*, 1996). They are best developed on the eastern side of the tor and are confined to soils developed on the metamorphic aureole. These mounds form a polygonal network but vary considerably in size and shape: some are elongated downslope, others show degraded profiles on their downslope side (Gerrard, 1983; Bennett *et al.*, 1996). Bennett *et al.* (1996) have classified the hummocks as: 1. 'single hummocks'

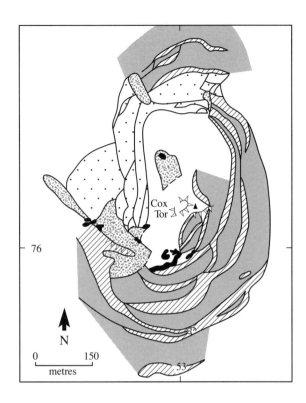

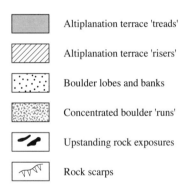

	Altiplanation terrace 'treads'
	Altiplanation terrace 'risers'
	Boulder lobes and banks
	Concentrated boulder 'runs'
	Upstanding rock exposures
	Rock scarps

Figure 4.7 The geomorphology of Cox Tor and adjacent areas. (Adapted from Gerrard, 1983.)

Weathering products

Two roadside exposures through slope materials occur on the lower slopes of Cox and Staple Tors, respectively (Gerrard, 1983). The exposure adjacent to Staple Tors shows up to 0.4 m of head comprising small weathered granite cobbles and more angular schorl underlain by a light-brown sandy gravel (lower head) with some very large boulders. A similar pattern, that is a coarse head overlain by a finer head, is shown by the Cox Tor exposure, although the materials are different. Here, the finer head exhibits a characteristic downslope orientation of particles (Gerrard, 1983; cf. Figure 4.3).

Interpretation

It has long been accepted that Dartmoor was never glaciated, but that it lay in the periglacial zone beyond the maximum limit of the ice sheets, and was subjected to cryonival processes (e.g. Waters, 1964; Gerrard, 1983). Most workers have therefore argued that significant landform elements and the superficial deposits at Merrivale, and elsewhere on Dartmoor, are primarily the result of periglacial, frost-assisted processes. Although the nature and origin of the decomposed granite (growan) are key factors in understanding the evolution of these landforms and deposits, and have been much disputed (see 'Kaolinization or Tertiary chemical weathering?'; this chapter), it is generally assumed that the tors were exhumed from beneath a weathered or 'softened' mantle mainly by Pleistocene periglacial activity. The close association between the tors, their clitter (blockfields and stone runs), local rock terraces and earth hummocks has been used as evidence that the features were all formed at a comparatively late stage of the Pleistocene by periglacial weathering (Waters, 1964, 1974; Green and Gerrard, 1977; Gerrard, 1983).

Tors

Both of the main models of tor formation explain the 'exhumation' of tors by periglacial stripping. Linton (1955; p. 476) defined a tor as ' … a residual mass of bedrock produced below the surface level by a phase of profound rock rotting effected by groundwater and guided by joint systems, followed by a phase of mechanical stripping of the incoherent products of chemical action … '. This two-stage mechanism first involved deep weathering under warm humid conditions (probably during the

(of various shapes, but mostly kite-shaped); 2. 'compound hummocks' (multi-peaked with variable morphology); and 3. 'rock-cored hummocks' (prevalent where clitter reaches the ground surface). The vast majority seem to be composed of a well-drained, fibrous brown loam derived from weathering of the diabase, although some are underlain by large boulders (Green and Gerrard, 1977; Bennett *et al.*, 1996). Whether boulder-cored or not, all exhibit considerable sorting of the soil (Gerrard, 1983).

Palaeogene) when a thick regolith developed with corestones occurring where the joint planes were most widely spaced. Second, the products of weathering (the regolith) were removed by mass wasting, leaving the corestones as upstanding tors. Linton proposed that the tors were exhumed under periglacial conditions in the Pleistocene when solifluction and meltwater would have been efficient agents in removing the regolith (Figure 4.2).

Alternatively, Palmer and Radley (1961) and Palmer and Neilson (1962) suggested that tors, such as those in the Pennines and on Dartmoor, had formed solely as a result of mechanical weathering under periglacial conditions. Indeed, Palmer and Neilson (1962) doubted the former existence of deep weathering at tor sites, and attributed the known occurrences of deep decomposition (e.g. Two Bridges Quarry) to the effects of pneumatolysis. Waters (1974) went further and suggested that the chronology of tor formation in the Dartmoor area could be related directly to local head sequences. In identifying two head facies (an upper coarse-grained head and a lower finer-grained head), he argued that the principal phase of tor exhumation had occurred during the Wolstonian (Saalian Stage) when a regolith of fine-grained material was stripped from around the tors and redeposited in valley-side and valley-bottom locations as the 'Main Head' (Figure 4.3). A second phase of periglacial activity, during the Devensian Stage, produced the 'Upper Head', much of the clitter now surrounding the tors and reducing the mass of the tors still further. All of these theories, however, involve stripping of weathering or alteration products from around the tors by periglacial mass wasting processes.

Gerrard (1983, 1988) argued that tors, such as those at Merrivale, formed where weathered material (whatever its origin) was removed faster than it was produced. Thus, the tors are found in 'high energy' locations such as steep valley-side slopes, at breaks of slope and on summits where the processes of removal are most efficient (Gerrard, 1983, 1988). He argued that a simple explanation, based on weathering followed by removal, is not adequate to explain the major elements of a granite landscape with tors, and that several possible relationships exist between the rates of accumulation and removal of material. For example, where the ground surface is stable, deepening of the regolith may occur. Where the ground surface is unstable, a gradual removal of regolith is likely. Alternatively, local conditions may promote the renewal and removal of material at similar rates. Thus, these

conditions may exist concurrently at different sites and vary in significance through time.

Such thinking is mirrored by Battiau-Queney's (1984, 1987) work in Wales, particularly on tor landscapes in the Preseli and Trefgarn areas (Campbell and Bowen, 1989). Like Linton, she argued that the tors had formed in response to two main factors. First, evidence, particularly from Trefgarn, showed that deep chemical weathering of the land surface had occurred in a hot humid environment (probably during the Palaeogene). Secondly, this weathered mantle had been stripped, but not solely by periglacial processes in the Pleistocene. Rather, the exhumation of the more resistant tors had occurred as the result of protracted uplift along old structural axes throughout the Cenozoic, and not simply because of changing climatic and environmental conditions. Battiau-Queney therefore suggested that the tors were formed in response to slow uplift where sub-aerial denudation had exceeded (perhaps only locally) the rate of chemical weathering. Consequently, a sharp deterioration of climate was not required to trigger stripping of the weathered regolith. Instead, a closely balanced relationship between persisting local uplift and erosion offered the most conducive conditions for tor formation (Battiau-Queney, 1984, 1987).

Green and Gerrard (1977) suggested that the pattern of vertical and horizontal jointing in the granite of Staple Tors (particularly Great Staple Tor) suggests that some form of 'unloading' mechanism has operated in the past, along a broadly whale-backed ridge of the granite. They argued that the centre of the ridge was probably its weakest part, accounting for the missing portion of the tor.

Clitter

It has been suggested that the Dartmoor clitter (blockfields, stone runs and stripes etc.) was derived from the periglacial demolition of the tors (e.g. Waters, 1964, 1974), and it has long been recognized that frost heaving and thrusting, as well as differential movement downslope, have been involved in its formation. However, the exact origin of the clitter is poorly understood (Gerrard, 1983).

The traditional view that tors and clitter are closely associated is partly borne out by the stone runs and garlands at Merrivale which lead directly from the base of some tors (Gerrard, 1983). However, other boulder accumulations appear to have originated *in situ*. For example, some of the

Figure 4.8 Great Staple Tor seen from Middle Staple Tor. The missing central portion or 'avenue' of Great Staple Tor can be seen clearly on the horizon. (Photo: S. Campbell.)

Figure 4.9 Looking west through the 'avenue' of Great Staple Tor. (Photo: S. Campbell.)

Figure 4.10 Great Staple Tor seen from Cox Tor, revealing a diverging anastomosing pattern of boulder runs on the west-facing slopes. (Photo: S. Campbell.)

stone stripes start mid-slope at a great distance from the nearest tor or outcrop. Similarly, many of the blockfields occur towards the base of slopes, suggesting that much of the clitter has not travelled very far, and has therefore originated *in situ* (Green and Eden, 1973; Green and Gerrard, 1977; Gerrard, 1983). The possibility that some of this clitter might resemble protalus ramparts or even rock glaciers must not be discounted (Harrison *et al.*, 1996).

Gerrard (1983) suggested that the size distribution of the boulders was clearly a function of the intensity of jointing in the granite, and therefore that the difference in pattern and arrangement of the boulders was likely to be the result of the relative abundance of blocks: where blocks occur in profusion, blockfields, boulder garlands, lobes and runs might be expected, whereas where fewer blocks are available, stone stripes may be the dominant feature.

Further support for the limited transport of blockfield material comes from the patterns of the constituent boulder long-axes. In the blockfields these show considerable variation, with many boulders being inclined at steep angles often into the slope. In contrast, boulders in the stripes show a dominant long-axis orientation parallel with the steepest local slopes, suggesting greater downslope movement and sorting (Gerrard, 1983). Gerrard tentatively suggested that slope angle was probably not a controlling factor in the distribution of the various boulder (clitter) patterns, since all types occur on slopes of similar angles. Recently, Bennett *et al.* (1996) have re-examined the boulder runs on the west-facing slopes beneath Great Staple and Middle Staple tors (Figure 4.10). Controversially, they have concluded that the runs originated from the erosion of soil in lines, perhaps by ancient springs, to reveal the clitter beneath. In this model, the boulder runs are regarded simply as a function of discontinuous soil cover.

Altiplanation terraces

The benched hillslopes or rock terraces described around Cox Tor (Te Punga, 1956, 1957; Waters, 1962; Brunsden, 1968; Green and Gerrard, 1977; Gerrard, 1983, 1988; Ballantyne and Harris, 1994) closely resemble altiplanation terraces described elsewhere in the world which have been attributed to periglacial conditions (e.g. Demek, 1968; Czudek and Demek, 1970). Altiplanation terraces in the

British landscape were first recognized by Guilcher (1950) who described several examples around the coasts of north Devon. The examples flanking Cox Tor were described in detail by Te Punga (1956) who argued that they closely resembled features currently forming in perennially frozen ground in Alaska (Eakin, 1916; Lewis, 1939). Te Punga suggested that the terraces had formed where snow patches, with their major axes lying transverse to the local pattern of drainage, had eroded backwards and downwards into the hillside by frost-action, leaving a series of pronounced 'treads' and 'risers' (Te Punga, 1956). There could be no doubt, he argued, that the formation of these terraces around Cox Tor (and elsewhere in southern England) was closely associated with perennially frozen ground (permafrost), developed under Pleistocene periglacial conditions. Comparable features were later recognized elsewhere on Dartmoor (Waters, 1962).

The lithological characteristics of the local bedrock were seen as important in the development of the features, particularly for maintaining a sharp shoulder at the margin of the terraces (Te Punga, 1956). Indeed, rock control appears to be crucial, since terraces such as those at Cox Tor and elsewhere on Dartmoor are always found on metamorphic rocks and not on the granite (Green and Gerrard, 1977; Gerrard, 1983): closely spaced joints, and the cleavage and bedding planes found in the former rocks, may have facilitated more effective frost-action. Similarly, Te Punga (1956) noted that variations in hardness of subhorizontally stratified rocks also favoured the development of altiplanation terraces: the features, however, are not restricted to such conditions for they commonly bevel steeply inclined strata and may also be cut in massive homogenous bedrock (Te Punga, 1956).

Te Punga also recorded that the surfaces of the altiplanation terraces in north Devon were closely associated with the occurrence of stone polygons. Indeed, stone polygons were so common on such terraces and adjacent summits elsewhere, that ' … it may be desirable to regard all areas covered by stone polygons, formed in material derived by periglacial robbing of the underlying bedrock, as altiplanation features' (Te Punga, 1956; p. 337). It was not made clear, however, whether the low earth mounds found on the terraces and flanks of Cox Tor were generically similar (see below).

Earth hummocks

Sharp (1942) first used the term 'earth hummocks' to describe patterned ground characterized by dome-shaped, apparently non-sorted nets or circles. Such features have a circumpolar distribution, and are found in northern Europe, Siberia, Greenland, Iceland and North America (Gerrard, 1983). Comparable features have been described on Ben Wyvis in northern Scotland (Ballantyne, 1986), in the Pennines (Tufnell, 1975) and Cumbria (Pemberton, 1980), and are broadly analogous to the 'thufur' of tundra areas (Ballantyne and Harris, 1994).

Te Punga (1956) likened the small, roughly circular mounds around Cox Tor (Figure 4.11) to the 'high-centred polygons' of high latitudes, and suggested that they had also been formed under periglacial conditions. Gerrard (1983, 1988) argued that there is no evidence to suggest that the features at Cox Tor are forming at the present time, and some are indeed being destroyed. Many of the mounds occur on top of clitter and only formed once the clitter had become stabilized, presumably after the Younger Dryas. Some have been removed by Bronze Age humans during the construction of circular huts in the area. It seems likely, therefore, that the mounds formed at some time between *c.* 9000 and 2000 BP (Gerrard, 1983), although Brunsden (1968) considered that they might be even younger, having formed as the result of spring action dissecting the thick soil cover.

There is considerable doubt, however, as to the mode of formation of the earth hummocks. Although two principal types occur around Cox Tor (boulder-cored and non-boulder-cored, with the latter being dominant), all demonstrate considerable sorting of the soil: the soil of the mounds is consistently finer than that at the same depth in the depressions alongside, and it is thought that frost-heaving is the primary process in their formation (Gerrard, 1983, 1988). Beskow (1935) has shown that frost-heaving is unlikely to occur in soils with less than 30% silt and clay (cf. Williams, 1957; Corte, 1963). Gerrard (1983, 1988) has argued that such requirements may have controlled the distribution of the earth hummocks at Merrivale, restricting the features to the fine silty loam soils derived from the weathering of the Cox Tor diabase, and excluding their development on adjacent granitic soils which are generally deficient in the silt and clay grades. Recent work by Bennett *et al.* (1996) has confirmed this fundamental relationship between soil type and hummock distribution. It is interesting that such hummocks occur all along the western edge of Dartmoor where similar rock types exist.

Most authors seem to agree that earth hum-

Figure 4.11 A profusion of earth hummocks on the east-facing slopes of Cox Tor, with Great Staple Tor and Middle Staple Tor on the horizon. (Photo: S. Campbell.)

mocks are formed by frost-heaving, caused by uneven ground freezing and thawing, although the specific mechanisms involved are disputed (Ballantyne and Harris, 1994). The main problem in understanding the genesis of these small-scale landforms, however, is establishing how the initial micro-relief forms. Once the mounds are created, their vegetation cover may afford better insulation than the intervening water-soaked areas where freezing is likely to occur first: this differential freezing may set up pressures forcing material inward and upward into the hummocks, causing them to 'grow' (Beskow, 1935; Gerrard, 1983, 1988; Bennett *et al.*, 1996). Although a variety of

mechanisms has been suggested which would produce the initial micro-relief, for example, hillwash, soil movement, wind deposition and differential vegetation growth in clumps (Gerrard, 1983; Bennett *et al.*, 1996), the exact mode of formation of the mounds is largely conjectural (Green and Gerrard, 1977; Ballantyne and Harris, 1994; Bennett *et al.*, 1996): all that can be said with any certainty is that the mounds are fossil features which appear unrelated to present conditions (Gerrard, 1983, 1988). Bennett *et al.* (1996) provide a comprehensive review of the earth hummocks at Merrivale, and discuss possible modes of formation.

Weathering products and slopes

The relationships between weathering products, solifluction deposits and resultant slope forms in the Merrivale area are complex. It has been suggested that two different head deposits occur in the Dartmoor area, namely a lower fine-grained deposit and an overlying coarse-grained sediment – both periglacial solifluction deposits (e.g. Waters, 1964, 1974; Mottershead, 1976). Such a sequence of head deposits has been taken to reflect an inversion of a normal granite weathering profile, that is, with finer-grained weathering products removed from upper slopes during an initial periglacial phase, and subsequently redeposited on lower slopes and in valley bottoms; and with a subsequent phase of periglacial activity removing large, sound blocks from tors and other surface exposures, and depositing them on top of the finer-grained head materials (Figure 4.3). Green and Eden (1973) challenged this view and demonstrated that much of the coarse debris actually occurs in the lower parts of the head sequences, and has been derived from proximal basal sources and not from more elevated and distant rock outcrops. It is likely, therefore, that the head deposits are derived from many parts of the slope and that an inversion of weathering profiles will only have occurred in localized situations (Green and Gerrard, 1977; Gerrard, 1983).

Gerrard (1982, 1983, 1989a) has presented detailed evidence from head sequences throughout Dartmoor to show that initial weathering profiles were probably more complex than hitherto thought, and that periglacial processes led to a substantial mixing of materials, rather than to simple re-sorting. Indeed, in 47 exposures across Dartmoor, Gerrard (1982, 1989a) has shown that the relationship between the coarse- and fine-grained facies is complex, with both layers often being intermixed and with large blocks of granite occurring throughout the beds. He argued that significant mixing of materials has therefore occurred, with gullies having been cut and infilled at different times, showing complicated sequences of slope modification rather than simple reworking of weathered granite. The Merrivale exposures may therefore be somewhat atypical in showing 'coarse' head overlain by 'fine' head, and exposures elsewhere show that the distribution and characteristics of solifluction deposits are far more complex.

In conservation terms, Merrivale provides an outstanding assemblage of the landforms characteristically developed on the granite intrusions and adjacent aureole rocks of upland South-West England. While the broad configuration of the area, with its large dome-shaped ridge of granite bordered by valleys, had probably been established by the mid-Tertiary, smaller details of the local slopes, altiplanation terraces, tors, clitter and slope deposits reflect the operation of periglacial processes throughout the Pleistocene. The Pleistocene periglacial legacy to the Dartmoor landscape has been substantial and Merrivale illustrates many of the key features of its periglacial geomorphology. The tors and clitter are some of the very finest examples anywhere in Britain: the Staple Tors and Roos Tor demonstrate, particularly clearly, the relationship between bedrock lithology and structure and tor morphology and development. The clitter shows a remarkable variety of forms (blockfields, stone stripes, stone runs and garlands), and its distribution and characteristics have a considerable bearing on processes of slope development, and on the origin of the tors themselves.

The Cox Tor area of Merrivale shows some of the finest altiplanation terraces in southern Britain and some of the most convincing evidence for cryoplanation: the Cox Tor examples are also of historical significance since they were the subject of the first detailed exposition of altiplanation terraces in the British landscape (Te Punga, 1956). The superb and profusely developed earth hummocks in this area also add significantly to its scientific interest, although their precise age and origin are still far from clear.

While the individual landforms are exceptional and worthy of special note, the landform assemblage as a whole is probably unparalleled elsewhere in Britain. In particular, Merrivale provides an outstanding assemblage of interrelated landforms which illustrate many of the most significant theories of long-term and, especially, periglacial landscape evolution in southern Britain.

Conclusion

Merrivale is one of Britain's most important sites for understanding the development of granite landscapes. When the Dartmoor granite was intruded into the surrounding 'country' rocks about 290 million years ago, it altered them profoundly, and the wide range of landforms now seen at Merrivale in part reflects differences in rock type and local geological structure: the Staple and Roos tors are granite landforms of textbook quality, showing clearly how tors have developed in relation to the pattern and density of jointing in the host rock. In contrast, the more angular outlines of Cox Tor reflect the nature of its constituent rocks, in particular the markedly different reaction of the diabase to protracted weathering. Cox Tor is surrounded by arguably the finest 'altiplanation terraces' in Britain, a 'staircase' of horizontal benches cut back into the rock by freeze-thaw processes at a time when perennial snow patches were present on Dartmoor. The siltier soils of the Cox Tor area have given rise to one of the best British examples of 'earth hummocks', controversial landforms of disputed age and origin, although almost certainly formed by frost-assisted processes in the Late Devensian. Many of the landforms at Merrivale – the large accumulations of loose rock or 'clitter', the altiplanation terraces and earth hummocks – owe their origin to a range of cold-climate processes which operated during the Quaternary. Frost-shattering of local rocks, the repeated contraction and heaving of the ground surface as it alternately froze and thawed, and the downslope movement (solifluction) of weathered materials over frozen ground (permafrost) all gave rise to characteristic landforms now seen in fossil form today. Merrivale is also important because some of the landforms, such as tors, may have begun to form well before the Pleistocene ice ages of the last two million years or so – perhaps as far back as the early and mid-Tertiary, when subtropical or even tropical conditions may have prevailed. Merrivale is unique in showing such a wide range of landforms in a small area, and provides important evidence for understanding how landscapes evolve over timescales of many millions of years.

BELLEVER QUARRY
S. Campbell

Highlights

A reference site demonstrating the relationship between slope deposits and granite weathering products on Dartmoor, Bellever Quarry provides particularly detailed evidence for the origin of 'bedded growan' deposits.

Introduction

Bellever Quarry is of considerable geomorphological interest for its related assemblage of periglacial slope deposits and granite weathering features, which are considered typical of many Dartmoor slopes. The generic relationship between the intact, relatively unaltered, granite here and the overlying weathered granite (growan), 'bedded growan' and periglacial head deposits is graphically illustrated, and has long attracted scientific interest (Waters, 1961; Brunsden, 1964, 1968). A detailed description and interpretation of the sequence at Bellever Quarry by Green and Eden (1973) has major implications for theories of slope development throughout the region (Te Punga, 1957; Waters, 1964; Mottershead, 1971; Green and Gerrard, 1977; Cullingford, 1982; Cresswell, 1983; Gerrard, 1983, 1989a).

Description

Bellever Quarry (SX 658763), sometimes known as Lakemoor or Laughter Quarry, lies on the lower slopes of the East Dart Valley in the Bellever Plantation, approximately midway between Riddon Ridge and Laughter Tor. It exposes granite, weathered granite and overlying slope deposits in one main face and several other less extensive and more overgrown faces. The stratigraphic sequence is as follows:

4. 'Head' consisting of granite clasts set in a coarse matrix of growan (up to 1.5 m)
3. Disturbed weathered granite ('bedded growan') (up to 1.0 m)
2. *In situ*, undisturbed, weathered granite or growan (> 2.0 m)
1. Intact, relatively unaltered granite

The head (bed 4) shows a concentration of clasts in its lower layers, but there is no indication that the bed should be divided on this basis (Green and Eden, 1973). The disturbed weathered granite (bed 3) shows colour bands which are conspicuously overturned in a downslope direction (Figure 4.12). The apparent layering is related to colour as well as textural variations, and can be traced into both beds 1 and 2. It is believed to be related to a zone or vein of tourmalinization (Green and Eden, 1973; Green and Gerrard, 1977). The amount of lateral displacement of individual layers relative to one another is only a few millimetres at *c.* 2 m depth below the top of the bedded growan (bed 3), but increases upwards in the profile (Green and Eden, 1973). The *in situ*, undisturbed, weathered granite or growan (bed 2) both overlies intact granite (bed

Figure 4.12 Granite alteration products and slope deposits at Bellever Quarry being examined during the 1977 INQUA trip to the South-West. (Photo: N. Stephens.)

1) and is juxtaposed between bosses or stacks of the relatively sound rock (cf. Two Bridges Quarry).

Interpretation

The first detailed work at this site is attributable to Brunsden (1968), who used evidence from various sites, including Bellever Quarry, to devise a classification of weathering zones found in the Dartmoor granite (see Kaolinization or Tertiary chemical weathering?; Two Bridges Quarry). The Bellever Quarry section showed most of the different weathering types proposed: from relatively intact granite with corestones, showing only partial decomposition along joints; through well-rotted, incoherent granite still showing details of the original rock structure; to undisturbed, stained, weathered granite containing much quartz, but displaying no detail of any previous structure. In addition, the sections showed evidence for soil creep (bed 3; the 'bedded growan' of later workers) as well as a capping layer of head, Brunsden's 'migratory layer'. Sites like Bellever Quarry were used by Brunsden to demonstrate that pneumatolytic alteration, deep chemical weathering and

physical, frost-assisted processes had all affected the Dartmoor granite, and that evidence for all three could be found in individual profiles.

Although disturbed weathered granite (bedded growan) is not always present between the *in situ* weathered granite and the overlying head on Dartmoor, it is common on many slopes and some interfluves (Green and Eden, 1973). Waters (1964) attributed its bedded appearance to downslope wash or creep, hence the term 'bedded growan': he appears to have favoured surface wash as the most likely agent in its formation (Waters, 1971; Green and Eden, 1973).

Green and Eden (1973) re-examined the bedded growan from a number of sections on Dartmoor, including those at Bellever Quarry. They showed that, in composition, the material is similar to the underlying undisturbed and *in situ* growan, despite the frequent colour banding and appearance of downslope bedding. They concluded that its general characteristics are not consistent with an origin as a surface-wash deposit. In studying the relationships of local head sequences and clitter patterns to the bedded and *in situ* growan deposits, Green and Eden surmised that the movement of the bedded growan had been contemporaneous with that of the overlying periglacial head: namely that the bedded growan was not a pre-existing surface deposit derived from upper slope source areas. In support of their claim they cited evidence for superficial layers having passed over and displaced underlying *in situ* material elsewhere (e.g. Penck, 1953; Fitzpatrick, 1963; Jahn, 1969).

The relationship between the periglacial slope deposits and the weathered granite established by Green and Eden (1973) at Bellever Quarry and other sites is significant. Formerly, slope deposits on Dartmoor were considered to have developed under periglacial climatic conditions, when successive layers of the pre-existing weathered profile were removed from the upper parts of slopes and deposited in the reverse order on lower slopes – the 'inversion' theory of Waters (1964) (Figure 4.3). Such a mechanism also found favour with other workers (e.g. Te Punga, 1957; Brunsden, 1964, 1968), and was the basis for a simple model of slope development. It was noted that

'The rapid wasting away of the land surface during periglaciation, due to the transportation of enormous quantities of material to lower levels, has produced a landscape of subdued aspect characterized by slopes that are convex near the top and concave near the bottom. The convex upper slope has been a zone of wastage and the concave lower slope has been a zone of deposition.' (Tricart, 1951; p. 196; Te Punga, 1957; p. 410).

The evidence presented by Green and Eden, derived substantially from Bellever Quarry, is vital for demonstrating that this slope transfer and inversion model is not generally applicable. These workers have argued that such inversions are rarely the norm, and that the processes responsible for the formation of slope deposits have included the erosion of substantial amounts of underlying material (ranging from large granite blocks to fine-grained growan) and its incorporation into the transported layer on all parts of the slope. Even on lower slopes, as at Bellever Quarry, the movement of slope deposits (head) appears to have been accompanied by the erosion of the underlying weathered granite.

On the basis of this evidence, it was possible to dispel two long-held notions: first, that slopes could be divided into simple 'source' and 'accumulation' areas; second, that there had been a widespread inversion of the pre-existing weathering profile (Green and Eden, 1973; Green and Gerrard, 1977). Waters' (1964) suggestion that transfers of material downslope during successive cold phases had led to a typical three-fold succession overlying the weathered granite, namely the bedded growan, the 'main head' and the 'upper head' (Figure 4.3), was therefore shown to be untenable on Dartmoor, and any widespread evidence for two separate layers of head was refuted (Green and Eden, 1973). A comparable study by Mottershead (1971), on head deposits overlying schist in south Devon, similarly failed to confirm the applicability of Waters' model. Recent studies by Gerrard (1982, 1983, 1989a) have further highlighted the complex relationships between the types of slope deposit and weathering products found in the region. He confirmed that considerable mixing of materials had taken place, with gullies being cut and infilled at various times, and he showed that complicated sequences of slope modification rather than a simple reworking of weathered granite had occurred.

Conclusion

Bellever Quarry provides particularly strong evidence to demonstrate that the bedded growan of Dartmoor did not accumulate at the ground surface, but instead formed beneath periglacial head deposits while they accumulated. Such evidence has profound implications for models of slope development in the region. Both Two Bridges and Bellever quarries provide significant evidence for the numerous arguments regarding the origin of the decomposed granite on Dartmoor, and its relevance to landscape evolution, and particularly tor formation. Whereas Two Bridges Quarry has become almost pre-eminent in such debates, Bellever Quarry provides complementary evidence, and is fundamental to understanding mechanisms of slope development, and especially the relationships of slope deposits (principally periglacial head) to the underlying granite weathering/alteration products.

TWO BRIDGES QUARRY
S. Campbell

Highlights

Two Bridges Quarry provides an excellent example of the 'decomposed' Dartmoor granite, and is one of the best sites in Britain for understanding the formation of tors.

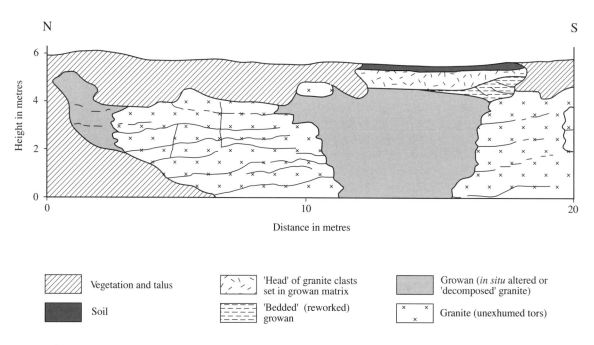

N   S

Figure 4.13 A cross-section through the granite and associated alteration products at Two Bridges Quarry, Dartmoor, adapted from Campbell (1991).

Introduction

Two Bridges Quarry is one of the most important geomorphological sites in South-West England, and is particularly noted for its association with D.L. Linton's classic theory of tor formation (Linton, 1955). The site shows heavily decomposed granite juxtaposed with masses of harder, less altered granite, and was used as a field model by Linton to illustrate the first stage of tor formation by differential weathering. The site has also featured widely in subsequent studies supporting, challenging or modifying Linton's model (e.g. Palmer and Neilson, 1962; Brunsden, 1964, 1968; Gregory, 1969; Green and Eden, 1971, 1973; Eden and Green, 1971; Green and Gerrard, 1977), and the roles of pneumatolysis (alteration of the granite by mineralizing fluids from deep within the Earth), subsurface chemical weathering and physical weathering by frost-action, in the formation of the altered granite (growan) have all been examined using evidence from this site (e.g. Doornkamp, 1974; Dearman and Baynes, 1978; Cullingford, 1982; Durrance and Laming, 1982). A summary of the site's scientific importance and conservation value was given by Campbell (1991).

Description

Two Bridges Quarry (SX 609751) lies near Princetown on Dartmoor, and *c.* 6 km east of Merrivale. The floor of this small disused quarry is now occupied by a car park, but the curved working face (*c.* 20 m long by 6 m deep) still provides an excellent section through the granite and associated weathering products (Figure 4.13). Both ends of the exposure are occupied by relatively solid and intact masses of grey, coarse-grained granite, including intrusions of fine-grained aplite (Green and Gerrard, 1977; Dearman and Baynes, 1978). The largest of these masses (to the north) measures some 6 m wide by 4 m deep: it shows both vertical joints (at *c.* 1–2 m intervals) as well as subhorizontal sheet jointing or 'pseudo-bedding' (at *c.* 0.5 m intervals).

Growan or decomposed granite, up to 6 m in thickness, occurs between the more coherent granite masses (Figures 4.13 and 4.14). Thin quartz-tourmaline veins are present in this material: the structure of the original granite is undisturbed and the primary rock-forming minerals are readily identifiable (Green and Gerrard, 1977). This pattern, however, breaks down within *c.* 1 m of the ground surface. Here, the growan is

Figure 4.14 The section at Two Bridges Quarry, showing large intact granite masses, the unexhumed tors and adjacent deeply altered granite. (Photo: S. Campbell.)

commonly mixed with large angular granite clasts (head) and, in one or two places, the growan shows colour banding which indicates downslope bedding.

Interpretation

Linton (1955) regarded the growan at this pit as the product of subaerial chemical weathering, and the masses of more solid rock as unexhumed tors. Indeed, he considered that the field evidence from Two Bridges Quarry showed that the Dartmoor tors, in general, had formed in response to differential sub-surface weathering, with the decomposed granite (growan) having formed as the result of deep weathering by percolating groundwaters under warm conditions in the late Tertiary (the Neogene). This, he argued, developed a thick regolith with corestones occurring only where joint planes were widely spaced and where the granite was least susceptible to chemical alteration (Figure 4.2). The products of this weathering (the regolith or growan) were subsequently stripped by mass-wasting, probably by periglacial

processes in the Pleistocene. Two Bridges Quarry was therefore fundamental for demonstrating an *in situ* example of the selective, deep chemical weathering necessary in Linton's model of tor formation: in most of the more exposed locations the weathering products had subsequently been removed, destroying the critical association of weathered and non-weathered rock.

Palmer and Neilson (1962) disagreed, and related the breakdown of the granite at Two Bridges Quarry (and elsewhere) to pneumatolytic processes, arguing that the decomposition was not a prerequisite for tor formation. They argued that the tors had evolved in response to Pleistocene periglacial processes, and that granite had been removed from around the tors by physical frost-shattering and solifluction. A principal line of reasoning was that the altered granite material was found in valley bottoms and not around the tors themselves.

Alternatively, Brunsden (1964, 1968), also using evidence from Two Bridges Quarry, argued that characteristics of all three forms of rock alteration and breakdown (subaerial chemical weathering, pneumatolytic alteration and physical breakdown) could be found together at individual sites on Dartmoor. He argued that the bulk of evidence from Two Bridges Quarry showed that the decomposed granite there had originated from chemical weathering, most of the weathered deposits belonging to the 'pallid zone' in his classification (see Kaolinization or Tertiary chemical weathering?). He cited, in particular, the pronounced eluviation of clay minerals (found coating joint faces lower in the profile) and the progressive stages of physical disintegration shown by the deposits, and the spheroidal weathering and grus formation found in the lowest parts of the section adjacent to the corestones and 'solid' rock (Brunsden, 1964). Since some of the joint faces are marked by thin veins of tourmaline (cf. Green and Gerrard, 1977), the sections at Two Bridges also probably show evidence for pneumatolytic alteration. However, Brunsden suggested that tourmalinization could only be confirmed if the zone of decayed granite increased and widened with depth, if a widespread cover of solid unaltered granite could be proven and if there was an upward decrease in the frequency of alteration products. Since the depth and extent of alteration at Two Bridges Quarry (and elsewhere) cannot be seen, the role of pneumatolysis remains uncertain in this instance.

The evidence for physical processes in the profile at Two Bridges Quarry, however, was believed to be more sound (Brunsden, 1964): the weathering profile was capped by a thin, 'migratory' layer of head (Brunsden, 1968), and the observed leaching of minerals in solution and the eluviation of clays was also seen as consistent with frost-assisted physical weathering (Brunsden, 1964).

Eden and Green (1971) undertook a detailed mineralogical and grain-size study of the growan at Two Bridges Quarry and from comparable sites elsewhere on Dartmoor. They concluded that the growan was in fact characterized by a relatively low silt and clay content and by a high feldspar residue. These findings were not in keeping with those of Brunsden (1964) who had argued, partly on the basis of the evidence from Two Bridges Quarry, that the granite was 'well rotted' and 'incoherent' and consisted of as much as 90% quartz residue. Neither was there evidence, according to Eden and Green, for the leaching and eluviation of weathering products claimed by Brunsden. Instead, they argued that the high feldspar content and persisting rock texture of the growan showed that there had been only slight chemical weathering and a limited removal of weathering products: the low clay content of the growan was attributed to

limited weathering rather than to leaching and eluviation. Although some translocated clay could be found in the profile near the ground surface and along joint planes, they argued that this had originated from the pedogenic zone since there was little evidence for its translocation deeper within the growan profile (Eden and Green, 1971). In conclusion, Eden and Green suggested that the growan at Two Bridges Quarry was only moderately decomposed, contrasting markedly with the alteration products caused by pneumatolysis (kaolinization) found elsewhere in the region. As such, they argued that it was unlikely that the growan had formed in the hot, humid environment perhaps implied by Linton (1955). Rather, it may have originated under conditions somewhat warmer than at present, perhaps akin to a 'meso-humid subtropical climate'.

Further detailed work was carried out at Two Bridges Quarry by Doornkamp (1974) and by Dearman and Baynes (1978). Doornkamp studied the micro-morphological characteristics of detrital quartz grains taken from head deposits, bedded growan and *in situ* growan at the site, and from other locations on Dartmoor. He demonstrated that only the growan material at Two Bridges Quarry showed significant evidence for chemical alteration, most of the deposits elsewhere were strongly affected by the processes of mechanical weathering. Indeed, the various facies of head deposits could not be distinguished on the basis of quartz grain micro-morphology, and even the bedded growan and *in situ* growan at most sites showed similar quartz grain surface features dominated by mechanical weathering. Quartz grains from the growan at Two Bridges Quarry, however, showed quite different characteristics, with solutional and etch features highly indicative of chemical weathering in a more humid and hotter environment than that found on Dartmoor today (Doornkamp, 1974).

He concluded that the evidence of chemically decomposed granite in Two Bridges Quarry alone could be construed to support Linton's hypothesis that a climate of a 'more tropical' nature had occurred on Dartmoor, and that it effectively produced a sub-surface differential weathering of the granite (Doornkamp, 1974; p. 81). Such evidence also supported Eden and Green (1971) who had concluded that there were only relatively few present-day remnants of any pre-existing widespread cover of a chemically weathered mantle. Either, most of this regolith had been removed during the Pleistocene or, more likely, the deep weathering described by Linton had been much more localized in the first place (Eden and Green, 1971; Doornkamp, 1974). A cautionary note should, however, be added: Doornkamp admitted that the evidence from Scanning Electron Microscopy (SEM) was probably insufficient to differentiate between the products of chemical weathering under a hot humid climate, and the effects of metasomatic (hydrothermal/pneumatolytic) alteration.

Dearman and Baynes (1978) further attempted to distinguish the relative effects of hydrothermal alteration, chemical weathering and frost-shattering on the formation of the rotten granite on Dartmoor, and also used evidence from Two Bridges Quarry. By mapping the distribution of equal intensities of granite decomposition at a number of sites, Dearman and Baynes constructed a model to allow differentiation between chemical weathering (characterized by an overall increase in intensity upwards to the ground surface) and hydrothermal alteration (characterized by an even distribution of alteration products with depth) at any given outcrop.

At Two Bridges Quarry, they demonstrated that only small proportions of the granite had been weathered to grade D in their engineering classification, most of the *in situ* growan belonging to their grade C. Since decomposition of the granite could therefore be demonstrated to increase in intensity upwards, they argued that the origin of the growan was attributable, at least partly, to chemical weathering. However, the precise structural controls on the extent and distribution of chemical weathering, like veins and joints, were also those which had controlled the original distribution of the hydrothermal alteration. Thus, although a dominant set of characteristics (chemical weathering or hydrothermal alteration) could be determined for any given profile, the precise contribution of each in exposures like those at Two Bridges Quarry, where both were believed to be present, was very difficult to ascertain (Dearman and Baynes, 1978). These workers also noted that nearly all of the granite weathering products (head, bedded growan and *in situ* growan) at the site had been substantially affected by frost-action, tending to support Brunsden's earlier argument.

Evidence from Two Bridges Quarry has important implications for landscape evolution in the Dartmoor region and, particularly, a major bearing on the genesis of granite landforms including tors. According to Eden and Green (1971) and Green and Gerrard (1977), it is clear that the growan at

Two Bridges closely resembles, and is indeed typical of, weathered granite found elsewhere on Dartmoor and in other parts of Europe. The material is quite unlike the products of pneumatolytic alteration also present on Dartmoor and which are quarried from the southern part of the moor as china clay (Green and Gerrard, 1977). It is significant that substantial depths of the growan are confined, apparently, to the present-day valleys: they are not found on interfluve summits, and this condition may be the principal determinant in the distribution and formation of the tors themselves which are also found in similar locations and not on plateaux surfaces or the higher interfluves (Eden and Green, 1971; Green and Gerrard, 1977).

Although the general principle of a two-stage mechanism in the formation of tors was accepted by Eden and Green, they argued that the weathering process on Dartmoor was likely to have been much less effective and widespread than previously envisaged. Linton (1955), for example, implied the previous existence of weathered granite (growan) up to 20–30 ft (6–9 m) in thickness on Dartmoor, from which the tors were subsequently exhumed. Since the extensive plateaux surfaces of the region are largely devoid of tors and growan, it would appear likely that the weathering process was indeed more localized. Eden and Green (1971) argued that the tors had been exhumed from a sandy and not clayey weathering zone located principally in or adjacent to the main river valleys. These authors also recognized that many of the Dartmoor tors had been exhumed and modified by periglacial processes.

Two Bridges Quarry has furnished evidence for all the theories put forward to explain the altered or decomposed granite or growan of Dartmoor. As such, it is a key site for the understanding of long-term landscape evolution in the region, and holds one of the principal keys to explaining the origin, long-debated, of local and other British tors. There is no doubt that the origin of the decomposed granite on Dartmoor is critical to understanding how tors, such as those at Merrivale, formed. Two Bridges Quarry shows a profile considered by some to be typical of the Dartmoor growan, and one which has been used by many workers to demonstrate proposed mechanisms for granitic decomposition including chemical weathering, pneumatolytic alteration as well as physical disintegration and weathering. Since the weathering products at Two Bridges Quarry are closely juxtaposed with more intact granite masses, which strikingly resemble unexhumed tors, the site has become a reference locality in theories of tor formation and granite landscape evolution. However, it must be stressed that the quarry does not provide a three-dimensional picture. The latest evidence from the site is probably the most convincing and suggests that weathering of the Dartmoor granite was selective and that it probably occurred under a warm and mildly humid climate. Major difficulties still remain, however, in assessing the relative effects of hydrothermal alteration and chemical weathering where they occur together in one section. Although the granite alteration scheme recently proposed by Floyd *et al.* (1993) for the St Austell Granite does not aid in this discrimination, it does provide an appropriate model and time-frame for geomorphologists working on the evolution of granite terrains and landforms.

Whereas nearby Merrivale exhibits one of the finest assemblages of tors (and associated periglacial landforms) anywhere in Britain, it does not provide the detailed juxtaposition of sedimentary evidence necessary to elaborate and test theories of the longer-term aspects of tor formation: this is provided at Two Bridges and Bellever quarries. Although these quarries share some common characteristics, the evidence is complementary, and Bellever Quarry provides an altogether different insight into the geomorphological evolution of Dartmoor, showing particularly detailed evidence for periglacial slope processes and head formation. Together, these sites form a network indispensable to any detailed reconstruction of long-term geomorphological evolution.

Conclusion

Two Bridges Quarry is one of the most important sites in Britain for understanding processes of granite alteration. The site shows relatively sound, unaltered masses of granite surrounded by softer, decomposed granite or growan. This association of rock and alteration products was central to D.L. Linton's proposal that many British tors had formed by differential chemical weathering of the granite in the Tertiary, followed by a stripping of the weathering products by periglacial processes in the Pleistocene. Many of the subsequent studies which have either challenged or modified Linton's classic theory, have also used Two Bridges Quarry as critical field evidence. In particular, the site has been central in establishing the relative rôles played in alteration of the granite by pneumatolysis (alteration by mineralizing fluids from deep within the

Earth) and chemical and physical weathering processes. The importance of Two Bridges Quarry, in conservation terms, can be summed up simply: its 'embryonic' tors juxtaposed with granitic alteration products will remain central to debates on the origin of British granite terrains and landforms.

DEVENSIAN LATE-GLACIAL AND HOLOCENE ENVIRONMENTAL HISTORY
J. D. Scourse

Introduction

This section introduces GCR sites located on the granite moorlands of Bodmin Moor (Hawks Tor and Dozmary Pool) and Dartmoor (Blacklane Brook and Black Ridge Brook). These four sites were selected because they demonstrate the most complete and detailed records of Devensian late-glacial and Holocene environmental history in these areas. They are effectively regional representatives in a national GCR site network designed to illustrate the most salient features of British Devensian late-glacial and Holocene environmental change.

With the exception of the Somerset Levels (Chapter 2), the environmental history of South-West England during the Devensian late-glacial and Holocene is perhaps less well understood than that of other areas in the British Isles. This partly reflects a lack of suitable depositional basins in which sediments from this time interval could accumulate. This is especially true of the granite moorlands, and is at least partly because large areas were never glaciated. In contrast, other areas of the British Isles north of the Devensian maximum glacial limit (Figure 2.3) are characterized by more profuse erosional (e.g. cirque) and depositional (e.g. kettle hole) basins in which both Devensian late-glacial and Holocene sedimentation occurred. Depositional basins yielding palaeoenvironmental data from this part of the Quaternary in South-West England, by contrast, consist generally of topogenous or soligenous mires, or are the sites where lakes developed in hollows on granitic bedrock. The most critical sites of this age in the region are, as a result, mostly confined to areas of granite bedrock on Bodmin Moor, Dartmoor and the Isles of Scilly. It should be noted that these areas are characterized by acidic soils, with Bodmin Moor and Dartmoor providing the highest relief of the region. The palaeoenvironmental records from sites in these areas should therefore be interpreted in the light of local edaphic and physical controls, and should not necessarily be regarded as being indicative of typical regional conditions.

Despite the uneven distribution and context of the most important localities, the GCR sites across South-West England together provide a comprehensive coverage of Devensian late-glacial and Holocene environmental history. These mire and lake sediments have yielded pollen and plant macrofossil data critical in vegetational reconstruction, and some have yielded diatom data which have enabled assessments of temporal changes in water chemistry. All the selected sites have been calibrated by the radiocarbon method. These data together enable a continuous record of climate, vegetational and environmental change to be reconstructed for the past 13 000 years.

No organic deposits are known in the region which date from between 21 000 to 13 000 BP, the full-glacial phase of the Late Devensian. The earliest record for the Devensian late-glacial is provided by Hawks Tor on Bodmin Moor (Conolly *et al.*, 1950; Brown, 1977, 1980) where the classic tripartite stratigraphy of the Devensian late-glacial was first identified in Britain. The study published by Conolly *et al.* (1950) was actually undertaken during the late 1930s and early 1940s prior to the identification of the Allerød event (Windermere Interstadial) at Windermere (Pennington, 1947) and at Flitwick (Mitchell, 1948): their study allowed correlation with better-established Devensian (Weichselian) late-glacial sequences in Ireland, Germany and southern Scandinavia, and made Hawks Tor a 'landmark' in the development of pollen biostratigraphy in Britain.

Although the Holocene record at Hawks Tor is incomplete, the Devensian late-glacial sequence there is undoubtedly the finest known in the region. Pollen and plant macrofossil evidence from the site indicates the presence of arctic-montane species prior to 13 000 BP (Older Dryas) at a time of active solifluction. The climatic amelioration of the Allerød (Windermere Interstadial) led to the development of birch woodland in sheltered valleys. The succeeding Younger Dryas (Loch Lomond Stadial) saw a reversion to active solifluction with arctic-montane species flourishing in an open, treeless environment. This record is important in demonstrating the relative paucity of the interstadial vegetation here in comparison with other sites in southern Britain at lower altitudes and on more base-rich soils, and in registering the severity of periglacial processes during the Younger Dryas: this implies a degree of continentality in marked contrast to the present oceanic climate.

More complete Holocene records are preserved in sediments at nearby Dozmary Pool (Conolly *et al.*, 1950; Brown, 1977) and at Blacklane Brook (Simmons, 1964a; Simmons *et al.*, 1983; Maguire and Caseldine, 1985) on Dartmoor. Both sites occur on upland granite characterized by acidic soils, but Blacklane Brook (457 m OD) lies at a considerably higher elevation than Dozmary Pool (265 m OD). In common with many other sites across southern England, these two localities demonstrate the major features of vegetational and environmental change characteristic of the Holocene: early Holocene open-grassland vegetation, gradual immigration of thermophilous arboreal species, a mid-Holocene forest stage, and then gradual clearance of woodland as a result of anthropogenic activity. There are, however, some important vegetational differences between the two sites. There is some evidence to suggest that Blacklane Brook lay close to the regional treeline during the phase of maximum forest cover during the mid-Holocene while the lower-lying environs of Dozmary Pool were dominated by open oak forest. This is in marked contrast to other areas of Britain which apparently demonstrate a much higher treeline at this time, and also contrasts with other sites in southern England which were dominated by other tree species (Bennett, 1989). The dominance of oak in the Holocene forest of South-West England is thought to be related to the dominantly acidic character of the soils. However, as indicated above, this may not hold true for the entire region and may simply reflect the acidic substrate of these particular sites. Whatever their spatial extent, these differences illustrate the significance of climatic and edaphic factors in determining the pattern of Holocene vegetational development.

Both Dozmary Pool and Blacklane Brook are extremely important in providing evidence of woodland clearance by fire as early as the Mesolithic. At Blacklane Brook, the evidence is based on changes in pollen assemblages at around 7500 BP, and at Dozmary Pool such changes are complemented by charcoal within the peat profile at around 7000 BP, and by a profusion of archaeological material close to the site. Both sites continue to be important in discussions regarding the relative significance of climatic and anthropogenic factors in disrupting woodland from the mid-Holocene onwards.

At around 6500 BP, many sites in the region indicate a change to a wetter climate. This is indicated by the development of raised and blanket bogs, and the rapid spread of alder at sites on Bodmin Moor and Dartmoor. It is perhaps significant that this coincides with the onset of peat accumulation at Higher Moors (St Mary's) on the Isles of Scilly (Chapter 8). Peat from the base of this mire has been dated to 6000 BP (Scaife, 1984), the oldest Holocene organic sediments yet reported from the islands. This site is important because its pollen evidence indicates that indigenous woodland was able to generate on the Isles of Scilly during the mid-Holocene: this has implications for studies of tree migration and dispersal given that the islands lie 45 km from the mainland against the direction of the prevailing south-westerly winds.

The Higher Moors site on Scilly, and the Bodmin Moor and Dartmoor sites, all demonstrate the progressive clearance of woodland by humans during the Neolithic and later. The Mesolithic clearance episodes were relatively minor and limited to the mainland, and most sites show evidence of post-clearance forest regeneration. The mainland Neolithic clearances, however, were more long-lasting, and pollen evidence, including pollen grains of cereals, weeds and ruderals, indicates active cultivation from this time onwards. On Scilly, regeneration occurred after an initial phase of Neolithic clearance, woodland only being irreversibly removed from the middle Bronze Age onwards.

HAWKS TOR
S. Campbell and N. D. W. Davey

Highlights

The most complete record of Devensian late-glacial vegetational and climatic changes in South-West England comes from Hawks Tor on Bodmin Moor. This site was central to the development of pollen biostratigraphy in Britain in the 1930s/1940s and still provides the regional 'standard' with which sequences elsewhere in Britain, Eire and continental Europe are compared.

Introduction

Hawks Tor provides the most detailed and extensive record of Devensian late-glacial conditions in upland South-West England, as well as a partial record of Holocene environmental changes. Pollen, plant macrofossils, diatoms and radiocarbon dates

from the site show unique evidence in the area for the Allerød Interstadial of the Late Weichselian (= Late Devensian). The kaolinized granite at Hawks Tor was discussed by Reid *et al.* (1910), and its origin and importance in landscape evolution were also mentioned in studies by Barton (1964) and Clayden (1964). The pollen biostratigraphy of the site has been studied in detail by Conolly *et al.* (1950) and Brown (1972, 1977, 1980), and radiocarbon dates for selected horizons were provided by Libby (1952) and Brown (1977, 1980). The site's importance is highlighted by its use as the reference locality for the Allerød Interstadial in the Geological Society's Quaternary correlation charts for South-West England (Stephens, 1973; Campbell *et al.*, in prep.), and by its continuing citation in both regional and national syntheses of Quaternary evidence (Te Punga, 1957; Pennington, 1974; Kidson, 1977; Caseldine, 1980; Bell *et al.*, 1984).

Description

Hawks Tor GCR site (SX 150749) lies at *c.* 220 m OD on Bodmin Moor, and consists of a small peat bog situated at the northern end of the lake which now occupies the disused pit of the Hawks Tor China Clay Works (Figure 4.15). The bog lies in a depression flanked to the east by the Warleggan River and to the west and north by the rising slopes of Menacrin Downs and Hawks Tor itself. Up to 2.5 m of organic sediment is exposed in the valley bottom, thinning to less than 0.5 m on the surrounding hillslopes. The upper layers of peat, however, have been reduced by cutting, and the bog surface is much disturbed by spoil, trackways, embankments and ditches. The Devensian late-glacial and Holocene deposits are exposed in sections along the lake edge, at the margin of the former kaolin workings. These sections also expose the underlying growan or grus (the kaolinized granite). Similar stratigraphic relationships occur to the south of the GCR site and were exposed, temporarily, during widening of the A30(T) in 1987.

Figure 4.16 shows the stratigraphic sequence exposed at the north-east end of the lake. This representation, based on the work of Brown (1977, 1980), is highly generalized: considerable lithological variations occur in the beds, the pattern of which is frequently disrupted by complex unconformities and cryoturbation structures. The sequence of vegetational development and inferred climatic changes given here is based principally on pollen analyses carried out on three separate

Figure 4.15 The disused china clay workings at Hawks Tor, Bodmin Moor. The altered granite or growan is seen in the foreground faces, with the cliffed, but somewhat degraded, Devensian late-glacial and Holocene sequence along the lake edge behind. (Photo: S. Campbell.)

monoliths by Brown (1977, 1980) (Figure 4.16) although a brief appraisal of the earlier work by Conolly *et al.* (1950) is also given.

Interpretation

Conolly *et al.* (1950) presented the results of pollen analyses carried out on the peat beds (beds 3, 5 and 6). Their 'upper peat' (beds 5 and 6) showed characteristic tree pollen assemblages from Pollen Zones V–VIII of the 'post-glacial' (Holocene), although there was some doubt as to whether deposits with a Pollen Zone IV flora (Pre-Boreal) were present at the site. With this exception, the 'upper peat' showed a pollen zonation similar to many other British tree pollen diagrams for the period.

The Pollen Zone V assemblage, found at the base of bed 5 in the wood peat, showed a preponderance of birch and hazel pollen with small amounts of pine, oak and willow. Such a pattern also characterizes the succeeding zone (also in bed 5), but with a decline in birch and hazel, and a corresponding increase in oak and alder. Pollen Zone VII

(bed 5) was characterized by a significant reduction in tree birch pollen and by the continued rise of oak. Alder and hazel pollen values are maintained at a steady level throughout this zone. The Pollen Zone VIII assemblage, found in the non-humified peats of bed 6, showed a renewed peak in birch, a sudden expansion in oak and a corresponding decline in alder. Beech makes its first significant appearance in this zone.

Conolly *et al.* (1950) noted that Pollen Zone IV (Pre-Boreal) pollen might be present in the upper layers of the underlying silty peats, sands and gravels (bed 4). However, these dominantly minerogenic beds had lithological and stratigraphical characteristics suggesting a Younger Dryas (Pollen Zone III) age (Conolly *et al.*, 1950).

These workers also identified pollen, fruits and seeds from the organic sediments in bed 3 which was sandwiched between two dominantly inorganic layers (beds 2 and 4). The fossil flora from bed 3 was characterized by plants now restricted to more northern parts of Britain – for example, *Betula nana*, *Salix herbacea* and *Thalictrum alpinum*. Pollen from this bed revealed an open-tundra or 'park-tundra' vegetation and, although

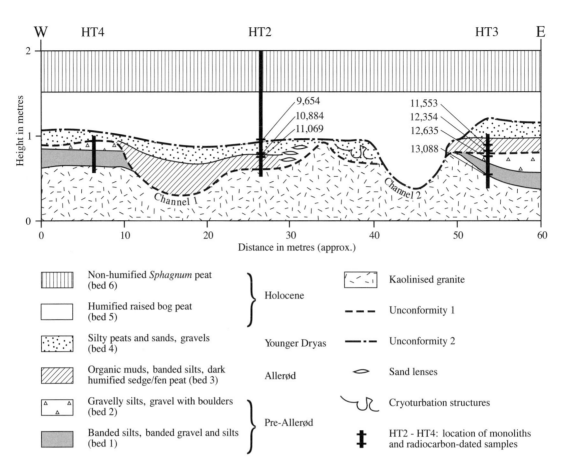

Figure 4.16 A simplified composite section of the north-east face of the exposures at Hawks Tor as exposed in 1970–1971. Adapted from Brown (1977, 1980).

there was no clear evidence for tree growth (but see below; Brown (1977, 1980)), conditions were clearly warmer than when the underlying and overlying, dominantly minerogenic, beds were deposited. Organic lake muds at the base of bed 3 yielded a diatom flora indicative of cool-temperate, moderately eutrophic conditions. On the basis of this evidence, Conolly *et al.* (1950) confidently ascribed bed 3 to the Allerød oscillation (= Pollen Zone II) of the classic Danish sequence.

The gravelly silt, which contains some large granitic boulders up to 1 m across (bed 2), was interpreted as a solifluction deposit, probably ascribable to cold conditions in Pollen Zone I times (Conolly *et al.*, 1950). Likewise, the dominantly minerogenic deposits of bed 4 were considered to have accumulated by solifluction during a deterioration of climate represented by Pollen Zone III (Younger Dryas): Conolly *et al.* argued that the disturbance, cracking and contortions found in the

underlying peats and lake muds (bed 3) had been caused by frost-action and cryoturbation at this time. The Hawks Tor sequence therefore provided a record showing the threefold subdivision of the Scandinavian Late Weichselian (= Late Devensian late-glacial), in addition to a comprehensive record of Holocene vegetation changes with the exception of detailed evidence in the earliest Holocene (the Pre-Boreal/Pollen Zone IV) (Conolly *et al.*, 1950). An early attempt by Libby (1952) to calibrate the Devensian late-glacial and Holocene sequence at Hawks Tor, using radiocarbon dating methods, provided ambiguous results (Libby, 1952; p. 75).

Brown (1977, 1980) re-investigated the pollen biostratigraphy of the Hawks Tor site, and provided radiocarbon dates from critical lithological and pollen assemblage boundaries (Figure 4.16). The oldest Quaternary sediments (bed 1) overlie kaolinized granite. These banded silts and alternations of silt and gravel appear to have resulted from cyclic sedimentation in still water. The pollen record shows that deposition occurred in a treeless, sparsely vegetated landscape, probably with hill-side snowbeds. The occurrence of pollen in these beds from the arctic-montane species *Artemisia norvegica* and *Astragalus alpinus* indicates low mean annual temperatures and a degree of base-element enrichment caused by local solifluction (Brown, 1980). The oldest radiocarbon-dated organic sediment (13 088 ± 300 BP (Q–979)) comes from banded silt in bed 1, near its junction with the underlying kaolinized granite. These sediments may have accumulated in a clear, shallow water, mud-bottomed channel. In places, they are succeeded by gravel which contains some large boulders (bed 2) reflecting continued slope instability and solifluction.

This cold channel environment with local solifluction gave way to warmer conditions by about 12 600 BP, when organic muds and a sedge mire (bed 3) started to accumulate. These dominantly organic sediments accumulated on an irregular channelled surface (Figure 4.16; unconformity 1) after a change in drainage had effected local removal of parts of beds 1 and 2 and, in places, the underlying bedrock. The pollen record from bed 3 shows that a tall herb fen and birch carr had developed by about 11 500 BP: these warmer conditions are correlated with the Allerød Interstadial (Brown, 1977, 1980). The Allerød deposits are succeeded unconformably in places by solifluction gravels (bed 4). Elsewhere, the peat is overlain by dominantly inorganic gravels, sands and silts, which contain bands of peat presumably reworked from bed 3. Radiocarbon dates show that this change from organic to dominantly minerogenic sedimentation occurred at around 11 000 BP. Pollen from these sediments (bed 4) shows a return to an open treeless vegetation with several cold-climate species. This climatic deterioration is correlated with the Younger Dryas event (Pollen Zone III), and was characterized by a return to solifluction, possibly wet-soil creep, in a cold, oceanic regime (Brown, 1977, 1980). Two notable botanic records occur in the Younger Dryas sediments: *Luzula arcuata* and *Epilobium alsinifolium* are arctic-montane species found now only at higher latitudes in northern England, Scotland and Scandinavia (Brown, 1980).

There is no pollen evidence for an unconformity between the solifluction/sheet-wash deposits of the Younger Dryas (bed 4) and the overlying Holocene peat (beds 5 and 6). The radiocarbon date of 9654 ± 190 BP (Q–1017) (Figure 4.16) from the base of the Holocene peat, however, does suggest an unconformity, much as Conolly *et al.* (1950) had suspected, since the date is too young to mark the start of the Holocene. The fact that the peat of bed 5 also directly overlies sediments of bed 3, and even granite in places (Figure 4.16; unconformity 2), shows that some erosion took place between deposition of beds 4 and 5.

The Holocene sediments consist of highly humified *Sphagnum* and sedge peats (bed 5) which grade into *Sphagnum/Eriophorum* peats (bed 6) of decreasing humification. This shows that the Late Devensian sediments were covered by a wet *Sphagnum* mire which dried out progressively to fen carr. This part of the Hawks Tor pollen sequence (covered by Pollen Zone V; Conolly *et al.* (1950)) is, however, discontinuous and there has been some erosion of the bed (Brown, 1977, 1980). The development of the fen carr was followed by a return to wetter conditions, impeded drainage and the development of raised bog (bed 6); blanket bog developed at the site by about 3100 BP, and the subsequent effects of pastoral and, later, arable agriculture are also clearly evident in the pollen record (Brown, 1977, 1980).

The Hawks Tor deposits are of considerable significance as they have yielded the most complete record of Devensian late-glacial conditions known from upland South-West England. Although incomplete, the Holocene pollen record from the site complements that from nearby Dozmary Pool, collectively providing some of the best evidence for the vegetation history and climate of this period on Bodmin Moor. (The Holocene pollen record for Bodmin Moor has been supplemented recently by data from nearby Rough Tor (Gearey and Charman (1996).)

Hawks Tor is also important in the historical development of pollen biostratigraphy in Great Britain. Conolly, Godwin and Megaw's (1950) study was undertaken at a time, in the late 1930s/early 1940s, when little evidence for Devensian late-glacial conditions in Britain was known. Although strong evidence for recognizing the Allerød Interstadial had been made at Windermere in the Lake District (Pennington, 1947) and at Flitwick in Berwickshire (Mitchell, 1948), other British sites with comparable sequences had only been described and interpreted more speculatively (Godwin, 1947). The study by Conolly *et al.* (1950) at Hawks Tor is therefore a significant 'landmark' in correlating the British Devensian late-glacial record with more comprehensively studied sequences in Ireland, Germany and southern Scandinavia.

The pollen and stratigraphic evidence, including periglacial structures in the beds, is also of importance for demonstrating the considerable severity of Devensian late-glacial climate at latitudes well south of the Late Devensian maximum ice limit (Pennington, 1974).

Conclusion

The sequence of Quaternary deposits at Hawks Tor includes frost-weathered, reworked granitic deposits, peats, lake muds and silts. It has yielded pollen, plant macrofossils and radiocarbon dates which have enabled one of the most detailed reconstructions of vegetational and climatic changes in South-West England during the Devensian late-glacial to be made. Particularly significant is its unambiguous evidence for the Allerød oscillation of the Late Devensian – a discrete interlude of relative warmth, lasting some 1500 years, sandwiched between periglacial phases (equivalent to the Older and Younger Dryas) when the local landscape resembled tundra. In conservation terms, Hawks Tor is important because it provides the most complete record in the South-West for the climatic and environmental conditions of this time interval and shows, uniquely, that trees were established on the Peninsula during the Allerød at high elevation. Although its record of Holocene environmental changes is incomplete, it complements the fuller record at nearby Dozmary Pool: both can be regarded as reference sites for understanding changing climatic and environmental conditions in the latest Quaternary. Hawks Tor has the further distinction of being a significant 'landmark' in the development of British pollen biostratigraphy.

DOZMARY POOL
S. Campbell

Highlights

Dozmary Pool is a key pollen site with a radiocarbon-dated record spanning almost the entire Holocene. Changes in the relative proportions of tree-shrub-herb pollen, in association with occurrences of charcoal in the peat layers, are central to the controversy surrounding the extent of human activity on Bodmin Moor during the Mesolithic.

Introduction

Organic deposits at Dozmary Pool preserve the most complete Holocene pollen sequence yet known from Bodmin Moor, and provide a key record of the vegetational history of South-West England for this period. Dozmary Pool is a reference site for interpreting more discontinuous Holocene pollen sequences found elsewhere in the region, for example, at Parsons Park, Stannon Clay Pit and Hawks Tor. The site was first referred to in an archaeological context by Whitley (1866), Brent (1886) and later by Robson (1944). The pollen biostratigraphy has been studied by Conolly *et al.* (1950), Brown (1972, 1977, 1980) and Simmons *et al.* (1987). Evidence from the site was also discussed by Caseldine (1980) and Bell *et al.* (1984) in reviews of environmental change in Cornwall during the last 13 000 years.

Description

Dozmary (Dozemare) Pool GCR site (SX 192743) lies on Bodmin Moor at *c.* 265 m OD, some 4.5 km east of Hawks Tor (Figure 4.1). The pool occupies a peaty depression to the north-east of Gillhouse Downs between the Fowey and St Neots rivers: the GCR site consists of a rectangular bog (*c.* 300 m × 150 m) situated at the south-west end of the lake where there is a small outlet stream (Figure 4.17). This raised mire has a vegetation community of *Eriophorum vaginatum*, *Rhynchospora alba*, *Erica tetralix*, *Calluna vulgaris*, *Molinia caerulea*, *Juncus effusus* and *Narthecium ossifragum*. The sediment sequence comprises *c.* 2.5 m of lake muds, sedge and fen peats topped by decreasingly humified raised bog peats. Conolly *et al.* (1950) recorded the following stratigraphy from a core of the deposits:

7. Non-humified fresh peat (0.72 m)
6. *Sphagnum–Eriophorum* peat with lake mud at base (0.23 m)
5. Moderately humified *Sphagnum–Calluna* peat with layers of ash and charcoal (0.90 m)
4. *Sphagnum–Eriophorum* peat and some organic mud (0.23 m)
3. Brown and black organic lake muds (0.23 m)
2. White lake mud (0.30 m)
1. Dark brown lake mud with *Phragmites* (0.10 m)

Brown (1977) recorded a deeper profile, with the following stratigraphy:

Figure 4.17 The peat bog at the south-west end of Dozmary Pool, Bodmin Moor. (Photo: S. Campbell.)

9. Coarse, non-humified *Sphagnum/ Eriophorum vaginatum* peat (0–64 cm)
8. Slightly humified *Sphagnum/Eriophorum vaginatum* peat (64–77 cm)
7. Coarse, non-humified *Sphagnum/ Eriophorum vaginatum* peat with angular quartz gravel (< 1 cm diameter) at 80–85 cm (77–93 cm)
6. Blackish, well-humified *Sphagnum/ Eriophorum vaginatum/Calluna* (raised bog) peat with carbonized material throughout, and charcoal fragments in distinct bands at 146, 154, 170 and 175–180 cm; large chunks of burnt peat at 145–150 cm. *Calluna* flowers, leaves and twigs frequent throughout (93–188 cm)
5. Highly humified monocotyledonous peat with fresh wood fragments at 190, 193 and 197 cm; thin black band at 198 cm; frequent minute carbonized fragments throughout, fungal perithecia frequent in the upper half of the bed (188–207 cm)
4. Highly humified sedge peat with abundant fresh *Salix* wood; *Carex* nutlets are frequent at the top and *Juncus* seeds at the base (207–215 cm)
3. Medium and fine muds with a fibrous layer at 220 cm and dark band at 223 cm; *Juncus* seeds frequent throughout. *Carex* nutlets, *Littorella uniflora* and *Hydrocotyle vulgaris* fruits, *Menyanthes trifoliata* seeds occasional towards the base (215–224 cm)
2. Fine, very silty mud; megaspores of *Isoëtes lacustris* and *Isoëtes echinospora* abundant; oospores of *Nitella* type, *Elatine hexandra*, *Luronium natans*, and *Juncus* seeds occasional; *Potamogeton natans* fruits rare; *Sphagnum* leaves present (224–235 cm)
1. Kaolin clay (below 235 cm)

Although Simmons *et al*. (1987) recorded a slightly deeper profile, their stratigraphy is essentially similar to that given by Brown (1977) (see above), although they do not record the highly humified sedge peat with abundant *Salix* wood (bed 4). Brown's stratigraphic sequence is given here in preference because his reconstructed pollen assemblage zones cover most of beds 2–6. Simmons *et al*. (1987), on the other hand, concentrated their pollen analyses on beds 5 and 6 only: the muds (beds 2 and 3), the sedge peat (bed 4) and the upper peats (beds 7–9) were not analysed. Brown's pollen analyses were supported by five radiocarbon dates (Q-1021 to Q-1025). Five additional radiocarbon dates (HAR–5077 to HAR–5080 and HAR–5083) for the profile were provided by Simmons *et al.* (1987).

Interpretation

Artefacts

First mentioned in the scientific literature by Whitley (1866), Dozmary Pool attracted archaeological attention: over 100 'very perfect flint-flakes' were recovered from around the site, mainly from the uppermost soil layers. Brent (1886; pp. 60–61) remarked that

'Although no traces of Lake Dwellings could be observed when Dosmare Pool was entirely dry in the summer of 1866, yet the presence of hut-circles, barrows, &c., on the surrounding moor; the five 'Kings' Graves', one since destroyed, on Bron Gilly; and the vast quantity of flakes, pieces, and some arrow-heads from the peat; would indicate that there was once a large population in this interesting district.'

Robson (1944) noted that 'Neolithic' flints occurred beneath the peat at Dozmary Pool, thereby proving the 'recent formation of the peat'. More up-to-date accounts of the microlith assemblages were given by Wainwright (1960) and Jacobi (1979). Wainwright described the microliths found at Dozmary (mostly during the nineteenth century) which have now been dispersed to various museum collections, including 2500 flints to Plymouth Museum alone: their original stratigraphic provenance is unknown. Brown (1977) recorded burnt peat within the sequence (his bed 6) immediately prior to the 'elm decline' (see below). He therefore argued that the microliths might be late Mesolithic in age. Jacobi (1979), on the other hand, has linked the microliths from Dozmary with an assemblage from Thatcham in the Kennet Valley, suggesting an early eighth millennium date, namely very early Mesolithic. Such an age would place the Dozmary flints in a period when pollen shows that *Empetrum* heath and juniper scrub grew on the gentle local hillslopes, and before the spread of birch woodlands. Such a view implies that the microliths either come from, or can be correlated with, a very much earlier deposit in the sequence than suggested by Brown (1977) (Simmons *et al.*, 1987).

Pollen biostratigraphy and radiocarbon dating

Conolly *et al.* (1950) provided an arboreal pollen diagram based on analyses in their beds 3–5 and from the lower layers of bed 6. The diagram follows the standard zonation with full detail for Pollen Zones VI–VII, and parts of Zones V and VIII. They noted that it was likely that the top metre of peat (beds 6 and 7) was formed in Pollen Zone VIII: there was also an ample depth of mud (beds 1 and 2) beneath the lowest counted sample to record the vegetation history in Pollen Zone IV and the earlier part of Zone V. The incompletely analysed record for Pollen Zone V (in bed 3) shows birch and hazel to be dominant: the latter, however, declines rapidly towards the end of the zone. The upper part of bed 3 (lake mud) and most of bed 4 (peat and lake mud) contain an arboreal pollen assemblage characteristic of Pollen Zone VI – with a rapid decrease in birch and the progressive invasion of oak. Alder first becomes continuously represented in this zone while hazel maintains steady values. A Pollen Zone VII assemblage occurs in the very lowest part of bed 4 and throughout bed 5 (moderately humified peat). This shows the progressive decline of birch to low but stable levels, the continued increase of oak and the steady maintenance of alder and hazel in the developing mixed deciduous forest. Conolly *et al.* (1950) further noted that the layer of ash and charcoal in bed 5 (Pollen Zone VIII) probably reflected the activities of prehistoric humans in clearing the local forests.

Brown (1972, 1977) reinvestigated the pollen biostratigraphy of the site, providing detailed evidence (absolute pollen frequencies) for the vegetation history of the beds which had previously yielded Pollen Zone VI and VII assemblages (Conolly *et al.*, 1950): a summary of this work was given by Brown (1980). His work covers an important gap in the existing site record – namely the early Holocene.

According to Brown (1977), the basal silty muds (his bed 2) started to accumulate at *c.* 9053 ± 120 BP (Q–1021), and are characterized by a local pollen assemblage zone dominated by grasses. Pollen and plant macrofossil evidence confirms that deposition took place in a shallow, clear-water, base-poor lake with a gravelly bed. The high grass and low tree pollen values, together with a rich herb flora, show that the early Holocene landscape was open, dominated by a short-turf grassland, with trees limited to only small areas or more continuously distributed at some distance (Brown, 1977).

A second pollen assemblage zone (DP2), found in beds 2 and 3, consists of an early *Corylus*-Gramineae zone with a subsequent Cyperaceae subzone (Brown, 1977). The first stage of a hydroseral succession is shown by increasing Cyperaceae values, the occurrence of *Carex* nutlets in the mud, the appearance of abundant *Sphagnum* spores and the disappearance of most aquatic pollen. This indicates that open-water conditions were gradually replaced by a *Carex/Sphagnum* mire which formed above the local water-table, and that Dozmary Pool itself was reduced in area shortly after 9000 BP in response to drier conditions (Brown, 1977). Elm pollen appears in the profile for the first time at a level dated by radiocarbon methods to 8829 ± 100 BP (Q-1022). A persistence of herb and grass pollen was taken to indicate the limited extent of the local woodland, which was perhaps restricted to sheltered valleys (Brown, 1977).

Local pollen assemblage zone DP3 (bed 4; humified sedge peat with *Salix* wood) is broadly similar to the preceding subzone, although *Salix* pollen reaches high values in the lower part of the wood peat: together with higher levels of birch pollen, and the occurrence of *Salix* wood itself, the evidence points to a mixed tree layer forming fen carr in the immediate vicinity (Brown, 1977).

The succeeding pollen biozone (DP4) is marked by high *Corylus*, with a *Calluna* subzone (DP4a): it spans the upper part of the wood peat laid down in a fen carr (bed 4) and the majority of bed 5. Following an initial persistence of birch/willow-dominated fen carr, dated to *c*. 7925 ± 100 BP (Q-1023), the carr was swamped by active mire growth as shown by high *Sphagnum* frequencies and the appearance of *Calluna* pollen (Brown, 1977). This zone is also characterized by a sparse but varied herb flora, fern spores and increasing Gramineae values. Oak and elm values are also higher and are taken as showing the continued spread of open woodland to well-drained hillside locations.

Zone DP5 and its subzone DP5a span the upper part of bed 5 (highly humified peat) and the base of the dark *Sphagnum/Eriophorum/Calluna* peat (bed 6). A radiocarbon date of 6793 ± 70 BP (Q-1024) from the upper layers of bed 5, provides a maximum age for the sediments covered by pollen assemblage zone DP5. This characteristic hazel/fern biozone sees a peak in grass and birch pollen early in the zone (DP5). Fluctuations in the pollen curve in this zone, and the appearance of

ferns, may indicate the restriction of the woodland on better soils by fire (Brown, 1977). Pollen assemblage zone DP5 may well cover the driest period in a climatic regime which had improved continuously since the opening of the Holocene some 3000 years earlier.

Calluna/Alnus subzone DP5a commences at 6451 ± 65 BP (Q-1025): the lower boundary of the zone corresponds with the base of the raised bog peat (bed 7). The appearance of alder for the first time in this zone reflects the development of alder carr around the mire, together with *Fraxinus* and *Tilia cordata*. Brown has argued that the presence of these trees shows the onset of a mild, but wetter, climate as does the ensuing development of the raised bog itself. Indeed, a general recession of woodland is indicated as waterlogging of local sites progressed (Brown, 1977).

Brown did not provide a detailed analysis of pollen present in the remainder of the beds (upper bed 6 and beds 7-9): the likely biozonation of these beds is, however, thought to conform with analyses carried out on comparable beds at nearby sites (e.g. Hawks Tor and Parson's Park) (Brown, 1977). Radiocarbon dates were not attempted from the upper peats at any of these sites due to the likely effects of modern rootlet contamination (Brown, 1977). It is assumed, on the basis of preliminary pollen work and on inter-site correlation, that the 'elm decline' is manifest in bed 6 in the Dozmary Pool record somewhere between 120-160 cm (Brown, 1977).

From the pollen and plant macrofossil evidence given by Brown (1972, 1977, 1980), the course of vegetation development and hydroseral succession, right up to and including the climatic optimum of the Boreal period, is clear. Initially, the Dozmary area was occupied by a base-poor clear lake. This was partially replaced by a sedge and *Sphagnum* mire, then by birch and willow carr and finally by raised bog. The latter may have been caused by the local development of fen carr near the present lake outlet. This would have impeded drainage from the lake (Thurston, 1930; Brown, 1977) which may indeed have been present continuously in the vicinity, albeit in varying size, since the Late Devensian (Brown, 1977).

Other salient features of Brown's pollen diagram, which covers the early and middle Mesolithic periods only (between *c*. 9000 and 6500 BP), include the maintenance of some open ground at this altitude throughout the period, and a tendency towards treelessness in the later part covered by the diagram. The latter coincides with increases in bracken spores and the pollen of grasses and *Potentilla*. He has argued that fluctuations in the pollen record at *c.* 6541 ± 75 BP (Q-1025) may indicate the restriction by fire of oak and elm, with a corresponding rapid spread of birch, ferns and grasses. He has correlated these changes with evidence from Dartmoor (see Blacklane Brook) where a record of forest recession, similar rises in fern spores and grass pollen, have been taken to indicate forest clearance by Mesolithic inhabitants (Simmons, 1964a). Brown has suggested, however, that such deliberate woodland clearance would have been unnecessary on Bodmin Moor where, at this time, the pollen evidence reveals an already open landscape.

It is against the background of Brown's work that Simmons *et al.* (1987) subsequently concentrated their pollen analyses on beds 5 and 6, where they judged the fluctuating conditions associated with Mesolithic activities to be present. All the vegetational changes recorded within this part of the sequence fall within a single local pollen assemblage zone (Simmons *et al.*, 1987).

Generally low total tree pollen percentages, together with sporadic occurrences of a range of herbs and open-land indicators, point to the persistence of some unwooded land in the vicinity of the site throughout the period (Simmons *et al.*, 1987). Some large fluctuations in the relative pollen contributions of trees, shrubs, small (ericaceous) shrubs and herbs do occur, with a major fluctuation between 209 and 203 cm in an equivalent of Brown's bed 6, coincident with much charcoal and mineral matter. Oak is dominant in the tree pollen record throughout: *Alnus* is lacking in the lowest samples (bed 5) but rises steadily before fluctuating and finally rising again to reach similar values. Pine is not judged to have been a constituent of the woods on Bodmin Moor during the period covered by the diagram constructed by Simmons *et al.* (1987). Elm pollen is generally low and is probably absent where charcoal is found in bed 4 at 205 cm (Simmons *et al.*, 1987).

Although the oldest radiocarbon date of 7590 ± 100 BP (HAR-5083) falls between Brown's (1977) dates of 7925 ± 100 BP (Q-1023) and 6793 ± 70 BP (Q-1024) derived from similar monocotyledonous peat (bed 5), covered by an equivalent pollen assemblage zone (Brown's DP4), the other radiocarbon dates given by Simmons *et al.* pose considerable problems of interpretation: these dates are not arranged in the expected chronological order in the profile. Since sample contamination is considered unlikely, the radiocarbon dates probably show that the site here was not disturbed by fire alone, but by the physical removal of peat and the inversion of the profile: this may have been on two separate occasions – at least as recently as *c.* 2740 BP in the case of the lower disturbance, and 510 BP for the upper. Alternatively, and more likely, both apparent disturbances may have happened simultaneously after the latter date. Considerable fluctuations in the pollen, charcoal and mineral contents at the 201–209 cm-levels lend much support to the latter view (Simmons *et al.*, 1987).

Dozmary Pool is a reference site for upland vegetation history in South-West England with one of the most extensive Holocene pollen records in the region. It is a member of a network of pollen sites in Britain which shows regional and altitudinal variations in the course of vegetation succession, and is particularly important for demonstrating the effect of oceanicity – that is, a local climate dominated by exposure to wind and rain – on vegetation development. In this respect, the vegetation record from Dozmary Pool shows strong similarities with sites on the western fringes of Britain and indeed areas on the seaboard of continental Europe, but contrasts with other inland and montane sites (Brown, 1977). In particular, the strong record of birch throughout the Holocene, as shown in part from the evidence at Dozmary Pool, contrasts markedly with other parts of southern England and even Dartmoor: its persistence here may be ascribed to its survival in a very open landscape, one maintained by exposure to the elements rather than by human activity (Brown, 1977).

Similarly, both on Bodmin Moor and on Dartmoor, oak is believed to have invaded rapidly, and although it never completely covered the uplands, being restricted by the lack of shelter, it soon reached its maximum extent to become the dominant tree species (Brown, 1977). Its rapid domination of the landscape may have been facilitated by the extensive areas of grasslands, open birch woods (see above) and immature hazel scrub which covered the uplands and which offered little competition for the advancing oak populations (Brown, 1977). Again, the open landscape is believed to have been strongly influenced by the oceanicity of the climate.

On the other hand, *Pinus* does not seem to have been a major constituent of the early forests of the South-West. Evidence from Dozmary Pool (and elsewhere) shows that it was absent on the moorlands of the South-West (Brown, 1977), although it may have been present at lower altitudes and in more sheltered southern coastal locations (Ussher,

1879c; Clarke, 1970). In this respect, the evidence from Dozmary Pool shows strong similarities with sites in Brittany (van Zeist, 1963; Brown, 1977). Elsewhere in the British Isles, *Pinus* spread rapidly north-westwards in response to the improving climate of the early Holocene. Where *Pinus* was not established, as in South-West England, hazel and oak expanded rapidly (cf. south-west Ireland and Northern Ireland). Brown (1977) has argued that the oceanicity of the Atlantic seaboard in these parts of Europe put *Pinus* at a disadvantage when in direct competition with *Quercus* as the climate improved. It is possible that light-demanding pine seedlings were quickly shaded-out when birch arrived (Brown, 1977), although the precise requirements of different genera are still poorly understood and must be established before definitive statements can be made about regional variations in forest composition (for discussion of these problems see Jessen, 1949; Iversen, 1954; Jones, 1959; Planchais, 1967; Carlisle and Brown, 1968; Bennett, 1989; Bennett and Birks, 1990).

Another important regional contrast exhibited by the pollen evidence from Dozmary Pool concerns the relative expansions of oak and elm in Britain during the early Holocene. Mitchell (1951) has shown that elm tended to expand in calcareous areas first, and this is clearly shown by pollen records from the calcareous areas of southern England where *Ulmus* either expanded before or simultaneously with *Quercus* (e.g. Seagrief, 1959, 1960; Seagrief and Godwin, 1960). Farther west in South-West England, on the other hand, oak expanded before elm, and this was clearly a result of the acidic upland soils, particularly those on Dartmoor and Bodmin Moor (Brown, 1977). The unbroken record of vegetation succession in Pollen Zone VII at Dozmary Pool is particularly important: it is almost certainly due to the continuously adequate water supply at the site, allowing anaerobic preservation throughout the climatic optimum of the Holocene (*c.* 7900–6500 BP). This contrasts with Hawks Tor and Parsons Park where the mires dried out at this time, fossil (pollen) preservation ceased, and where erosion probably occurred (Brown, 1980). Retention of water sufficient to form a lake and support the growth of a mire at Dozmary Pool during the driest (Boreal) period, may have been promoted by the restricted drainage outlet (see above), and may also have been a function of the greater annual rainfall which would have been received at these higher altitudes (Shorter *et al.*, 1969). In any event, Dozmary Pool contains arguably the most complete Holocene pollen sequence on Bodmin Moor with an unbroken pollen/sediment sequence through Pollen Zones V and VI, although this evidence has been supplemented recently by archaeological and palynological data from Rough Tor on Bodmin Moor (Gearey and Charman, 1996). The evidence from Dozmary Pool has been instrumental in revealing the existence of unconformities in other regional profiles which date approximately from this time (see Hawks Tor and Blacklane Brook). The pollen stratigraphic record from Dozmary Pool is therefore a key to the interpretation of more abbreviated sequences found elsewhere in upland South-West England.

The potential of the site for assessing the extent of Mesolithic activities in the area has not, unfortunately, been realized from the pollen evidence, although it will remain central to resolving the controversy, and may provide complementary evidence to the Rough Tor site in this respect (Gearey and Charman, 1996). The apparently continuous pollen record from a peat sequence containing Mesolithic flints and charcoal, provides great scope for elaborating the patterns and timings of various postulated Mesolithic clearances in the area. The most recent study of these phenomena was severely hampered by unforeseen disturbances to the peat profile at the chosen sample point. Although positive evidence of Mesolithic clearance, as interpreted from pollen sequences on Dartmoor (Blacklane Brook and Black Ridge Brook), cannot yet be demonstrated at Dozmary Pool, the evidence afforded from the most recent study here has wide implications. The evidence from Dozmary Pool suggests that stratigraphical and pollen data from cores alone, especially where no peat faces are available, may provide misleading results: the disturbances and inversion of parts of the Dozmary profile were not visible in the sample core. Only the radiocarbon evidence reveals the scope of the problem and highlights the need for detailed radiocarbon calibration for all pollen biostratigraphic studies carried out on core material. It is possible that the disturbances noted at Dozmary Pool are localized: further pollen work with additional radiocarbon dates may well prove the extent of disturbance and reveal the true importance of anthropogenic factors on the regional vegetation history.

Conclusion

Peat and lake muds at Dozmary Pool preserve the fullest record of changing Holocene environmental conditions on Bodmin Moor. This radiocarbon-dated sequence, and that of Blacklane Brook on Dartmoor, together provide some of the best available Holocene vegetational and inferred climatic records for upland South-West England. The conservation value of this GCR site stems not only from the unusual combination of archaeological and stratigraphic evidence, but from the apparently continuous sedimentary record which spans the driest (Boreal) part of the Holocene, a time when water supplies at other sites were inadequate to allow continued peat growth and pollen preservation. The pollen record demonstrates very clearly that the vegetation of Bodmin Moor developed throughout the Holocene in response to exposure to the elements, and a generally open, sparsely wooded landscape is indicated. Although there is clear archaeological evidence for anthropogenic activities at the site, the relative rôles of climatic change and human interference in vegetational development are not entirely clear from the site evidence. Recent studies here, using radiocarbon dating methods, demonstrate very clearly the potential pitfalls of interpreting pollen data from borehole cores without the use of such calibration.

BLACKLANE BROOK
S. Campbell

Highlights

A reference site for Holocene vegetational history on Dartmoor, Blacklane Brook provides exceptionally detailed evidence for a pre-*Ulmus* (elm) decline in forest cover, attributed to the activities of Mesolithic people.

Introduction

Relatively few detailed studies have been carried out on the peats of South-West England, and Blacklane Brook preserves one of the most extensive Holocene vegetational records yet known from the region. First studied in detail by Simmons (1964a), the site has become a cornerstone for studies of Holocene vegetational and environmental history (e.g. Simmons, 1962, 1964b; Smith, 1970; Stephens, 1973; Caseldine and Maguire, 1981, 1986; Cullingford, 1982; Simmons *et al.*, 1987; Caseldine and Hatton, 1996; West *et al.*, 1996). Recent accounts of the site's pollen stratigraphy and plant macrofossils were given by

Simmons *et al.* (1983) and Maguire and Caseldine (1985), respectively.

Description

The Blacklane Brook pollen site (SX 627686) lies on southern Dartmoor at *c.* 457 m OD, and consists of a series of shallow peat sections exposed along this tributary of the southward-flowing River Erme (Figure 4.18). The site lies in an area covered mostly by blanket peat. Simmons (1964a) originally described a peat section some 1.2 m deep, but a more complete 2.2 m-deep section was later described by Simmons *et al.* (1983). The following stratigraphy is reconstructed here using data from two separate, but overlapping, monoliths (Simmons *et al.*, 1983).

7. Red-brown, fibrous *Eriophorum* peat, becoming increasingly humified towards base (0–110 cm)
6. Dark brown, well-humified amorphous peat (110–174 cm) with wood fragments at 142 cm and 169 cm
5. Wood layer of *Salix* and *Betula* (174–182 cm)
4. Dark brown amorphous peat with scattered wood fragments (182–202 cm)
3. Dark brown, pseudo-fibrous laminated peat with occasional lighter bands of more fibrous peat (202–217 cm)
2. Dark well-humified peat with increasing mineral matter with depth (217–222 cm)
1. Grey-brown silty clay with granite gravel (below 222 cm)

Three radiocarbon age determinations (HAR–4460 to HAR–4462) were obtained from materials within the section (Simmons *et al.*, 1983).

Interpretation

The original pollen spectrum (Simmons, 1964a) was thought to cover the period from the end of the Devensian late-glacial to the opening of the Sub-Boreal or Pollen Zone VIIb of Godwin's scheme. A phase of open-country conditions dominated by shrubs, birch and pine (corresponding to Godwin Pollen Zone IV) gave way to a phase characterized by the immigration of trees, particularly *Corylus*, which displaced the open-ground species (= Pollen Zone V and early and mid-Zone VI). Finally, the pollen record showed a rapid 'clear-

Figure 4.18 Blacklane Brook pollen site, southern Dartmoor. (Photo: S. Campbell.)

ance' phase followed by the stabilization but not regeneration of many taxa (= late Zone VI and early Zone VIIb; Simmons, 1964a; Simmons *et al.*, 1983). The main characteristics of the clearance phase were: 1. a slight reduction in the frequency of *Quercus* pollen; 2. the appearance of *Fraxinus* and *Prunus-Sorbus* type pollen; 3. a fall in *Corylus/Myrica* pollen; followed by 4. a peak in grass pollen and fern spores. This clearance was judged to have taken place late in Pollen Zone VI, well before the Neolithic, leading to the speculation, now more widely accepted, that woodland clearance had been initiated in this area by Mesolithic people.

From the evidence presented by Simmons

(1964a), Simmons *et al.* (1983) and Maguire and Caseldine (1985), the following updated sequence of vegetational, climatic and environmental changes can be interpreted from the peat sections at Blacklane Brook. The pollen record here commences in the early Holocene at an estimated 10 200 BP: it has been divided into six local pollen assemblage zones (BLB1–BLB6) from which the mire's history and local forest development can be reconstructed (Simmons *et al.*, 1983). From the pollen contained in beds 1 and 2, the initial vegetation of the site and local area (pollen assemblage zone BLB1) appears to have been predominantly dry grassland or meadowland, with perhaps some localized willow and birch, *Empetrum* heath and juniper scrub. The succeeding pollen assemblage zone BLB2 (occurring in most of bed 3) indicates the continuation of fairly open vegetation, but with a persistence of local birch and willow stands and juniper scrub. The site itself consisted of a sedge and *Sphagnum* mire at this time. Local pollen zone BLB3 (upper bed 3; lower bed 4), dominated by birch and grasses, shows a distinct shift with the mire changing from sedge- to grass-dominated. At the same time, birch and oak woodlands also became established within the pollen catchment, perhaps at lower elevations, on drier sites or even on the hillsides around the mire. Simmons *et al.* interpreted this evidence as indicating a change to more acid conditions prior to the development of blanket peat.

Pollen zone BLB4 (upper bed 4; bed 5; lower bed 6) indicates vegetation succession to the deciduous forest stage (mid-Holocene). Although dominated by shrub pollen, this *Quercus–Corylus/Myrica* zone suggests that forest covered much of the local area, perhaps encroaching locally on the site. However, pollen suggestive of open communities persists throughout the zone, and it is possible that the immediate environs remained as a bog, with perhaps some willow carr (Simmons *et al.*, 1983). A radiocarbon date of 7660 ± 140 BP (HAR–4462) from the lowest part of bed 6 gives a minimum age for the layer of *Salix* and *Betula* wood beneath (bed 5).

Although not radically different to its predecessor in its tree pollen content, it is within the same context of deciduous forest cover in zone BLB5 (bed 6) that human modification to the vegetation can first be detected. The distinguishing feature of this zone is the onset at *c.* 7660 radiocarbon years of higher pollen and spore frequencies usually associated with the opening of the forest: *Pteridium* spores, and the pollen of Rosaceae and that of a number of herbaceous types are all found within the early part of this zone. These floral changes are also accompanied by increased amounts of charcoal within the sediment profile. A radiocarbon date of 6010 ± 90 BP (HAR–4461) within zone BLB5 marks the rise of *Alnus*. The evidence from this pollen zone was interpreted by Simmons *et al.* as indicating the activities of Mesolithic people, and their use of fire to create grassy clearings (see below).

Local pollen assemblage zone BLB6 (upper bed 6; bed 7) (*Quercus–Alnus–Corylus* and *Myrica–Calluna*) reveals a decline in human pressure upon the local forest cover, although the maintenance of a weak herbaceous flora indicates some contemporary human activity. There is no direct evidence for agriculture and no prehistoric remains have yet been discovered from the site or its immediate environs (Simmons *et al.*, 1983). An important feature in this part of the Blacklane pollen record is the 'elm decline' dated to *c.* 4260 ± 90 BP (HAR–4460), and occurring right at the beginning of local pollen assemblage BLB6. Another is the development of a distinctive weed flora including *Plantago lanceolata*, interpreted as evidence for Neolithic activity (Simmons *et al.*, 1983): as in the previous zones, there is no direct evidence for agricultural activity, for example, cereal growing. During this biozone, soil acidity rose and bog growth probably continued.

Relatively few sites have yet been studied in this region for their Holocene (and Devensian late-glacial) pollen biostratigraphy. Blacklane Brook not only provides a key record of Holocene vegetation changes on Dartmoor but, in addition, forms a vital element in a national network of pollen sites which shows major regional variations in the vegetational history of the British Isles. In common with other sites in southern Britain, Blacklane Brook reveals that the periglacial conditions of the Younger Dryas (= Loch Lomond Stadial) were replaced at *c.* 10 000 BP by the gradually ameliorating climate which marks the onset of the Holocene and the progressive development of vegetation to mixed deciduous forest by mid-Holocene times. The record from this site provides important details of this succession, from the colonization of open heathland to the deciduous forest stage, and gives important insights regarding the composition of the forest and the height of the regional tree line. There seems little doubt that *Quercus* sp. was the dominant tree in the forest: its early arrival relative to other trees is a characteristic feature of pollen diagrams from South-West and south-central England (Conolly *et al.*, 1950; Seagrief, 1959, 1960; Simmons, 1964a).

The occurrence of wood remains (*Salix, Betula* and *Sorbus aucuparia* (rowan)) in the peat of bed 5 at Blacklane Brook is significant (Simmons, 1964a; Simmons *et al.*, 1983; Maguire and Caseldine, 1985). This, together with pollen evidence from elsewhere on Dartmoor, enabled Simmons (1962, 1964a, 1969) to estimate that the tree-line lay in the range between 427–457 m. This led him to argue that all of upland Dartmoor had been wooded during the so-called 'forest maximum' (Simmons, 1964a), a possible exception being the very exposed summits which may have remained bare, with waterlogged hollows and only thinly dotted with birch or oak-hazel scrub (Simmons, 1969). With the exception of some sites in Wales (Taylor, 1980) and on Bodmin Moor (Brown, 1977), this is one of the lowest published tree-line estimates in Britain for the period. Nowhere else is the tree-line fixed below 650 m; even in the Cairngorms and the Lake District it is reputed to have exceeded 760 m (Maguire and Caseldine, 1985). This has led to the suggestion that all of Dartmoor was in fact wooded at the time of maximum forest development, although, from macrofossil evidence alone, it is only certain that trees actually grew up to altitudes between *c.* 497 and 547 m OD (Maguire and Caseldine, 1985).

Equally significant is the record from Blacklane Brook of the rôle of prehistoric people in modifying the regional vegetation cover. Although many examples of forest clearance by Mesolithic people in the British Isles have now been reported (e.g. Dimbleby, 1962, 1963; Smith, 1970; Jacobi *et al.*, 1976; Mellars, 1976; Simmons, 1979; Simmons and Innes, 1981), and the evidence from Blacklane Brook shown to be by no means unique even on Dartmoor (Caseldine and Maguire, 1981; Hatton, 1991; Caseldine and Hatton, 1993), the sections here are important for they provided the first firm scientific basis for connecting small-scale fluctuations in the local pollen record with the activities of Mesolithic people: this was at a time when little such evidence had been adduced elsewhere (e.g. Dimbleby, 1962). The importance of the early pollen biostratigraphical work at Blacklane Brook (Simmons, 1964a) in this context has been enhanced by additional and more detailed pollen and macrofossil work and by the application of radiocarbon dating methods (Simmons *et al.*, 1983; Maguire and Caseldine, 1985).

The relative responsibility of climatic change and Mesolithic humans to account for these small-scale fluctuations in the pollen record has not been unequivocally determined (Cullingford, 1982). However, it is now widely accepted that Mesolithic alterations did occur to the woodland fringe areas (Caseldine and Maguire, 1986; Hatton, 1991; Caseldine and Hatton, 1993, 1996). There is no evidence at Blacklane Brook to support repeated burning of the area, as occurred in the southern Pennines and the North York Moors (cf. Jacobi *et al.*, 1976; Simmons and Innes, 1981). However, there is no reason why such local clearances were not effected by the use of fire to maintain open ground and scrub in a predominantly mature forest landscape: there is evidence (see above) that some areas of Dartmoor remained unwooded even at the

peak of forest development. Alternatively, the Blacklane Brook evidence may suggest clearance as a reaction to the rapid encroachment of forest – which perhaps was less rich in necessary animal food resources than the more open and ecotonal systems which it replaced (Simmons *et al.*, 1983).

The appearance of a 'weed flora' subsequent to the 'elm decline' at *c.* 4260 BP, and perhaps after a temporary cessation of pressure on the local forests, was taken by Simmons (1964a) to be consistent with the archaeological picture of a very sparse Neolithic occupation of Dartmoor. However, the quality of evidence for the elm decline on Dartmoor is very poor (Caseldine and Hatton, 1996) and the interactions between early Neolithic activity and climatic/vegetational change are still poorly understood (Fleming, 1988; Caseldine and Hatton, 1996).

Conclusion

Blacklane Brook provides an important radiocarbon-dated record of vegetational and climatic changes on Dartmoor during the Holocene. This record charts the development of vegetation from the early Holocene colonization by open grassland and heathland through until the attainment of a mixed deciduous woodland dominated by oak in mid-Holocene times: the latter forest may indeed have spread to all but the most exposed summits of Dartmoor, and the wood remains from Blacklane Brook have long centred in a controversy regarding the height of the regional tree-line. Fluctuations in the pollen record thereafter provide crucial evidence for determining the relative effects of climatic change and anthropogenic activities on regional vegetation development: from the Blacklane Brook evidence, there seems little doubt that Mesolithic inhabitants were having an important impact on the forest cover, perhaps even as early as the eighth millennium BP. The details and precise durations of these anthropogenic activities have yet to be determined, and Blacklane Brook will undoubtedly play a key rôle in resolving many of the outstanding questions. Although elsewhere there is much evidence to suggest that Neolithic people had a much greater impact on the forest cover, Blacklane Brook shows no direct evidence of cereal cultivation or other agricultural practices at this time. This has been used to support the view that Dartmoor may have been only sparsely populated by Neolithic people. The pollen biostratigraphic evidence from this site complements that from Black Ridge Brook on northern

Figure 4.19 Shallow peat sections exposed along Black Ridge Brook, northern Dartmoor. (Photo: C.J. Caseldine.)

Dartmoor, where an extensively radiocarbon-calibrated profile provides particularly detailed evidence of changing conditions at the Devensian/Holocene transition.

BLACK RIDGE BROOK
S. Campbell and R. Cottle

Highlights

Black Ridge Brook GCR site provides an important record of Devensian late-glacial and early Holocene environmental conditions on northern Dartmoor. It shows that human modification of the natural mid-Holocene forest cover occurred as early as Mesolithic times. The site has also yielded important information regarding the extent of woodland on Dartmoor during the 'forest maximum'.

Introduction

Black Ridge Brook has yielded one of the most extensive records of Holocene environmental change in South-West England. The human rôle in

modifying the natural vegetation cover can be shown, on the basis of radiocarbon-dated pollen evidence from this site, to have begun as early as the Mesolithic. The pollen record has also been seen as central to establishing the extent of tree cover ('tree-lines') on Dartmoor during the forest climax in the mid-Holocene. Referred to by Caseldine and Maguire (1981), the site has since been studied in detail by Maguire (1983), Maguire and Caseldine (1985) and Caseldine and Maguire (1986). Pollen studies at this and adjacent sites have allowed a comprehensive picture of vegetational and environmental changes to be reconstructed for the area (Simmons, 1964a; Simmons *et al.*, 1983; Hatton, 1991; Caseldine and Hatton, 1993, 1996).

Description

Black Ridge Brook rises in the northern part of Dartmoor on the south-west-facing slopes of Black Ridge (573 m OD), an interfluve between the northerly draining West Okement river and the south-west-flowing Amicombe Brook. The Black Ridge Brook pollen site (SX 579842) is located

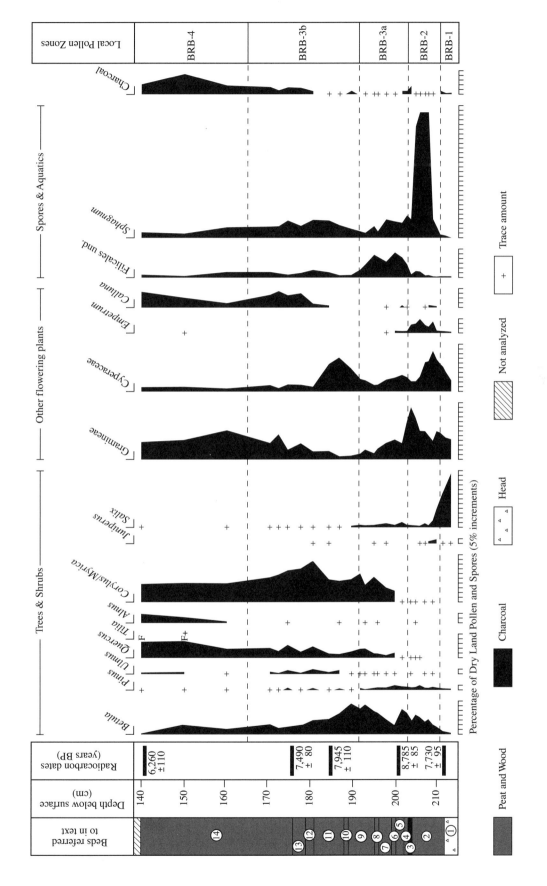

Figure 4.20 Simplified pollen diagram for Black Ridge Brook, adapted from Caseldine and Maguire (1986).

approximately at the confluence of the Black Ridge and Amicombe brooks (Figure 4.19). The surrounding interfluve areas, including nearby Little Kneeset (507 m OD), are characterized by gently undulating topography, largely blanketed with peat and occasionally punctuated by tors. Several pollen profiles have been described from different locations along this reach of Black Ridge Brook: the main site, however, is that referred to by Maguire and Caseldine (1985) as Black Ridge Brook (lower).

The most detailed description of the site was provided by Caseldine and Maguire (1986) who recorded a 2.12 m-thick sequence from a river section. The following detailed stratigraphy was provided for the lower part of this section:

14. Black pseudo-fibrous peat (140-176 cm)
13. Wood (176-179 cm)
12. Wood / well-humified brown peat (179-181 cm)
11. Grey/dark brown well-humified peat (181-188 cm)
10. Wood (188-189 cm)
 9. Black well-humified peat (189-195 cm)
 8. Wood (195-196 cm)
 7. Black well-humified peat (196-199 cm)
 6. Wood (199-200 cm)
 5. Black well-humified peat (200-202 cm)
 4. Wood (202-203 cm)
 3. Charcoal (203-204 cm)
 2. Well-humified grey/brown peat (204-212 cm)
 1. Head (mineral matter) (> 212 cm)

The pollen and radiocarbon evidence derived from this sequence is shown in simplified form in Figure 4.20: five radiocarbon dates are available, and the pollen spectra have been divided into five local pollen assemblage biozones (BRB1 to BRB4).

Interpretation

The basal sediments (beds 1 and 2) contain a *Salix*-Gramineae-Cyperaceae pollen assemblage (BRB1), indicating open-ground conditions with very few trees. High *Salix* values are taken as denoting the presence of *Salix herbacea* rather than local stands of more shrubby forms of *Salix* (Caseldine and Maguire, 1986). The dominantly minerogenic nature of bed 1 and the pollen evidence are strongly suggestive of conditions at the Devensian late-glacial/Holocene transition, similar to those described throughout the British Isles and north-west Europe (Caseldine and Maguire, 1986).

The radiocarbon date of 7730 ± 95 BP (GU-1606), from sediments covered by the lowest part of the biozone, is thought to be too 'young', perhaps due to groundwater seepage along the peat/mineral boundary between beds 1 and 2 or from rootlet contamination (Caseldine and Maguire, 1986).

Local pollen assemblage zone BRB2 spans beds 2-4 (well-humified peats with charcoal and wood layers). This *Juniperus-Empetrum-Pinus-Betula*-Gramineae zone signals the establishment of an early Holocene birch woodland. The fall in *Salix* pollen, the continued presence of Gramineae and Cyperaceae pollen and successive peaks in the pollen of *Empetrum*, *Juniperus* and *Betula*, which characterize the biozone, reflect a pattern of plant succession with an initial gradual development of dwarf birch scrub and the eventual immigration and establishment of tree birches. A radiocarbon date of 8785 ± 85 BP (GU-1700) from the base of bed 5 suggests that the disappearance of dwarf scrub elements of the vegetation and the establishment of birch woodland, at best only scrubby and open in form, was a slow process. Certainly, in comparison with other areas of the British Isles, where birch spread rapidly after the climatic amelioration following the Younger Dryas (Smith and Pilcher, 1973; Huntley and Birks, 1983), the late spread of birch to Dartmoor and other upland areas of South-West England seems anomalous. Caseldine and Maguire (1986) suggested that the delay in the development of birch woodland could have been due to the soil properties of the area. After the close of the Devensian, Dartmoor would have been dominated by large areas of bouldery clitter and very coarse periglacial detritus rather than the thick glacial sediments present over much of Scotland, northern England and Wales. Where fine fractions were largely absent from the soil, it is thought that the reduction in water retention capacity may have limited birch establishment. Other factors such as exposure and winter temperature regimes may also have played a part in the observed delay in woodland development. The prominent charcoal layer (bed 3) suggests that fire (naturally rather than human-induced) may have been instrumental in retarding woodland succession at this early stage of the Holocene (Caseldine and Maguire, 1986).

The succeeding pollen biozone (BRB3a) covers beds 5-8 and the lower part of bed 9, a series of alternating well-humified peat and wood layers. This *Corylus/Myrica-Quercus-Betula*-Filicales assemblage denotes the establishment of birch-hazel woodland on Dartmoor after *c.* 8785 BP. *Corylus* becomes co-dominant in the pollen record by *c.* 8500 BP, a relatively late date. Caseldine and Maguire (1986) have argued, on the basis of arboreal and shrub pollen frequencies, that for a time birch-hazel woodland reached close to, if not over, the summit areas, largely clothing Dartmoor, with perhaps only pockets of more open ericaceous heath on the very highest areas. The pollen frequencies at Black Ridge Brook certainly demonstrate the existence of woodland around the site, some 120 m below the nearby summits. During this period of woodland expansion (roughly 8700–8500 BP), *Quercus* would have become the dominant tree in more sheltered valleys. Changes in local conditions are also denoted by pollen zone BRB3a: at *c.* 8785 BP shallow *Sphagnum* bog was replaced by *Salix* carr which lasted until *c.* 7490 BP – well into the succeeding pollen assemblage zone BRB3b (Caseldine and Maguire, 1986).

Local pollen assemblage BRB3b spans the upper part of bed 9, beds 10–13 and the lower part of bed 14, and provides evidence of vegetational changes between *c.* 8000 and 7500 BP. It shows that woodland communities still remained at high altitudes after *c.* 8000 BP, but that locally, oak and hazel developed at the expense of birch; the evidence from this and other Dartmoor sites shows that hazel was dominant in the woodland above *c.* 400 m (Caseldine, 1983; Maguire, 1983; Simmons *et al.*, 1983; Caseldine and Maguire, 1986). In common with other areas of the South-West at this time, *Pinus* was sparse, and elm appears never to have become a significant element of the upland woodland community, having been restricted to more sheltered lowland locations (Caseldine and Maguire, 1986). The pollen evidence shows that woodland retreat from the highest areas of Dartmoor may have begun at *c.* 7700–7600 BP.

The succeeding local pollen assemblage zone (BRB4) occurs in the upper part of bed 14 and in the sediments above it (not analysed in detail; Figure 4.20). This zone opens at *c.* 7490 ± 80 BP, and demonstrates colonization of the area by *Alnus* (alder), which displaced birch especially in local valley floors and areas marginal to the bog (Caseldine and Maguire, 1986). A general reduction in woodland cover is indicated by increases in the percentages of Gramineae and *Calluna* pollen at the expense of *Corylus/Myrica* and *Betula*. Although these changes may partly reflect variations in local conditions – from *Salix* carr to blanket peat – higher percentages of charcoal in the profile after *c.* 7000 BP suggest that vegetational development was being governed by widespread external influences.

Caseldine and Maguire (1986) have argued that these changes (a fall in arboreal and shrub pollen, rising amounts of charcoal in the profile and the extension of blanket peat on relatively level ground and gentle slopes) were brought about by the recurrent burning of woodland by Mesolithic hunting populations. This burning may have led to soil deterioration and the onset of blanket peat formation (cf. Jacobi *et al.*, 1976; Simmons *et al.*, 1983). The continuous presence of charcoal in the peat profile at Black Ridge Brook up to and after *c.* 6260 BP is not matched by a further decline in tree and shrub pollen after *c.* 7000 BP. Caseldine and Maguire (1986) suggested that once the higher areas had lost their woodland cover, and blanket peat was established, an equilibrium was achieved with fire maintaining the open character of the newly developed moorland communities, and suppressing woodland there, but not further diminishing areas of upland woodland.

The pollen diagram from Black Ridge Brook (Figure 4.20) provides some measure of changes after 6260 BP, but the nature of the record and the lack of radiocarbon dates makes detailed interpretation unwise (Caseldine and Maguire, 1986).

Conclusion

Black Ridge Brook GCR site provides a detailed pollen record, calibrated by radiocarbon dating methods, which covers the Devensian late-glacial/early Holocene time interval. Sites providing environmental data from this time interval are not only rare on Dartmoor, but there are relatively few in South-West and central southern England as a whole. The record from Black Ridge Brook is perhaps most notable for showing unequivocal evidence for a significant delay in vegetational development in the early Holocene, especially in comparison with other areas of the British Isles. This has been attributed to a variety of environmental factors including the coarse nature of local parent materials and extreme climatic exposure. The site also provides some of the best evidence in Britain for the influence of Mesolithic people in reducing woodland cover by fire, and for providing a link with these activities and the widespread initiation of blanket peat. The site's pollen record is one of the most comprehensive in South-West England and is central to reconstructions of regional vegetation history. It is intimately linked to human influence on the natural landscape.

Chapter 5

Pleistocene cave sequences

INTRODUCTION
S. Campbell

South-West England contains some of the most important Pleistocene cave sequences in Britain. The Carboniferous Limestone of the Somerset and Avon districts (particularly the Mendips), and the Devonian limestones of south Devon (Figures 2.1 and 5.1), host numerous caves and cave systems, many of which contain important sediment sequences, faunal remains and artefacts. As a result, the caves have long attracted scientific interest and several locations remain the subject of intensive modern research. This chapter reviews evidence from four of the stratigraphically most significant localities – Kent's Cavern (Torquay), Tornewton Cave (Torbryan), Chudleigh Caves and Joint Mitnor Cave (Buckfastleigh).

The importance of the region for Pleistocene cave studies stems from a number of circumstances. First, the region was situated overwhelmingly, if not entirely, beyond the southern limits of Pleistocene glaciation (Figure 2.3), and lay centrally in a zone across which a whole range of Pleistocene mammal species would have migrated in response to climatic and environmental changes (Sutcliffe, 1969; Cullingford, 1982).

Second, the Devonian and Carboniferous limestones of the region exhibit significant karst development, providing a profusion of suitable caves, fissures and shafts where sediments, faunal remains and artefacts have accumulated by a variety of agencies. In some locations, animals fell into caves via open shafts and became trapped (e.g. Joint Mitnor Cave and Banwell Bone Caves; Cullingford, 1953; Sutcliffe, 1960; Macfadyen, 1970); other locations were used as natural refuges or dens, for example by bears or hyaenas, in which the animals died, leaving their own remains as well as those of their prey (e.g. Tornewton Cave; Sutcliffe and Zeuner, 1962; Sutcliffe, 1974). Occasionally, fossil bones are preserved in cave stream deposits (e.g. Eastern Torrs Quarry Cave, near Plymouth; Sutcliffe, 1969, 1974). Particularly significant are the remains of rodents, which evolve more quickly than the larger mammals and which can provide important palaeoenvironmental evidence (Sutcliffe, 1974; Sutcliffe and Kowalski, 1976). Bone preservation, even of microfaunal remains, has been aided in many circumstances by alkaline conditions. Thirdly, the South-West, and Devon in particular, was the scene for much of the pioneering work carried out in the nineteenth century.

Geological Conservation Review site selection

The large numbers of caves and fissures in South-West England where sediments, faunal remains and artefacts have been described, provide significant difficulties with respect to selecting the 'best' or most appropriate sites for conservation. Following the guidelines outlined in Chapter 2 (Ellis *et al.*, 1996), four sites believed to contain the most important stratigraphical and palaeoenvironmental evidence (and with some of the most extensive histories of research) were selected, and are described in this volume (this chapter). A further five sites in Somerset and Avon (Sun Hole (Cheddar Complex); Bridged Pot and Savory's Hole (Ebbor Gorge); Badger Hole and Rhinoceros Hole (Wookey Hole); Picken's Hole and Beeche's Hole (Crook Peak–Shute Shelve Hill); Banwell Bone Caves) are to be described in a separate volume of the GCR Series – *Fossil Mammals and Birds* (in prep.) – in the *Pleistocene Vertebrata* section. Tornewton Cave and Joint Mitnor Cave (this volume; Chapter 5) and Brean Down (this volume; Chapter 9) are described again in the latter volume by virtue of their palaeontological importance, which was assessed independently.

The caves of south Devon

The south Devon caves have been studied for nearly 200 years (e.g. Polwhele, 1797). In fact, the first bone cave to be studied scientifically in Britain was excavated by Joseph Whidbey at Oreston, near Plymouth, in 1816 (Sutcliffe, 1969; Cullingford, 1982). In 1825, the Reverend J. MacEnery began his excavations at Kent's Cavern, later to be continued by William Pengelly (e.g. Pengelly, 1868b, 1869, 1871, 1878, 1884). William Buckland also visited many of the Devon caves during the 1820s, including Kent's Cavern, Pixie's Hole (Chudleigh), Oreston and Anstey's Cove Cave among others (Figure 5.1; Sutcliffe, 1969). Importantly, Pengelly's excavations in the Brixham Cave in 1859 provided the first published account to demonstrate that humans had occupied the region before the extinction of the cave mammals, a fact later to be reinforced by his work in Kent's Cavern (note also the work of MacEnery; this chapter) (Sutcliffe, 1969).

Of the early excavations, J.L. Widger's in the Torbryan Caves from about 1865 to 1880, are particularly notable. From the 1920s until the

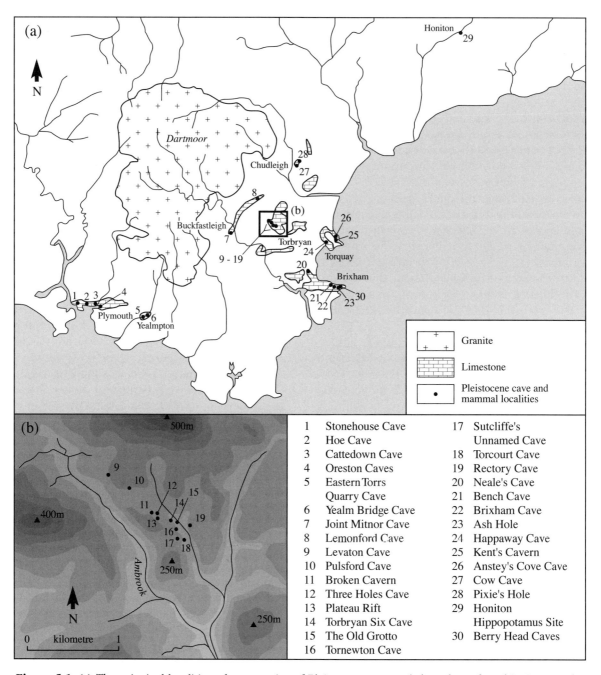

Figure 5.1 (a) The principal localities where remains of Pleistocene mammals have been found in Devon, after Sutcliffe (1969). (b) Excavated caves in the Torbryan Valley, after Roberts (1996). The location of Berry Head 'sea caves' (Proctor, 1994, 1996) is also shown.

beginning of World War Two, the Torquay Natural History Society excavated in Cow Cave (Chudleigh), Kent's Cavern, Tornewton Cave and Joint Mitnor Cave. Since the War, excavations have been carried out by Almy, Cheesman, Neale, Rosenfeld, Sutcliffe, Zeuner and others in Tornewton, Three Holes and Levaton caves (Torbryan); Joint Mitnor Cave; Eastern Torrs Quarry

Cave, Yealmpton (discovered by quarrying in 1954); and Neale's Cave, Paignton (discovered in 1958) (Sutcliffe, 1969, 1974, 1977).

The principal localities where remains of Pleistocene mammals have been found in south Devon are shown in Figure 5.1. With the exception of the Honiton hippopotamus site, where a peat deposit with an Ipswichian fauna was excavated during roadworks in 1964 (Turner, 1975), the caves are all located in Devonian limestone below *c*. 100 m OD (Sutcliffe, 1969, 1974, 1977).

Kent's Cavern, Tornewton Cave and Joint Mitnor Cave have long been regarded as the most important sites, both stratigraphically (in terms of length and detail of palaeoenvironmental record) and palaeontologically. However, Chudleigh Caves also contain significant sediment sequences, fossils and artefacts which have been investigated systematically by Collcutt (1984, 1986); Pixie's Hole is particularly notable in containing the greatest demonstrable extent of Later Upper Palaeolithic deposits surviving in any known British cave site (Collcutt, 1996). Recent Uranium-series work on speleothem in the Berry Head Caves (Figure 5.1) confirms their potential for calibrating marine Pleistocene events (Proctor and Smart, 1991; Baker, 1993; Proctor, 1994; Baker and Proctor, 1996) and necessitates their future assessment for GCR status.

Kent's Cavern is a site of considerable historical importance in British Pleistocene studies (e.g. Pengelly, 1884; Campbell and Sampson, 1971; Proctor and Smart, 1989; Proctor, 1994). It contains a sequence of sediments, with fossil fauna and artefacts, which extends well back into the Middle Pleistocene. The interpretation of the sequence has been hampered both by Pengelly's removal of most of the cave sediments and by an inability, until recently, to date the earliest deposits in the cave: the latter have appeared to most workers to be significantly older than deposits in the upper part of the sequence. Some have speculated that the controversial Breccia, with its *Ursus deningeri* von Reichenau fauna (with *Homotherium latidens* Owen), could date from the Cromerian Complex (e.g. Hinton, 1926; Campbell and Sampson, 1971; Sutcliffe and Kowalski, 1976; Straw, 1983), but recent Uranium-series and Electron Spin Resonance (ESR) dates associated with the breccia (Proctor, 1994, 1996) suggest an age of *c*. 300–400 ka BP for the deposit, and a possible correlation with Oxygen Isotope Stages 9 or 11 of the marine record. Proctor's work, however, confirms earlier views that the Crystalline Stalagmite represents a lengthy hiatus between deposition of the Breccia and the overlying Cave Earth. Controversially, Straw (1995, 1996) has suggested that the Crystalline Stalagmite was broken up by an earthquake. The traditional interpretation of the Cave Earth as a Devensian deposit is upheld by U-series dates which invite a correlation with Oxygen Isotope Stages 4 to 2: the deposit contains a typical and diverse Devensian mammal fauna dominated by spotted hyaena, which evidently used the cave as a den (Proctor, 1994, 1996).

Tornewton Cave, in the Torbryan Valley (Figure 5.1), is of considerable importance for a sequence of richly fossiliferous deposits spanning the entire Late Pleistocene and extending well back into the Middle Pleistocene (Roberts, 1996). Sutcliffe and Zeuner (1962) interpreted the sequence as an Ipswichian interglacial layer (Hyaena Stratum) sandwiched between Saalian (Glutton Stratum) and Devensian (Reindeer Stratum) cold-climate deposits, and this view has since been reiterated in a number of publications and wider correlations (e.g. Sutcliffe, 1969, 1974, 1977; Stephens, 1973). Recent multi-disciplinary work, sponsored by the British Museum between 1989 and 1992 at Torbryan, has confirmed the palaeoenvironmental value of the caves, especially Three Holes Cave, Broken Cavern and Tornewton Cave, and led to considerable re-evaluation and elaboration of the sequences (e.g. Proctor, 1994; Barton, 1996; Berridge, 1996; Cartwright, 1996; Caseldine and Hatton, 1996; Currant, 1996; Debenham, 1996; Gleed-Owen, 1996, 1997; Irving, 1996; Price, 1996; Proctor and Smart, 1996; Seddon, 1996; Stewart, 1996). Although the full results of the Torbryan Caves Research Project await publication (Roberts *et al.*, in prep.), it is likely that the sequence at Tornewton Cave spans from Oxygen Isotope Stage 7 (Otter Stratum), through Stage 6 (minor debris flow), to the warmest parts of Stage 5 (Bear Stratum, Hyaena Stratum and 'Dark Earth') to Stage 4 (isolated deposits with a *Rangifer* and *Bison* fauna) and Stage 3 (material at the cave mouth) to Stage 2 (Glutton Stratum – a huge mass-flow deposit), thus providing one of the most complete Pleistocene cave sequences known in Britain.

Although a comparable reappraisal of the sequence at Joint Mitnor Cave at Buckfastleigh is not available, the site remains one of the most important localities for Ipswichian mammal remains in Britain (Stuart, 1982b), and is included in this volume on that basis.

The caves of Somerset and Avon

The Carboniferous Limestone of the Bristol, Bath and Mendip areas hosts a significant number of caves, fissures and shafts which contain Pleistocene sediments, fossils and artefacts (e.g. Balch, 1947, 1948; Donovan, 1954, 1964; Ford, 1968; Macfadyen, 1970; Drew, 1975; Campbell, 1977; Hawkins and Tratman, 1977; Irwin and Knibbs,

1987; Waltham *et al.*, 1996). The caves have long been the subject of scientific study, but overwhelmingly (with the notable exception of Westbury-sub-Mendip) the evidence dates largely from the Devensian Stage, particularly Oxygen Isotope Stages 4 and 2, and does not provide a sufficient length or detail of palaeoenvironmental record to have merited selection within the subject 'blocks' of the GCR described in this volume (Chapter 2). However, a number of the Somerset and Avon cave sites are of considerable palaeontological significance, and are thus described elsewhere in the GCR Series of volumes (see above; Geological Conservation Review site selection). Excellent reviews, with comprehensive bibliographies, of the cave and fissure deposits of these areas are provided by Donovan (1954, 1964), Macfadyen (1970) and Hawkins and Tratman (1977), and only brief background comments are provided here on the more significant localities for Pleistocene studies.

The longest and most significant record from this region comes from Westbury-sub-Mendip, where a fossil mammal fauna (Westbury 1 fauna) may provide evidence of conditions in Cromer-Complex times (Heal, 1970; Bishop, 1974, 1982; Stringer *et al.*, 1996). The later Westbury faunas (Westbury 2 and 3 faunas) appear to be post-Cromerian (cf. West Runton) but pre-Hoxnian *sensu stricto* (Stringer *et al.*, 1996), and thus represent an important part of the Pleistocene record only poorly known elsewhere in the region. Historically, the Westbury deposits have proved difficult to conserve effectively, although the site's adjacent karstic features are included within a GCR site selected for the 'block' *Karst and caves of Great Britain* (Ellis *et al.*, 1996; Waltham *et al.*, 1996). Continuing use of the locality for stratigraphic and palaeontological studies, however, undoubtedly warrants reconsideration of its GCR status for the site 'blocks' – *Quaternary of South-West England* and *Quaternary of Somerset and Avon* – described in this volume, and also for the *Pleistocene Vertebrata* section of the GCR.

Evidence for conditions in the Ipswichian Stage comes from relatively few cave localities in Somerset and Avon (Chapter 2; Ipswichian Stage). Particularly significant are the sequences at Badger Hole and Rhinoceros Hole (Wookey). The former contains *in situ* deposits which contain mammal remains of Middle and Late Devensian (pre-Devensian late-glacial) age in association with Early Upper Palaeolithic artefacts. Significantly, Rhinoceros Hole contains a sequence of deposits

containing a characteristic Ipswichian interglacial fauna including hippopotamus, as well as younger deposits with contained faunas probably dating from the Middle Devensian (Stuart, 1982b).

Banwell Bone Caves are significant for having yielded a rich Pleistocene bone assemblage, mainly bison and reindeer (Balch, 1948; Cullingford, 1953; Macfadyen, 1970; Stuart, 1982b). Like Joint Mitnor Cave in south Devon, the animals here appear to have fallen to their deaths through a large hole in the cave roof. Extensive deposits remain *in situ*, but further work is necessary to establish the age and full composition of the fauna, and the stratigraphical relationships of deposits within the cave (Stuart, 1982b).

Bridged Pot, in the Ebbor Gorge, is notable for one of the best (presumed) Late Devensian small-mammal assemblages in Britain (Stuart, 1982b). The fauna from this small cave includes steppe pika *Ochotona pusilla* Pallas, arctic lemming, Norway lemming, various voles, red deer and reindeer: largely undisturbed deposits at nearby Savory's Hole are likely to yield a similar assemblage (Stuart, 1982b).

Picken's Hole (Crook Peak–Shute Shelve Hill) is of considerable importance for its clear, well-stratified sequence of deposits and faunas, all dating from within the Devensian Stage. The rich 'Layer 3' fauna (radiocarbon dated to 34 265 + 2,600/− 1,950 BP) includes spotted hyaena, lion, arctic fox, mammoth, woolly rhinoceros, horse, reindeer, suslik and northern vole *Microtus oeconomus* (Pallas), and provides a major source of information for the Middle Devensian. It is also notable for being the most carefully excavated hyaena-den site in Britain (Stuart, 1982b). The nearby Beeche's Hole is likely to contain deposits and faunal remains of comparable age (Stuart, 1982b).

Sun Hole, Cheddar, provides a varied fauna which has been radiocarbon dated to the end of the Late Devensian. The fauna includes both arctic and Norway lemming, various voles, steppe pika, brown bear, wolf, horse, reindeer and, of particular interest, saiga antelope – the only well-dated record of this species in Britain (Stuart, 1982b).

KENT'S CAVERN
D. H. Keen

Highlights

Kent's Cavern contains a wide range of Quaternary deposits, the earliest of which date from the Middle

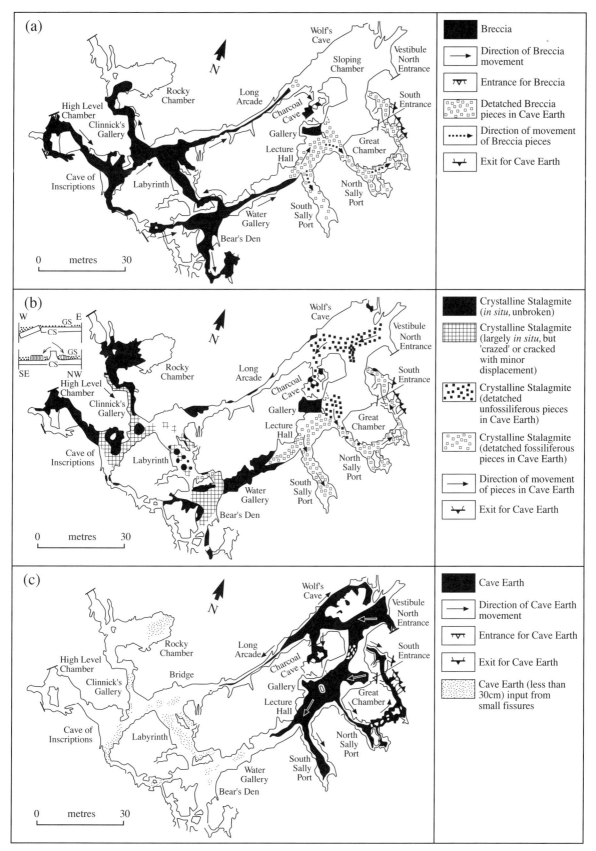

Figure 5.2 Kent's Cavern, after Straw (1996). Distribution of: (a) Breccia; (b) Crystalline Stalagmite; (c) Cave Earth. (a)–(c) are shown as indicated in Reports to the British Association by W. Pengelly, 1865–1880. Cave outline is based on the survey by Proctor and Smart (1989).

Pleistocene. Bones and human artefacts from these deposits were central to nineteenth century controversies about the antiquity of human beings, and the site provides one of the longest records of Pleistocene events in South-West England.

Introduction

Kent's Cavern contains one of the most important Pleistocene sequences in Britain. Its evidence of Middle Pleistocene conditions is unique in the South-West and the site has one of the most protracted histories of research of any British Quaternary locality. Excavations in the cave date from as early as 1824 (Northmore, 1868; Kennard, 1945; Campbell and Sampson, 1971; Straw, 1995, 1996). The earliest excavations by Northmore and Dean Buckland (1824–1825) were shallow and did not penetrate the stalagmite floor below which the majority of bone- and artefact-bearing sediments occur. Work later in 1825 and in 1826, by Reverend J. MacEnery, penetrated farther into the cave and managed to break through the stalagmite to expose softer deposits beneath. In these sediments were found the bones of hyaena and woolly rhinoceros, together with human artefacts, thus demonstrating the contemporaneity of human beings and extinct animals. Because of views prevailing in the 1820s about the age and origin of humans and of geological phenomena, MacEnery did not reveal his findings, and his notes were only published after his death by Vivian (1856) and Pengelly (1869).

Small excavations were also carried out between 1830 and 1850, but most of these were non-systematic and poorly documented (Campbell and Sampson, 1971). The most major excavations in the cave were conducted by William Pengelly between 1865 and 1880. In contrast to previous excavators, he dug the cave painstakingly layer by layer, using a grid system to establish the three-dimensional context of 'finds' and sediments. The results were published both in monthly and annual reports (Pengelly, 1868b, 1869, 1871, 1878). A final report on the excavations (Pengelly, 1884), however, did not include full sections of the deposits, which remained in Pengelly's excavation diary. Although the finds of bones and artefacts were well recorded by Pengelly, no further work or any illustration of them was published except for a small series by Evans (1897). Further limited excavations were conducted in the 1920s and 1930s (Dowie, 1928; Beynon *et al.*, 1929; Smith, 1940). Reviews of all earlier work were published by Kennard (1945) and Campbell and Sampson (1971). The latter authors published a composite stratigraphy partly based on Pengelly's notes, and also reviewed the industries and fauna found in the cave. A short summary by Straw (1983) drew largely on the work of Campbell and Sampson (1971). Lister (1987) examined some aspects of the mammalian fossil record from the site and radiocarbon dates were provided by Campbell and Sampson (1971) and Hedges *et al.* (1989). Proctor (1994, 1996) applied U-series and Electron-spin Resonance (ESR) dating techniques to stalagmite from the cave.

Description

Kent's Cavern (SX 93456415) is cut in Middle Devonian Torquay Limestone and lies on the west side of the dry Ilsham Valley. It comprises a series of large solution cavities linked by smaller passages cut along joints and bedding planes in the limestone. The largest caverns occur at the intersection of major joints or other partings in the bedrock (Figure 5.2). The infill of the cave passages resulted from a range of interlinked processes common to many karst cave systems, the most important being roof collapse, precipitation of stalagmite and fluvial deposition. Kent's Cavern has two small entrances on the valley side and these have been used by humans and animals to gain access to the cave passages.

The stratigraphy of Kent's Cavern is complex. Different sequences of deposits occur in adjacent parts of the cave and their formation has been controlled by local factors of sedimentation such as flow of water or proximity to the cave walls or roof. The following generalized stratigraphy is recorded by Campbell and Sampson (1971) and is based largely on Pengelly's notes:

7. *Black Mould*: silt and vegetable matter with artefacts ranging from Mesolithic to Mediaeval in age (0.3 m) *F/D

6. *Granular Stalagmite*: stalagmite with Neolithic and Mesolithic fauna and artefacts (1.5 m) *C2

5. *Stony Cave Earth*: limestone fragments in a light red silt/sand matrix with Upper Palaeolithic faunas, artefacts and hearths (2.0 m) *B2

4. *Loamy Cave Earth*: light red silt/sand with a few limestone fragments, some rounded. Upper Palaeolithic artefacts and faunas in the

top of the deposit and Middle Palaeolithic artefacts and faunas through the main body of the deposit (10.0 m) *A2

Erosion level

3. *Crystalline Stalagmite*: stalagmite intermittently present (4.0 m) *C1
2. *Breccia*: angular and rounded limestone fragments in a red sand/silt matrix. Massive concentration of bear remains and Lower Palaeolithic artefacts (3.0 m) *B1
1. *Red Sand*: dark red sand/silt with few artefacts or bones *A1

    * bed notations given by Campbell and Sampson (1971)

    (maximum bed thicknesses in parentheses)

Older crystalline stalagmite and laminated silts may be present in patches below beds 1–7 and over the bedrock. Detailed differences in this general stratigraphy are noted by Campbell and Sampson (1971).

The principal faunal remains were recovered from three units – the Breccia (bed 2), the Loamy Cave Earth (bed 4) and the Stony Cave Earth (bed 5). A summary of the fauna given by Campbell and Sampson (1971) was compiled from notes and publications of MacEnery, Pengelly and Evans. The fauna of the Breccia (bed 2) is composed overwhelmingly of cave bears (referred to *Ursus spelaeus* Rosenmüller and Heinroth by Campbell, but noted as *U. deningeri* by Bishop *in* Straw (1996)). Bed 2 also contains remains of the sabretooth *H. latidens* and extinct voles *Arvicola greeni* Hinton and *Pitymys gregaloides* Hinton. The latter specimens, recovered by MacEnery between 1825 and 1829, are not well provenanced and may even be derived from older sediments. The Loamy Cave Earth (bed 4) has a profuse fossil fauna dominated by remains of spotted hyaena (*Crocuta crocuta* Erxleben), woolly rhinoceros (*Coelodonta antiquitatis* Blumenbach) and horse (*Equus* sp.), but also including giant deer (*Megaloceros giganteus* Blumenbach), mammoth (*Mammuthus primigenius* Blumenbach) and brown bear (*Ursus arctos* Linné). The fossil fauna in the Stony Cave Earth (bed 5) is less profuse, but similar in composition, with a dominance of horse and brown bear remains, but also with some of hyaena and woolly rhinoceros.

Campbell and Sampson (1971) and Campbell (1977) recorded sparse pollen (two grains only) from the basal Red Sand (bed 1). The Loamy Cave Earth (bed 4) and the Stony Cave Earth (bed 5),

however, yielded pollen in abundance, both assemblages being dominated by the pollen of herbs (68% and 61%, respectively) but also with *Salix* and *Juniperus*. The contemporaneity of the pollen and sediment, however, is doubtful and the interpretation of such obviously derived plant fossils very difficult.

The cave has long been famous for its Palaeolithic artefacts. These were a focus of the nineteenth century excavations of MacEnery and Pengelly, and the occurrence of Palaeolithic material has been reviewed by Campbell and Sampson (1971). The earliest industry in the cave is represented by artefacts of Acheulian typology, probably derived from the Breccia (bed 2): modern analysis is difficult because only 29 of the 116 tools recovered by Pengelly between 1872–1900 have survived to be examined by recent workers (Campbell and Sampson, 1971). The tools are mostly of flint and consist of crude hand-axes and flakes with rare choppers and cleavers.

The Loamy Cave Earth (bed 4) yielded most of the other artefacts recovered from the site. Pengelly retrieved about 1000 pieces from the cave, but in Campbell and Sampson's (1971) reassessment only 33 of these could be traced. A further 12 specimens from Ogilvie's excavation (1926) are also described by Campbell and Sampson. These artefacts are of flint and Greensand Chert in about equal quantities, and Campbell and Sampson (1971) recognized seven different tool types characteristic of a Mousterian industry. The upper part of the Loamy Cave Earth yielded artefacts indicative of an Early Upper Palaeolithic industry, made largely of flint, and comprising 18 recognizable tool types. Later Upper Palaeolithic or Creswellian artefacts have been recovered from a level in the Stony Cave Earth known as the 'Black Band'. As with the other industries in the cave, most material was obtained by Pengelly between 1865 and 1880, and has since been lost. The surviving Creswellian material comprises 16 tool types, mostly of flint, as well as needles, awls and harpoon points of bone (Campbell and Sampson, 1971).

Interpretation

The deposits of Kent's Cavern provide a palaeontological and archaeological record for a major section of Pleistocene time which is otherwise poorly recorded in the South-West. The faunal assemblage of the Breccia (bed 2) contains no cold-climate species, and a temperate climate, perhaps towards the end of an interglacial, is indicated (Straw, 1983). The remainder of the faunal remains in the cave indicate open vegetation conditions typical of cool or cold steppe. The ages of these faunas and their associated archaeological assemblages are uncertain. Radiocarbon results reported by Campbell and Sampson (1971) from bone of bear and rhinoceros, probably from the Loamy Cave Earth (bed 4), found in association with earlier Upper Palaeolithic artefacts, gave ages of 28 160 ± 435 and 28 720 ± 450 BP. Bones of bear and giant deer from later Upper Palaeolithic contexts yielded dates of 14 275 ± 120 and 12 180 ± 100 BP. A radiocarbon determination (8070 ± 900 BP) on human bone collected by Pengelly (Hedges *et al.*, 1989) suggests that deposits above the Granular Stalagmite floor (bed 6) are of Mesolithic age. A fragment of human bone, probably of *Homo sapiens* Linné, collected during a 1920s' excavation and probably from the Loamy Cave Earth (bed 4), gave a date of 30 900 ± 900 BP, indicating a relatively early occupation of Britain by *H. sapiens*: the Mousterian artefact assemblage in the lower levels of the Loamy Cave Earth (bed 4) must be older than this, but specific dates are as yet unavailable.

The oldest deposits in the cave which contain faunal remains are difficult to date. The Breccia (bed 2) contains the bones of *Homotherium latidens* and perhaps *Ursus deningeri*: these species, and the accompanying voles, are regarded by most authorities to have become extinct in early Middle Pleistocene times (Stuart, 1982a). The occurrence of this cave bear, sabre-tooth and vole fauna, might indicate a broadly 'Cromerian' age for the Breccia (Straw, 1983). Certainly, the association of these bones with Acheulian artefacts is no longer problematical and it is now widely recognized that humans gained access to Britain relatively early in the Middle Pleistocene (Bishop, 1975; Wymer, 1985; Roberts, 1986; Shotton *et al.*, 1993). However, the possibility exists that the bones of the sabre-tooth and voles were reworked from an older deposit, and the precise age of the Breccia remains far from certain. Dates by the U-series and ESR methods (Proctor, 1994, 1996) suggest ages of 300 to 400 ka BP for the Breccia (bed 2) which would place it in Stage 9 or 11 of the oceanic Oxygen Isotope record. Such dates, particularly the latter, would confirm the early Middle Pleistocene age suggested by the mammalian evidence, and indicate that formation of the Crystalline Stalagmite (bed 3) and the erosion phase occupied a long period of Middle Pleistocene time.

The Crystalline Stalagmite is, in many places

within the cave, severely broken so that sharp-edged blocks of it have become incorporated in the Loamy Cave Earth of bed 4 (Figure 5.2). Pengelly suggested that water pressure was responsible for the destruction of this crystalline layer, but Straw (1995, 1996) states that the uniform nature of the fracturing throughout the cave may indicate a more general cause – a seismic event? The youngest parts of the Crystalline Stalagmite have been dated by U-series to *c.* 100 ka BP (Proctor, 1994), thus suggesting that the inferred seismic disturbance occurred in the latter part of Oxygen Isotope Stage 5.

Conclusion

The deposits of Kent's Cavern provide an extensive record of sedimentation which spans the Middle and Upper Pleistocene. Evidence from the site supports both an early date for the occupation of the British Isles by humans and for the entrance of *Homo sapiens* into Britain. The richness of the faunal and archaeological remains makes Kent's Cavern one of Britain's most important Pleistocene sites. Its famous record of *Homotherium latidens* is unique in this part of the South-West: together with faunal remains at Westbury-sub-Mendip, it may provide evidence for temperate conditions in part of the Cromerian Stage. The remaining deposits at Kent's Cavern, together with the surviving museum specimens, provide great potential for elaborating conditions during a part of the Pleistocene which is otherwise only very poorly represented in South-West England.

TORNEWTON CAVE
A. P. Currant

Highlights

Tornewton Cave contains one of the most complete Upper Pleistocene sequences in Britain. Its deposits include the famous 'Glutton Stratum' and 'Hyaena Stratum', and recent studies confirm that the biostratigraphical succession spans at least two major interglacials – equivalent to Oxygen Isotope Stages 7 and 5. The cave provides the only record of the clawless otter in the British Pleistocene.

Introduction

The Torbryan Caves were discovered by James Lyon Widger (1823–1892) who referred to them as the 'Alexandra Caves'. The first excavations in the group of cave passages now known as Tornewton Cave were undertaken by him around 1877. His only personal reference to the caves was published posthumously (Widger, 1892). Widger's large collection of Pleistocene vertebrate remains and human artefacts from the caves was sold to the London dealer F.H. Butler and subsequently dispersed. Some 600 specimens labelled merely 'Torbryan Caves', many of which may have come from Tornewton Cave, are now in the Natural History Museum, London. Other material, similarly labelled, is widely scattered in public and private collections.

Re-examination of Widger's description of the caves and of a longer manuscript version of the same account (Torquay Natural History Society archives) shows him to have been a perceptive and skilful interpreter of the complex history of events represented by the deposits and structures which he examined. Regrettably he was not successful in attracting the interest of the scientific establishment of his day and much valuable information was irretrievably lost on his death. Early accounts of Widger's life and work were given by Lee (1880) and Lowe (1918), and these have since been supplemented by Walker and Sutcliffe (1968): some of his finds were also figured by Reynolds (1902, 1906, 1909, 1922).

Although further excavations were undertaken in the cave by A.H. Ogilvie, the results were not published and the main account of the site remains that of Sutcliffe and Zeuner (1962). These workers tentatively ascribed the principal deposits at Tornewton, on the basis of their contained faunal remains, to the 'Penultimate Glaciation' (Saalian), 'Last Interglacial' (Eemian) and 'Last Glaciation' (Weichselian). However, distinctive species groupings within the Eemian (Ipswichian) mammal remains at Tornewton, and at other British sites, led Sutcliffe (1975, 1976) to the far-reaching conclusion that the Ipswichian Stage comprised two separate temperate phases. Rodent remains from the site were described by Kowalski (1967) and Sutcliffe and Kowalski (1976). The site is also widely referred to in accounts of both national and regional Pleistocene history (e.g. Sutcliffe, 1969, 1995; Macfadyen, 1970; Kidson, 1977; Cullingford, 1982; Lowe and Walker, 1984). The importance of the Torbryan Cave sequences has been confirmed by recent detailed reinvestigation (Willemsen, 1992; Proctor, 1994; Barton, 1996; Berridge, 1996; Cartwright, 1996; Caseldine and Hatton, 1996; Currant, 1996; Debenham, 1996; Gleed-Owen,

1996; Irving, 1996; Price, 1996; Proctor and Smart, 1996; Seddon, 1996; Stewart, 1996; Roberts *et al.*, in prep.).

Description

The Torbryan Caves (Figure 5.1) occur in an outcrop of Devonian limestone on the south-west side of the Torbryan Valley, near Ipplepen, Devon. Tornewton Cave (SX 81726737) is an ancient feature bearing no clear relationship to present topography. The cave comprises two sub-vertical phreatic rifts with associated horizontal passages. The larger rift is called the Main Chamber and now has three separate connections to the outside: the Upper, Middle and Lower entrances. The Upper Entrance is a rock arch and was Widger's original access to the cave. The Middle Entrance connects with the Main Chamber via a narrow passage called the Middle Tunnel. The Lower Entrance is partly artificial and has been enlarged to give access into the Lower Tunnel, a structural extension of the Main Chamber in its lower part. The smaller rift is known as Vivian's Vault. It is a structural continuation of the Middle Tunnel, but the interconnecting passageway is narrow and totally blocked by stalagmite. A hole high in the wall towards the back of the Main Chamber now allows access to Vivian's Vault. Excavations outside the cave show that it once extended farther out into the Torbryan Valley.

An adaptation of Widger's description of the Tornewton Cave deposits (Widger, 1892; Widger, ms.) is repeated here because it provides the only firsthand account of several of the younger units and remains valuable for interpreting surviving deposits:

6. The 'Black Mould' – of unspecified thickness, containing flints, shells, pottery, pebbles, charcoal, a Roman coin and the remains of rodents. The pre-excavation floor of the cave was covered by slabs of angular stones.

5. 'Diluvium' (Widger believed in the Biblical deluge) – 5 feet (1.5 m) of an unspecified deposit containing a few worked flints, charcoal, and the remains of rodents and bats. This unit was capped by a white stalagmite floor about 1 foot (0.3 m) thick.

4. The 'Reindeer Stratum' – 6 feet (1.8 m) of red earth containing abundant reindeer antler, the ribs, vertebrae and limb bones of large animals and remains of a large species of bear.

The Reindeer Stratum is covered by a stalagmite floor 'a few inches thick'.

3. The 'Dark Earth' – 2 feet (0.6 m) thick and emitting an unpleasant smell when first dug and containing jaws and teeth of different animals, mainly hyaena. The Dark Earth contained a scatter of well-rolled quartz pebbles on its upper surface.

2. The 'Great Bone Bed' – 3 feet (0.9 m) thick and containing 'most of the British cave fauna'.

1. The 'Bear Deposit' – originally 7 feet (2.1 m) thick but mostly washed out. The remains of the 'smaller species' of bear were found here.

Widger's excavations at Tornewton Cave appear to have been extensive but cannot be delimited with any certainty due to later digging, mainly by Ogilvie and others in the late 1930s. Although these later excavations are unpublished and undocumented, extensive collections from them are held by the Torquay Natural History Society.

From 1944 until the early 1960s, the cave was excavated still further by A.J. Sutcliffe and colleagues. His findings elevated Tornewton Cave to international fame. He recovered a series of discrete mammal faunas in apparent stratigraphic superposition which were believed to span the period from the penultimate cold stage (in modern terms, Oxygen Isotope Stage 6), right through the last interglacial (Stage 5) and including much of the last cold stage (Stage 4 to Stage 2). The published stratigraphy and associated faunas (Sutcliffe and Zeuner, 1962) are summarized below with an updated taxonomy:

6. Hyaena Stratum – equated with Widger's 'Great Bone Bed'. Clearly the product of a prolonged phase of denning by the spotted hyaena *Crocuta crocuta* Erxleben. Much of this unit, which was also present throughout the Main Chamber and Lower Tunnel, consists of teeth, bones and bone debris, hyaena coprolites and fragmented coprolitic material. As in the underlying Bear Stratum, rock clasts were few in number. Some 1300 isolated hyaena teeth were recovered representing a minimum of 76 adult and 41 juvenile animals. (Widger claims to have recovered around 20 000 teeth from the same deposit.) Hyaenas of all ages were represented. The body-part representation from this unit is highly biased in favour of teeth and foot bones, other skeletal elements being quite rare. It would appear

that the hyaenas had consumed all but the least digestible parts, even of their own kind. Species other than hyaena were quite rare given the abundance of bone within this unit, and in all cases they were represented by either teeth, foot bones or very heavily gnawed limb bone fragments. The following animals are listed: spotted hyaena, wolf *Canis lupus* Linné, fox *Vulpes vulpes* (Linné), lion *Panthera leo* (Linné), bear *Ursus* sp., narrow-nosed rhinoceros *Stephanorhinus hemitoechus* (Falconer), hippopotamus *Hippopotamus amphibius* Linné, fallow deer *Dama dama* (Linné), red deer *Cervus elaphus* Linné, a bovid (larger than that from the Glutton Stratum), hare *Lepus* sp., vole *Arvicola* sp., and undetermined species of bird. This fauna was taken to represent a considerable climatic amelioration and is assigned to the last interglacial (Stage 5/Ipswichian Stage).

5. Stalagmite Floor – forming a thin but fairly continuous sheet across the Main Chamber but represented by individual stalagmites in the Lower Tunnel.

4. Bear Stratum – described as immediately overlying the Glutton Stratum and extending the full length of the Main Chamber and Lower Tunnel. The authors maintain that there was no faunal distinction between the Bear Stratum and the underlying Glutton Stratum, but the nature of the deposit was apparently quite distinct. Although much of the remaining Bear Stratum was secondarily cemented, by infiltrating stalagmite, where it was not cemented the deposit was much less compacted than the Glutton Stratum and showed signs of faint internal stratification. There were occasional groups of related bones, and the very few contained rock clasts were of limestone rather than shattered stalagmite. No separate faunal list was given for this unit.

3. Glutton Stratum – a compacted, unstratified cave earth containing abundant teeth and bone fragments which represent a variety of vertebrate species, primarily bears. This deposit was believed to have accumulated by 'sludging' during periglacial conditions. The deposit was thickest (over 4.5 m) at the back of the cave, thinning out entirely at the mouth of the Lower Entrance. Other than abundant brown bear *Ursus arctos* Linné remains, those of wolf, fox and lion were recorded as common, while those of other species, such

as glutton *Gulo gulo* (Linné), reindeer *Rangifer tarandus* Linné, a small bovid and hare *Lepus* sp. were said to be rare. Other animals, often represented by single specimens, included horse *Equus ferus* Boddaert, a rhinoceros, badger *Meles meles* (Linné) and clawless otter *Cyrnaonyx antiqua* Blainville. The remains of rodents, bats, shrews, birds and fish were listed without further identification.

2. Stalagmite Formation – although listed as a separate bed by Sutcliffe and Zeuner (1962), this material comprises a large quantity of broken stalagmite which is found in the overlying bed. No such material is found in the underlying Laminated Clay and evidently this speleothem once capped the clay as part of a substantial and probably continuous floor/wall dripstone layer. Its break-up and redistribution into the Glutton Stratum were attributed to periglacial processes.

1. Laminated Clay – unfossiliferous waterlain clays and silts, excavated down to the watertable. The upper part of this unit was much disturbed and deeply incised by the overlying deposit (in effect bed 3; see note in bed 2).

Sutcliffe and Zeuner also described higher units within the cave, but these descriptions are based on interpretations of Widger's report and so do not represent new data. They do, however, describe a series of deposits outside the cave which adds to its palaeontological interest. At the bottom of their excavations, heavily disturbed laminated clays similar to those found inside the cave could be traced beyond the present extent of the cave system, but no direct equivalents of the Glutton Stratum, Bear Stratum, Hyaena Stratum or Widger's 'Dark Earth' were present. A series of very poorly fossiliferous 'head' or talus deposits, apparently containing rare finds of reindeer, had accumulated against the limestone cliff outside the cave, up to the level of the Middle Entrance. It would appear that the Middle Entrance and the Middle Tunnel provided the main access route to the Main Chamber, at least until the end of the last interglacial, at which time the inner end of the Middle Tunnel appears to have become blocked by rocks (Widger's observation).

Immediately outside the Middle Entrance, on top of the poorly fossiliferous head deposits, was a unit which Sutcliffe and Zeuner (1962) named the 'Elk Stratum'. This material, which was of very limited extent, nonetheless produced quite a rich faunal assemblage, including spotted hyaena, woolly rhinoceros *Coelodonta antiquitatis* (Blumenbach), horse and reindeer. Overlying deposits, believed by the excavators to represent an external equivalent of Widger's 'Reindeer Stratum', contained less abundant remains of the same faunal assemblage, but with the interesting inclusion of a human lower incisor and several flint implements. This faunal grouping was attributed to an interstadial phase within the Devensian Stage. The 'Elk' from which this deposit took its name was later re-identified as a red deer (A. Lister, pers. comm.).

Subsequently, Sutcliffe extended his excavations into Vivian's Vault. Here, he removed a large quantity of spoil dumped by earlier excavators and discovered a previously unrecorded deposit containing abundant microvertebrate remains, including those of a white-toothed shrew *Crocidura* sp. (Rzebik, 1968), a large number of bird bones and the teeth and bones of the clawless otter *C. antiqua*. These discoveries are mentioned by Sutcliffe and Kowalski (1976) under the designation of finds from the Otter Stratum. The rodent remains from this deposit were listed by Sutcliffe and Kowalski (1976) as follows:

Cricetus cricetus Linné	
European hamster	2 specimens
Lagurus lagurus (Pallas)	
steppe lemming	1 specimen
Dicrostonyx torquatus (Pallas)	
collared lemming	12 specimens
Lemmus lemmus (Linné)	
Norway lemming	1 specimen
Microtus nivalis (Martins)	
snow vole	4 specimens
Microtus oeconomus (Pallas)	
northern vole	82 specimens
Apodemus sylvaticus (Linné)	
wood mouse	2 specimens
Clethrionomys glareolus (Schreber)	
bank vole	19 specimens
Microtus agrestis (Linné)	
short-tailed vole	5 specimens
Arvicola sp.	
water vole	8 specimens

The authors noted that this stratum contained a mixture of two possibly discrete faunal assemblages. Remains from a warm period predominated, and contained clasts of stalagmite yielded only temperate forms such as *M. agrestis*, *M. oeconomus*, *Arvicola* sp., *C. glareolus* and *A. sylvaticus*. This temperate fauna was attributed to the last interglacial (Ipswichian Stage), while species such as *L. lagurus*, *C. cricetus* and *M. nivalis* were taken to indicate cold conditions during the penultimate cold stage (Oxygen Isotope Stage 6). The colder elements found in the deposits of the Otter Stratum were correlated with similar finds described from the Glutton Stratum in the same publication. Such an interpretation appeared to be secure, given the fact that the Glutton Stratum was stratified beneath the Hyaena Stratum and intervening Bear Stratum, and the Hyaena Stratum contained a large-mammal fauna characteristic of the Ipswichian.

Interpretation

Recent reappraisal of the Tornewton Cave collections now housed in the Natural History Museum, London, and selective excavation and sampling within the cave as part of the British Museum's Torbryan Caves Research Project, have together led to a partial reinterpretation of the Tornewton Cave sequence (Currant, 1996; Roberts *et al.*, in prep.).

Existing collections of material, recorded as having come from the Glutton Stratum, contain a much wider range of species than originally listed by Sutcliffe and Zeuner (1962). Newly identified species include: hedgehog *Erinaceus europaeus* Linné, marten *Martes* sp., narrow-nosed rhinoceros, hippopotamus, fallow deer, red deer, roe deer *Capreolus capreolus* (Linné) and wild boar *Sus scrofa* Linné. Together, these animals add a strongly 'temperate' component to what is supposed to be a cold, Oxygen Isotope Stage 6, assemblage. Samples taken from the Glutton Stratum during the British Museum excavations by Lorraine Higbee were examined for signs of internal stratification. She confirmed the complete admixture of temperate and boreal indicators noted above, and recovered the conjoining halves of a dentary of a glutton from two adjacent samples. This find was notable for its relative completeness and excellent state of preservation. Following a lead suggested by Widger's original description of the cave, in which he indicates the existence of some kind of partially blocked tunnel running down the back of the Main Chamber and through his 'Bear Deposit', this specimen yielded an Accelerator Mass Spectrometry (AMS) radiocarbon age estimate of 22 160 ± 460 BP (OxA-4587).

This dating has helped considerably in interpreting the controversial Glutton Stratum. The deposit would appear to have been emplaced, possibly under periglacial conditions, close to the Devensian stadial maximum (Oxygen Isotope Stage 2). It contains material from older units within the cave, including material which must have come from Widger's 'Reindeer Stratum' and from his 'Dark Earth', the Hyaena Stratum and the Bear Stratum. It may also include material from even older deposits which once adhered to the back wall of the cave. Sutcliffe and Zeuner (1962) described a 'cave within a cave' feature in the lower part of the Main Chamber, caused by the collapse of partly consolidated sediments which belonged to both the Bear Stratum and Hyaena Stratum. It seems likely that this collapse took place at the time of, or shortly after, the emplacement of the Glutton Stratum beneath them.

This might explain the curious absence of Middle Devensian deposits inside the cave. It seems likely that most of the cave system was completely filled during the earlier part of the Devensian (Oxygen Isotope Stage 4) and remained so throughout the Middle Devensian (Stage 3). During this phase, the only access appears to have been to the Middle Tunnel, where the Elk Stratum and associated deposits contain the only characteristically Middle Devensian fauna. It seems likely that the Elk Stratum deposits are themselves part of a small debris flow issuing from the mouth of the Middle Tunnel. Only after material filling the upper part of the cave had become mobilized and flowed down the back of the Main Chamber, was the main cavern space partially reopened. It is possible that the dated glutton jaw belonged to an animal that was using the cave while the debris flow process was still active.

Excavations in the entrance to Vivian's Vault in 1991 and 1992 located a small remnant of *in situ* Hyaena Stratum overlying quite a large fragment of *in situ* Bear Stratum against the wall of the vault, confirming Widger's description of the sequence here. Immediately beneath the Bear Stratum, where the walls of the vault converge, a thin unit of barren clay overlies a partly broken stalagmite floor. This stalagmite contained pockets of sediment rich in *Crocidura* remains and had bones of clawless otter adhering to some of its broken surfaces. This stratigraphic relationship, between the Bear Stratum and the deposits of the Otter Stratum, would seem to confirm that the latter represents a pre-Stage 5 temperate phase. *Crocidura* has not been found in any Stage 5 contexts in Britain, but appears to be a characteristic element of what are believed to be Stage 7 faunas from other British sites such as Aveley and Orsett Road, Essex and Itteringham, Norfolk (Currant, 1989). Detailed examination of blocks of partly cemented sediment suggests a very complex history of deposition, but most of the deposits excavated in Vivian's Vault appear to represent material which originally accumulated under temperate conditions. At present there is no reason to suppose that the deposits belong to more than one temperate stage. The vault is less than 0.3 m wide in places, so digging conditions are extremely difficult, but it is believed that the exact area where Sutcliffe recovered the cold elements of his Otter Stratum rodent fauna has been located. Immediately beneath the entrance to Vivian's Vault from the Main Chamber there is a 'pipe' of sediment running down into the disturbed sediments of the Otter Stratum, apparently associated with the broken edge of the stalagmite floor mentioned above. A palate and dentary of a European hamster has been recovered from a sample of this deposit, which is tentatively interpreted as a smaller-scale debris flow, similar in character to the Glutton Stratum in the Main Chamber, though possibly much older. Sadly, the deposits immediately overlying this feature had already been excavated, so it was not possible to establish whether this feature was sealed by the Bear Stratum and Hyaena Stratum, or whether it cut through them.

An interesting microvertebrate assemblage has recently been recovered from a complex sequence of stalagmite deposits which block the passage between the inner end of the Middle Tunnel and Vivian's Vault beyond. Sandy partings in the stalagmite have yielded *Lemmus*, numerous teeth of *M. oeconomus* and as yet unidentified bird remains. This cold-stage assemblage superficially resembles some pre-Stage 5 faunas like that from Crayford, Kent, but its age remains uncertain.

In 1992 a large stalagmite mass blocking the back of the Main Chamber, just by the entrance to Vivian's Vault, was drilled through and access was gained to a small cavern space beyond. This has been named the Main Chamber Extension. The extension contained an undisturbed sequence equivalent to at least the upper part of the Bear Stratum, the Hyaena Stratum and Widger's 'Dark Earth'. These deposits were sampled for microvertebrate fossils and surprisingly the Hyaena Stratum yielded quite profuse remains of northern vole. This species, which is usually found in wet grasslands, tends to be absent from deposits representing the warmest part of the Ipswichian Stage (Stage 5e), yet there is good reason to suppose that this is the only period in which the hippopotamus was present in Britain during the later part of the

Pleistocene. Stalagmite containing hippopotamus remains from Victoria Cave in North Yorkshire has been dated to around 125 000 BP (Gascoyne *et al.*, 1981). At Bacon Hole, West Glamorgan, a series of mammal faunas recovered from Stage 5 sediments shows that *M. oeconomus* is not present in local faunas during the earlier part of the stage, but is quite common during subsequent sub-stages of the interglacial (Currant *in* Sutcliffe *et al.*, 1987).

These findings have major repercussions for re-evaluating the Tornewton Bear Stratum. Sutcliffe and Kowalski (1976) described a small assemblage of rodent remains found close to the entrance to the Lower Tunnel which they interpreted as being contemporary with the Bear Stratum. The assemblage includes *D. torquatus*, *L. lemmus*, *C. glareolus*, *Arvicola* sp., *M. agrestis*, *M. nivalis*, *M. oeconomus* and *L. lagurus*, and was originally correlated with the Glutton Stratum as the equivalent of a Stage 6 assemblage. However, this collection of material appears to be far more closely connected with deposits immediately outside the originally narrow Lower Tunnel Entrance than it does with anything inside, and examination of Sutcliffe and Zeuner's own section through the cave deposits (Sutcliffe and Zeuner, 1962; plate 28) shows this to be the case.

It should be noted that the snow vole *M. nivalis* is not now believed to have been present in Britain during the Quaternary, and that material ascribed to this species actually represents a minor dental variant of *M. oeconomus* (A.J. Stuart, pers. comm.). Moreover, the biostratigraphic significance that has previously been attached to the occurrence of this form (Sutcliffe and Kowalski, 1976) is equally unfounded, similar morphologies having been found in faunas of widely different age. It is possible that the records of such species as *L. lagurus* at Tornewton Cave represent new elements of a poorly known Early Devensian (Oxygen Isotope Stage 4) fauna rather than being part of an even more poorly known Stage 6 fauna. Even such well-known temperate species as *M. agrestis* and *C. glareolus* may be making spurious appearances in faunal lists, particularly when unequivocally associated with more boreal species. The teeth of the narrow-skulled vole, a boreal species, often closely resemble those of *M. agrestis*, and the more boreal species of *Clethrionomys*, such as *C. rutilus* and *C. rufocanus*, may well have gone unnoticed: the identity of all of this material has to be re-checked.

Returning to the Bear Stratum, examination of the collections from Sutcliffe's excavations arouses the suspicion that it was not always certain which deposit was being excavated. The only species which appears to have been common in the Bear Stratum was brown bear. In fact, there is nothing else in the collection which is definitely said to have come from the Bear Stratum, which is particularly odd given the observation that, on faunal grounds, the Bear Stratum could not be differentiated from the Glutton Stratum (Sutcliffe and Zeuner 1962; p. 131). There is a varied group of material marked up as 'H.S./B.S. undifferentiated' (including at least one specimen of *Hippopotamus*) and an even larger group marked up as 'B.S./S.B. undifferentiated' (S.B. = sub-Bear, the original name for the Glutton Stratum). The only firm information available on the fauna from the Bear Stratum comes from a series of samples collected inside the entrance to Vivian's Vault in 1991 by Cornish who was able to demonstrate that the contained microvertebrate fauna was consistently interglacial in character.

The Bear Stratum, therefore, accumulated under warm temperate conditions, most likely the early part of the last interglacial (Stage 5e). This fits in well with the abundance of *M. oeconomus* in the overlying Hyaena Stratum which can probably be placed in Stage 5c, and with the survival of interglacial fauna into the overlying 'Dark Earth' which probably represents Stage 5a, but does not square particularly well with the reported occurrence of *Hippopotamus* in the Hyaena Stratum, especially as the two deposits are separated by a minor phase of stalagmite formation, suggesting the passage of a significant period of time between their deposition.

Hippopotamus fossils are not common at Tornewton Cave, and nearly all of the individual specimens present in the collections at the NHM have something odd about them. Several are from poorly stratified contexts, one in the Glutton Stratum has been derived, and most of the remainder have clear signs that their original field data have been erased or altered. The largest piece, a fragment of canine tooth with a dark, heavy staining quite unlike all other Hyaena Stratum specimens, was found on a spoil heap (A.J. Sutcliffe, pers. comm.) even though it is marked '(H.S.)'. Sadly, it would appear that little faith can be placed on the provenance of a number of biostratigraphically significant specimens which have been attributed to the Hyaena Stratum. One should not be too critical of such discoveries. Interpretation of cave deposits is often difficult, particularly when dealing with huge amounts of compacted spoil from earlier excavations and isolated blocks of *in situ* sediments, themselves partly

disturbed by collapse. Given the then-prevailing belief that the Bear Stratum represented a non-temperate environment, one can understand the collector's doubts about the provenance of earlier finds which originally may have been attributed to this unit but which were themselves temperate indicators.

Fortunately the new finds from the Main Chamber Extension are from unambiguous contexts. It is unlikely that some of the rarer species will be represented in the samples taken from the 1992 excavations, but there should be enough environmental data to resolve the exact placement of the various Stage 5 deposits. Particular hope is pinned on obtaining Uranium-series dates from a clean, crystalline stalagmite boss which was found growing on the Bear Stratum and sticking up through the Hyaena Stratum and the overlying 'Dark Earth', with 'skirts' marking the tops of each of these units.

The lateral equivalent of Widger's 'Reindeer Stratum' has been found in a previously unexplored horizontal 'tube' (Price's Passage) leading outwards from the Main Chamber just below the Upper Entrance. It may have had an entrance just above the Middle Entrance in what is now a mass of shattered rock and tree roots, but it would be unsafe to explore this further from the outside. Price's Passage is partly choked with soft, dark red loam which contains angular limestone fragments and a fauna with reindeer, bison and wolf. It appears to have been a wolf den for a significant time during the earlier part of the Devensian (Stage 4).

Conclusion

An interglacial vertebrate fauna with *C. antiqua* and abundant *Crocidura* has been recovered from the Otter Stratum of Vivian's Vault and is correlated with Oxygen Isotope Stage 7. A minor debris flow penetrating the Otter Stratum deposits contains a cold fauna with *Lagurus* and *Cricetus*. Its age is uncertain, but it could belong to Stage 6. The Bear Stratum, representing a major phase of denning by the brown bear, the Hyaena Stratum, representing a major phase of occupation by the spotted hyaena, and the 'Dark Earth', representing a less intense phase of hyaena usage, each contains interglacial faunas which appear to represent the warmer episodes of Stage 5. Overlying deposits, now found only as isolated remnants, contain what appears to be a Stage 4 fauna dominated by *Rangifer* and

Bison. A coherent Stage 3 assemblage, with *Coelodonta*, *Equus* and *Crocuta*, is represented outside the cave by material at the mouth of the Middle Tunnel. At some time during Stage 2, a huge mass of cave deposit formed a debris flow which moved down the back of the cave and beneath the remaining stratified sequence. This constitutes the deposit known as the Glutton Stratum. It has been seen to fill deep, steep-sided cavities in the underlying waterlain sediments and contains a highly mixed and derived fauna. Later deposits have been entirely removed by previous excavators.

CHUDLEIGH CAVES
S. Campbell and S. N. Collcutt

Highlights

Deposits containing rich archaeological and palaeontological remains make Chudleigh Caves important for reconstructing late Middle and Upper Pleistocene environmental conditions in South-West England.

Introduction

Chudleigh Caves, and Pixie's Hole in particular, have long been important for Pleistocene studies. The caves were extensively excavated in the nineteenth century by Buckland, Ackland, Trevelyan, Northmore and MacEnery among others. Pengelly (1873b) provided a useful account of these early explorations and gave details of the artefacts and bones recovered as the cave deposits were quickly sifted and removed. More recent accounts of the caves and their deposits were provided by Beynon (1931), Shaw (1949a, 1949b) and Simons (1963), and the site has been placed in a wider context by Cullingford (1953, 1962), Sutcliffe (1969, 1974, 1977), Macfadyen (1970), Campbell (1977) and Selwood *et al.* (1984). Detailed accounts have been provided of the cave sediments and fauna (Collcutt, 1984, 1986) and the Palaeolithic and Mesolithic artefacts (Rosenfeld, 1969; Collcutt, 1984). Holman (1988) described a herpetofauna from Cow Cave.

Description

Some 20 caves, mostly small, are located in the Chercombe Bridge Limestone around Chudleigh in Devon (Selwood *et al.*, 1984; Figure 5.1). The best

known are Pixie's Hole, Chudleigh Cavern and Cow Cave which are located in or around Chudleigh Rocks (SX 865787): several fissures in the immediate vicinity also contain important deposits. Chudleigh Rocks is part of a restricted outcrop of the Devonian limestone (Selwood *et al.*, 1984) which has been bisected at Chudleigh by the Kate Brook; the resulting gorge is known as the Glen. The Kate Brook rises on Great Haldon some 7 km to the north-east, and joins the Teign *c.* 0.5 km downstream of Chudleigh Rocks. The geology around the Rocks is complex, and the lithologies traversed by the Kate Brook in its upstream catchment are diverse. The latter, which include Carboniferous shales and sandstones, Devonian limestone, Lower Cretaceous pebble and shell beds and, notably, the Upper Carboniferous/Lower Permian Teignmouth Breccia, are important constituents of the cave sediments (Collcutt, 1984). The GCR site consists of Chudleigh Rocks themselves (containing the caves of Pixie's Hole and Cow Cave), Chudleigh Cavern a short distance to the north and part of the gorge (containing Tramp's Shelter). The site includes slope and terrace deposits which may eventually be linked with the history of the caves themselves.

Cow Cave

Cow Cave (SX 86467867) is located on the south side of the Rock. Its entrance, some 6 m wide by 4.5 m, high leads into a spacious cavern about 18 m long (Shaw, 1949b). This cave receives no mention in Pengelly's (1873b) account of the Chudleigh Caves, and it is not clear if any of the early cave explorers, such as MacEnery or Buckland, visited it. The first account of the cave, its layout and deposits, appears to be that by Beynon (1931) who summarized the results of excavations by the Torquay Natural History Society between 1927 and 1931. Unfortunately, none of the mammal remains, including bear, wolf, fox, hyaena, deer, wild cat, Irish deer and ox, nor the eight flint implements, comes from a known stratigraphic context. Beynon (1931) alludes to a stratigraphic sequence of surface 'limestone' soil, overlying stalagmite and a series of cemented and uncemented breccias, but even these details are unclear. Great numbers of well-rounded pebbles were also described from the cave, and were believed to have been washed in by great floods during a 'long pluvial period' (Beynon, 1931). Collcutt (1984) comments that the cave deposits described by Beynon appear to bear similarities with the Siliceous Group described from Pixie's Hole (Collcutt, 1984; see below). The Torquay Natural History Society excavations continued at Cow Cave until at least 1935 (Alexander, 1933, 1934, 1935), although no further detailed publication followed (Collcutt, 1984). Human remains, and those of rhinoceros and beaver were added to the faunal list and one flint of Aurignacian type was recovered (Alexander, 1935).

The most detailed accounts of this cave, however, are those by Simons (1963) and Rosenfeld (1969). Simons recorded the following stratigraphy:

6. Stalagmite floor
5. Frog stratum
4. Stalagmite floor
3. Reindeer stratum
2. Broken stalagmite floor
1. Stream deposit

Based, unsatisfactorily, on the state of their preservation, Simons assigned bones to the various stratigraphic levels. The earliest deposit (stream deposit; bed 1) was deemed to have yielded bear, lion, hyaena, red deer and a rodent. The reindeer stratum (bed 3) is considered to have contained bear, hyaena, bison, reindeer, hare, red deer, roe deer and water vole (Simons, 1963). From the most recent clastic sediment, presumably the Frog stratum (bed 5), are believed to have come the remains of hare, rabbit, dog, hedgehog, sheep, pig, horse and humans among others (Simons, 1963). The occurrence of *M. nivalis* among the various rodent remains excavated from the cave by Simons, led Sutcliffe and Kowalski (1976) to speculate that a pre-Ipswichian deposit may be present at the site. However, the tooth in question is non-diagnostic of the species, and *M. oeconomus* (common in the Late Devensian) is a likely alternative.

Rosenfeld (1969) recorded that six flakes of Middle Palaeolithic age had been recovered from the cave. One flake, from Simons' (1963) investigations, had purportedly come from bed 1. Since the Reindeer stratum (bed 3) lying above this deposit and above a layer of fractured stalagmite (bed 2) was believed to be of 'last glacial' (Devensian) age, it has been speculated that this artefact may have been derived from an interglacial deposit of pre-Devensian (possibly even pre-Ipswichian?) age (Rosenfeld, 1969; Sutcliffe, 1969). The flakes were described by Rosenfeld as coming from prepared cores, some having faceted platforms, but none being of distinctive types. Campbell (1977), on the other hand, regarded the artefacts as of Earlier Upper Palaeolithic age. In addition to the Middle Palaeolithic (Rosenfeld, 1969), Aurignacian (Alexander, 1935) and 'Azilian' (Beynon, 1931) age interpretations, it is even unclear whether the authors are referring to the same or to different objects (Collcutt, 1984). However, the 'Azilian' piece referred to by Beynon is extant (a curved-back point), and almost identical to those found by Collcutt in Pixie's Hole: these tools are indicative of a date late in the Devensian Stage.

Collcutt (1984) has argued that Cow Cave and Pixie's Hole were probably once connected; the westernmost extension of the latter consists of a sediment-choked passage at the same level, and only *c*. 12 m from the easternmost extension of Cow Cave, which is similarly choked with deposits (Collcutt, 1984).

Tramp's Shelter

Tramp's Shelter lies on the south-east side of the Kate Brook, almost above the waterfall (Collcutt, 1984). It occurs in a steep bluff of limestone and has a south-westerly-facing entrance at *c*. 51 m OD. This deep shelter penetrates about 10 m inwards from a 6 m-wide entrance. The history of excavation at this cave is far from clear; certainly much material has been removed from the cave and at its entrance (Collcutt, 1984). Rosenfeld (1969) refers to work by Miss M. Collins and a cave earth which had yielded bones of *Bos*, *Equus* and *Cervus elaphus* Linné. The same deposit also apparently contained a 'backed blade industry with obliquely blunted blades' (Rosenfeld, 1969). Campbell (1977) refers to excavations by Smith, apparently in 1968, and refers to 10 artefacts, possibly attributable to the late Upper Palaeolithic.

Collcutt (1984) also briefly examined Tramp's Shelter, recording at least 2 m of homogeneous deposits (limestone scree and blocks set in a dense matrix of red gritty clay), and recovered several flint artefacts (certainly of Later Upper Palaeolithic age).

Chudleigh Fissure

The location of this feature is not clear. It is recorded as a palaeontological site which has yielded the remains of small mammals (Sutcliffe and Kowalski, 1976) and birds (Bramwell, 1960) indicative of cold conditions probably during the Devensian Stage. An early reference to the site by Hinton (1926), cites Kennard as the original excavator (Collcutt, 1984).

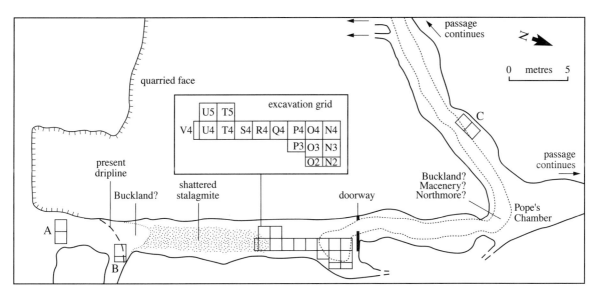

Figure 5.3 Plan of Pixie's Hole, after Collcutt (1984).

Pixie's Hole

This cave is located on the south side of Chudleigh Rocks (at SX 86547867), and passes through the remaining limestone which has been left by quarrying on both its northern and southern sides (Collcutt, 1984; Figure 5.3). It is undoubtedly the most important of the Chudleigh Caves. It is a relatively small but complex cave system, the full extent and layout of which are not fully known. Shaw (1949a, 1949b) provided descriptive details of the cave and passage morphology. He recorded a vertical network with 265 m of passages and with four openings; Selwood *et al.* (1984), however, refer to only three openings. The areas relevant to the recent investigations are clearly marked on a plan given by Collcutt (1984) (Figure 5.3). The most important area of the cave currently known is the relatively simple eastern section, although farther west the cave develops on at least four different levels, and choked passages lead off perhaps linking up with other named entrances elsewhere in the Chudleigh Rocks (Collcutt, 1984). Much of the long eastern passage, which runs northwards into the Pope's Chamber, is developed in a near-vertical fault (Collcutt, 1984). The cave formerly contained fine karst features including stalactite curtains (Shaw, 1949b), but these have been seriously degraded by visitors. (The entrances are now gated.)

The earliest known reference to the cave is that in Ridson's *Survey of Devon* (carried out between 1605 and 1630; Ridson, 1911), although it was probably first mentioned in a scientific context by Paddon (1797) who estimated the extent of the cave. Buckland is the first recorded excavator: according to Pengelly (1873b), he probably dug there in 1824–1825, finding flints, pottery, domestic animal bones and charcoal. The results of these excavations were never published by Buckland, although brief mention is given to the site in his 1824 *Reliquae Diluvianae*. Again, according to Pengelly (1873b), Buckland visited the cave in 1825 with Sir Thomas Acland, finding the remains of 'antediluvian animals' such as hyaena, deer and bear. Excavations were continued between about 1829 and 1841 by MacEnery, but little extra faunal material was recovered (Pengelly, 1873b). From the notes and letters left by these early workers, including Northmore, Pengelly (1873b) summarised that the remains of rhinoceros, ox, hyaena, deer, bear, elephant and hippopotamus had been recovered from the caves at Chudleigh.

Collcutt (1984) excavated at four locations within the eastern part of Pixie's Hole, proving depths of cave sediment in excess of 5 m. The stratigraphy is extremely complex and variable across the cave: Collcutt provides a comprehensive lithostratigraphic classification of the deposits. These are, from top to bottom, as follows:

17. Stalagmitic floor (STF)
16. Silty Clay (SCL)
15. Diamicton (DIA)
14. Stony Cave Earth (SCE)
13. Stony Silt (SST), Dark Silt (DST) and Light Silt (LST)

12. Dark Earth (DET)
11. Grey Clay and Silts (GCS)
10. Red Clayey Sand (RCS)
9. Red Sand (RSD)
8. Red Sandy Gravel (RSG)
7. Friable Speleothem (FSP)
6. Concreted Stones (CST)
5. Silty Fine Gravel (SFG)
4. Cemented Gravel (CGL)
3. Loose Gravel (LGL)
2. Dense Gravel (DGL)
1. Silt, Sand and Gravel (SSG)

Interpretation

Units 1–10 were classified by Collcutt (1984) as the Siliceous Group. Bedding structures in these sediments suggest that they were emplaced largely by water, although at various levels there are also signs that deposition occurred by mass movement (debris flow). The overall impression of the sediments which comprise this group, however, is that they are similar to normal subaerial alluvial fan deposits (Collcutt, 1984). Within this part of the sequence, the carbonate-rich units (6 and 7) are clearly identifiable as a cave breccia capped by a stalagmitic floor (Collcutt, 1984). Most of the bones recovered from unit 6 are those of bear. Unit 7 is almost certainly an interglacial deposit. Collcutt favours an Ipswichian age, although the material (recrystallized) is undateable.

The succeeding Grey Clay and Silts (unit 11) appear to have entered the cave via a fissure in its roof: the overlying Dark Earth (unit 12) is composed of similar material, and its convoluted lower boundary may suggest subsequent mass movement (Collcutt, 1984). The overlying silty units (unit 13) appear to have been emplaced by wash processes; fine lag gravels, 'puddle' deposits and micro-deltas also occur in these units, providing evidence for a complexity of depositional environments within the cave system at this time. Taking the sequence as a whole, there appears to be evidence for a reduction in energy levels up through the deposits. Some of the sediments within unit 13 may have been derived from loessic material outside the cave: remains of woolly rhinoceros *C. antiquitatis* in this unit have been seen as compatible with a 'Main Devensian' age (Collcutt, 1984). These secondary loess deposits are closely comparable with deposits in the platform sequence at Tornewton Cave and may be of a similar age.

Unit 14 is a typical cave earth composed largely of limestone debris. The lack of non-calcareous silt in its matrix may suggest that external sources of silty drift (loess?) had been depleted by this time (Collcutt, 1984). The succeeding sediments (unit 15) were undoubtedly emplaced by mass movement, being derived largely from sediments elsewhere within the cave system. The sharp transition between this and the underlying bed probably represents an erosion surface (Collcutt, 1984).

The overlying Silty Clay (unit 16) is predominantly a fine-grained sheet-wash deposit. It is capped by the Stalagmite Floor (unit 17). The calcareous deposits from unit 13 upwards contain common faunal material, both megafauna and microfauna (Collcutt, 1984). Preliminary examinations of the recovered material suggest that the bones of bear, hyaena, wolf, fox, reindeer, red deer, *Bos* sp., woolly rhinoceros and horse are present (Collcutt, 1984). Initial interpretations of the fauna indicate that it forms a 'cold' assemblage, perhaps of later Devensian age (Collcutt, 1984). The presence of *Bos* at this high stratigraphic level might suggest an interstadial, possibly the Windermere Interstadial (see below).

Collcutt's (1984) excavations also yielded archaeological material, some 400 artefacts largely of flint, all referable to the Later Upper Palaeolithic (*sensu* Campbell, 1977). Dense spreads of redeposited charcoal occur between units 15 and 16, and some flints also show signs of having been burned, indicating that there is evidence for at least one hearth to have been present within the cave (Collcutt, 1984). Collcutt plotted the positions of the recovered artefacts and concluded that the principal source was the Stony Cave Earth (unit 14), which may have carried a 'living floor'. The distribution of the finds suggests that in places this floor may remain undisturbed. Nonetheless, elsewhere, the artefacts have been reworked and incorporated into the overlying mass flow deposits (Collcutt, 1984). There appears to be no typological difference between the artefacts found in these units. Although the artefacts and hearth feature of unit 14 may amount to an 'archaeological occupation', there is no direct evidence to relate the artefacts to the faunal remains (Collcutt, 1984). Much faunal material does occur at this level in the sequence, and while there are many herbivore bones which it is tempting to relate to the archaeological material (butchered?), the occurrence of carnivore bones suggests that animals such as wolf, may have been responsible for the accumulations (Collcutt, 1984).

Conclusion

The Chudleigh Caves and the landforms and deposits of the adjacent area contain valuable information regarding the Pleistocene (and earlier) evolution of the region. Unfortunately, much of the evidence presented by early workers was gleaned by excessively destructive methods: the stratigraphical context of most early finds is unknown, and their value for interpreting events therefore limited. Recent excavations at Pixie's Hole, in particular, have revealed a complex sequence of sediments largely attributable to fluvial, mass flow and freeze-thaw processes in the Devensian, and clearly demonstrate the value of the site for reconstructing Pleistocene palaeoenvironments. Even so, there is no absolute timescale for events here as yet, and much of the archaeological and faunal material remains to be published in detail.

Firm evidence for a Later Upper Palaeolithic occupation of Pixie's Hole can, however, be demonstrated. Not far from the main entrance occurs a 'living floor' consisting of a rudimentary hearth (charcoal, burnt limestone and flint, underlain by baked fine sediment). Elsewhere, farther into the cave system, the tools from this occupation have been dispersed by debris flows.

Pixie's Hole, in particular, provides intriguing evidence for the activities of Upper Palaeolithic humans in Devon. It is clear that flint was knapped at the site; debitage, retouched tools and cores have been recovered. It is likely that the occupation occurred in respect both to the shelter afforded by the cave and with regard to local supplies of flint. Although the local flint gravels are not totally unsuitable for tool manufacture, the presence of good quality flint farther afield in the Kate Brook catchment may well have been a critical factor in the selection and occupation of Pixie's Hole (Collcutt, 1984, 1986).

Dating this occupation, beyond a general ascription based on the faunal evidence, to the Late Devensian is problematical. None of the bone or antler recovered from Pixie's Hole carries unequivocal traces of human modification. Indeed, there is ample evidence for the presence at Chudleigh of denning carnivores. Therefore it is not possible to link the archaeological and faunal evidence directly in time. The industry on its own, however, is fully compatible with a Later Upper Palaeolithic date; a Federmesser industry slightly younger than the classic Creswellian seems likely (cf. the 'Final Palaeolithic' of Barton (1996)).

The evidence from the other Chudleigh caves and deposits is even more difficult to interpret. Rosenfeld's (1969) archaeological and Sutcliffe's (1969) faunal evidence from Cow Cave, raise the interesting possibility of Ipswichian (and perhaps even pre-Ipswichian) interglacial deposits being present at the site. The rather disparate interpretations of the ages of the artefacts from the latter site, however, throw doubt on to this claim (see above). Nonetheless, Cow Cave and Pixie's Hole lie at about the same altitude. Collcutt (1984) has interpreted units 1–10 in Pixie's Hole (Siliceous Group) as having been emplaced largely by water, with unit 7 (speleothem) almost certainly being an interglacial (?Ipswichian) deposit. Sutcliffe (1966) argues that bed 1 in Cow Cave is almost certainly a stream deposit: it now lies high above the modern stream, making it unlikely that the intervening downcutting could have taken place within the timespan of the Devensian alone.

The full potential of the Chudleigh Caves for elaborating regional Pleistocene conditions has not yet been realized: their remaining artefact- and bone-bearing beds (Pixie's Hole has the most extensive Later Upper Palaeolithic deposits known to survive at any British cave) will undoubtedly contribute significantly to the growing knowledge of Pleistocene humans and their environment in Devon.

JOINT MITNOR CAVE
S. Campbell and A.J. Stuart

Highlights

Joint Mitnor Cave has yielded one of the richest fossil assemblages of Ipswichian age yet known in Britain. Much of the fossiliferous deposit is preserved *in situ*.

Introduction

Joint Mitnor Cave is one of several caves opening into the disused Higher Kiln Quarry, at Buckfastleigh in Devon. Although long known as a cave, it was discovered to contain bones in 1939, and was named after W. Joint, W. Mitchell and F.R. Northey. Three main deposits can be distinguished in the cave: 1. a basal, unfossiliferous waterlain deposit; 2. an overlying, highly fossiliferous cave earth, which has yielded a profusion of mammal bones and teeth characteristic of the Ipswichian Stage; and 3. a capping stalagmitic floor. The deposits were excavated between 1939 and 1941 by A.H. Ogilvie and other members of the Torquay Natural History Society (Anon., 1948), and most of the finds now lie in the Torquay Natural History Society Museum and in the British Museum (Natural History), London. A permanent demonstration section of the deposits, with bones *in situ*, has been preserved at the site. The fauna was re-examined in detail by Sutcliffe (1960) and Sutcliffe and Kowalski (1976), and the site has been referred to widely in regional and national syntheses of Pleistocene history (e.g. Sutcliffe, 1969, 1974, 1977; Macfadyen, 1970; Stephens, 1973; Cullingford, 1982; Stuart, 1982a, 1982b, 1983, 1995).

Description

Joint Mitnor Cave (SX 744665) is part of a large cave system which has been intersected by two large quarries, Baker's Pit and Higher Kiln Quarry, excavated close to Buckfastleigh Church. The system includes Baker's Pit Cave (opening in Baker's Pit), and Reed's Cave, Disappointment Cave, Rift Cave, Spider's Hole and Joint Mitnor Cave – the latter all opening into Higher Kiln Quarry. The system contains over 2100 m of passages lying between *c.* 61–82 m OD (Sutcliffe, 1960, 1977; Macfadyen, 1970). The caves lie some 250 m to the west of the River Dart which flows in a generally south-east direction at a level of *c.* 40 m OD. The river is bordered by a low terrace, and has a well-developed terrace on the east bank at *c.* 60 m OD and another on the west bank at *c.* 90 m OD (Sutcliffe, 1977).

Joint Mitnor Cave, which extends only about 20 m from the disused quarry face, lies in faulted and jointed Devonian limestone, and the cave system is underlain by a basin-like deposit of green volcanic tuff (Sutcliffe, 1977). Access to the cave is via the disused Higher Kiln Quarry which now houses the William Pengelly Cave Studies Centre.

Sutcliffe (1960) recorded the following stratigraphy from Joint Mitnor Cave:

3. Stalagmite floor, shattered and partly re-cemented (up to *c.* 0.5 m);
2. Loosely packed earth and stones, with numerous mammalian remains (cave earth and breccia) (up to *c.* 1.5 m);
1. Sterile waterlain sediments, much disturbed and locally consisting of finely laminated clay and silt (up to *c.* 1.5 m).

Sutcliffe (1960) notes that deposits similar in composition and texture to those in bed 1 were encountered in each of the trial pits excavated within Joint Mitnor Cave, as well as in similar locations elsewhere within the cave system, particularly within Baker's Pit Cave. Locally, these deposits are brecciated and overlie bedrock which shows no sign of speleothem growth; their maximum altitude is consistently at *c.* 64 m OD (Sutcliffe, 1960). Bed 2 is recorded as being thickest beneath the boulder-choke at the highest point of the Entrance Chamber, but thinning to < 0.3 m near the cave entrance. Sutcliffe describes this highly fossiliferous deposit as part of a talus cone, similar cones occurring elsewhere within the cave system. He notes that numerous waterlain pebbles occur in both beds 1 and 2, some having originated from Dartmoor.

Sutcliffe (1960) and Sutcliffe and Kowalski (1976) recorded the following mammalian remains from the cave earth/talus cone (bed 2) (nomenclature updated):

MAMMALIA
    Lagomorpha
    Lepus sp., hare
    Rodentia
    Arvicola terrestris cantiana (Hinton), extinct
      water vole
    Microtus agrestis (Linné), field vole
    Carnivora
    Canis lupus Linné, wolf
    Vulpes vulpes (Linné), red fox
    Ursus cf. *arctos* Linné, brown bear
    Meles meles Linné, badger
    Crocuta crocuta (Erxleben), spotted hyaena
    Panthera leo (Linné), lion
    Felis sylvestris Schreber, wild cat
    Proboscidea
    Palaeoloxodon antiquus (Falconer &
      Cautley), straight-tusked elephant
    Perissodactyla
    Stephanorhinus (Dicerorhinus) hemitoechus
      (Falconer), extinct rhinoceros
    Artiodactyla
    Sus scrofa Linné, wild boar
    Hippopotamus amphibius Linné,
      hippopotamus
    Megaloceros giganteus (Blumenbach), giant
      deer
    Dama dama (Linné), fallow deer
    Cervus elaphus Linné, red deer
    Bison priscus (Bojanus), extinct bison

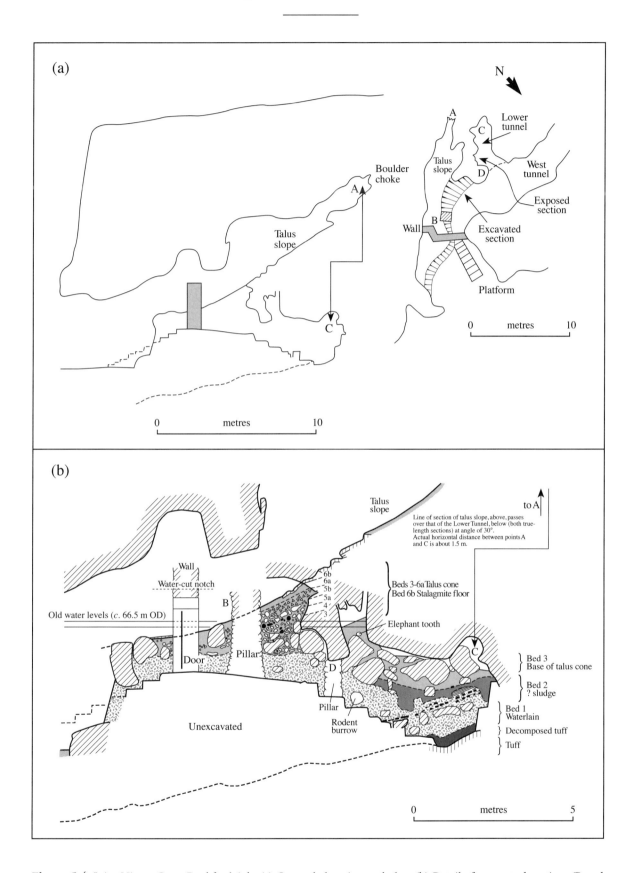

Figure 5.4 Joint Mitnor Cave, Buckfastleigh: (a) General elevation and plan. (b) Detail of excavated section. (Based on the work of A.J. Sutcliffe and adapted from Sutcliffe's original drawing and Sutcliffe's (1974) simplified section.)

Interpretation

Probably the first record of the Buckfastleigh Caves was provided by Polwhele in 1797, and although the caves were visited both by MacEnery in 1829 and Pengelly in 1859 (MacEnery, 1859; Pengelly, 1873c), they failed to yield fossils until 1936 when bones were noticed in an unnamed cave (later to be called Joint Mitnor) in Higher Kiln Quarry (Cheesman, 1959; Sutcliffe, 1960). In 1939, systematic investigations of the Buckfastleigh Caves were started by members of the Devon Spelaeological Society, and resulted first in the discovery of Reed's Cave (Hooper, 1950) and then the re-discovery of the cave at the southern end of Higher Kiln Quarry (named Joint Mitnor after its discoverers).

Excavations in Joint Mitnor between 1939 and 1941 were supervised by A.H. Ogilvie, who obtained 4307 specimens of mammal bones and teeth, now known to comprise at least 127 individual animals of 16 species (Ogilvie, 1939–1941; Sutcliffe, 1960). A preliminary account of these excavations (Anon., 1948) assigned the remains loosely to an 'interglacial period', and noted that the cave probably had not been used as a hyaena den, since gnawed bones appeared to be absent.

A.J. Sutcliffe re-studied Joint Mitnor Cave and the area around it, dug three trial pits in the cave during 1954–1955 (Sutcliffe, 1957), and re-examined the mammalian remains collected by Ogilvie. The most significant phase of excavation, however, dates from about 1959 onward, when the cave was set up as a demonstration. This work had already been started before the purchase of the quarry in 1961 and the establishment of the Pengelly Centre (Sutcliffe, 1966). During these excavations, a huge rock was removed with explosives and the roof reinforced, and the demonstration section established. Many bones were found at this stage (now housed in the Natural History Museum), but no new species were added except for the two rodents. Although largely unpublished, important documentation was undertaken, including drawing a detailed section of the deposits (Sutcliffe, 1966; Figure 5.4).

Sutcliffe (1960, 1977) regarded the relationship between the Higher Kiln Quarry Caves and the terraces of the River Dart as instructive in determining the history of development of Joint Mitnor Cave. The highest local terrace (the Upper Ambersham Terrace of Green (1949)) overlies the caves at a level of *c.* 90 m OD. The adjacent 60 m-level (and lower) terraces of the Dart, lie beneath the general level of the cave system and have been correlated with the Boyn Hill and Taplow terraces, respectively, of the Thames (Green, 1949). Irrespective of the validity of these correlations, Sutcliffe (1960, 1977) has suggested that the geomorphological evidence shows that the caves were formed beneath a water-table which would have existed until at least 90 m-terrace times. As the Dart downcut its bed, the water-table was lowered and the caves were drained, probably quite quickly. Sutcliffe (1960, 1977) therefore concluded that the sterile water-lain sediments (bed 1), found throughout the cave system, had accumulated at a time when the caves lay, at least partially, beneath the water-table; there was no evidence to suggest that they had been laid down in a limited vadose channel. The sediments of bed 1 were therefore believed to post-date the 90 m-terrace materials lying above the cave system, but to pre-date the lower Dart terraces (Sutcliffe, 1960).

Sutcliffe (1960) interpreted the materials of bed 2 (the source of the mammal remains) as part of a talus cone which had accumulated beneath a fissure in the cave roof. He argued that the faunal assemblage had resulted from animals falling to their deaths through a former opening, perhaps concealed by bushes, in the cave roof, and then becoming incorporated in the talus deposits beneath. The cave is thus thought to have operated as a natural pitfall trap in this respect, with the opening becoming blocked later by roof-fall materials.

Regarding the faunal remains, Sutcliffe (1960) concluded that the assemblage was composed only of mammals characteristic of warm climatic conditions, or of those with no specific climatic significance; cold-climate species were absent. The assemblage was therefore regarded as having accumulated during an interglacial, its homogeneity suggesting relatively rapid accumulation (Sutcliffe, 1960). Sutcliffe has argued that the faunal remains probably date from the warmest part of the Ipswichian (Eemian) Stage, citing the abundance of hippopotamus, *Stephanorhinus (Dicerorhinus) hemitoechus* and spotted hyaena, and the lack of horse and *Stephanorhinus (Dicerorhinus) kirchbergensis* (Jäger), as the principal evidence. He also notes that the remains of herbivorous mammals far outweigh those of carnivores, and that the cave does not appear to have been a hyaena den, the presence of the bones being best explained as the result of an 'accidental process of accumulation' (Sutcliffe, 1960).

Sutcliffe (1977) correlated the faunal remains

from Joint Mitnor Cave with those from the Upper Flood Plain Terrace of the River Thames (Ipswichian), and noted that similar assemblages could be recognized at widely distributed localities throughout England and Wales, at least as far north as Yorkshire. The fauna from Joint Mitnor in fact also compares very closely with 'hippopotamus faunas' reported from a number of other open sites (e.g. Barrington, Cambridgeshire; Swanton Morley, Norfolk) which have been correlated by pollen biostratigraphy with substages Ip IIb, and early Ip III of the Ipswichian (Stuart, 1982a, 1982b, 1995).

No detailed analysis of the stalagmite floor (bed 3), which overlies the fossiliferous sediments at Joint Mitnor Cave, has yet been undertaken, although Sutcliffe (1960) favoured that the surface had been frost-shattered by periglacial activity in the Devensian. He subsequently revised this interpretation (Sutcliffe, 1966) to include the possibility that the fracturing could have been caused by an earthquake (cf. Straw, 1995, 1996).

Conclusion

Joint Mitnor Cave is significant for providing one of the richest known faunal assemblages of Ipswichian age in Britain. The fauna, of 18 species, includes hippopotamus, straight-tusked elephant, wild boar, fallow deer, spotted hyaena, lion, bear and some small mammals, making Joint Mitnor one of Britain's outstanding Pleistocene mammal localities. The cave contains a lower waterlain deposit which probably formed while the cave was at least partly below the water-table, and an overlying talus deposit rich in the bones and teeth of mammals. The animals are thought to have fallen to their deaths through a fissure in the cave roof. A permanent section through the deposits and fossils is preserved in the cave, which now forms part of the William Pengelly Cave Studies Centre. The importance of the cave deposits is heightened by their geomorphological relationship with the adjacent terraces of the River Dart.

Chapter 6

The Quaternary history of the Dorset, south Devon and Cornish coasts

INTRODUCTION
D. H. Keen

Marine and periglacial deposits of the Dorset, south Devon and Cornish coasts

The Pre-Mesozoic rocks of South-West England, together with the harder Jurassic formations, provide the platforms which support intermittent outcrops of raised marine sediments and associated periglacial ('head') deposits. These deposits have an extremely protracted history of research with notable sites, such as Godrevy, having been described early in the nineteenth century (De la Beche, 1839). Although most coastal exposures comprise a simple sequence of raised beach deposits covered by head, in sheltered localities virtually throughout the Peninsula, 'sandrock', consisting of ancient dune sediments, accompanies the marine deposits. Periglacial 'heads' and related sediments occur widely throughout the region and testify to the effectiveness of frost-dominated and other (e.g. slope-wash) processes between the various phases of marine action.

At Trebetherick Point on the north Cornish coast, the occurrence of erratic boulders between the sandrock and the bulk of the head has caused some authors to consider the possibility of a glacial encroachment on to the Peninsula at some stage of the Pleistocene. The isolated gneiss boulder, known as the Giant's Rock, at Porthleven in south Cornwall also seems to suggest the former presence of glacier ice in the region (see Early glaciation; Chapter 2).

The generally acid rocks of most of South-West England provide superficial formations which are deficient in calcium. However, at a few locations (e.g. Godrevy) the calcium content is sufficient to preserve shell. Together with shells found in raised marine deposits on Devonian and Jurassic limestones at Torbay and Portland, these have provided the raw material for amino-acid age determinations which have gone some way to providing a geochronological framework for the raised beach deposits of the South-West.

The raised marine deposits

Pleistocene marine deposits found around the Peninsula occupy low topographic positions and rest on platforms cut between modern mean sea level and *c.* 5 m above it. Overwhelmingly, the beach deposits consist of gravel and cobbles, with the spectacular west-facing boulder beds of Porth Nanven, near Land's End, providing the coarsest grade material. Prior to the advent of amino-acid geochronology, the beach sediments had been regarded mostly as a coherent unit of similar age around the coast. The age of these deposits, however, was not certain, with some authors (Arkell, 1943; Mitchell, 1960; Stephens, 1970a) regarding the beaches as Hoxnian in age, on the basis of perceived stratigraphies in the overlying head, and others (Zeuner, 1959; Bowen, 1973b; Kidson, 1977) preferring an Ipswichian age.

With the revisions of chronology spearheaded by the amino-acid method, current opinion is that the sea has reoccupied the same platform on a number of occasions and that the currently visible raised beach deposits, although at roughly the same altitude and occupying the same geomorphic position in the landscape, date from at least two separate stages of the Pleistocene: Davies (1983) regarded these as being equivalent to Oxygen Isotope Stages 5 and 7 of the deep-sea record, and further detailed work by Bowen *et al.* (1985) has confirmed these ascriptions.

The general lack of fossils, in all but the beaches on the limestones, precludes any detailed comment on climatic conditions at the times the beaches were formed. Despite early statements of a 'glacial' or at least cold-climate origin for the beach deposits (Reid and Flett, 1907), there is now wide agreement that the beaches were formed by high sea levels during temperate interglacial conditions. The exact temperatures and conditions prevailing during deposition of the beaches can be determined more precisely by shells recovered from the beach deposits at Torbay (Hope's Nose and Thatcher Rock) and Portland Bill. Authors in the 1930s (Baden-Powell, 1930) still regarded the beaches here as having accumulated under cool climate conditions and cited the occurrence of molluscs, also currently found in Scottish waters, as evidence of temperatures several degrees below those of the present Channel. More quantitative analyses of raised beach molluscs at Portland Bill (Davies and Keen, 1985) and at Torbay (Mottershead *et al.*, 1987), however, show that sea temperatures during both Stage 7 and Stage 5 were no cooler than today. Indeed, the Portland Bill fauna provides some evidence that during Stage 5 the sea was warmer than now by two or three degrees.

The sediments contained in the raised beaches also reflect deposition under conditions of wave approach and energy little different to those of the present, with the calibre of the beach gravels being similar to those of the modern beaches in the same area. This is even true for the spectacular raised beach deposit at Porth Nanven which comprises water-worn boulders commonly up to 0.5 m in diameter: this west-facing site on the exposed Penwith Peninsula appears to have been subject to the same, extremely high energy conditions during both present and past interglacials.

Blown sand and fossil dunes

The common occurrence of fossil dune material overlying the raised beach deposits around the coasts of the Peninsula is exemplified by the sequence at Trebetherick Point, but similar sequences can also be found where there has been a suitable coastal configuration and sediment supply. These sands are generally lightly cemented with iron and calcium carbonate and were probably deposited during times of falling sea level. This mechanism was suggested by Arkell (1943) for the dune sands at Trebetherick Point, although he also thought that the sands were a product of a warming climate. Current views would suggest that deposition at a time of climatic cooling is more consistent with falling sea level.

The specialized nature of modern dune habitats might suggest that ancient dune deposits would be poor in fossils. However, in the dunes with a moderate calcium carbonate content, as at Godrevy and especially on the Devonian limestone at Torbay, shell does occur. This consists mostly of comminuted marine shell debris blown from the strandline, but occasionally land shells also occur, as at Hope's Nose, Torbay (Mottershead *et al.*, 1987), thus confirming the terrestrial origin of some of the sand in which they are found.

Head and related sediments

From the earliest descriptions by De la Beche (1839), it was recognized that the deposits overlying the marine horizons were of terrestrial origin and their vernacular name in the South-West, 'head', is now widely adopted. Perhaps the most representative section in the South-West is that described by Mottershead (1971) between Start Point and Prawle Point. The vital characteristics of head development are all demonstrated by this section, with easily broken bedrock to provide the

clasts in the sediment, a wide marine platform to provide a foundation for head accumulation and a sheltered aspect (on the east side of Bolt Head) which has preserved the deposits from excessive stripping by modern westerly waves. Where these three criteria are met, the thickest head sequences occur. However, even at exposed places such as Porth Nanven, some head has been preserved because the original sediment bodies ('fans' or 'aprons') were large and have not therefore been removed totally by Holocene marine action.

Although numerous facies of head can be recognized in the coastal sections, many workers have divided the head deposits into a Lower or Main Head and an Upper Head (e.g. Stephens, 1970a; see Chapter 2): traditionally a 'Wolstonian' (Saalian) age has been preferred for the lower and a Devensian age for the upper. At many locations, however, the head deposits contain lenses and layers of finer-grained sediments including silts and muds. Pollen recovered from organic-rich muds in the head sequence at Boscawen, on the south Cornish coast, indicates that the periglacial processes involved in head formation may have ceased, temporarily, when tundra vegetation colonized pools on the surface of the solifluction deposits: radiocarbon dating of these organic-rich sediments suggests that the upper head here can be no older than *c.* 30 ka BP (Scourse, 1985a). Recent lithostratigraphical comparisons of head sequences in the region suggest that this twofold division is in fact widespread (Scourse, 1985a), and a Late Devensian age for the upper head is supported by thermoluminescence (TL) dating of aeolian sediments ('sandloess') intercalated with it (Wintle, 1981). However, the precise age of the lower head, at most localities, is unknown: continuing uncertainty surrounding the age of raised beach deposits which underlie the head precludes precise dating. Detritus from a number of Pleistocene cold stages (equivalent to Oxygen Isotope Stages 4 and 6) may be present, but such ascriptions must remain provisional. In this context, it is worth noting that head deposits at least as old as Stage 8 have been recognized in the similar Pleistocene sequences of the Channel Islands and Normandy (Keen, 1993; Keen *et al.*, 1996; Chapter 2).

Further subdivision of regional head sequences has, however, been possible at Portland Bill, where the thin head overlying the West Beach (Davies and Keen, 1985) is divided in two by a marked palaeosol. Below this horizon, intense weathering has caused decalcification of the head. Above it, silty head with a high shell content is indicative of sub-tundra conditions and its upper layers are highly cryoturbated suggesting a further intense periglacial phase. The evidence from Portland Bill therefore suggests three separate periglacial phases after deposition of the West Beach during Oxygen Isotope Stage 7: tentatively, these could be ascribed to cold Oxygen Isotope Stages 6, 4 and 2 (Keen, 1985).

Evidence of glaciation

In the area under consideration there is little evidence for glaciation, but at Trebetherick Point the erratic boulders, perhaps separating the two heads, may be an exception. It is nonetheless difficult to see how a glaciation extensive enough to deposit erratics at Trebetherick could avoid leaving them elsewhere (but see Chapter 7) and none has been found in the area between Trebetherick Point and Land's End. If the Trebetherick erratics were brought by ice, it is perhaps more probable that it was in the Middle or Early Pleistocene and that only remnants of the tills survived to be reworked into the head.

Further evidence for an incursion of glacier ice on to the South-West Peninsula may be provided by the Giant's Rock at Porthleven. The provenance of this famous gneiss boulder is far from certain, but the highly metamorphosed terrain necessary to provide such a rock-type can lie no nearer than Scotland or Brittany, and indeed its source may be much farther afield (see Chapter 2). A Scottish source would fit with a glacial origin, but virtually no theoretical reconstruction of ice sheets, except that of Kellaway *et al.* (1975), would allow a Breton source. Whatever its provenence, it is difficult to imagine any agency other than ice which could have been responsible for its emplacement. The type of ice is a further uncertainty. The lack of any till in south Cornwall has led to the suggestion of icebergs as the mechanism of deposition (Flett and Hill, 1912), and this still seems the most probable explanation, although a sea cold enough to carry icebergs of the size capable of floating the Giant's Rock should have occupied a much lower sea level than that of the modern intertidal platform where the rock is stranded. For the Giant's Rock alone to have survived from the glacial phase which deposited it, the glaciation would need to have been of Middle or Early Pleistocene age to allow the time for the destruction of all other glacial material. There is no other trace of such an episode in the South-West.

Geochronology

Prior to the 1980s, the age of the raised marine deposits was variously determined from their altitude (and inferred related sea level) or from the lithostratigraphy of the overlying head(s) and associated deposits. According to Mitchell (1960) and Stephens (1970a), nearly all raised beach deposits in the region were of Hoxnian age: a Lower or Main Head was of 'Wolstonian' age, the last interglacial (Ipswichian) was represented by an unconformity or weathering horizon in the head sequence, and the last major cold phase of the Pleistocene, the Devensian, saw deposition of the uppermost head layers. A different scheme proposed by Bowen (1973b), which also used evidence from Wales and southern Ireland, suggested that the raised beach deposits everywhere in the region were of the same, Ipswichian, age. The raised beach deposits and overlying head, therefore, could be accommodated by a single interglacial/glacial cycle which spanned the Ipswichian (warm) and Devensian (cold) stages.

The application of amino-acid, thermoluminescence (TL), Uranium-series and radiocarbon dating techniques has not solved all the problems, but has provided much clarification. The principal discovery, revealed by amino-acid analyses on marine mollusc shells, is that the raised beach sediments were deposited in two or more episodes (Davies and Keen, 1985; Bowen *et al.*, 1985). The most likely ascriptions for these marine phases are Stages 7 and 5 of the oceanic oxygen isotope record. An alternative interpretation of the data involves deposition of marine sediments in more than one of the sub-stages of Stage 5 (Bowen *et al.*, 1985).

Although not obtained from the GCR sites described in this volume, the Uranium-series dates derived from the caves on Berry Head, Torbay, by Proctor and Smart (1991) and Baker and Proctor (1996) are significant because they date three major high sea-level events equated by the authors with Oxygen Isotope Stages 5e, 7 and perhaps 9, and provide the potential to calibrate the amino-acid chronologies obtained from shell carbonate outside caves (Bowen, 1994b). Although it has been suggested that the amino-acid ratios obtained from shells from raised beaches below 10 m OD may indicate ages in either Stage 5e or 7 (Bowen *et al.*, 1985; Davies and Keen, 1985; Bowen, 1994b), or within Stage 5e (Bowen *et al.*, 1985), the evidence from the speleothem dates from Berry Head reinforces the former chronology of two separate high sea-level events, and adds a third series of dates around 328 ka BP for a sea level at *c.* 25 m above mean sea level (see above; Chapter 2).

The thermoluminescence and radiocarbon dating of the uppermost parts of the head and related sediments to the last 30 000 years (Wintle, 1981; Scourse, 1985a), suggests that the lower parts of head sequences could be either Early Devensian in age (Oxygen Isotope Stage 4), or date from Stage 6 (see above; Head and related sediments). Unless the underlying marine deposits are dated (see Portland West), the age of any individual head deposit cannot be determined except within these broad limits.

In the following account, the selected GCR sites are described in order around the coast from Portland Bill to Trebetherick Point (Figure A; Preface).

PORTLAND BILL
D. H. Keen

Highlights

Portland Bill is one of the most important raised beach sites in Britain. It shows evidence of pre-Stage 5 shoreline deposits and terrestrial sediments, a Stage 5 raised beach deposit and Devensian head. The beach sediments and Devensian head have yielded mollusc faunas which have been used to reconstruct former environments and for amino-acid dating.

Introduction

Portland Bill is one of the few places on the south coast where two separate raised marine deposits, in association with terrestrial sediments, can be demonstrated. The site was first described by De la Beche (1839) who noted the existence of marine deposits. Later authors mentioned aspects of the succession and its fauna (Bristow, 1850; Weston, 1852; Damon, 1860; Whitaker, 1869; Prestwich, 1875; Baden-Powell, 1930; Arkell, 1943, 1947; Green, 1943; Carreck, 1960; Pugh and Shearman, 1967; Macfadyen, 1970; Mottershead, 1977b). Of these, Prestwich's (1875) description was the most comprehensive and has been used by most later writers as a stratigraphical basis. The marine fauna was summarized in detail by Baden-Powell (1930). In recent times, full descriptions of the deposits and their included faunas have been published by Keen (1985) and Davies and Keen (1985). The latter authors also

presented preliminary isoleucine epimerization data to provide a geochronology for the deposits. This method was also used by Bowen *et al.* (1985) who suggested alternative dating schemes.

Description

The full Quaternary sequence at Portland Bill (SY 677681) does not occur in superposition in any single section, and separate localities and deposits are needed to describe the sequence adequately. The key elements comprise the Portland West Beach and the Portland East Beach (Davies and Keen, 1985), and the Portland Head and Loam (Prestwich, 1875; Arkell, 1947; Keen, 1985) which are described separately below and shown in Figure 6.1.

Portland West Beach

The earliest comprehensive description of this part of the site was by Prestwich (1875). The main section today stretches for 40 m south of SY 67506860 in the grounds of the Admiralty Underwater Weapons Establishment (AUWE). The widespread occurrence of loose beach gravel, for 200 m south of the current section, probably indicates the former extent of the raised beach deposits. The beach deposit also crops out at SY 68056885 in a disused quarry 100 m south-west of the Old Lower Lighthouse (now the Portland Bird Observatory). The beach deposits rest on a lightly wave-smoothed surface cut in Jurassic limestone. In the main AUWE section, the beach consists of up to 2.5 m of well-sorted sandy gravel arranged into as many as seven fining-up units, each grading from pebbles to coarse sand (Figure 6.1). The deposits are planar-bedded and cemented by calcium carbonate, although considerable voids also occur. The pebbles of the beach comprise a variety of lithologies: flint makes up between *c.* 85–90%, with the remaining clasts comprising chert and limestone from the local Portlandian, and quartzite, chalk, and Greensand. In addition to these pebble-types, Prestwich (1875) noted ferruginous grit (Tertiary), micaceous sandstone (Devonian), red and purple sandstone (Triassic), red feldspar porphyry and red granite. The source of much of this non-local material was probably the South-West Peninsula. The Portland West Beach is largely devoid of shelly fossils, except in its basal layers, but enough fragments of rocky shore gastropods – *Nucella lapillus* (Linné), *Littorina littorea* (Linné),

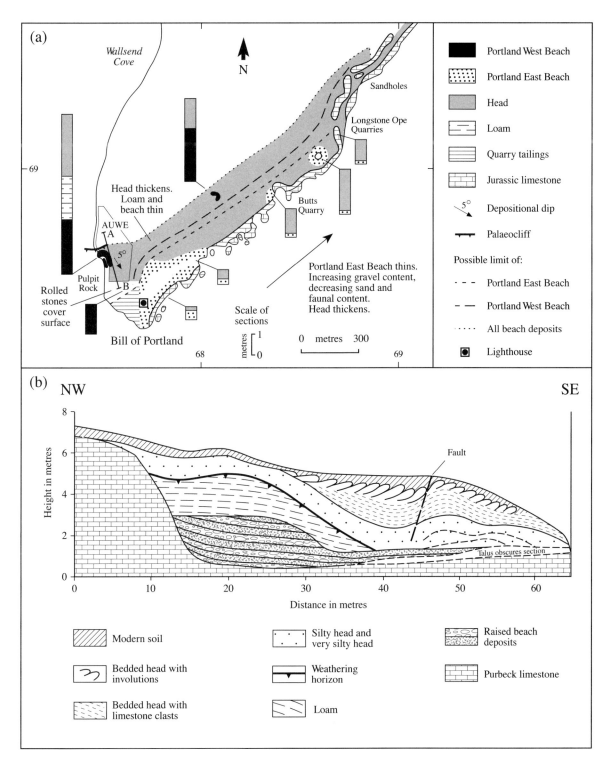

Figure 6.1 (a) Quaternary deposits at Portland Bill, adapted from Davies and Keen (1985). (b) The Quaternary sequence at AUWE, adapted from Keen (1985). The cross-section follows line A–B shown in plan above.

Patella spp. and bivalves *Cerastoderma* spp., *Mytilus edulis* (Linné) – were recovered for amino-acid analyses (see below).

Portland Loam and Head

The Portland Loam and Head were also first described in detail by Prestwich (1875). The two units rest on the cemented shingle of the West Beach and are best seen in the AUWE section (Figure 6.1). The deepest part of the section shows the following sequence:

4. Head with limestone clasts and small quantities of mollusc shell (1.2 m)
3. Silty head with numerous shells (1.4 m)
2. Silty head with few clasts and very numerous shells (0.3 m)
1. Loam with calcareous pellets and topped with a weathering horizon (0.5 m)

The loam is devoid of shell, despite Prestwich's (1875) assertion that both loam and head are shelly. Keen (1985) showed that the calcium carbonate content of the loam had been completely reworked so that no shell remains were present. This reworking has allowed the development of calcareous concretions in the loam which is otherwise silty in texture and devoid of coarser material, except for a few pebbles near the base derived from the underlying raised beach deposits.

The head (beds 2, 3 and 4) overlies the loam with a sharp boundary (Figure 6.1). The matrix of this deposit also consists largely of silt, but limestone clasts, up to 0.25 m long, are also present. Both the loam and the head are crudely bedded, with the former dipping south at 3°, with individual beds being picked out by lines of calcareous pellets. The head is more steeply inclined at 5–10° to the south. The top metre of the head is disrupted by periglacial structures similar to the festoons described widely in southern England (Williams, 1975). Deeper disruption occurs near the south-eastern end of the section where an inclined zone of disturbance reflects reverse faulting under periglacial conditions (Keen, 1985; Figure 6.1). At its eastern extreme, the entire thickness of the head deposit is disrupted by periglacial structures which extend into the Jurassic bedrock (Pugh and Shearman, 1967). The head is not decalcified and contains abundant land shells. The total fauna comprises fourteen species of which only three – *Pupilla muscorum* (Linné), *Lymnaea truncatula* (Müller) and the slug genus *Deroceras* sp. –

are numerous (Keen, 1985). The head also contains fossil ostracods which may have lived in small brackish pools on the land surface where the head was accumulating (Keen, 1985).

Portland East Beach

The deposits of this beach crop out north-eastwards for 1.5 km between Portland Bill and Longstone Ope Quarries (SY 688691; Figure 6.1). They consist of subangular clasts of Portland and Purbeck limestone with a few pebbles of flint and chert and calcareous fossil debris in a sandy matrix. The largest clasts are *c.* 60 cm in diameter and in places these represent the whole thickness of the beach. Elsewhere, the beach deposit is less than 45 cm thick and consists of shell which infills interstices between the larger clasts. The deposits of the beach are structureless, probably due to post-depositional cryoturbation (Pugh and Shearman, 1967). Unlike the West Beach, the East Beach is almost entirely uncemented and is richly fossiliferous: a 2 kg sample from the most fossiliferous part of the exposure, 200 m north-east of Portland Bill, yielded 6670 individual shells (Davies and Keen, 1985). The fauna is dominated by rocky shore gastropods – *Littorina* spp., *Gibbula* spp., *Patella* spp., *Nucella lapillus* (Linné), *Rissoa* spp., and the bivalve *Turtonia minuta* (Fabricius). A total of 34 gastropod taxa, one chiton, and 17 bivalve taxa were recorded from the East Beach by Davies and Keen (1985). A further seven species of gastropod and four species of bivalve were recorded by Baden-Powell (1930). Other faunal remains include foraminifera, *Balanus* spp. plates, and crab and echinoderm fragments (Davies and Keen, 1985).

Interpretation

The sequence of deposits at Portland Bill provides evidence for two high sea levels and two phases of terrestrial deposition in the Upper Pleistocene. Amino-acid ratios (D-*alloisoleucine* : L-*isoleucine*) derived from fossil shells in the sequence, enable these events to be placed in relative stratigraphic order and provide a tentative chronological framework.

Portland West Beach

Planar bedding in this sediment indicates that the Portland West Beach was deposited under high energy conditions. The few shell fragments it contains are suggestive of sea temperatures no colder than now and a sea level perhaps 10 m higher (Davies and Keen, 1985). Amino-acid D/L ratios suggest the beach is older than the East Beach and Davies and Keen (1985) concluded that an age of 200 ka ± 30 ka BP was likely. The West Beach could therefore have been deposited during Oxygen Isotope Stage 7 of the oceanic record.

Portland Loam

The decalcified nature of this deposit makes it difficult to determine its environment of formation and its age. However, it may be a slope deposit and, since it overlies the West Beach, it must be younger than Oxygen Isotope Stage 7. A Stage 6 age for the loam seems probable and thus the weathering horizon which separates it from the head may have formed in the temperate conditions of Stage 5 (see below).

Portland East Beach

The thin deposits of the East Beach (< 0.6 m) give little sedimentological evidence for its conditions of deposition. However, its fauna, more extensive than that from any other raised beach deposit on the south coast, gives comprehensive details of the contemporary marine environment and, through amino-acid D/L ratios, of its age. The earliest detailed work on its fauna was by Baden-Powell (1930). He concluded that sea temperatures were approximately 4°F (2.2°C) colder than those of the current Channel, because 'northern' species of mollusc, such as *Margarites helicinus* (Fabricius), were present. Recent work by Davies and Keen (1985) shows that most of the fauna comprises species still found today in the Channel. Molluscs with a restricted northern range, such as *Tricolia pullus* (Linné), outnumber those with a northern distribution today, thus indicating that sea temperatures were no colder than at present. These conclusions are supported by fossil foraminifera recovered from the East Beach. These comprise, exclusively, modern-day English Channel-types, including one species – *Elphidium crispum* (Linné) – which is now found no farther north than the Channel (Davies and Keen, 1985).

Amino-acid ratios presented by Davies and Keen (1985) suggest that the Portland East Beach accumulated during Oxygen Isotope Stage 5e. The ratios were calibrated with a Uranium-series date obtained from travertine in similar raised beach deposits at La Belle Hougue Cave, Jersey, some 130 km to the south (Keen *et al.*, 1981). Further amino-acid D/L ratios derived from raised beach shells elsewhere in western Britain (Bowen *et al.*, 1985) suggest that the Portland East Beach could be younger than Oxygen Isotope Stage 5e, perhaps dating from Stage 5a. These amino-acid ratios were obtained by a different preparation method and gave slightly lower, thus younger, ratios than those obtained from other raised beach sediments in South Wales: the latter were dated by Bowen *et al.* (1985) to their Pennard, or 5e, Stage.

Portland Head

The sediments and included fauna of the Portland Head together are indicative of a terrestrial origin. Because the species of mollusc found in the head can still be found in Britain today, both Prestwich (1875) and Arkell (1947) assumed that it had formed under conditions like those of the present. However, the deposit is ill-sorted, contains angular clasts and has a restricted fossil mollusc fauna indicative of open-ground conditions. Together, these characteristics point to the deposit having accumulated under cold conditions (Keen, 1985). Its age, however, is uncertain. The fauna it contains is unlike those of Devensian late-glacial deposits described from Kent (Kerney, 1963; Keen, 1985), and the occurrence of the relatively thermophilous mollusc, *Helicella itala* (Linné), appears also to rule out a Middle Devensian age. These comparisons led Keen (1985) to propose that the head accumulated during the Early Devensian and was then cryoturbated in the Middle or Late Devensian.

Bowen *et al.* (1989) presented a small number of amino-acid ratios derived from *Trichia* shells in the head: the deposit was considered to have accumulated during Oxygen Isotope Stage 7. This is at variance with Davies and Keen's (1985) suggestion that the West Beach is of Stage 7 age. If the Portland Head indeed dates from Stage 7, then the age of the West Beach must be much older, Oxygen Isotope Stage 9 at youngest. The Oxygen Isotope Stage 7–equivalent amino-acid ratios reported from the head by Bowen *et al.* (1989) are also inconsistent with its origin as a cold-stage, periglacial, deposit. It is, however, possible that the *Trichia* shells were reworked into the head from a 'temperate' Stage 7 terrestrial deposit, but no such source has so far been identified at Portland. At present, the least problematical interpretation of the sequence is to regard the West Beach as an Oxygen Isotope Stage 7 deposit, and the overlying loam and head as cold-climate materials formed later in the Pleistocene.

Conclusion

The Pleistocene deposits on Portland Bill present a fascinating association of terrestrial and marine sediments ranging from perhaps 200 ka BP. The faunal content of the marine units is unrivalled along the south coast and allows a palaeoenvironmental reconstruction of considerable detail and value. The use of amino-acid geochronological techniques has enabled this complex of marine deposits to be dated to Oxygen Isotope Stages 7 and 5 of the deep-sea record: this preliminary ascription might suggest that the cold-climate deposits and structures at the site date from Oxygen Isotope Stages 6, 4 and 2. The evidence of two raised beach deposits, of different ages, confirms Portland Bill as a critical reference site with regard to interpreting coastal Pleistocene sequences throughout south-west Britain.

HOPE'S NOSE AND THATCHER ROCK
D. H. Keen

Highlights

The Torbay raised beach deposits are among the most fossiliferous of their kind on the south coast. Together with shelly raised beach sediments at Portland Bill, they provide a cornerstone for Pleistocene palaeoenvironmental and stratigraphic studies in the South-West.

Introduction

The raised marine deposits in Torbay have received considerable attention since their earliest description by Austen (1835). Much work was done on the sequences and their faunas in the nineteenth (De la Beche, 1839; Ussher, 1878; Hunt, 1888; Prestwich, 1892) and early twentieth centuries (Hunt, 1903, 1913a, 1913b; Ussher, 1904; Jukes-Brown, 1911; Shannon, 1927, 1928; Lloyd, 1933; Green, 1943). In recent times, Orme (1960b) provided new material, while reviews by Zeuner (1959), Macfadyen (1970) and Mottershead (1977b) drew substantially on earlier data, especially those of Hunt and Lloyd. Amino-acid ratios derived from shells in the beaches were presented by Davies (1983) and Bowen *et al.* (1985): a recent detailed study by Mottershead *et al.* (1987) uses these and other palaeoenvironmental data to establish a geochronological framework for the sequence.

Description

The raised beach deposits in Torbay occur on both of the two limestone promontories which enclose the bay. In the south, around Shoalstone Point (SX 939568), 1.3 m of beach sediment is overlain by *c.* 2 m of head. The exposures here extend nearly 300 m from Shoalstone, east towards Berry Head. On the north side of the bay, Pleistocene sediments are best seen at Hope's Nose (SX 949637) where the total thickness of marine and terrestrial deposits is around 9.5 m. Offshore, the islet of Thatcher Rock shows a thin development of both head and raised beach sediments. The maximum thickness of Pleistocene deposits here is *c.* 5 m. All of the major beach exposures are fossiliferous, principally with gastropods of rocky shorelines, but also with bivalves from a range of habitats, and a restricted microfauna. Early work on the molluscan fauna (Hunt, 1888, 1903; Lloyd, 1933) suggested that the beach was deposited by a sea colder than the present. Modern work (Mottershead *et al.*, 1987) regards both the molluscan fauna and the microfauna as indicating water temperatures no colder than in the Channel today.

The GCR site known as Hope's Nose and Thatcher Rock includes three main elements: (a) Shoalstone; (b) Hope's Nose; and (c) Thatcher Rock which are described below.

(a) Shoalstone

The deposits at Shoalstone consist of gravel and cobble beach deposits overlain by head. Mottershead *et al.* (1987) describe the following sequence at SX 939568 (maximum bed thicknesses in parentheses):

6. Sandy loam with pebbles (0.7 m)
5. Head of angular cobbles of limestone in a loamy matrix and with apparent downslope bedding (2.0 m)
4. Coarse sand with shell debris (0.3 m)
3. Cobble beach of limestone, flint and slate without sand and with little shell (0.6 m)
2. Beach gravel with abundant shells, principally of oysters, but also *Cerastoderma edule* (Linné) and gastropods (0.2 m)
1. Angular cobbles and boulders in a sandy matrix (0.2 m)

The beach deposits rest on a wave-smoothed platform of Devonian limestone and are intermittently cemented with calcium carbonate. The shells are often fractured.

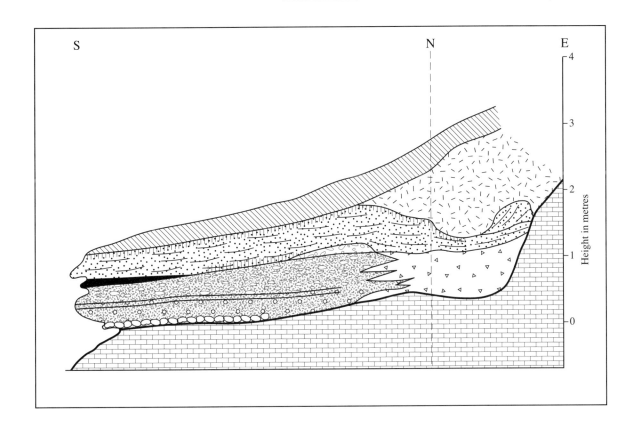

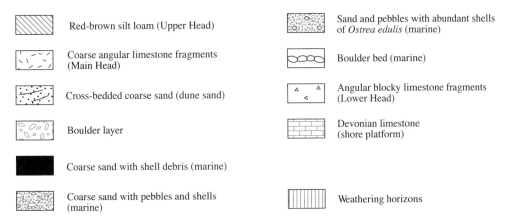

Figure 6.2 The Quaternary sequence at Hope's Nose. (Adapted from Mottershead *et al.*, 1987.)

(b) Hope's Nose

The best section occurs at the eastern extremity of the exposure, at SX 949637, and is shown in Figure 6.2. The sequence, simplified from Mottershead *et al.* (1987), is as follows:

4. Head of coarse limestone debris in a loamy matrix (4.0 m)
3. Dune-bedded sand with occasional cobbles of limestone. The boundary of the sand and head is gradational (1.5 m)
2. Coarse sand with intermittent boulders towards the top and interbedded with silt layers lower down. Shelly throughout, but with bedding planes crowded with oysters in the lower levels of the deposit (2.85 m)
1. Boulder bed resting on Devonian limestone and interdigitated with 1.0 m of angular head close to the fossil cliff (0.3 m)

The whole sequence of deposits rests against a

fossil cliff and on a platform cut into Devonian limestone. The marine units (bed 2) in general fine-upwards and are well cemented by calcium carbonate. Shells are mainly fragmented, but whole shells occasionally occur. Especially abundant are valves of *Ostrea edulis* (Linné) although other bivalves and gastropods are also present. Overlying the beach deposits is a dune-bedded sand (bed 3) which contains a sparse fauna of land gastropods – *Pupilla muscorum* (Linné) – and derived marine shell debris together with foraminifera (Mottershead *et al.*, 1987). The beach deposit (bed 2) and blown sand (bed 3) are weathered and stained with iron oxide. The overlying head (bed 4), which forms the uppermost unit of the succession, contains much dune sand in its basal metre. Higher up, it is composed of angular limestone fragments, up to 0.5 m long, in a silt and sand matrix.

(c) Thatcher Rock

The Pleistocene deposits on Thatcher Rock occupy two small and unconnected areas on the islet. In the north (SX 944629), patches of marine and terrestrial sediment total up to 1.8 m in thickness. In the south-east (SX 944628), two smaller outcrops exhibit up to 5.2 m of head and beach sediment. The following sections were described by Mottershead *et al.* (1987):

North Beach
3. Head with angular blocks of limestone in a red silty sand matrix (0.4 m)
2. Sand, reddened at the top and cemented, with locally abundant shells (1.2 m)
1. Pebbles and cobbles of flint, limestone, and igneous rock. Cemented in places and penetrated by weathering (0.2 m)

East Beach
2. Head deposit of angular, crudely bedded limestone blocks in a sandy matrix (5.0 m)
1. Beach deposit of well-rounded pebbles with shells in a clay/silt matrix (0.2 m)

Both beach deposits rest on a platform and against a cliff cut in Devonian limestone. The elevation of the beach fragments, between 7.8 and 10.3 m OD, has allowed considerable recent erosion of the beach and head deposits by waves: only small areas of sediment remain on an increasingly exposed platform. The higher elevation of the Hope's Nose deposit, between 9.1 and 12.1 m OD, has preserved this beach remnant from present-day marine action.

The shell content of the Thatcher Rock raised beach sediments is described by Mottershead *et al.* (1987). The fossil fauna consists of numerous gastropods, mainly *Littorina* spp. and *Patella* sp., but with 20 other species noted, and also 21 species of bivalve, mostly represented by single shells. The fauna is most numerous in the north outcrop and the East Beach has a fauna composed of the most common species found in the North Beach. Accompanying the fossil molluscs in the North Beach were four species of foraminifera, barnacle plates, fish vertebrae, crab and echinoid debris and rolled 'pebbles' of mammal bone.

Interpretation

The local palaeoenvironment of the raised marine deposits around Torbay can be determined from their mollusc and other fossils. Earlier workers recorded shells from all the outcrops described above and these records were consolidated by Lloyd (1933): the general conclusion was that sea temperatures at the time of deposition were lower than those of the modern Channel. In particular, the occurrence of species such as *Trophon truncatus* (Ström), with a mainly Scottish distribution at present, was used to support this interpretation. Mottershead *et al.* (1987) counted over 1000 individual fossil shells from the three major beach localities, and concluded that sea temperatures were no colder then than now. This assessment is most secure for the Thatcher Rock beach deposit, where faunal remains are most extensive. For these raised beach sediments, the fauna is closely comparable to marine faunas of today, and there is no reason to suggest temperatures different from those of the modern seas around Torbay. The evidence from the Hope's Nose and Shoalstone beach deposits is less clear due to a more restricted fauna, but the inference from these localities is also consistent with a climate no cooler than today's.

Because the raised beach deposits occur in similar positions and at approximately the same height above sea level, previous authors (see Mottershead, 1977b) have assumed a similar age for them. The application of amino-acid epimerization relative dating techniques (Davies, 1983; Bowen *et al.*, 1985; Mottershead *et al.*, 1987) and further detailed stratigraphic studies, however, have established a more complex chronology. The amino-acid ratios of Davies (1983) and Bowen *et al.* (1985) were obtained by slightly different analytical methods. The data sets, however, are 'internally' consistent and therefore allow comparisons with other sites in south-west Britain. Ratios from the Hope's Nose beach deposits, derived by both analytical methods, are consistently higher (indicating a greater age) than those for the other sites in Torbay. The ratios obtained from the Thatcher Rock fauna were compared by Davies (1983) with those obtained by Keen *et al.* (1981) from similar raised beach sediments at La Belle Hougue Cave, Jersey. The latter have yielded, from travertine, a Uranium-series date of 121 ka + 14 ka/− 12 ka BP, leading Davies to conclude that the Thatcher Rock raised beach deposits accumulated during Oxygen Isotope Stage 5e (Ipswichian Stage). Although the few ratios from Shoalstone noted by Davies were higher than those from Thatcher Rock, these beach deposits also were ascribed to Stage 5. The Hope's Nose raised marine sediments, therefore, were considered to be older than Oxygen Isotope Stage 5 and were ascribed to Stage 7 (*c.* 180–220 ka BP). The results of Bowen *et al.* (1985) confirm that the Thatcher Rock beach deposit probably dates from Oxygen Isotope Stage 5 (Pennard Stage) and that the Hope's Nose beach deposit dates from Stage 7 (Minchin Hole Stage). However, the slightly higher ratios derived from shells in the Shoalstone deposits (lying on a scale somewhere between those of the Hope's Nose and Thatcher Rock deposits) suggest an intermediate age between Oxygen Isotope Stages 5 and 7 (equivalent to the 'unnamed' Stage of Bowen *et al.*, 1985). In this context, the latter may equate either with the end of Stage 7 or the beginning of Stage 5, since no discrete marine deposits can be recognized in the site's stratigraphy.

The geochronological dating of these marine events also establishes the relative age of the terrestrial deposits in Torbay. The head, which underlies or interdigitates with the Hope's Nose beach deposit, must date from at least Oxygen Isotope Stage 7. The dune sand which overlies the Hope's Nose raised beach deposit has yielded amino-acid ratios indicative of a Stage 5 age (Bowen *et al.*, 1985) and its snail fauna, and the weathering it has undergone, also suggest interglacial conditions. The head, which overlies the raised marine and/or dune sediments at all three sites, is therefore likely to have accumulated during the Devensian Stage: greater precision than this is not yet possible.

Conclusion

The Pleistocene deposits exposed in Torbay record an impressive variety of marine and terrestrial

events dating back to at least 200 ka BP. Amino-acid geochronological techniques have shown that raised beach deposits of two distinct ages are present. These have been correlated with Oxygen Isotope Stages 7 and 5 of the deep-sea record. Palaeontological analyses at this site provide vital evidence to show that sea temperatures during both of these marine phases were similar to today's. Dune sand overlying the raised beach deposits at Hope's Nose can be ascribed to warm conditions in Oxygen Isotope Stage 5 (Ipswichian Stage), while terrestrial deposits which underlie it must be at least as old as Oxygen Isotope Stage 7 (*c.* 200 ka BP). Head deposits which 'cap' the raised beach and dune deposits throughout the Torbay sections are likely to have accumulated under periglacial conditions in the Devensian Stage. Very few raised beach localities in Britain have yielded so much palaeoenvironmental information. Together with the Pleistocene sequence at Portland Bill, the Torbay deposits provide an important stratigraphic and chronological model with which other coastal Pleistocene sequences throughout the South-West can be compared.

START POINT TO PRAWLE POINT
D. H. Keen

Highlights

This 6 km-section of coast provides some of the finest exposures of periglacial deposits in Britain. They demonstrate superbly the salient characteristics of coastal head deposits.

Introduction

The coast between Start Point (SX 830370) and Prawle Point (SX 773350) is extensively mantled with Pleistocene, periglacial slope deposits. These deposits are especially well developed here because the schist bedrock is particularly prone to destruction by frost-action, and because most of this coast faces south or south-east and is thus protected from westerly and south-westerly waves. Alternatively, on the exposed sides of headlands, such as the section between Bolt Head (SX 725359) and Hope Cove (SX 673398), head deposits have been stripped by marine erosion (Mottershead, 1971). Authoritative descriptions of the coastal head deposits of South-West England were first made by De la Beche (1839) who introduced the

Figure 6.3 Coastal head deposits overlying a raised shore platform at Great Mattiscombe Sand (SX 816369), 1.2 km west of Start Point. (Photo: D.H. Keen.)

term 'head'. The sections between Prawle Point and Start Point received particular mention by Ussher (1904), Steers (1946), Masson-Phillips (1958) and Orme (1960b), but the most comprehensive sedimentological and morphological descriptions were given by Mottershead (1971). The quantitative characterization of the head deposits along this stretch of the coast is detailed by Mottershead (1972, 1976, 1982b). Estimates of current rates of bedrock weathering and erosion along the shoreline are provided by Mottershead (1981, 1982a, 1982c). Morawiecka (1993, 1994) described palaeokarstic features in 'sandrock' at Prawle Point.

Description

The sections range up to 33 m in height and are banked against a fossil cliffline up to 125 m high. The head rests on former wave-smoothed surfaces at the foot of the cliffs. The maximum thickness of the head is seen only at localities such as Mattiscombe Sands (SX 816369) where subsequent erosion has cut deeply into the deposit (Figure 6.3). At other localities, such as Langerstone Point

(SX 782354), sections are as little as 2 m high: these exposures occur up to 270 m away from the fossil cliff. In contrast, the thickest sections occur only a short distance from the former cliffline. The head deposits fill the mouths of coastal valleys, as at Mattiscombe Sands and are of variable thickness where the bedrock floor is gullied or irregular. They are banked against the fossil cliff in a series of 'fans' or 'aprons', the surfaces of which form concave slopes of 10–15° near the bedrock slope, declining to 2–3° away from it (e.g. Langerstone Point) (Mottershead, 1971).

The head is poorly sorted and is composed of all particle sizes from boulders to clay (Mottershead, 1971). Sediments nearer to the ancient cliff are generally coarse, often containing boulder-sized material, whereas those away from it (and the source of sediment) are generally finer-grained, consisting largely of pebble- and granule-size material.

The lithology of clasts in the head reflects the nature of the bedrock found upslope, namely a combination of quartz-mica schist and green (chlorite/hornblende) schist. Mottershead (1971) noted a general increase in the ratio of quartz to schist in the sections farthest from the fossil cliffline, which he suggested was due to comminution of the more friable schist-types during transport, and perhaps even to the selective transport of smaller clasts. All particles, however, are angular or subangular, even those found farthest away from the old cliff. The clasts in the head show a strong unimodal orientation downslope, reflecting their direction of movement away from the cliffline and sediment source. The head sometimes shows crude stratification (e.g. at Rickham Sand; SX 755367) with thin layers picked out in places by the different colours of the constituent mica and hornblende schists. Most of the head, however, is structureless.

Towards the base of many sections, the head also contains sand and rounded pebbles of marine origin. In places, possibly *in situ* raised beach deposits are covered by the head. At others, the former presence of a beach deposit is indicated by the occurrence of rolled stones within the head: some, such as those of flint, are clearly of non-local derivation.

Interpretation

Most deposits exposed in the coastal cliffs between Start Point and Prawle Point, are believed to have been derived by the action of frost on the ancient cliff which now lies upslope (Mottershead, 1971). They arrived on the ancient marine platform by mass movement processes, principally solifluction, under periglacial conditions. The morphology of the sediment bodies ('fans' or 'aprons' of head), and the sedimentological and lithological characteristics of the deposits are entirely consistent with such an origin. Alternating layers of different schist material within the head may indicate successive shallow solifluction 'flows', each derived principally from a particular bedrock lithology: other massively bedded and unstructured parts of the head sequence may denote the arrival of large thicknesses of saturated debris *en masse* (Mottershead, 1971). Whatever process variations the head may denote, there is general agreement that they indicate mass wasting under periglacial conditions, and are the products of degradation of the coastal bedrock slope (Mottershead, 1977a). These head sequences find close parallels elsewhere around the coast of South-West England (Mottershead, 1977b) although they are rarely, if ever, better developed.

Whereas their origin is not disputed, the age of the deposits is far from certain. Mottershead (1971) divided the sequences found along this coastline into a variety of facies: a Main Head (forming the bulk of the sequence); an Upper Head (distinguished by a silty texture, perhaps indicating the incorporation of loess); and a sporadic Lower Head (separated from the Main Head by raised marine sediments, for example, at Sharpers Cove, SX 786357). Although Stephens (1966b, 1970a) had earlier effected a similar division of head deposits in north Devon, he had assigned them to the 'Wolstonian' (Lower/Main Head) and the Devensian (Upper Head) because the underlying raised beach deposits were believed to be of Hoxnian age. Bowen (1973b) regarded this differentiation as unrealistic and ascribed the different head facies to various stadial phases of the Devensian Stage. Mottershead (1971) instead followed the thinking of Masson-Phillips (1958) and Orme (1960b) and suggested that the raised beach deposits had accumulated during milder, interstadial, phases of the Devensian, with head accumulating during colder periglacial phases.

Modern work has shown that none of these schemes is universally applicable. Detailed regional geochronological studies (Keen, 1978; Lautridou, 1982; Davies and Keen, 1985; Bowen *et al.*, 1985; Mottershead *et al.*, 1987) have shown that the raised marine deposits of the western Channel date from at least two high sea-level stands in the Middle and Upper Pleistocene – probably equivalent to Oxygen Isotope Stages 7 and 5 of the deep-sea record. At sites where an Oxygen Isotope Stage 7 raised beach deposit lies beneath head, the possibility clearly exists for head facies to have accumulated during a number of Pleistocene cold stages (e.g. Oxygen Isotope Stages 6, 4 and 2). Where the head overlies an Oxygen Isotope Stage 5 marine deposit, on the other hand, a Devensian age (equivalent to either or both of Oxygen Isotope Stages 4 and 2) must be assumed. With the absence of datable materials, the ages of the head and raised marine sediments between Start Point and Prawle Point remain unknown.

Conclusion

Start Point to Prawle Point GCR site provides some of the finest sections through periglacial slope deposits (head) anywhere in Britain. Although their precise age is unknown, these deposits exhibit all the characteristic features of coastal head deposits, and are believed to have accumulated during a variety of cold stages in the Pleistocene when periglacial conditions prevailed. Alternating facies of head, comprising different clast lithologies, point to deposition by a variety of individual, shallow solifluction flows, with different bedrock layers in the old cliffline successively succumbing to the effects of frost-action. Elsewhere, massively bedded head deposits suggest downslope movement of substantial quantities of probably saturated debris *en masse*. In conservation terms, the sections from Start Point to Prawle Point provide a wide variety of slope deposits and related landforms of textbook quality, against which other, less well-developed examples, can be compared and interpreted.

PENDOWER
S. Campbell

Highlights

This site provides a textbook example of a compound shore platform. Its Pleistocene succession of raised marine deposits and periglacial head and loess is one of the finest in Cornwall.

Introduction

Coastal exposures at Pendower typify the south Cornish Pleistocene succession, and show one of the finest examples of a compound shore platform overlain by raised beach sediments anywhere in south-western Britain. First described by Boase (1832) and later by Whitley (1866) and Reid (1907), the site has also featured in studies by Rogers (1909, 1910), Davison (1930), Robson (1944), Everard *et al.* (1964), Mottershead (1977b), James (1974, 1994) and Stephens and Sims (1980). The most detailed accounts of the site are those of James (1981b, 1994) and Scourse (1985a, 1996c).

Description

Pleistocene sediments are exposed more or less continuously for a distance of 8 km in Gerrans Bay, on the south Cornish coast. Perhaps the best exposures, however, are found between SW 899381 and *c.* 910380, a stretch of coast along Pendower Beach, through Spire Point to beyond Giddleywell in the east. The sequence at the western end of these exposures is shown in Figure 6.4. The following succession can be generalized from field observations and the descriptions given by Stephens and Sims (1980), James (1981b, 1994)

and Scourse (1985a, 1996c) (maximum bed thicknesses in parentheses):

4. Silty sand (loess) (0.5 m)
3. Breccia (head of various facies) (*c.* 12 m)
2. Sand (coastal dune sand) (2.5 m)
1. Pebbly conglomerate (raised beach deposits) (3.0 m)
   Compound shore platform

The shore platforms in Gerrans Bay can be traced almost continuously from St Anthony Head in the south-west to east of Pendower Beach (James, 1974, 1981b). They cut across south-east-dipping slates, shales and sandstones with interbedded quartz veins of the Portscatho and Veryan series. Stephens and Sims (1980) and James (1974, 1981b) divided the platform into a lower surface (coincident with the modern beach) and an elevated surface of limited extent, frequently obscured by Quaternary sediment.

The ubiquitous lower platform is notched into a small cliff below the higher platform surface at a maximum level of 4.4 m OD (James, 1974, 1981b; Mottershead, 1977b), and extends to below high water mark. The upper platform also notches a fossil cliff which, in many places, is obscured by Pleistocene sediments. The lower platform is at times more than 10 m wide (James, 1974, 1981b) and both platform surfaces are remarkably flat. Large boulders are stranded on the platform surfaces (Stephens and Sims, 1980): some may be erratics (Stephens, 1970a), but others consist of local quartzite, chert, conglomerate and spilitic lava.

Deposits immediately overlying the lower raised shore platform consist of a clast-supported conglomerate of pebbles and cobbles (bed 1) with occasional, larger, and more angular blocks of local derivation (Figure 6.5). For the most part, this bed consists of small rounded quartz pebbles with some of slate in a sandy matrix. It is strongly cemented with oxides of iron and manganese. In places, fragments of material from bed 3 (head) are also incorporated into the upper layers of the raised beach deposit. Although considered unfossiliferous (Reid, 1907; James, 1981b), shells from the raised beach deposit were reported by Rogers (1910). Scourse (1985a, 1996c) and James (1994) record that bed 1 is overlain by cross-bedded sands (coastal dunes), although James (1994) notes that this unit is locally obscured by slope-wash materials.

The succeeding breccia (bed 3) is a variable deposit consisting almost entirely of angular clasts of local shales, slates, grits and vein-quartz frag-

Figure 6.4 Coastal exposures at the western end of Pendower Beach, showing a compound shore platform cut across steeply dipping slates, overlain by cemented raised beach deposits and head. (Photo: S. Campbell.)

ments. In places, this bed comprises a breccia of fine shale fragments showing faint stratification: in others, it is a blocky deposit consisting of large, angular cobble- and boulder-sized fragments. Finer lenses of silt and sand also occur within this bed. In most of the sections at Pendower, the head is banked against a steep cliff and, where the deposits have fallen away, the old, apparently water-worn slate cliff (Reid, 1907) is seen. The upper layers of the head (between *c.* 0.5–1.0 m) consist of fine, sharply angular, comminuted shale and slate fragments; cryoturbation features (involutions) and fossil ice-wedge casts are also common in the

upper parts of this bed (Stephens and Sims, 1980; James, 1981b, 1994). Stephens and Sims recorded one locality where the head (bed 3) was divisible into two beds, separated by sand containing only occasional rock fragments. For much of the site, however, they recognized a coarse blocky head overlain directly by a finer head.

Finally, the Pleistocene sequence is capped by a thin (up to 0.5 m) silty sand with occasional pebbles, either a loess (Scourse, 1985a) or a slope-wash deposit (Stephens and Sims, 1980; James, 1981b). Submerged forest beds are also present along this stretch of the coast (Robson, 1944).

Figure 6.5 A striking unconformity between the shore platform and overlying, cemented raised beach deposits at Pendower, viewed by members of the Quaternary Research Association in 1980. (Photo: S. Campbell.)

Interpretation

Boase (1832) referred to the 'head' in Gerrans Bay as 'Diluvium'. Reid (1907) regarded the head (bed 3) as ' ... a mass of angular rubble shattered by frost, swept down the slopes, and deposited on the lowlands when arctic conditions prevailed in Cornwall ... ' (Reid, 1907; p. 58). The climatic conditions under which the raised beach deposits accumulated were not known, the bed being unfossiliferous and there being no clear evidence for penecontemporaneous ice-carried erratics – unlike other examples of raised beach sediments elsewhere in Cornwall (Reid, 1907).

Although referred to briefly by Davison (1930), James (1974) and Mottershead (1977b), the sections at Pendower were not studied again in any detail until the work of Stephens and Sims (1980), James (1981b) and Scourse (1985a). The beds were described in detail by Stephens and Sims (1980) who, apart from interpreting bed 1 as a raised beach deposit formed during temperate, high sea-level conditions in the Pleistocene, and the head (bed 3) as a periglacial solifluction deposit, were noncommittal about the possible chronostratigraphic attribution of the beds.

James (1981b) used the thermoluminescence (TL) evidence of Wintle (1981) to provide a relative dating scheme for the beds at Pendower. Wintle (1981) gave TL determinations which supported a Late Devensian age for loess deposits (stratigraphically equivalent to bed 4 at Pendower) at locations in southern England, including south Cornwall and the Isles of Scilly. James (1981b) argued that the various facies of head (bed 3) underlying the silty sand (bed 4) at Pendower, were thus probably Devensian in age, with the upper layers and cryoturbation structures almost certainly belonging to the coldest part of that stage – the Late Devensian. He argued that the raised beach conglomerate (bed 1) was therefore most probably ascribable to the Ipswichian Stage. Where raised beach (bed 1) and head (bed 3) material was mixed at the site, he argued that marine and terrestrial deposition had taken place in latest Ipswichian or earliest Devensian times. The underlying shore platform was believed by him to have been re-trimmed, if not comprehensively formed, during the Ipswichian Stage.

Scourse (1985a, 1996c) correlates bed 1 at Pendower with the raised beach deposits at Godrevy (Godrevy Formation). He regards bed 2 as an aeolian coastal dune deposit. A TL date of between 165 ka and 252 ka BP (QTL-466; Southgate, 1984; unpublished data) from sand in bed 2 was used as evidence by Scourse to show that the underlying raised beach conglomerate was older than Ipswichian in age (Oxygen Isotope Stage 5). He regarded the head (bed 3) as a periglacial solifluction deposit (Penwith Formation), formed substantially, although not necessarily exclusively, during the Devensian Stage (Scourse, 1985a, 1996c).

In summary, the elevated shore platform is likely to be composite in age (e.g. Mitchell, 1960; Kidson, 1971; Stephens, 1973), although it was probably re-trimmed during the Ipswichian (James, 1981a, 1994). Likewise, the precise age of the raised beach sediments at Pendower is not established. The TL dating from the site, although tentative, might suggest that the raised beach deposit here may be older than many of the other examples in South-West England and Wales which have been ascribed to the Ipswichian Stage (Oxygen Isotope Stage 5e – c. 125 ka BP). It is relevant that preliminary amino-acid data (Bowen *et al.*, 1985) similarly indicate a pre-Ipswichian age for the raised beach deposits at Godrevy: these sites may therefore provide crucial evidence to demonstrate that not all of the raised beach deposits in the region are ascribable to the Ipswichian Stage. Whether they can be ascribed to Oxygen Isotope Stage 7 is unproven at present, although a distinct possibility (see Portland Bill; Hope's Nose and Thatcher Rock). The firm attribution of these raised beach beds to a pre-Ipswichian high sea-level event would be crucially important for interpreting Pleistocene successions throughout the region: some elements of the complex head and solifluction sequences, therefore, may have accumulated in cold conditions prior to the Devensian Stage. A cautionary note, however, must be added. The *Patella vulgata* shell from Godrevy which yielded the Stage 7 age, came from an isolated conglomerate 'bridge' some distance from the main Godrevy raised beach exposure (James, 1997). James regards this 'bridge' as a Stage 7 remanié section, but notes that the rest of the Godrevy raised beach section is likely to date from Stage 5e. Indeed, a recent reassessment of the geochronological evidence (James, 1995) suggests that many of the raised beach deposits in South-West England contain remanié material of earlier interglacial events (e.g. Stage 9 and/or 7), but that most of the locations were last occupied during Stage 5e.

Conclusion

Although the precise ages of the beds at Pendower are far from certain, the site provides one of the most detailed Pleistocene records in Cornwall. It is particularly notable for the fine development of the compound shore platform with its associated marine cliffs and notches. These features, together with the closely associated Pleistocene sediments (large erratics, raised beach, head and possible loess deposits), reveal an unusually complete record of the complex and protracted evolution of the south Cornish landscape.

PORTHLEVEN
S. Campbell

Highlights

One of Cornwall's most controversial geomorphological localities, Porthleven provides possibly the oldest evidence of Pleistocene glacial conditions in the South-West. Its famous 50-ton erratic of gneiss, known as the Giant's Rock, could have arrived here on floating ice.

Introduction

Porthleven has long been famous for the large stranded erratic known as the 'Giant's Rock'. Its exact origin and mode of emplacement have been much debated but never satisfactorily solved: both, however, have significant repercussions for regional Pleistocene conditions. The Giant's Rock was first described in detail by Flett and Hill (1912). Subsequently, it has figured prominently in the scientific literature (e.g. Davison, 1930; Robson, 1944; Flett and Hill, 1946; Mitchell, 1960, 1965; Everard *et al.*, 1964; Stephens, 1966b, 1970a, 1973, 1980; Stephens and Synge, 1966; Kidson, 1971, 1977; Hall, 1974; Scourse, 1985a, 1996c; Holder and Leveridge, 1986; Todd, 1987; Goode and Taylor, 1988; Bowen, 1994b).

Description

The Giant's Rock (SW 623257) lies some 400 m north-west of Porthleven harbour on a wide shore platform, locally named Pargodonnel Rocks (Figure 6.6). It is fully exposed only at low tide and rests

Figure 6.6 The Giant's Rock at Porthleven, the South-West's most famous erratic – author for scale. (Photo: S. Campbell.)

in a large rock pool on the abraded platform surface from which, significantly, it is never shifted even in the heaviest storms (Flett and Hill, 1912; Hall, 1974). The erratic measures 3 m in length and weighs an estimated 50 tons. It is highly polished, brown in colour, and is composed of garnetiferous microcline gneiss, with garnet crystals up to 1 cm across (Hall, 1974; Stephens, 1980). On the remainder of the platform there are other numerous smaller boulders including those of slate, granite, gabbro and vein quartz. Towards the base of the steps, which lead down into the small cove from the cliff top, are patches of iron-stained and cemented sand and gravel (the latter quite well rounded in places) which adhere to the slate bedrock. These are probably the remains of a raised beach deposit from which other small boulders, elsewhere on the platform, may have been washed. Stephens (1973) noted that the platform extended to substantial notches (which contain the cemented raised beach gravels) in the base of a rock cliff now being re-exposed and re-trimmed by wave action.

Interpretation

Flett and Hill (1912) provided one of the most significant accounts of the erratic, establishing that the rock-type could not in fact be matched with any other British example. They later argued that it had been stranded ' … by the ice floes of the Glacial Period.' (Flett and Hill, 1946; p. 168).

Modern interpretations have centred on whether the Giant's Rock was deposited directly by glacier ice or was floated into position on pack ice or on a massive iceberg. Mitchell (1960, 1965) argued that, in view of the widely held relationship between Pleistocene glacial stages and low eustatic sea levels, the erratic could only have been borne to its present position directly by an ice sheet; he used the Porthleven erratic on the south Cornish coast and the beds at St Erth (part of which he then regarded as till) to define his southernmost limits for an ice sheet of Lowestoft (Anglian) age. The St Erth Beds, however, have since been securely re-established as marine in origin (Mitchell *et al.*, 1973a; Jenkins *et al.*, 1986), and Mitchell's proposed Anglian ice limit in this region would appear to have no foundation.

Kidson (1971, 1977) propounded that large erratics found on shore platforms around the Croyde–Saunton Coast in north Devon, and indeed those along the south coast as far east as Prawle Point (including, by implication, the Giant's Rock?), were also emplaced directly by an ice sheet. He argued that the erratics at Croyde and Saunton had been derived from Wolstonian (Saalian) Irish Sea glacial sediments (which include the Fremington Clay; see Chapter 7) and then incorporated into raised beach deposits during the Ipswichian Stage. Although this Wolstonian ice sheet may also have impinged on the Cornish coast at Trebetherick Point and in the northern Isles of Scilly (Kidson, 1977), there are no coherent glacial sediments anywhere else along the south Cornish and Devon coasts which suggest a more extensive inundation by ice at this time (Stephens, 1966b; Kidson and Bowen, 1976). Recent amino-acid dating studies of raised beach deposits in the region (e.g. Andrews *et al.*, 1979; Davies, 1983; Bowen *et al.*, 1985; Davies and Keen, 1985) have shown that raised beach sediments from at least two separate interglacial phases of the Pleistocene are present. In many cases, however, the raised beach deposits are unfossiliferous and cannot be dated; the true age relationship of the erratics to the local raised beach deposits therefore remains uncertain in the vast majority of cases.

In contrast, many authors have favoured ice-rafting as the most likely mechanism for emplacement of the Giant's Rock and similar large erratics in the South-West. Indeed, Tricart (1956) also favoured this mechanism to explain the presence of large erratics on the French Channel coast. However, if floating ice carried the erratics to the south coast, then problems arise regarding contemporary Pleistocene sea levels. Mitchell (1972) sidestepped the problem of low sea levels during glacial stages by arguing that these erratics were rafted into position at the beginning of the Saalian Stage when the level of the sea might still have been relatively high after the preceding, warm, Hoxnian Stage. Similarly, Stephens (1966b) argued that the large erratics could have been emplaced by pack-ice and icebergs during the waning of an early pre-Saalian (Anglian?) glacial period when world sea level would have been high enough to allow the erratics to be 'floated' into position (Fairbridge, 1961). As an alternative hypothesis, Stephens suggested that towards the end of the Saalian ice-sheet glaciation, isostatic depression of the land had allowed the sea

to move icebergs against these coasts despite a generally low eustatic sea level. Such a mechanism is similar to recently proposed models of Late Devensian glaciomarine sedimentation in the Irish Sea Basin (e.g. Eyles and McCabe, 1989, 1991). Bowen (1994b) recently suggested that the Giant's Rock could have originated in Greenland and then been transported to the South-West on ice-floes from a disintegrating, Early Pleistocene, Laurentide ice sheet.

In support of the ice-rafting hypothesis, the most convincing evidence is that the erratics are very largely confined to a narrow coastal zone, invariably below 9 m OD, and within the reach of storm waves today (Stephens, 1966b). Also, comparable very large erratics are not known from inland locations in the region, with the exception of those recorded from the Fremington Clay in the low-lying Barnstaple Bay area (Stephens, 1966b). Smaller examples, however, are found at various levels around the south coast, and some may have sources in Brittany, the Channel Isles and Côtentin (Kellaway *et al.*, 1975; Mottershead, 1977b); they may have been worked from sea-bed deposits, carried up and finally emplaced by successive transgressive Pleistocene high sea levels. Such a mechanism is not plausible for emplacement of the Giant's Rock which does not shift at all even in the heaviest of storms and wave regimes of the present day. This has led some authors to consider the possibility that the Giant's Rock is not a glacial erratic: it may have been derived from the Normannian High by means of slumping during the deposition of the Mylor Slates in Devonian times (Holder and Leveridge, 1986; Goode and Taylor, 1988). Such a view is not, however, widely held.

Conclusion

The Giant's Rock is the most impressive and intriguing of the large erratics found around the south and west coasts of Britain. Despite having attracted scientific interest for nearly a century, its exact origin and mode of emplacement are still unknown and it remains the subject of much controversy. Although some workers have maintained that the 50-ton erratic was emplaced directly by glacier ice, most believe that it was delivered to its present location on floating ice. One recent interpretation invokes Greenland as a possible source and a disintegrating Laurentide ice sheet as a transport mechanism. The Porthleven erratic and the classic examples of the Croyde–Saunton Coast (some of which are found in a stratigraphic context) are central to the reconstruction of earlier Pleistocene events in the region: they have major implications for the extent of pre-Devensian ice sheets, for Pleistocene sea levels and, more controversially, for a possible mechanism involving isostatic depressions of the land's crust proximal to ice-sheet margins.

BOSCAWEN
S. Campbell

Highlights

Organic sediments found in the head sequence here provide unique evidence for environmental conditions in Cornwall during the Devensian Stage. Pollen and radiocarbon-dated evidence shows that the bulk of head deposits at Boscawen accumulated after *c.* 29 ka BP, during the Late Devensian.

Introduction

Boscawen GCR site comprises St Loy's and Paynter's coves, and is an important Pleistocene stratigraphic locality. Organic beds within the head sequence have yielded pollen and have also been radiocarbon dated. They provide a rare opportunity to subdivide and interpret head sequences in the region. The sections were first described by De la Beche (1839), and subsequently by Reid and Flett (1907), Rogers (1910) and Davison (1930). A contemporary assessment of the deposits is provided by Scourse (1985a, 1987, 1996c) and the site is also mentioned in the BGS memoir (Goode and Taylor, 1988).

Description

Good sections in Pleistocene sediments occur between Merthen Point in the west (SW 418227), through St Loy's and Paynter's coves, almost to Boscawen Point in the east (SW 430230). The Pleistocene sequence is relatively straightforward with a rocky shore platform, cut entirely in granite, overlain by a sequence of raised beach and then solifluction deposits. The raised beach deposit overlies the shore platform at about 3.5 m OD, and reaches a maximum thickness of 1.8 m (Scourse, 1985a). For the most part, it consists of a matrix-supported deposit of granite pebbles and cobbles,

although erratics are also common (including greenstone, hornfels, chalk flints, Greensand chert and purple sandstone): in places, more than half the pebbles are non-granitic (Reid and Flett, 1907).

The overlying beds (up to 12 m in thickness) consist mainly of a breccia made up of angular and subangular granite clasts (Figure 6.7). Considerable facies variations, however, occur both vertically and laterally, and sandy silt lenses can be seen particularly towards the top of the sections (Scourse, 1985a). In places, there are stratified sandy and gravelly bands, and throughout the sections slump and flow structures occur. Near the top of the sections, there is a coarse deposit of granite boulders. Towards Boscawen Point, the raised beach is underlain by a thin breccia made up of granite clasts.

At *c.* SW 423230, Scourse (1985a, 1996c) records the following sequence (maximum bed thicknesses in parentheses):

5. Breccia of granite fragments (0.5 m)
4. Sandy silt with scattered granite clasts (0.75 m)
3. Breccia of granite fragments (2.0 m)
2. Granular humic silt (0.5 m)
1. Breccia of granite fragments (0.5 m)

The silty humic sediments range from black through chocolate-brown to ochre in colour. They have yielded pollen, and have been radiocarbon dated to 29 120 + 1690/− 1400 BP (Q-2414) (Scourse, 1985a, 1996c).

Interpretation

The raised beach deposits at Boscawen were noted by De la Beche (1839) but were first described in detail by Reid and Flett (1907). They recorded a Pleistocene marine-cut platform overlain by raised beach and head deposits. The erratic content of the raised beach deposit was used as evidence that it (like the overlying head) had accumulated under cold, glacial, conditions with the erratics having been derived from grounded ice-floes (Reid and Flett, 1907). This interpretation was also followed by Davison (1930).

Scourse (1985a, 1996c) recognized the same basic stratigraphy but interpreted the sequence in far greater detail. He correlated the raised beach deposit with that at Godrevy, classifying it lithostratigraphically as the Godrevy Formation. Despite this correlation, the age of the raised beach deposit

Figure 6.7 Boscawen GCR site (St Loy's Cove): solifluction deposits interbedded with organic sediments towards the base of the section. (Photo: J.D. Scourse.)

is far from certain; Scourse (1985a, 1996c) has tentatively ascribed it to a pre-Oxygen Isotope Stage 5 event (see Godrevy and Porth Nanven). The overlying beds were interpreted as consisting mainly of solifluction deposits (Penwith Formation) (Scourse, 1985a, 1996c).

The organic sediments within the head sequence at Boscawen are fundamental to Scourse's (1985a, 1996c) interpretation of the beds and the cornerstone of his scheme for Pleistocene subdivision in Cornwall. The organic sediments are arranged in a number of beds, the upper ones apparently having been soliflucted. The pollen contained in the lower, *in situ*, beds, although not zoned in the conventional manner, does however indicate a tundra vegetation and an Arctic climate: the deposits appear to have accumulated in a small pool on the surface of a solifluction flow (Scourse, 1985a, 1996c). The fossil flora is dominated by grasses. The spectrum also indicates that sedges were present in wetter areas, and the herb pollen are those from species which would have colonized the

poor, disturbed minerogenic soils of a periglacial landscape (Scourse, 1985a, 1996c).

A radiocarbon age determination of 29 120 + 1690/− 1400 BP (Q-2414) from these humic sediments, if reliable, provides a maximum, Middle to Late Devensian, age for the bulk of the Penwith Formation at the site. The *in situ* organic bed, regardless of its age, also shows that the breccia accumulated on at least two separate occasions, and although a Late Devensian (Oxygen Isotope Stage 2) age for the upper part of the sequence seems likely, there is no way at present of knowing the age of the underlying head, particularly in view of the lack of a reliable age estimate for the raised beach deposits beneath. Similarly, the duration of the climatically slightly less severe interlude in which the organic sediments accumulated is not known.

Discontinuous sand lenses in the upper part of the head sequence are interpreted by Scourse as periglacial wind-blown sand ('sandloess') and correlated with the Lizard Loess (Roberts, 1985). The latter has been dated on the Lizard Peninsula by the TL method to 15.9 ± 3.2 ka BP (QTL 1e) (Wintle, 1981).

Boscawen is primarily a reference site for the interpretation of solifluction deposits in South-West England. It also provides a valuable record, in the form of raised beach deposits, for fluctuations in relative sea level during the Pleistocene. At present, the age of the raised beach deposit here is unknown. The dated organic deposits from within the solifluction beds show clearly that most of the breccia, including the lenses of wind-blown material, is probably Late Devensian in age. They also show that head deposits, indistinguishable from those above them and classified, lithostratigraphically, as part of the Penwith Formation, formed at some stage prior to *c.* 29 ka BP (Scourse, 1985a, 1996c). An Ipswichian age for the raised beach deposit (equivalent to Oxygen Isotope Stage 5e) would confine this early period of head formation to Early or Middle Devensian times. Such an age, however, is far from certain, and if the raised beach deposit proves to be older (e.g. Stage 7), then this head could equally well date from a pre-Devensian cold stage (e.g. Stage 6). The precise character and duration of the period in which the organic sediments at Boscawen accumulated has yet to be determined. It is not clear, for example, if the surviving organic beds represent the early part of an interstadial phase of the Devensian, or simply a brief, even seasonal, respite from active solifluction. Boscawen is therefore important for establishing that the head and solifluction deposits fringing the Cornish coast were not necessarily deposited in a single, synchronous, Late Devensian event or cycle.

Conclusion

Boscawen demonstrates important evidence for Pleistocene climatic and environmental conditions. Its raised beach and head deposits are in many ways typical of those along the south Cornish coast. However, organic sediments found within the head here have allowed a more detailed reconstruction of events than is usually possible. Solifluction appears to have been interrupted, c. 29 000 years ago, by a period of less harsh conditions when the land surface was colonized by tundra vegetation. A return to more extreme periglacial conditions and renewed solifluction is shown by thick head deposits which overlie the organic beds. Whereas the age of head deposits in most coastal sections can be determined only very broadly, the bulk of those at Boscawen can be shown to have accumulated after c. 29 ka BP, probably during the coldest part of the Devensian Stage (Late Devensian/Oxygen Isotope Stage 2). Head deposits below the organic sediments here must therefore be older, although like the raised beach deposits beneath them, their age is unknown.

PORTH NANVEN
S. Campbell

Highlights

Possibly the most spectacular section through raised beach deposits in Britain, the raised 'boulder' bed at Porth Nanven comprises a thick development of granite clasts, some in excess of 0.6 m diameter.

Introduction

Porth Nanven provides one of the finest examples of a raised 'boulder' beach in Britain. This classic raised beach deposit was first noted by Borlase in 1758, and was subsequently included in De la Beche's (1839) treatise on the geology of Cornwall. The site has also featured in studies by Austen (1851), Whitley (1866), Ussher (1879a), Reid and Flett (1907), Robson (1944), Guilcher (1949), Savigear (1960), Everard *et*

al. (1964), Stephens (1973), Todd (1987), James (1994) and Scourse (1996c). The most detailed descriptions are provided by Scourse (1985a) and Goode and Taylor (1988).

Description

Particularly fine exposures of raised beach and head deposits occur between SW 355310 and SW 356308 in the small bay of Porth Nanven, 0.75 km south-east of Cape Cornwall (Figure 6.8). Here, a straightforward sequence occurs on top of a shore platform (Reid and Flett, 1907; Scourse, 1985a):

2. Coarse breccia of angular granite clasts (up to 4 m)
1. Large rounded cobbles and boulders, principally of granite (up to 8 m)

The shore platform, lying at c. 8.5 m and developed across the Land's End Granite, terminates inland at a nearby vertical granite cliff. The latter, according to Reid and Flett (1907), shows signs of being 'water-worn' almost up to 12 m above OD. Overlying the platform, and banked up against the fossil cliff, is a deposit comprising mainly large granite cobbles and boulders. This bed (bed 1 – raised beach deposit) is clast-supported, and discernibly coarser in the upper 2 m of the bed where boulders in excess of 0.6 m in diameter occur (Scourse, 1985a). The basal 4–6 m of the bed consist of cobble- and gravel-sized clasts. The bed contains much Killas and greenstone gravel near its base (Reid and Flett, 1907).

The overlying breccia or head (bed 2) consists of angular granite fragments which fine-upwards within the bed. The base of the bed consists of coarse angular boulders of granite. The sequence is capped by a thin soil developed directly on the head. The latter can be traced inland via stream sections which have been comprehensively worked for 'stream tin' (Reid and Flett, 1907; Scourse, 1985a).

Interpretation

The remarkable development of the raised beach deposits at Porth Nanven was first noted in Borlase's treatise on the natural history of Cornwall in 1758; he also referred to the overlying head as ' … a load of rubbish … ' (Borlase, 1758; p. 76). The spectacular development of the bed, and its unusually large cobbles also ensured that it was

Figure 6.8 Porth Nanven, west Cornwall: the spectacular 'raised boulder beach' overlain by solifluction deposits. (Photo: D.H. Keen.)

referred to in De la Beche's (1839) classic work on Cornish geology. The best early description of the Porth Nanven sequence was, however, given by Reid and Flett (1907). They regarded the raised platform as a marine surface of Pleistocene age cut into an older, Pliocene, surface. The water-worn granite face on the landward side of the platform was regarded as an ancient Pleistocene cliffline (Reid and Flett, 1907).

Although regarded as the oldest Pleistocene deposit of the region, the precise age of the raised beach deposit was not discussed. However, on the basis that similar beds elsewhere contained erratics (e.g. a pebble of 'biotite-trachyte' from nearby Priest's Cove), Reid and Flett argued that the raised beach had formed in one of the 'Glacial Stages' of the Pleistocene. Such contemporary thinking was also mirrored in other parts of Cornwall (e.g. Hill and MacAlister, 1906; Reid and Scrivenor, 1906; Reid *et al.*, 1910; Flett and Hill, 1912). The overly-

ing head (bed 2) was also attributed to cold conditions in the Pleistocene (Reid and Flett, 1907), frost-shattering of the local bedrock having been followed by the redistribution of material along the coastal slopes and local valleys by meltwaters during the 'Glacial epoch'. It was during this latter phase that stream tin was washed from the shattered debris and deposited, selectively, within the gravel and head sequence by virtue of its high specific gravity (Reid and Flett, 1907).

Savigear (1960) briefly referred to Porth Nanven in a study of the morphology of the local coastal slopes and cliffs. He identified four narrow and fragmentary benches below the main granite cliffs at Porth Nanven, the uppermost of which was overlain by the raised storm beach and head sequence. He concluded that the main shore platform, the beach deposits, cliffs, spurs and bevels of the coastal zone all, therefore, antedated the whole, or part of the last glaciation, since they were blanketed by head up to a height of some 30 m OD.

The most recent interpretation of the Pleistocene sequence at Porth Nanven was given by Scourse (1985a, 1996c) and James (1994). Scourse also interpreted the beds as a raised beach deposit (bed 1; Godrevy Formation) overlain by solifluction deposits (bed 2; Penwith Formation). He commented that the raised beach deposit was the thickest and coarsest in Cornwall, its pronounced upward coarsening perhaps showing a regression of the high Pleistocene sea level.

Despite these preliminary lithostratigraphic correlations, the age of the Porth Nanven raised beach deposit is far from certain. Preliminary amino-acid data from fossil shell material in the Godrevy raised beach deposit indicate that it was perhaps not formed in the Ipswichian Stage (Oxygen Isotope Stage 5e) (Bowen *et al.*, 1985). If the present correlation of the Porth Nanven and Godrevy raised beach deposits (Scourse, 1985a, 1996c), based entirely on lithostratigraphy, is correct, then the Porth Nanven example too may date from a pre-Ipswichian (pre-Oxygen Isotope Stage 5e) high sea-level event. Such an ascription, however, is unproven (Pendower; this chapter).

Complementing the stratigraphic importance of the site, the nature of the raised beach deposit itself is of great interest. Its perceived coarsening-upward sequence and upper, substantial, boulder layer is unusual, perhaps reflecting a progressive regression in relative sea level. The coarseness of the bed may reflect both the nature of the sediment source (coarse, widely jointed granite) and the prevailing wave energy; the site faces due west and lies only a few kilometres from Land's End, one of the most exposed coastal locations in southern Britain today. As such, the site provides a fine example of a raised 'boulder' beach, a facies variation rarely recorded elsewhere at other British Pleistocene localities.

On the basis of a radiocarbon date (*c.* 29 ka BP) from organic sediments interbedded in the head sequence at nearby Boscawen, Scourse (1985a) has attributed the bulk of head deposits there, and other examples of the Penwith Formation, to the Late Devensian (see Boscawen). Problems in estimating the age of the underlying raised beach deposit (bed 1), however, make estimates of the age of the overlying head at Porth Nanven tenuous. The possibility clearly remains that if the raised beach deposits accumulated during a pre-Ipswichian (Oxygen Isotope Stage 5e) high sea-level event, then solifluction deposits from several subsequent cold phases of the Pleistocene may be present. It seems likely, however, that a substantial part of the head sequence accumulated in the Late Devensian (Scourse, 1985a). To some extent, the nature of the head deposits at Porth Nanven may throw light on the relative dating of the beds. The well-developed and clear fining-upward head sequence may show that in fact there was only one major phase of periglacial activity, and a single cycle of head formation at the site, the coarsest material representing the products of a previous weathering cycle (interglacial or interstadial?), and the progressive fining of the beds reflecting the diminution in availability of weathered products for solifluction.

Conclusion

Porth Nanven is an important Pleistocene stratigraphic locality providing fine examples of a shore platform, raised beach and head deposits. The raised 'boulder' beach is particularly unusual and was probably formed as a storm beach during a fall in relative sea level. Its coarseness reflects a high energy wave regime promoted by the particularly exposed westerly location of this part of the Cornish coast. Porth Nanven is also notable as one of the first raised beach localities to have been described in Britain.

GODREVY
S. Campbell

Highlights

Excellent sections through interglacial marine deposits, blown sand and periglacial head, make Godrevy one of the South-West's most important Pleistocene sites. Part of its raised beach deposit is tentatively ascribed to Oxygen Isotope Stage 7, leaving the possibility that some overlying head deposits and associated periglacial structures were formed during several Pleistocene cold stages.

Introduction

Godrevy has long been regarded as a reference site for raised beach, blown sand and head deposits. The site was mentioned in early studies by De la Beche (1839), Whitley (1866, 1882), Ussher (1879a), Reid and Flett (1907), Rogers (1910) and

Davison (1930). It has featured in more recent studies of Pleistocene chronology by Robson (1944), Stephens (1961a, 1966a, 1970a), Everard *et al.* (1964) and Mitchell (1972). The sediments and stratigraphy have also been described by Hosking and Pisarski (1964), James (1975b, 1994, 1995), Hosking and Camm (1980), Sims (1980), Scourse (1985a, 1996a, 1996c), Goode and Taylor (1988) and Morawiecka (1993, 1994). Bowen *et al.* (1985) provided amino-acid ratios from shell in the raised beach deposits.

Description

Continuous sections through Quaternary sediments extend from Magow Rocks (SW 582423) almost to Godrevy Point (SW 581430). The Quaternary sequence rests on a well-developed shore platform cut across Devonian (Mylor) slates (James, 1975b), mainly highly contorted blue (hard) and yellow (soft) slates intersected by quartz veins (Figure 6.9). The platform lies between 4 and 10 m OD (Scourse, 1996c) and in places is as much as 100 m wide. It is notched at a level of *c.* 8 m OD (James, 1975b) and, locally, its surface is fragmented (Stephens, 1966a). The following succession can be generalized from the descriptions given by Stephens (1966a), James (1975b) and Scourse (1985a, 1996c):

5. Soil
4. Silty sand (up to 1.0 m) (Holocene)
3. Head of various facies (up to 5.0 m)
2. Sand, cemented in places, with clay bands – 'sandrock' (up to 3.5 m)
1. Basal pebbles consisting mainly of local rocks, but with greenstone and chalk flints and mixed with sand in places. Occasionally cemented by iron and manganese oxides (up to 2.0 m)
   Shore platform

Clay bands are common in the lower part of bed 2, especially at its junction with bed 1. They vary in thickness from 2–5 cm, form an undulating pattern and fill troughs between sand ripples (James, 1975b). In places, thicker and more numerous clay bands occur in the sand, especially below the junction with the overlying head (bed 3), which interdigitates with the non-indurated sand of bed 2 (James, 1975b, 1994). This junction is sometimes marked by a convoluted band of manganese-cemented sand. Where the sand is cemented, as at

Godrevy Point, it is usually made up of pale yellow, shelly, medium to coarse sand, more than 60% of which consists of shell. Some of these sediments are characterized by large-scale, planar cross-beds (James, 1975b; Scourse, 1985a, 1996c).

In places, the sand (bed 2) is penetrated by vertical pipes (cf. St Agnes Beacon and Trebetherick Point) with an average diameter of 12.5 cm and a depth of over 1 m. Many of these pipes are now empty but others, filled with brown uncemented sand, have been observed (James, 1975b). These dark sands also form a thin layer on top of the cemented sands and beneath the overlying head. Texturally and petrologically, this sand is identical to that elsewhere in bed 2; it differs only in being decalcified (James, 1975b).

The head (bed 3) consists of a highly variable breccia of slate and quartz fragments and can be subdivided, in places, into a lower festooned bed and an upper less disturbed bed (Stephens, 1966a, 1970a). James (1994), however, notes that nearly all of the clearly observed quartz clasts are vertically aligned, indicating significant post-depositional disturbance to the upper layer. Occasionally, some stratification is present within the head (Scourse, 1985a, 1996c). Stephens (1966a) noted that the raised beach deposits, at one locality, were also underlain by a thin development of head.

Interpretation

Whitley (1882) was one of the first to describe the sections at Godrevy in any detail. He regarded bed 1 as an alluvial deposit, since it was thickest where exposed in the central part of the valley mouth, and claimed that ' … it becomes obvious that the gravel bed is not a sea beach, but a valley deposit, cut back and exposed by the action of the sea.' (Whitley, 1882; p. 134). Reid and Flett (1907), however, regarded the Godrevy sections as showing raised beach deposits overlain by blown sand and head. Although most subsequent authors (e.g. Robson, 1944; Everard *et al.*, 1964; Stephens, 1966a) have accepted this broad classification, significant differences of opinion have arisen regarding the age of the beds.

Stephens (1961a, 1966a, 1970a) used Mitchell's (1960) framework of Pleistocene events in southwest Britain to interpret the sediment succession at Godrevy. He argued that the lowest head deposit (recorded at one locality beneath raised beach deposits) and the frost-shattered surface of the

Figure 6.9 Coastal exposures near Godrevy Cove, showing shore platform overlain by raised beach cobbles, sand and various head facies. (Photo: S. Campbell.)

silty sand of bed 4 (a mixture of blown sand and slope-wash deposits) was thought to have accumulated in Holocene times. This chronology was also adopted by Sims (1980).

James (1975b) provided new evidence for the depositional environment of the sand (bed 2). In places, the sand differs in grain size through the bed, reflecting that the basal layers may be marine, the upper aeolian (James, 1975b): occasional well-rounded pebbles are found at least 1 m above the base of the sand bed. James suggested that the clay bands found towards the base of the bed had been deposited by slow-moving water, the clay having infilled ripples on the surface of the underlying marine sand. This material was believed to have been washed from surrounding slopes prior to the deposition of wind-blown sand. James (1975b) assigned bed 4 to the Holocene but made no age ascriptions for the other beds.

Hosking and Camm (1980) suggested that the iron oxide cementing agent of the Godrevy raised beach conglomerate showed that swampy conditions had prevailed in the adjoining valley prior to the accumulation of solifluction deposits. It is likely that the iron and manganese oxides were leached from the valley bedrock and transported to the sites of oxide deposition, in the raised beach sediments, as organic complexes (Hosking and Camm, 1980).

Bowen *et al.* (1985) presented three amino-acid ratios derived from a single specimen of *Patella vulgata* Linné collected from the raised beach deposits (bed 1) by H.C.L. James. This shell yielded an average ratio of 0.175 ± 0.021, a value similar to those derived from shells in the Inner Beach at Minchin Hole Cave, Gower (Bowen *et al.*, 1985; Bowen *et al.*, 1989; Campbell and Bowen, 1989). Similar results were provided for the Godrevy raised beach deposits by Davies (1985). Such ratios, and others from raised beach deposits elsewhere in southern Britain (e.g. Butterslade and Horton Upper Beach, Gower; Saunton, north Devon; and Fistral Beach, north Cornwall), have been correlated with high relative sea levels in Oxygen Isotope Stage 7 (Bowen *et al.*, 1985) and calibrated by Uranium-series techniques (e.g. Davies, 1983) to around 210 ka BP. These shells are therefore believed to be significantly older than those characteristic of raised beaches ascribed to the Pennard D/L Stage (Ipswichian Stage/Oxygen Isotope Stage 5e) (Bowen *et al.*, 1985; Bowen and Sykes, 1988) (see Portland Bill). James (1995), however, notes that the preparation methods used by Davies (1983, 1985) and earlier

shore platform recorded a periglacial event of unspecific, but broadly 'Early Pleistocene' age. The raised beach deposits (bed 1) were believed to have accumulated on the shore platform (and locally above head deposits) at a time of high relative sea level and during temperate conditions in the Hoxnian Stage. The overlying dune sand (bed 2) was believed to indicate falling sea level and the onset of cold conditions at the beginning of the Wolstonian (Saalian Stage). Stephens (1966a) divided the overlying head (bed 3) into two distinct facies: an upper, relatively undisturbed head; and a lower, thicker, much weathered and cryoturbated head. The lower unit was interpreted as a typical periglacial solifluction deposit which was both deposited and cryoturbated during the Saalian. Temperate conditions in the ensuing Ipswichian Stage were invoked to account for weathering of the bed. The upper part of bed 3 (Stephens' upper head) was assigned to a separate phase of periglacial conditions in the Devensian, while the workers, produced amino-acid ratios consistently higher (and therefore 'older') than the methods currently in use. He raises the possibility, therefore, that Davies' ratios from Godrevy could be taken to indicate a Stage 5e age for the raised beach deposits, and cites preliminary Infra-red Stimulated Luminescence (IRSL) dates (Richardson, 1994) in support of this Stage 5e ascription (see also Pendower; this chapter).

Scourse (1985a, 1996c) re-examined the Godrevy sections in detail. The excellent development of raised beach deposits and associated dune sands led him to use Godrevy as a regional type-site for such sediments and the deposits were classified as the Godrevy Formation. Scourse used palaeocurrent vectors derived from foreset laminae to show that the sands (bed 2) had been deposited by north-westerly winds. The overlying head (bed 3) was confirmed as a typical periglacial solifluction deposit (assigned to the Penwith Formation; Scourse, 1985a, 1996c).

Morawiecka (1993, 1994) presented details of 'pipe' phenomena developed in the calcareous sandstone (bed 2) at Godrevy. Most of the pipes occur in the upper and middle parts of the 'sandrock' profile, but none extends down to the level of the rock platform. Morawiecka concludes that the pipes developed after the overlying head (bed 3) had been deposited (cf. West, 1973). She raises the possibility that solution of the 'sandrock' occurred, perhaps quasi-catastrophically, under cold conditions close to the Devensian/Holocene transition, thus making this the youngest palaeokarst demonstrated in the UK.

Although the models of Pleistocene chronology applied to sequences in the South-West by Stephens (1966a, 1970a, 1973) and Mitchell (1960, 1972) have been criticized as unnecessarily complex (e.g. Kidson, 1971, 1977; Bowen, 1973b), recent amino-acid measurements show clearly that raised beach deposits of significantly different ages are present around the coast of south-west Britain. As far as the limited amino-acid data from Godrevy can show, the site provides a fine example of a raised beach deposit that has been ascribed to Oxygen Isotope Stage 7 (but see James, 1995). If this dating is confirmed, then the overlying head could have accumulated during several different Pleistocene cold stages, perhaps equivalent to Stages 6, 4 and 2 of the deep-sea record. James (1994, 1995), however, believes that the bulk of raised beach deposits at Godrevy date from Stage 5e, implying that the overlying head is entirely Devensian in age.

The sequence also provides a comprehensive record of Quaternary environments. Whereas the raised beach conglomerate and the head facies show extremes of climatic change in the region (from temperate high sea-level conditions, to cold periglacial environments), the intervening sands probably reflect conditions towards the end of an interglacial and before the establishment of a full periglacial regime. These sediments clearly record a fall in relative sea level and deflation of sand, probably from an exposed sea floor. The clay bands found in the sand bed may show that sheet-washing of sparsely vegetated hillslopes occurred before the onset of more severe (periglacial) conditions when the overlying head(s) accumulated. (Similar sands at Belcroute and Portelet, Jersey, contain palaeosols indicative of temperate Stage 5c- and 5a-type conditions (Keen *et al.*, 1996).)

Importantly, the Pleistocene sediments exposed at Godrevy show no direct evidence for glacial activity in this area. Some reworked erratics found in the raised beach deposit may attest to an earlier glacial event of considerable age, but otherwise the sections demonstrate that the site, and probably its immediate environs, lay in the periglacial zone during a variety of Pleistocene cold stages. The site therefore shows no tangible evidence to corroborate the reconstructed ice-sheet limits proposed by a variety of workers (e.g. Mitchell, 1960, 1972; West, 1968, 1977a; Jones and Keen, 1993).

Conclusion

Godrevy provides one of the most informative Pleistocene sequences in Cornwall. Its raised beach sediments, blown sand and solifluction deposits (head) are referred to in many important texts on the Pleistocene. Fossil shell material in the raised beach deposits here has been dated by amino-acid geochronology; a tentative correlation with Oxygen Isotope Stage 7 (warm) of the oceanic record is suggested. A fall in relative sea level and cooler climatic conditions are recorded by the overlying sand, some of which was blown inland from an exposed sea floor by north-westerly winds. The head deposits at Godrevy record a further deterioration of climate. These deposits are arranged in distinct layers which may have accumulated during different cold, periglacial phases of the Pleistocene – possibly equivalent to one or more of Stages 6, 4 or 2 of the deep-sea record. The site shows that this part of Cornwall was probably not overrun by glaciers.

TREBETHERICK POINT
S. Campbell

Highlights

This classic locality has featured in numerous reconstructions of the Pleistocene history of South-West England. Its highly controversial 'boulder gravel' has been used as evidence by some for glaciation of the north Cornish coast.

Introduction

Sections at this site reveal a complex sequence of raised beach, blown sand, 'boulder gravel' and head deposits. The controversial 'boulder gravel', in particular, has attracted much interest and debate, and has been interpreted variously as a beach deposit reworked from glacial sediment, as a head or solifluction deposit, as river or outwash gravels and as till *in situ*. The precise origin of this bed and its stratigraphic relationship to sediments exposed elsewhere are of prime importance in establishing the sequence and nature of Pleistocene events in the region. The site was described in early studies by Ussher (1879a), Rogers (1910) and Dewey (1913, 1935). It also featured in more detailed studies by Reid *et al.* (1910), Clarke (1965a, 1965b, 1969, 1973), Stephens (1966a) and West (1973): the classic account is that of Arkell (1943). It has also been referred to in regional evaluations of Pleistocene history by Robson (1944), Arkell (1945), Balchin (1946), Clarke (1962, 1968), Mitchell and Orme (1967), Macfadyen (1970) and Edmonds *et al.* (1975), and in more extensive correlations (Green, 1943; Mitchell, 1960; Bowen, 1969; Stephens, 1970a, 1973; Kidson, 1971, 1977). More recently, the site has featured in studies by Clarke (1980), Sims (1980), Campbell (1984), Bowen *et al.* (1985) and Scourse (1985a, 1985b, 1987, 1996a, 1996c). A detailed description of the site is also provided by James (1994).

Description

Trebetherick Point (SW 926779) is a promontory of Devonian rocks on the east side of the Camel Estuary. Pleistocene sediments are well exposed for *c.* 0.75 km around the Point between the southern end of Greenaway Beach (SW 928783) and the dunes at the north end of Daymer Bay (SW 928777). They fringe an ancient rock cliff and rest on a conspicuous shore platform of Upper Devonian purple and green slates which varies in height between *c.* 1–7 m OD. Stephens (1966a) divided the platform into upper and lower elements: the Pleistocene deposits rest on the higher of these.

The Pleistocene sequence is both laterally and vertically extremely variable. Arkell (1943) figures three sections and Stephens (1966a) provides eight separate stratigraphic sections to illustrate this variability. Figure 6.10 shows Arkell's 'Section A' and a revised stratigraphy for the principal exposure of the boulder gravel at *c.* SW 927781. The following succession can be generalized from the descriptions given by Arkell (1943), Stephens (1966a) and Scourse (1985a, 1996c) and from field observations (maximum bed thicknesses in parentheses):

9. Soil
8. Sand (dune sand) (2.0 m)
7. Stony loam/pebbly clay (colluvium) (0.5 m)
6. Breccia (upper head) (1.0 m)
5. Diamicton ('boulder gravel') (2.0 m)
4. Breccia (lower or main head) (3.0 m)
3. Sand (dune sand) (6.0 m)
2. Boulders, gravel and sand (raised beach deposits) (2.0 m)
1. Breccia (lowest head) (0.3 m)
   Shore platform

The lowest breccia (bed 1) is not widely developed at the site and consists predominantly of fine slate fragments. Stephens (1966a) states that it rests directly on the shore platform and beneath raised beach deposits (bed 2), but Arkell (1943) shows the material to interdigitate with raised beach sediments towards the fossil cliff (Figure 6.10).

The overlying bed comprises a highly variable deposit of boulders, gravel and sand. Some of the sand is bedded and contains stringers of shingle, fine slate and occasional pebbles (Arkell, 1943). In other places, the bed consists of a poorly cemented, clast-supported deposit of pebbles and cobbles (Scourse, 1985a, 1996c). Clasts are subangular to rounded and of both local and non-local origin. Arkell (1943) recorded quartz, grit, slate, elvan (dyke rock of granitic composition), dolerite and flint from the bed as well as some very large boulders of 'greenstone' (sill material).

Bed 3 consists overwhelmingly of sand which exhibits large-scale cross-beds and is unevenly cemented. The latter gives rise to alternating hard and soft layers which make bedding conspicuous. Stephens (1966a) and subsequent authors have

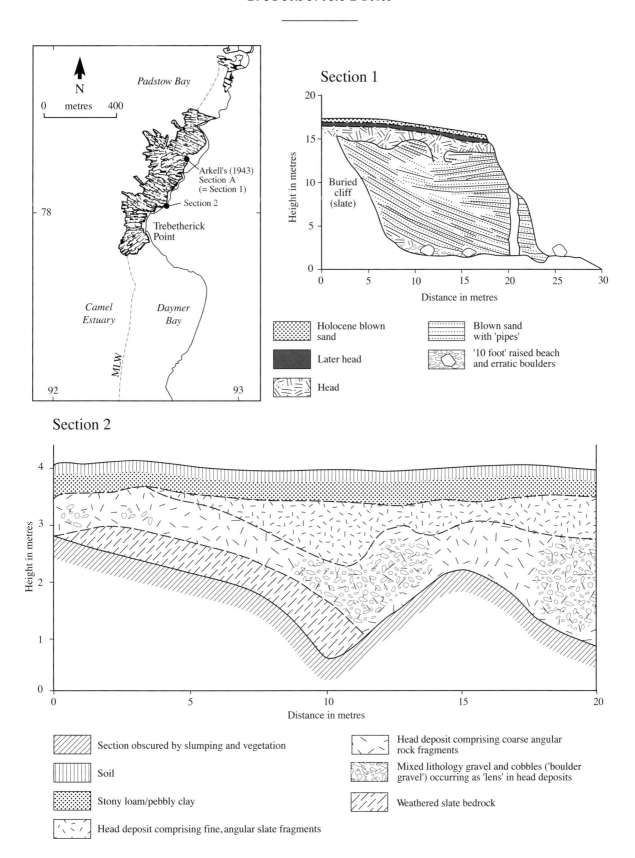

Figure 6.10 The Quaternary sequence at Trebetherick Point: Section 1 after Arkell (1943); Section 2 compiled by S. Campbell.

referred to the material as 'sandrock'. It is most steeply bedded (28°) at its junction with the fossil cliff, but more gently bedded to the west (Figure 6.10). The sand contains impersistent layers of slate and greenstone fragments, broken limpet shells, whole crab (*Cancer pagurus* Linné) carapaces (Arkell, 1943) and comminuted shell material (Rogers, 1910). Large solution pipes run from the top of this bed to its base (Figure 6.10).

The overlying breccia (bed 4) has sometimes cut-out or 'channelled' the sands beneath (Arkell, 1943; Stephens, 1966a; Scourse, 1985a), but elsewhere it interdigitates with the sand (Stephens, 1966a). It consists of angular slate fragments with some larger pebbles of quartz and other rocks (Arkell, 1943) and is generally coarse and 'blocky' in nature: it sometimes has a sand-clay matrix (Stephens, 1966a). The deposit exhibits cryoturbation structures, fossil ice-wedge casts and, according to Stephens (1966a), is highly weathered. In places, it directly overlies the shore platform.

The stratigraphic relationships of the boulder gravel (bed 5) are not entirely clear, although most authors record that it overlies bed 4. This diamicton is restricted in outcrop; its main exposure is at SW 927781, although Scourse (1985a, 1987, 1996c) also records it at Tregunna (SW 960740) and other localities around the Camel Estuary. Figure 6.10 (section 2) shows that the deposit extends for *c.* 20 m (Scourse has traced its lateral extent for *c.* 60 m) and that it grades laterally into a coarse breccia of angular slate and quartz fragments. It may, therefore, be a facies variation of bed 4 (the lower or main head). It is a largely matrix-supported, mixed lithology gravel with cobbles and occasional boulders. Many of the clasts are locally derived (purple and green slate, vein quartz, phyllite), but others appear to have come from farther afield. The latter include sandstone, conglomerate, granite, porphyry, dolerite, basalt, ironstone and flint (Arkell, 1943), mica-schist (Clarke, 1965b; Scourse, 1996c) and chert (Scourse, 1996c). Some clasts are in excess of 0.5 m diameter and many of the larger ones are subrounded to rounded. In places, the matrix consists of a breccia of very fine slate fragments (Figure 6.10). Arkell (1943) described the sediment as conspicuously bedded in its lower layers, but Stephens (1966a) contradicts this, and today there is little evidence of stratification.

The overlying breccia (bed 6) is also laterally impersistent, and elsewhere rests directly on the lower head (bed 4). It always has a sharp junction with the lower head (bed 4) and is sometimes separated from it by a thin sand layer (Stephens, 1966a). It is a highly variable deposit consisting mostly of fine angular slate fragments, but with some larger slate blocks, quartz and other pebbles. Locally, it appears to contain material similar to, and probably derived from, the boulder gravel (Stephens, 1966a). Bed 6 sometimes grades into a stony loam. This dominantly fine-grained and matrix-supported deposit contains occasional slivers of slate and pebbles of quartz. Arkell (1943) notes that its upper surface is flat and even. It is overlain in places by partially cemented dune sand which contains abundant land snails (*Pomatias elegans* (Müller)), shells of mussels, and Mesolithic flint flakes (Arkell, 1943).

Interpretation

Ussher (1879a) was the first to establish the importance of the sections at Trebetherick. He remarked that at no other site in South-West England has ' … so interesting a collection of Pleistocene, or Post-Tertiary, phenomena been observed within so small a space.' (Ussher, 1879a; p. 6). He recorded erratic stones set in clay (presumably the boulder gravel – bed 5), and interpreted the deposit as fluviatile in origin. In contrast, Reid *et al.* (1910) suggested that the boulder gravel was a head deposit and that the underlying sand (bed 3) was a 'cemented sand reef' and, by implication, waterlain. Dewey (1913) fired the controversy over the boulder gravel by hinting at a glacial origin. He later referred to it unequivocally as 'boulder clay' and correlated it with the beds of clay and striated erratics at Croyde Bay and Fremington in north Devon (Dewey, 1935; p. 67) (Chapter 7).

Arkell's (1943) description of the sequence at Trebetherick is still one of the best, although some of his interpretations have since been revised. He regarded beds 2 and 3 as raised beach deposits and aeolian sand respectively, and correlated them with comparable deposits at Saunton in north Devon. He assigned the beds to the 'Boyn Hill or Middle Acheulian Interglacial' (equivalent to the Hoxnian of later terminology). He argued that the absence of shells in the raised beach deposits (bed 2) indicated deposition under relatively cold but high sea-level conditions at the beginning of an interglacial, with the constituent erratic boulders having been transported on ice-floes. He refuted the suggestion of Reid *et al.* (1910) that bed 3 was waterlain. Instead, he suggested it was an aeolian deposit derived from the north-west, and cited the inverted arrangement of crab carapaces within the deposit as evidence for wind action. Arkell argued that the deposit showed a fall of sea level, while the profusion of comminuted marine molluscs in the bed was taken to indicate a concurrent marked climatic improvement.

Arkell (1943) interpreted the overlying breccia (bed 4) as a solifluction deposit formed under periglacial conditions during the 'Cornovian Glaciation' (= Wolstonian Stage). He did not concur with a periglacial or glacial origin for the boulder gravel (bed 5) which he interpreted as either a river gravel or, more probably, a raised beach deposit. In favouring the latter hypothesis, he argued that the bed had accumulated during warm conditions and high sea levels (*c.* 17 m OD) in his 'Wolvercote or Micoquian Interglacial' (= Ipswichian Stage) and that the constituent erratics had all been derived from relatively local sources within the catchment of the River Camel, flints and Tertiary pebbles having been reworked from inland plateaux (Arkell, 1943). In the discussion following Arkell's paper, Bull claimed that the boulder gravel could equally well be a solifluction deposit, while George noted that its erratic content was ' ... quite different from the Irish Sea drift.' (George *in* Arkell, 1943; p. 148). Arkell regarded the pebbly clay (bed 7) as a mixture of solifluction and sheet-wash deposits reworked from the boulder gravel. Together with head deposits found elsewhere around Daymer Bay (= bed 6), these deposits were believed to represent a return to periglacial conditions during the 'Cymrian Glaciation' (= Devensian Stage). Arkell noted that the evenly planed upper surface of bed 7 supported a fossil soil which contained Mesolithic artefacts and which was believed to represent an ancient land surface.

The overlying sands (bed 8) were regarded as wind blown in origin and of Holocene age. Their contained fauna of indigenous land snails and derived marine molluscs was taken as indicating temperate conditions subsequent to a period of vigorous forest growth around the coastal margin, as attested by a now submerged forest bed in parts of Daymer Bay (Arkell, 1943).

Stephens (1966a) regarded the raised beach deposits (bed 2) at Trebetherick as Hoxnian in age and suggested that the overlying sands (bed 3) had been blown inland from an exposed sea bed as sea level fell at the beginning of the Wolstonian (Saalian Stage). The lower or 'main' head (bed 4) was interpreted as a solifluction deposit which had both accumulated and been cryoturbated during periglacial conditions in the Wolstonian. Stephens (1966a) groups the boulder gravel (bed 5) with the main head (bed 4), but implies that it may originally have been deposited as outwash from an Irish Sea ice sheet of Wolstonian age (the same ice sheet was believed to have been responsible for depositing the Fremington Clay; Chapter 7). Whether it lies *in situ* or has been soliflucted is not made clear (Stephens, 1966a). Stephens also draws attention to the very different nature of the principal solifluction deposits at Trebetherick (beds 4 and 6), arguing that the upper (Devensian) is relatively 'fresh' and unweathered, whereas the lower (Saalian) is much disturbed by frost-action (Saalian and Devensian) and significantly weathered (Ipswichian).

Sections at Trebetherick Point and elsewhere around the Camel Estuary have also been described and interpreted in a series of papers by Clarke (1962, 1965a, 1965b, 1969, 1973) who proposed a variety of mechanisms to explain the boulder gravel (bed 5). In 1962, he refers to it as 'head', but hints that it may have originated as a glacial deposit. Clarke's (1965a, 1965b) papers return to the thinking of Arkell (1943), by proposing that the boulder gravel is an Ipswichian raised beach deposit derived by ' ... marine erosion of a moraine in the Bristol Channel ... ', but then reworked by solifluction processes in the Devensian (Clarke, 1965b; p. 274).

His later papers (Clarke, 1969, 1973) reflect the influence of a growing body of evidence for the presence of an ice sheet in the Bristol Channel and Western Approaches during Wolstonian (Saalian) times (e.g. Mitchell, 1960, 1968, 1972; Stephens, 1966b; Mitchell and Orme, 1967). His 1969 paper attempts to cover several eventualities by stating ' ... a tongue of this ice invaded the north projecting Camel estuary, gathering beach pebbles in its progress and leaving a patch of till.' (Clarke, 1969; p. 90). He argued that this explanation of the boulder gravel was consistent with Mitchell's (1968) evidence that Saalian Stage (Wolstonian) Irish Sea ice had surrounded Lundy Island to a height of 105 m, and also with Stephen's (1966b) evidence for Irish Sea till in the Barnstaple Bay/Fremington area, and finally with evidence for an incursion of the same ice sheet on to the Isles of Scilly (Mitchell and Orme, 1967).

In 1973, Clarke reported a further exposure of the boulder gravel west of Tregunna House on the south side of the estuary (SW 960740), and alluded to the possibility that the material was deposited by a glacier originating on Bodmin Moor. In his 1980 paper, however, he returns to the Irish Sea ice sheet hypothesis and explains the boulder gravel as part of a recessional moraine, occurrences of material at Tregunna and Little Petherick being used to define lateral margins of the moraine. This ingress of ice into the estuary was believed to have impounded the proto-Camel, forming a 'lake flat' at Trewornan (SW 988743). However, Scourse (1985b) has since shown that this feature comprises Holocene estuarine sediments.

Mitchell (*in* Mitchell and Orme, 1967) regarded Arkell's pebbly clay (bed 7) as a weathered facies of the north Devon Fremington Clay – a till believed by him to be of Wolstonian age: the boulder gravel (bed 5) beneath was regarded either as an outwash gravel or raised beach deposit. Stephens (1970a, 1973) disputed this interpretation, concluding that the boulder gravel was a mixture of head, glacial outwash and Irish Sea till, subjected to later frost-action and weathering.

Kidson took quite a different view of the evidence, rejecting a glacigenic origin for any of the deposits at Trebetherick. He noted that the boulder gravel (bed 5) graded laterally into head and might therefore be soliflucted river gravels – forming one element in a multiple facies head sequence. Moreover, George (*in* Arkell, 1943) had shown that the erratic content of the boulder gravel and pebbly clay was quite different from that of glacial deposits elsewhere which had been derived from the Irish Sea basin, the Cornish origin of many of the clasts supporting the idea that they had been derived by solifluction from within the catchment of the River Camel itself (Kidson, 1977). In refuting the presence of an *in situ* Wolstonian glacial deposit in the sections at Trebetherick, Kidson followed Zeuner (1945, 1959), Arkell (1945) and Bowen (1969) in assigning the raised beach sediment and blown sand (beds 2 and 3) to the Ipswichian Stage, and the overlying beds of head to the Devensian.

Although recent studies have clarified several important issues posed by the succession at Trebetherick, they have failed to reach a firm conclusion regarding the origin of the boulder gravel. Campbell (1984; *in* Scourse, 1996c) analysed quartz sand grains from the boulder gravel using Scanning Electron Microscopy (SEM). He concluded that the constituent grains showed features characteristic of a subaqueous origin (unspecified) but none indicative of a glacial or glaciofluvial environment of deposition.

Scourse (1996c) assigns the raised beach deposits and overlying sands (beds 2 and 3) to the Godrevy Formation of his lithostratigraphical classification, and the overlying head deposits (bed 4) to the Penwith Formation: the boulder gravel is afforded special status as the Trebetherick Boulder Gravel Member of the Tregunna Formation (the latter includes head deposits overlying bed 5 which locally contain materials reworked from it). He interpreted the bed as a partially soliflucted outwash gravel or ablation till, and provided fabric data in support of this conclusion. Its lithological content, however, was regarded as typical of the Variscan outcrops which occur both inland from the Camel Estuary and seaward of it. In determining the height of various boulder gravel outcrops around the margins of the estuary, Scourse (1996c) has concluded that only the lower-level occurrences, such as those exposed at Trebetherick Point, have been soliflucted, whereas those at higher level lie *in situ* and were the source of the reworked material. Additional clast lithological analyses from the material now point strongly to an inland source, perhaps as far afield as Bodmin Moor. The mechanism for emplacement, however, is still uncertain, although Scourse cites river ice as a possibility. Although this mechanism would explain its lithological content, Scourse concedes that the reconstructed gradient of the various boulder gravel outcrops around the Camel would be more consistent with a glacigenic (outwash) origin.

Bowen *et al.* (1985) provided an amino-acid ratio of 0.113 from a specimen of *Littorina saxatilis* (Olivi), collected by H.C.L. James, from the raised beach deposits at Trebetherick (bed 2), and ascribed the beach to Oxygen Isotope Stage 5e (Ipswichian) of the deep-sea record. Scourse (1996c) uses this geochronological evidence as well as stratigraphical evidence from the Isles of Scilly (Chapter 8) to assign a Devensian age to the boulder gravel whatever its origin.

Scourse (1996a) described trace fossils (burrows of the talitrid sandhopper) from the sand beds (bed 3) at Trebetherick. He argued that they provide independent evidence of a backshore-frontal dune environment of deposition. He notes that the ecological requirements of *Talitrus saltator* (Montagu) – the most likely burrowing agent – support an interglacial depositional environment.

Morawiecka (1993, 1994) studied the palaeokarstic 'pipes' found in the upper part of the 'sandrock' profile at Trebetherick (cf. Arkell, 1943; Figure 6.10, section 1). She concluded that they had been formed beneath a cover of head deposits under cold climatic conditions during end-Pleistocene times (cf. Godrevy).

Conclusion

Trebetherick Point is one of the most controversial Quaternary sites in South-West England. It shows a sequence of raised beach deposits, wind-blown sand and periglacial 'head' deposits. Of particular interest is the deposit known as the 'boulder gravel' which occurs within the head sequence. This highly controversial deposit has been interpreted by some as a glacial sediment and used as evidence to reconstruct the southern margin of a 'Wolstonian' Irish Sea ice sheet. Others have interpreted the material as fluvial sediment or raised beach deposits. Whatever its original mode of deposition, most agree that it was finally moved into place by periglacial (solifluction) processes. The latest suggestion is that the boulder gravel was derived entirely from within the catchment of the River Camel (some rock-types may have come from Bodmin Moor), having been transported to Trebetherick on floes of river ice similar to those seen in present-day Arctic Canada. Preliminary geochronological evidence from the site shows that the raised beach sediments were probably formed during Oxygen Isotope Stage 5e (Ipswichian) and that the overlying head deposits, including the boulder gravel, accumulated during the Devensian.

Chapter 7

The Quaternary history of north Devon and west Somerset

INTRODUCTION
N. Stephens

The Quaternary deposits and associated landforms of north Devon and west Somerset have been investigated for over 150 years. Some of the most significant early contributions were made by Sedgwick and Murchison (1840), who studied the raised beach and associated deposits between Saunton and Baggy Point (Croyde Bay), and De la Beche (1839) who established the term 'head' for a variety of slope deposits developed under non-temperate freeze-thaw conditions. Maw (1864) identified and mapped the controversial Fremington Clay for which he proposed a glacial origin. Dewey (1910, 1913) considered the possible source of various erratic and striated pebbles and boulders and investigated the stratigraphical relationships of the local raised beach deposits and the Fremington Clay.

Since this early work, numerous examinations of Quaternary deposits in north Devon have been carried out (e.g. Mitchell, 1960, 1972; Stephens, 1961a, 1961b, 1966a, 1966b, 1970a, 1973, 1974, 1977; Churchill and Wymer, 1965; Wood, 1970, 1974; Kidson, 1971, 1977; Edmonds, 1972; Kidson and Wood, 1974; Gilbertson and Mottershead, 1975). Attention has also been given to a variety of landforms including hogback cliff profiles (E. Arber, 1911; Steers, 1946; M. Arber, 1949, 1974), dry valleys, including the Valley of Rocks (Simpson, 1953; Stephens, 1966a; Pearce, 1972; Mottershead, 1977c; Dalzell and Durrance, 1980), and the buried rock channel of the Taw-Torridge Estuary (McFarlane, 1955). Erratic-rich gravels on Lundy Island have been investigated (Mitchell, 1968) as well as the curious flint-bearing gravel at Orleigh Court near Bideford (Rogers and Simpson, 1937).

Numerous attempts have been made to date and correlate deposits in the region, but as yet there is little agreement on many aspects of its Quaternary history. However, it is generally agreed that the bulk of South-West England was not overrun by Pleistocene ice sheets. Fragmentary evidence for the encroachment of an ice sheet along the north coast, from the Isles of Scilly to north Devon, is, however, widely recorded, taking the form of giant erratics stranded on shore platforms (e.g. Saunton), postulated glacigenic gravels (Isles of Scilly, Trebetherick and Lundy Island) and possible outcrops of till (Fremington) (Stephens, 1973; *in* Mitchell *et al.*, 1973b). This has led to the proposition that the most extensive of the Pleistocene ice sheets reached its southernmost limit at or near the north Devon and Cornish coasts (Figure 2.3). Traditionally, this glacial event has been regarded as 'Wolstonian' in age (Mitchell, 1960, 1972; Stephens, 1966a, 1966b, 1970a; Mitchell and Orme, 1967; Kidson, 1977). However, recent intense scrutiny of the status of 'Wolstonian' deposits in Britain (e.g. Sumbler, 1983a, 1983b; Rose, 1987, 1988, 1989, 1991) has necessitated a radical revision of the concept of the Wolstonian. Best estimates now ascribe glaciation of the northern Isles of Scilly to the Late Devensian (Scourse, 1985a, 1987, 1991), refute evidence of glacial activity at Trebetherick (Kidson, 1977; Scourse, 1996c), and assign the Fremington Clay or Till of north Devon, albeit tentatively, to a glaciolacustrine event of Anglian age (Croot *et al.*, in prep.; Campbell *et al.*, in prep.). Evidence of an earlier glacial event comes from the Bristol district where marginal marine deposits, dated by amino-acid geochronology to Oxygen Isotope Stage 15 (Cromer Complex), are underlain by glacigenic sediments (Gilbertson and Hawkins, 1978a; Andrews *et al.*, 1984; Bowen, 1994b). An attribution of the glacial beds to Stage 16, a major ice-volume episode, is possible, and may also be appropriate for some of the large stranded erratics such as the Giant's Rock at Porthleven (Bowen, 1994b).

Large erratic boulders weighing up to *c.* 50 tons are found around the shores of Barnstaple Bay and Croyde Bay. Some authorities have linked them with similar erratic material recovered from the Fremington Clay (Kidson, 1977), but in reality the relationship is far from clear. Moreover, large erratic boulders are also found on the south coast of Cornwall (the Giant's Rock at Porthleven; Flett and Hill, 1912), along the English Channel coast as far east as Sussex and even in Brittany (Tricart, 1956).

The presence of glacier ice in the Bristol Channel has also been used to account for numerous dry valleys, channels and cols found in north Devon especially near Lynmouth and Hartland Point. Stephens (1966b, 1974) raised the possibility that these were cut by pre-Devensian, probably Wolstonian, meltwater. Others have interpreted them as remnants of dismembered fluvial valley systems formed by protracted coastal erosion and cliff retreat (Steers, 1946; Dalzell and Durrance, 1980).

Throughout the Peninsula there is widespread evidence of former periglacial activity. The development of large coastal 'terraces' of head deposits testifies to the mobility of slope deposits, including weathered residues and frost-shattered rock. These sediments bury ancient cliff notches, some of the marine-planated rock platforms and raised beach deposits, fossil sand dunes (the sandrock of the Croyde–Saunton coast; Stephens, 1966b, 1970a; Kidson, 1977) and some of the large erratic boulders. On the higher slopes the head deposits are generally thinner, and, where sections exist, display a more coarse-textured and blocky nature than at the foot of the coastal slopes, but there is considerable sedimentological variation.

Above about 175 to 200 m no genuine erratic material has been recorded from the mantle of head deposits, although detailed mapping is far from complete. Investigations of erratic limits and erratic-free areas have proved difficult, largely because of human activity (e.g. Madgett and Inglis, 1987). For example, quantities of Carboniferous Limestone from South Wales were off-loaded at local quays and beaches and then transported inland for spreading on cultivated fields. Such activity undoubtedly accounts for the presence of numerous small 'foreign' pebbles such as quartz and flint and some Palaeozoic rocks found in ploughed fields for several kilometres inland from the coast. The modern sandy shingle beaches are rich in quartz and flint pebbles together with erratic material, but the former inland extent of any ice from the Bristol Channel and Irish Sea is unknown.

Thus not only the extent of former glaciation, which most authorities have argued must be pre-Devensian, but the age of the Fremington Clay (till?) and its relationship to the arrival of the large erratic boulders, and the fitting of these events into a timetable which also encompasses deposition of the raised beaches and the various head deposits are far from clear. This is reflected by widely disparate views and correlation tables which have been published of the Pleistocene history of the area (e.g. Edmonds, 1972; Stephens, 1973; Kidson, 1977; Bowen *et al.*, 1985, 1986; Campbell *et al.*, in prep.).

The Holocene is represented inland principally by sand and silt deposits which make up the present floodplains of the Taw–Torridge rivers and their tributaries. Holocene gravel and boulder debris is associated with the steeply falling north coast rivers such as the East and West Lyn, which are deeply incised in the coastal plateau. On slopes, variable amounts of hillwash and colluvium can be observed and, on Exmoor, pollen-bearing peats provide evidence of Holocene climatic and environmental changes (Merryfield and Moore, 1974; Moore *et al.*, 1984). Coastal peat, organic clays and trees from the submerged forest testify to the pro-

gressive drowning of the coast by the Holocene sea.

The coastal deposits and the sequence of events at Westward Ho! have been described in detail by Rogers (1908), Churchill (1965), Churchill and Wymer (1965) and Balaam *et al.* (1987). Sea level is considered to have been 4–6 m below present some 6–7 ka BP, thus allowing peats and peaty muds to accumulate and trees to grow on what is now the modern foreshore. Radiocarbon dating of the peat places it at 6585 ± 130 BP and a Mesolithic kitchen midden was recorded from the same set of deposits. Holocene sea-level rise is also well documented from other parts of the Bristol Channel and the Somerset Levels (Kidson and Heyworth, 1973, 1976; Heyworth and Kidson, 1982; Chapter 2). At Stolford, a submerged peat with tree stumps and fallen trees is seen in much the same position as the Westward Ho! deposits (between MLWOT and MHWOT) and radiocarbon dates indicate a very similar timescale of events, culminating in the sea returning to approximately its present level.

Thus in north Devon and the Somerset Levels, and probably also for much, if not all, of the South-West, the Holocene transgression appears to have been a continuous process, implying crustal stability over a wide area (Kidson and Heyworth, 1973; Kidson, 1977). In Barnstaple Bay the last few thousand years have seen construction of the enormous shingle and cobble ridge which extends northwards from Westward Ho!; the south-western end of this structure is estimated to have moved inland at least 60–70 m in the last 150 years (Stuart and Hookway, 1954). Massive sand dune systems have also formed at Northam Burrows, Braunton Burrows, in Croyde Bay and Woolacombe Bay; these dunes mask parts of the foot of the coastal slope in each of these localities (Kidson *et al.*, 1989; Chisholm, 1996).

The evidence

Rock platforms

In north Devon, Quaternary deposits constitute a wedge of sediments up to 18–20 m thick and mask the mainly convex coastal slopes. These deposits overlie a series of marine-planated rock platforms which lie between *c.* − 1.5 m to + 15 m OD. Cliffs cut in the Pleistocene sediments vary in height from *c.* 2 to 20 m and it is clear that the extensive rock platforms now visible were once buried and have been exhumed by recent Holocene marine erosion. Although there has been some trimming of the old platforms, the evidence indicates that the greater part of the exposed platforms are ancient features.

At Saunton the platform at 5 m OD is most prominent. However, Kidson (1971) has shown other platforms also occur, extending from a few metres below OD and the modern beach to + 8 m OD. In Croyde Bay a series of platforms can be observed between Middleborough House and Pencil Rock, the 8-10 m OD platform being the dominant feature. A higher platform at 12-16 m OD is seen near Freshwater Gut and Middleborough House, rising to 14-15 m OD at Pencil Rock. Although marine processes undoubtedly account for the erosion of the platforms, it is not clear whether this took place entirely under interglacial climatic conditions; some parts of the platforms may have been cut by the sea under periglacial conditions (Dawson, 1990).

At Westward Ho! several platforms are superbly exposed. The present tidal platform extends for some 200 m at low tide with a height range of 2-3 m OD; a lower rock surface at about − 1.05 m OD disappears northwards below the modern beach. The 2-3 m platform terminates in a near-vertical rock cliff above which the 8-9 m OD platform is seen. Isolated remnants of a platform at 5 m OD rise locally above the 2-3 m surface (Stephens, 1970a; Kidson, 1971, 1977; Figure 7.9). The generally ragged appearance of the wide 2-3 m platform gives way to more smoothed surfaces at the cliff-foot where wave action has constructed an impressive storm beach of sandstone pebbles and cobbles. These appear to have been derived from raised beach material lying on the 8-9 m platform. The seaward edge of the 8-9 m platform lies 3-4 m above the highest storm beach ridge and about 6 m above the back of the 2-3 m surface. The notch of the 8-9 m platform is estimated to be at about 13-14 m OD although it is completely hidden by a suite of superficial deposits.

Buried rock channels

McFarlane (1955) demonstrated a buried rock channel at the confluence of the Taw-Torridge rivers, lying at a depth of about − 24 m OD and buried by *c.* 30 m of sediment in the estuary. He considered that both rivers formerly graded to a level of − 24 m OD north of Appledore. Similar flat-bottomed buried channels elsewhere in South-West England may have formed when sea level was lower by up to 45 m, and the shoreline some 15 km off the present coast.

The age of the channels has not been determined, but they are likely to have been formed during a variety of Pleistocene cold stages when sea levels were low. Similar buried channels have been described elsewhere in South-West England by Codrington (1898), and comparable features have been traced and recorded in South Wales and the Severn Estuary (Anderson, 1968). The sediment content of the channels varies from till, sand and gravel to a variety of Holocene deposits, and much of the sedimentary infill has been reworked.

Erratics and glacigenic materials

Large erratic boulders on the coasts of north Devon and Cornwall, on the English Channel coast as far east as Sussex and in Brittany, together with significant quantities of pebble-sized erratics and some other glacially derived sediments have long provided interesting problems for Quaternary researchers (Williams, 1837; Pengelly, 1867, 1873a; Hall, 1879b; Dewey, 1910; Taylor, 1956; M. Arber, 1964; Stephens, 1966b; Kidson, 1971; Edmonds, 1972; Madgett and Inglis, 1987; Hallégouët and van Vliet-Lanoë, 1989; Sims, 1996). Virtually all the large coastal erratics are found within a narrow range of height (0-10 m OD) and resting upon one of the several marine-planated rock platforms (Stephens, 1970a, 1974; Kidson, 1971, 1977). A few have allegedly been recorded at higher altitudes, as for example an epidiorite block on Baggy Point at + 80 m OD (Madgett and Madgett, 1974), but some erratic material may owe its position to anthropogenic activity. Madgett and Inglis (1987) have made a comprehensive survey of erratics, of all sizes, found between Baggy Point and Saunton, and Sims (1996) reviews the possible mechanisms for their emplacement.

The wide distribution of the large erratics has been accounted for by the movement of land-based ice masses (Mitchell, 1960; Kidson, 1971; Kellaway *et al.*, 1975) or floating icebergs and floes (Synge and Stephens, 1960; Stephens, 1970a, 1974; Bowen, 1994b). However, deposition of the large erratics has also been linked directly to the emplacement of the Fremington Clay (Kidson and Wood, 1974; Kidson, 1977) where the erratic suite is similar to that on the open coast at Croyde and Saunton. The rejection of the proposition (Kellaway *et al.*, 1975) that land-based ice could have advanced sufficiently far eastwards into the English Channel to deposit large erratics from south Cornwall to Sussex and in Brittany (Gibbard,

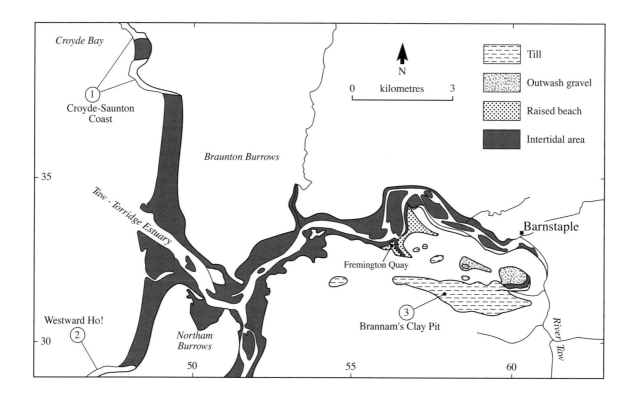

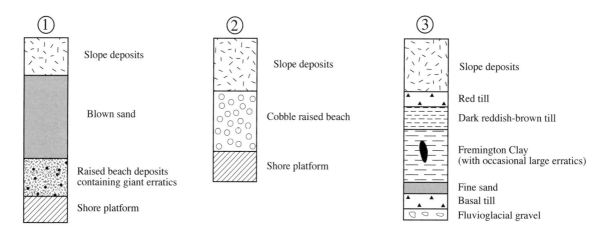

Figure 7.1 The distribution and proposed stratigraphical relationships of Quaternary deposits around the Taw-Torridge Estuary. (After Kidson and Heyworth, 1977.)

1988), suggests that, at least for these coasts, floating ice was probably involved. In north Devon and north Cornwall the problem is more difficult because there is good evidence for incursions of Irish Sea ice into the Bristol Channel during some stage or stages of the Early and Middle Pleistocene to deposit till and related sediments as far east as Somerset (Hawkins and Kellaway, 1971; Hawkins, 1977; Andrews *et al.*, 1984). Thus two mechanisms

of deposition may be involved in the South-West Peninsula, although it is also possible that the wide geographical distribution of the large erratic boulders was accomplished during a single glacial event of pre-Devensian age.

Erratic emplacement by floating ice requires a delicate balance between sea level and climate to allow the stranding of ice-rafted material. If isostatic adjustments were also involved then a complex relationship between land and sea level must be envisaged for which, at present, there is little or no evidence during the Early and Middle Pleistocene. Oscillations of the position of the North Atlantic Polar Front have been postulated by Ruddiman and McIntyre (1981) during the Devensian. If similar movements occurred during earlier Pleistocene glacial events then climatic conditions may well have allowed substantial ice-floes to stream southwards to latitudes of the Bristol and English channels while sea level remained sufficiently high to allow the floes to be stranded on the lower of the rock platforms.

The Fremington Clay and related deposits

An important suite of Pleistocene deposits is known from a small area between Fremington and Barnstaple. The description and analysis of these deposits began in the mid-nineteenth century but their interpretation and dating remains difficult and contentious (Maw, 1864; Dewey, 1910; Taylor, 1956; Mitchell, 1960; M. Arber, 1964; Stephens, 1966b, 1970a, 1974; Wood, 1970, 1974; Edmonds, 1972; Kidson and Wood, 1974; Kidson, 1977). The Fremington Clay in fact comprises a complex and variable sequence of clays, silts, sands and stony clays overlying gravels (Stephens, 1966b, 1970a; Croot, 1987; Croot et al., 1996). Up to 30 m of clay have been recorded and erratics, including some large examples, and striated stones are abundant in certain horizons (but see Brannam's Clay Pit). The sub-clay gravels exposed in the base of the clay pits, at Fremington Quay, Fremington Railway Cutting and at Lake Quarry provide further problems of interpretation and correlation (Figure 7.1).

Sections at Fremington Quay, the Railway Cutting, Penhill Point and Lake Quarry show gravels which have been considered to represent a raised beach (usually equated with that exposed in the Croyde–Saunton coastal section) resting on bedrock and overlain by erratic-bearing stony clays (Figure 7.1). Stratigraphical variation over several hundred metres of section, however, has made interpretation difficult and dating controversial (Dewey, 1913; Stephens, 1966b, 1970a, 1974; Wood, 1970, 1974; Kidson and Wood, 1974; Kidson, 1977; Croot et al., in prep.; see Fremington Quay).

Several metres of sand and gravel cap the Bickington-Hele ridge which rises to 55 m OD and stands above the main body of the Fremington Clay (at 30 m OD). These deposits, which contain erratic stones, may represent ice-marginal or sub-

ice sediments associated with the Fremington Clay, which until recently has been regarded as, at least in part, a till. Edmonds (1972) re-mapped this area and reaffirmed the presence of till, a view accepted by most authors but recently questioned by Croot (1987) and Croot et al. (1996) who have proposed a glaciolacustrine origin for the bulk of the Fremington Clay (see Brannam's Clay Pit). Edmonds also mapped river terraces in the lower Taw Valley and between Barnstaple and Swimbridge. These terraces were related to a series of events involving both deposition of the Fremington Clay and the raised beach of the outer coast, and were considered to have accumulated during the Wolstonian and Ipswichian stages.

Foundation trenches for a housing development in Croyde Village revealed a red-brown clay containing erratics and striated stones. This too may be related to the Fremington Clay and other stony clays exposed in low cliffs below modern sand dunes in Croyde Bay (Madgett and Inglis, 1987). Although a glacial origin has traditionally been proposed for the Fremington and Croyde clays, there is still considerable doubt concerning the precise origin and age of these deposits and estimates of both vary widely. Most authorities have invoked a land-based ice sheet of Wolstonian (Saalian) age to account for the deposits, but this now seems highly unlikely (see Chapter 2; Anglian and Saalian events). Some workers have suggested the possibility of a Devensian age for the Fremington Clay; Eyles and McCabe (1989) postulated that the deposits could have accumulated as a glaciomarine mud drape from a disintegrating, floating, Devensian Irish Sea ice sheet. Such a possibility was also given credence by Bowen et al. (1989) and Campbell and Bowen (1989), and also by Synge (1977, 1979, 1981, 1985) who has argued consistently for extensive floating Devensian ice off the south-east Irish and south-west Wales coasts. Although such dating of the Fremington Clay fits neatly with Scourse's (1985a, 1986) proposal that land-based Devensian ice reached the Isles of Scilly, it runs counter to the view of Kidson (1977) and others that there is no evidence of a glacial event post-dating the raised beaches at Croyde, Saunton and Westward Ho! The most recent attempts to date the Fremington Clay using OSL techniques (Croot et al., 1996; Gilbert, in prep.) have provided ambiguous results. An age greater than c. 400 ka BP, however, is indicated suggesting that the deposits are at least of Anglian age. Evidence from the Bristol district suggests that glacigenic sediments there may be as old as Oxygen Isotope Stage

16 (Gilbertson and Hawkins, 1978a; Andrews *et al.*, 1984; Bowen, 1994b). If a correlation can be proved between these deposits and the Fremington Clay, then the Stage 16 ice limit in South-West England (Figure 2.3) proposed by Bowen and Sykes (1988) may appear increasingly realistic.

Lundy Island

Further evidence for the former presence of glacier ice in north Devon occurs on Lundy Island (Mitchell, 1968). The island is composed mainly of Tertiary granite except for a small area in the extreme south-east where Upper Devonian slates (Upcott Slates) crop out. There are also numerous dykes of basic and acid rocks. Pebbles of sandstone, flint, chert and greywacke up to 10 cm diameter are widely scattered over a considerable area at the northern end of the plateau between *c.* 84 and 107 m OD, and occasional erratics are also found in head which mantles coastal slopes in the east. Ice-moulded bedrock, demonstrating possible evidence for a WNW to ESE direction of ice movement, is recorded on the west side of the island north of St James's Stone. In the north of the island, small, deep dry valleys slope eastwards and trench the coastal slope: these have been interpreted as glacial meltwater channels (Mitchell, 1968). Although the precise age of the Pleistocene sediments and landforms is unknown, the evidence appears to confirm the presence of an ice sheet in this vicinity to a minimum height of 115 m OD: both Anglian and Wolstonian glacial events have been suggested (Mitchell, 1960, 1968, 1972).

The Doniford gravels

The Doniford gravels crop out for approximately 2 km along the Somerset coast near Watchet. According to Gilbertson and Mottershead (1975), they appear to have no direct connection with the glacial deposits and raised beaches of north Devon nor the glacial deposits reported from the Somerset Levels at Kenn (Gilbertson and Hawkins, 1978a) and in the Bristol area (Hawkins and Kellaway, 1971).

The Doniford gravels have a maximum thickness of about 5 m and consist mainly of slates, sandstones, grits and chert derived from the Devonian rocks of the Bredon Hills, with subordinate blocks of local Liassic shale. Both the gravels and the underlying shale bedrock are severely disturbed by periglacial structures which extend upwards into the lowest member of a series of pebble-bearing

loams; no far-travelled erratics have been detected in either set of sediments.

The gravels contain remains of *Mammuthus primigenius* (Blumenbach) (woolly mammoth) and well-rolled Palaeolithic implements (hand-axes and flakes) of Acheulian typology (Wedlake, 1950; Wedlake and Wedlake, 1963) which might appear to raise the possibility of a pre-Devensian age. However, since the uppermost loam, which caps the gravel sequence, contains Mesolithic artefacts and because there is no marked discontinuity between the lower loams and the gravels, Gilbertson and Mottershead (1975) have assigned the bulk of the Doniford sequence to the Devensian, arguing that the gravels accumulated as a mixture of periglacial fluvial and slope deposits.

Raised beach, sand and head deposits

Slope deposits of varying composition and thickness mantle most of the landscape of north Devon. They range from weathered clasts supported in a sandy-silt matrix to coarse, loosely packed angular rock fragments. At the foot of some coastal slopes the deposits form 'terraces' of considerable extent and thickness. The term 'head' has been universally adopted to describe the bulk of these deposits since it was first used by De la Beche (1839).

It is accepted that non-temperate freeze-thaw processes were responsible for the break up of the regolith and underlying bedrock and, together with some water-action, enormous quantities of material were moved downslope. The terraces of head are particularly well developed around Croyde Bay, at Saunton, Lee Bay and Westward Ho!, and in each case the head overlies extensive raised beach deposits (Figure 7.1). In Croyde Bay and at Saunton, cemented dune sands rest directly on raised beach sediments and these in turn are covered by head, which in places interdigitates with dune sand. At the base of the sections, and overlain by either raised beach deposits, cemented dune sand or head, are found some of the large erratic boulders; in every known case these are in contact with the bedrock of the shore platforms.

The head deposits are clearly younger than the raised beach sediments and the large erratic boulders. The cemented sands (sandrock) rest directly on the raised beach sediments, which are pebbly towards their base at Saunton but pass upwards into current-bedded (marine) sand. The latter then gives way, often imperceptibly, to cemented dune sand (Greenwood, 1972; Gilbert, 1996) which accumulated, at least in part, when periglacial processes began to produce the extensive head deposits, thus accounting for the interdigitation of the two very different sediments.

Attempts to recognize divisions within the head deposits (Stephens, 1966a, 1966b, 1970a) have been based on the presence of a coarse, angular and blocky layer up to 2 m thick – the Upper Head – which forms the upper surface of the head terrace. The terrace was considered to comprise a Lower Head, a relatively consolidated deposit with well-weathered clasts set in a matrix of sand- and silt-sized material, while the Upper Head consists mainly of angular rock fragments, little weathered and without a substantial matrix. Occasional ice-wedge casts and other infilled (with silty clay) cracks disturb the bedding of the Upper Head. However, such a division of the head deposits is not universally accepted and Kidson (1971, 1977) regards the head as a single, if very variable, sedimentary unit. Thus, Kidson would regard the head as the product of a major periglacial phase during the Devensian cold period while Stephens suggested that at least two separate periglacial phases are represented. Aeolian sand, which underlies and interdigitates with the head along the Croyde–Saunton coast (e.g. Greenwood, 1972; Gilbert, 1996), has been dated recently by Optically Stimulated Luminescence (OSL) techniques (Gilbert, 1996). The preliminary results show that the aeolian sediments were deposited during Oxygen Isotope Stage 4, around 70 ka BP, and that the considerable thicknesses of overlying head must be Devensian. Precise subdivision and dating of the head are still, however, tenuous. There is no evidence that the surface of the coastal terrace here, or at other localities, was fashioned by marine activity.

The raised beach deposits would therefore appear to represent an interglacial sea-level event, or events, before at least one, and possibly more than one, major periglacial phase. Exposures of raised beach deposits vary in height and it is only in Croyde Bay, at Pencil Rock on Baggy Point and at Middleborough House at about 10 to 14 m OD that notches in bedrock can be observed. Raised beach deposits are seen at lower levels towards Saunton but here the rock notch is obscured.

Attempts to resolve the age of the raised beach deposits between Croyde and Saunton have been made using amino-acid analyses of fossil marine shells (Andrews *et al.*, 1979; Davies, 1983; Bowen *et al.*, 1985). The results show that most of the raised beach deposits can probably be correlated with high relative sea levels during Oxygen Isotope Stage 7. However, the presence of some shells with amino-acid ratios typical of Stage 5e (Ipswichian) and Stage 9 provides major problems of stratigraphic interpretation: Bowen *et al.* (1985) have tentatively suggested that locally Stage 5e deposits may be banked against an older (Stage 7) raised beach deposit (see Croyde–Saunton Coast). Gilbert (1996) assigns the bulk of raised marine sediments at this locality to Stage 5e.

The raised cobble beach deposits at Westward Ho! have been ascribed both to the Ipswichian (e.g. Kidson, 1977) and to the Hoxnian (e.g. Stephens, 1970a, 1973). However, the deposit lacks shells and cannot therefore be dated by amino-acid techniques. Comparison based upon position, stratigraphy and height range in fact suggests that the raised beach deposit at Westward Ho! is possibly equivalent to the Croyde–Saunton raised beaches, although such evidence alone is insufficient to justify correlation. The raised beach deposit at Lee Bay (see the Valley of Rocks), sealed below massive head deposits, may also be related. Unfortunately, there is no section known in north Devon where two undoubted raised beach deposits of different ages occur in clear superposition. Only in South Wales, at Portland Bill and in the Isles of Scilly (Mitchell and Orme, 1967) have two superposed beaches of different age been identified, although in the latter locality Scourse (1986) interprets the higher (younger) deposit as having been derived by solifluction from the older (Chapter 8).

The evidence presented by Bowen *et al.* (1985, 1989) and Bowen and Sykes (1988) has been summarized admirably by Jones and Keen (1993). It appears that for the present, a number of permutations for the ages of the various outcrops of raised beach deposits in southern Britain are possible. If the Saunton and Croyde raised beach sediments are not Ipswichian (the Pennard Stage of Bowen *et al.* (1985)) but belong to some pre-Ipswichian event (the Unnamed Stage and/or the Minchin Hole Stage of Bowen *et al.* (1985)), then many questions still arise concerning the age of the overlying head deposits, the Fremington Clay and the raised beach sediments reported from Fremington Quay, Fremington Railway Cutting and at Penhill Point.

Attention must also be given to the substantial patches of well-rounded pebbles and cobbles exposed at low tide on the foreshore at Westward Ho! These deposits are exposed sporadically beneath a suite of Holocene sediments. The pebbles and cobbles vary in size, many exceeding 30 cm in length. Most are embedded in a head deposit and stand on-end, and some appear to have suffered post-depositional fracturing, perhaps by freeze-thaw processes. Rough polygonal patterns are displayed by the upstanding cobbles, suggesting an exposure of relic patterned ground in material almost identical to that of the modern storm beach ridge and the 8–10 m OD raised beach deposit (see Westward Ho!).

The lower intertidal rock platforms (− 1.5 m and/or 2–3 m OD levels of Kidson, 1971) must surely pass below these deposits. Dating and correlation, however, remain very difficult although it has been suggested tentatively that these foreshore deposits constitute a periglacially disturbed beach of different age from the 8–10 m OD raised beach deposit (Stephens, 1974). Such an explanation is rejected by Kidson (1974) as an unnecessary division of the same raised beach. However, it can be argued that these periglacially disturbed beach deposits were closely associated with the cutting of the 2–3 m OD rock platform and removal of some of the head deposits originally covering its surface. Although the sea is at present trimming this low platform it seems clear that it is a relic feature cut below the higher raised beach platform at 8–10 m OD. There remains, therefore, the possibility that two different phases of rock platform cutting and beach deposition are preserved at Westward Ho!

The relationship of the raised beach deposits to the Fremington Clay and related sand and gravel deposits also remains highly contentious. The raised beach sediments exposed at Fremington Quay, the Railway Cutting and at Penhill (resting on bedrock at 10 m OD) are considered by most authors to post-date the Fremington Clay and to represent an Ipswichian beach deposit (Kidson, 1971; Edmonds, 1972; Kidson and Wood, 1974; Wood, 1974; Kidson, 1977). However, alternative explanations of the stratigraphy at these sites have been proposed (see Fremington Quay) and the age and origin of a sub-Fremington Clay gravel remains crucial to establishing the relative age of the raised beach (Stephens, 1966b, 1974). Dewey (1913) considered that at Fremington Quay raised beach gravels were overlain by weathered stony clay containing erratics and striated pebbles, while Stephens (1966b) recorded contorted beach gravels overlain by till in the nearby Railway Cutting; whether or not these stony, erratic-bearing clays overlying the raised beach rest *in situ* or not requires further investigation.

It is thus of critical importance that the age of the raised beach deposits on the outer coast at Croyde and Saunton be determined, that there should be a fresh examination of the Westward Ho! beach deposits (both in cliff and foreshore exposures), and that a major effort be made to establish the precise stratigraphical relationships of various deposits in the Fremington area and their relationship to the sedimentary sequence on the outer coast. At present it is not possible to assign any of the beach deposits, the sandrock, the very large erratic boulders, the Fremington Clay and related sands and gravels at Bickington and Hele to a precise Quaternary timescale with any confidence.

Holocene deposits at Westward Ho!

The finest coastal exposures of Holocene deposits in north Devon crop out on the foreshore at Westward Ho! The sediments consist of peat and organic-rich clays with tree stumps and blue-grey clays. The sequence contains mammal remains and Mesolithic artefacts (I. Rogers, 1908; E.H. Rogers, 1946). Detailed excavation of the intertidal site was carried out by Churchill (1965), and Churchill and Wymer (1965) provided a radiocarbon date of 6585 ± 130 BP (Q–672) for a fen peat overlying a Mesolithic kitchen midden. A comprehensive reassessment of the site's palaeoenvironmental and archaeological history is provided by Balaam *et al.* (1987).

The site is one of many localities in southern Britain where former land surfaces (the 'submerged forests') can be shown to have been transgressed by the rising Holocene sea. At the time of the formation of the peat at Westward Ho!, sea level is considered to have been 4–6 m below that at present, which is consistent with the evidence from the Somerset coast at Stolford and South Wales (Kidson, 1977). The Holocene deposits rest upon a sterile blue-grey clay which in turn overlies the well-rounded pebbles and cobbles of periglacially disturbed beach deposits and head (Stephens, 1966b, 1970a).

Dry valley systems in north Devon

Dry valleys, dry cols and related features are widely recorded in north Devon. These include the Valley of Rocks near Lynmouth, the Hartland and Damehole Point channels south of Hartland Point, and some dry valleys and 'through' valleys near Bideford and Newton Tracey.

Controversy surrounds the origin of the Valley of Rocks, the associated dry col at Lee Abbey and a series of flattened spurs at Duty Point, Crock Point and Woody Bay, extending in all some 5 km westwards along the high cliffed coast from the joint mouth of the East and West Lyn rivers. The Valley of Rocks is incised over 100 m below the northern edge of the Exmoor plateau and hangs 120 m above the sea. It is a dry valley trending east-west, cut in the sandstones and slates of the Devonian Lynton Beds. It appears to occupy an anomalous position in relation to the mainly north-flowing drainage off the Exmoor plateau, except for segments of the East Lyn river.

Stephens (1966a, 1966b) suggested that if large ice masses had once occupied the Bristol Channel, there was a distinct possibility that conditions would have favoured the development of ice-marginal meltwater channels between the ice and the northern edge of Exmoor and even, locally, subglacial channels. However, other mechanisms involving coastal retreat and river capture have been suggested (E. Arber, 1911; Steers, 1946; Mottershead, 1964, 1967, 1977c; Pearce, 1972, 1982; Dalzell and Durrance, 1980). Pearce (1972, 1982) outlined a series of possible drainage diversions involving the East and West Lyn rivers, and the Lee stream, which he argued resulted from the retreat of the cliffed coastline.

Mottershead (1977c) invoked coastal erosion to account for the diversion of the East and West Lyn rivers to their present outlet at Lynmouth and the consequent abandonment of the Valley of Rocks. Such an hypothesis does not, however, take into account the enormous thickness of superficial (mainly head) deposits obscuring the rock floor of the Valley of Rocks and the Lee Abbey col. Similarly, the projection of river profiles from the East Lyn River to fit the Valley of Rocks and the Lee Abbey col also ignores the fact that thick superficial deposits overlie the rock floor (Simpson, 1953; Stephens, 1966b).

Dalzell and Durrance (1980) confirmed the existence of considerable thicknesses of superficial material, including thick head deposits, in the Valley of Rocks and the Lee Abbey col. Rockhead was shown to fall steeply westward to Wringcliff Bay and, in the Lee Abbey col, stands well above any westward projection of its profile under the Valley of Rocks. In a carefully argued account they envisaged that the East and West Lyn rivers and the Lee stream formerly joined at Wringcliff Bay and then flowed westwards along the coast to Duty Point and Crock Point, where the flattened spurs represented remnants of the old valley floor. This extended river system was then dismembered by coastal erosion and substantial cliff retreat during the Ipswichian.

The well-argued case that marine erosion and considerable cliff retreat is the most likely agent responsible for the capture and diversion of the East and West Lyn drainage is attractive and reasonable. Indeed, examination of the East Lyn Valley indicates that if very rapid cliff retreat took place about 4 km east of Foreland Point, near Glenthorne and County Gate, to reach an elbow of the river at Southern Wood and Malmsmead, where the Badworthy Water joins the East Lyn, then another gently sloping dry valley might be created between Southern Wood and Leeford. The dry valley would be about 2 km long. The lower gorge section of the East Lyn Valley between Leeford and Barton Wood, where it is joined by the Farley Water, would also form part of the abandoned system but would not be completely dry because of the input from several small streams. The implication of such an hypothesis is that cliff retreat of about 1.5 km would be necessary through ground rising to over 300 m OD. If a similar explanation is accepted for the evolution of the Valley of Rocks then the rate of cliff retreat must have been exceptionally fast to have been completed during the Ipswichian, or even during the entire Pleistocene Period.

The presence today of long, subaerially and periglacially modified slopes above a limited development of vertical cliffs, to form the hogback cliffs (E. Arber, 1911; Steers, 1946; Stephens, 1990) suggests that marine erosion along the north Devon coast has in fact been relatively slow. Consequently, the efficiency of wave attack to bring about river capture of the kind proposed must be questioned, unless special geological conditions were present. These might include the presence of particularly weak strata immediately offshore and systems of fault lines along the northern edge of the Exmoor plateau. While little is known about the rate of marine erosion on hard rocks in southern Britain, there seems insufficient evidence from north Devon to indicate that cliff recession is taking place at a rapid rate. Furthermore, the existence of a raised beach deposit (Ipswichian or older?) deeply buried by head deposits in Lee Bay indicates that at least some of the existing crenulations in the coastline have been in existence for some considerable time.

Consequently, the possible role of meltwater associated with an ice mass pressing against the north Devon coast during a pre-Devensian glacial event cannot be completely dismissed. Although there is no unanimity of agreement as to which glacial event (Oxygen Isotope Stage 16, Anglian or later?) may have been involved, there can be no doubt as to the former existence of ice masses in the Bristol Channel between the Isles of Scilly, Somerset and the Bristol area (Jones and Keen (1993) provide a summary of the evidence.). Erratic pebbles have been recovered from all the modern and raised beaches in north Devon and there is some evidence provided by erratic material that ice extended to about 150–175 m OD on the western plateau behind Ilfracombe and Berrynarbour. Meltwater erosion may have occurred with or without the development of ice-dammed lakes and the channels could have operated with steep gradients and very variable directions of flow, both parallel to the coast and transverse towards the Bristol Channel, perhaps contributing to the breaching of the seaward rim of Hollerday Hill at Wringcliff Bay. The matter clearly requires further investigation.

A series of dry channels has also been identified at Hartland Quay, Damehole Point and Speke's Mill Mouth in north Devon (Steers, 1946; Stephens, 1966b, 1974). The wide flat-floored valleys 'hang' above the sea near Hartland Point and Damehole Point and effectively isolate the prominent St Catherine's Tor; they continue as flattened spurs at Hartland Quay and Speke's Mill Mouth, which are similar features to those at Duty Point and Crock Point. Accumulations of coarse blocky head up to 3 m thick emphasize the flat floors and their accordance of level at about 25–30 m OD. Some parts of the channels contain small streams which plunge by waterfalls to the sea, as for example on the north side of St Catherine's Tor.

Steers (1946) regarded these features as part of a system of small valleys which were dismembered by cliff retreat, drawing on the evidence of the prominent vertical cliff profiles seen along the coast from Hartland Point southwards. The explanation could be accepted without question if it were not for two factors. The first is the considerable disparity between the size of the channels and the very small streams now flowing in some parts of them, the discharges of which appear to do little more than cut modest 'gutters' in the head which occasionally reach rockhead. The second is the recorded presence of glacial erratics on Lundy Island (up to 107 m OD) some 30 km to the north-west, and of the general acceptance that ice reached this coast during the Early or Middle Pleistocene. Thus, it is suggested that meltwater associated with an ice-front may well have played a part in the formation of these channels.

There are also low-level dry valleys near Bideford, which have been described by Edmonds (1972) as possible drainage channels resulting from the incursion of Wolstonian ice into Barnstaple Bay and as far inland as Barnstaple and Fremington. Edmonds provided a possible sequence of events involving drainage diversions resulting from the blocking of the outlets of the Taw and Torridge rivers and their tributaries by ice occupying the estuary, the deposition of the Fremington Clay (till and lake clay) and a series of terraces as multiple ice advances and retreat took place (Figure 7.4). Among the valleys that may have been used, and enlarged, by meltwater, is the dry valley extending from the Torridge at Bideford to the coast at Cornborough, 1.2 km south of Westward Ho! The valley floor rises from about 10 m OD in Bideford to about 30 m OD at its western end in the Cornborough col and was formerly used by the railway line to Westward Ho!

Another 'through' valley which he identified extends from the east side of the Torridge at Bideford eastwards to the Taw Valley. The linked valleys are flat-floored and both contain tiny streams: the highest point of a possible glacial drainage channel is 55 m OD at Newton Tracey (Edmonds, 1972). Edmonds also envisaged another such channel extending from the Torridge Valley at Landcross, via the River Yeo valley to Yeo Vale and then north-west towards Ford and the coast.

The Orleigh Court flint-bearing gravels

A deposit of flinty gravels and sands covers some 0.75 km^2 on a low plateau above the valleys of the River Yeo and River Duntz at Orleigh Court, 4 km west-south-west of Bideford. The gravels rest on Carboniferous sandstones and were recorded by Vancouver (1808), De la Beche (1839) and Rogers and Simpson (1937). The deposit is estimated to be about 8 m thick and consists of:

3. Upper gravels: flint-rich (flints up to 39 cm) with indurated ferruginous grit, much clay and silt
2. Lower gravels: relatively few flints, mostly small, with 'clean' sand
1. Yellow clay of unknown thickness

The most abundant flints are seen in ploughed fields over a rather smaller area of about 0.3 km^2 and there has clearly been downslope movement of the gravels off the plateau, which ranges in height from about 61–80 m OD. The deposits have been considered to represent a Pliocene outlier with affinities to the Lower and Upper Greensand and with fossils derived from the Upper Chalk

(Chapter 3). However, the origin of the gravels is unknown. What does appear to be important is the lack of carriage of large flints to adjacent areas and the survival of about 8 m of sediments in an exposed position. There is no indication that the gravels represent a glacial deposit for no erratics or striated stones have been recovered, but it is suggested that the existence of the deposit may indicate that there has been no ice movement across this part of north Devon to reach a height of 80 m OD. This in no way precludes ice from having crossed Lundy Island (84–107 m OD), entered Barnstaple Bay to reach Fremington, and pressed against the north Devon coast.

BRANNAM'S CLAY PIT
S. Campbell and D. G. Croot

Highlights

Brannam's Clay Pit is one of South-West England's most important Pleistocene sites, providing the best exposures of the enigmatic Fremington Clay, the age and origin of which have been the subject of scientific debate for over 130 years. The material has been interpreted variously as the product of land-based ice, as a glaciolacustrine deposit and as glaciomarine muds. Estimates of its age have varied widely between Oxygen Isotope Stage 16 (pre-Anglian) and the Late Devensian.

Introduction

The Fremington area has long been known for a contentious sequence of deposits which has a major bearing on Pleistocene reconstructions in South-West England. Part of the sequence (often referred to as the Fremington Clay or Till) has been interpreted as the product of a pre-Devensian (Saalian Stage) glaciation, but recent evidence suggests that the deposits may not have been laid down directly by glacier ice and that they are older. The site was noted in an early study by Maw (1864) and since by Ussher (1878), Prestwich (1892), Dewey (1910, 1913), Balchin (1952), Taylor (1956), Zeuner (1959), Mitchell (1960, 1972) and Everard *et al.* (1964). Detailed accounts were provided by Stephens (1966a, 1966b, 1970a, 1974), Wood (1970, 1974), Edmonds (1972), Kidson and Wood (1974), Kidson (1977) and Kidson and Heyworth (1977). Recent excavations were undertaken by Croot (1987), Croot *et al.* (in prep.) and Gilbert (in prep.). The site has also been mentioned widely elsewhere (e.g. Waters, 1966b; Stephens, 1961a, 1961b, 1973; Gregory, 1969; Cullingford, 1982; Hunt, 1984; Campbell and Bowen, 1989; Eyles and McCabe, 1989; Bowen, 1994b; Campbell *et al.*, in prep.).

Description

Maw (1864) demonstrated that the Fremington Clay or 'boulder clay' extended in an oval area some 5.6 km long by 0.8 km wide (Figure 7.2), from Lake to Mullinger. Several small outliers of the clay were also noted. (The distribution of the Fremington Clay and related sediments has since been revised by Croot *et al.* (1996), although much of Maw's original mapping still holds good.) The clay reached a maximum thickness of 27 m and was described by Maw as a tough, homogeneous, smooth brown clay with stones towards the top and with blackened wood fragments at the base. Near Bickington the clay overlay a gravel similar to that exposed on the coast at Fremington. A comparable succession was also described by Ussher (1878).

Brannam's Clay Pit (SS 530316), sometimes known as Higher Gorse Clay Pits, lies off Tew's Lane near Bickington and is located almost centrally in the known extent of the Fremington Clay. The stratigraphic succession exposed in Brannam's Clay Pit over the years has revealed both lateral and vertical variation. However, most authors have recognized a sequence of basal gravel, capped by stoneless clay overlain by stony clay and head over quite wide areas. Figure 7.3 depicts the sequences described by Stephens (1966a, 1966b, 1970a), Croot (1987) and Croot *et al.* (1996). Stephens' sequence can be summarized thus:

9. Soil
8. Solifluction deposits with frost-cracks
7. Sand
6. Weathered till
5. Fresh till
4. Stoneless clays
3. Sand
2. Fresh till
1. Pebbles (raised beach deposits)

Broadly comparable sequences were described by Edmonds (1972) and Kidson and Wood (1974), although the interpretation of the origin and ages of individual beds has varied considerably. Croot *et*

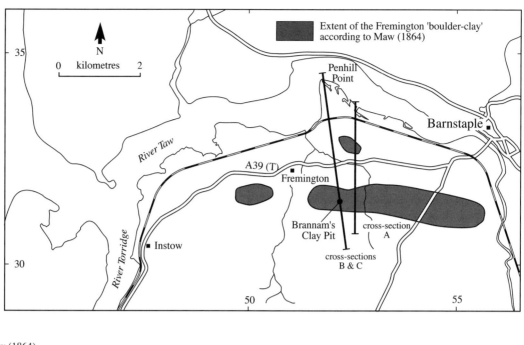

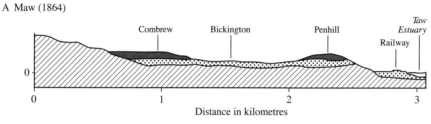

A Maw (1864)

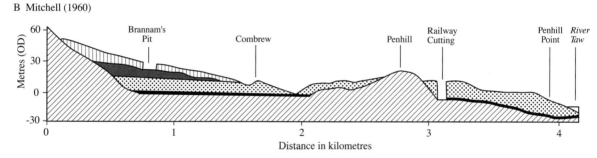

B Mitchell (1960)

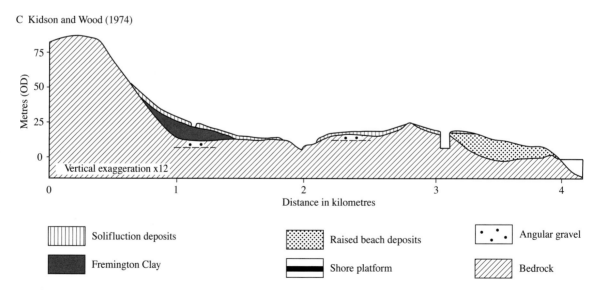

C Kidson and Wood (1974)

Figure 7.2 The extent of the Fremington 'boulder-clay' according to Maw (1864), and proposed stratigraphical relationships in the Fremington area. (After Maw, 1864, Mitchell, 1960 and Kidson and Wood, 1974.)

al. (1996) recorded the following sequence in Higher Gorse Clay Pits, within 50 m or so of the sections previously described by Stephens (1970a) and Croot (1987) (Figure 7.3c):

E. Gravelly sand and clay (head). A matrix-supported deposit with angular and subangular clasts of local bedrock set in a matrix of coarse yellow/brown clay. The deposit is uniformly 1.5 m thick across the exposed section and its contact with unit D is irregular and gradational (Figure 7.3b). (1.5 m*)

D. Clast-rich weathered clays. The material comprises more than 50% of clasts in a matrix of red clayey silt. Constituent clasts comprise mostly small gravels of lithologies identical to those in unit B. The matrix contains ill-defined and deformed silt-rich horizons. The unit is disrupted by irregular compressional and extensional stress patterns and clasts exhibit no consistent fabric. (1.0 m*)

C. Sand lenses within unit B. This unit comprises irregularly shaped lenses of sand and silt. The deposit consists mainly of medium-grained sand grading into coarse sand and medium silt. Grains are mainly of quartz but some of haematite and magnetite are present, together with clasts of sandstone, mudstone and, especially, lignite. The lenses exhibit no evidence of bedding or grading structures, and their contact with the surrounding clays of unit B is sharp and irregular. The sand bodies have no clearly defined channel form. (2.5 m*)

B. Dark brown clay. Contains occasional subrounded clasts and buff-coloured, irregularly spaced silt-rich beds. The proportion of clasts rises from 5% near the base of the unit to *c.* 40% where unit B grades into unit D. Contact between the silt bands and clay matrix is diffuse. Clasts within the unit lack any distinct fabric (orientation or dip) and do not appear to disturb the bedding of clays and silts. All but one of the clasts (sample size > 1500) are derived from bedrock lithologies found within 10 km of the site. The single exception is a cobble-sized, well-striated clast of microdolerite found *c.* 4 m below the surface of unit B. (9.0 m*)

A. Gravels. Comprises gravels in a sandy-silt matrix. The component clasts are dominantly subangular and comprise very locally derived sandstones, siltstones and shales of the Crackington Formation (Prentice, 1960); they exhibit a weak fabric with a north-west to

south-east orientation but no discernible dip. (2.0 m*)

(*) denotes maximum bed thickness

Erratics associated with the Fremington Clay were described by Maw (1864), Dewey (1910), Taylor (1956, 1958), Vachell (1963) and M. Arber (1964). Maw (1864) recorded a large boulder of 'basaltic trap' which he believed had originated from the middle of the clay bed (probably unit B; see above). Dewey (1910) recorded boulders of hypersthene andesite and spilite which had apparently been derived from the 'till'; he suggested that the former had originated from the west coast of Scotland, the latter from north Cornwall. Taylor (1956, 1958) recorded boulders of quartz porphyry, quartz dolerite and olivine dolerite from the clay, and additional erratics were recorded by M. Arber (1964). Taylor noted that the erratic recorded by Maw was striated. However, Croot *et al.* (1996) argue that the microdolerite clast from unit B represents the first reported find of an ' ... unequivocally glacially-transported *in-situ* clast at the Brannam's clay pit site.' (Croot *et al.*, 1996; p. 20). This assertion, however, is strongly refuted by Stephens who cites his own observations and those of the old clay pit foreman, as evidence of *in situ* erratics in the clay (Stephens, pers. comm.).

Edmonds noted that the upper part of the sequence (beds 6–8; Stephens' description) contained numerous rock fragments and pebbles, mainly sandstone and quartzite, but also slate, vein quartz, chert, granite, dolerite and flint. Stephens (1966a, 1970a) recorded that the beds he interpreted as till (his beds 2, 5 and 6) and the stoneless clays (his bed 4) contained pieces of lignite, shell fragments (especially abundant according to Waters (1966b)), erratics and striated stones. Croot *et al.* (1996) record $CaCO_3$ values of between 10% and 20% for units B and D; sand from unit C, surprisingly, has a lower value (5.5%), but contains numerous small (+ 1ϕ to 0ϕ) shell fragments of undetermined species. Units B, C and D contain abundant pollen and spores but no material uniquely of Pleistocene age. Damaged palynomorphs of Mesozoic and Palaeozoic age are also present. Hunt (1984) identified and listed Carboniferous, Jurassic, Cretaceous and Palaeogene palynomorphs in samples from these units. The nannofossil assemblage is dominated by reworked Cretaceous forms (Croot *et al.*, 1996) although Wood (1970) records a varied fauna of foraminifera which he suggests rules out a freshwater origin.

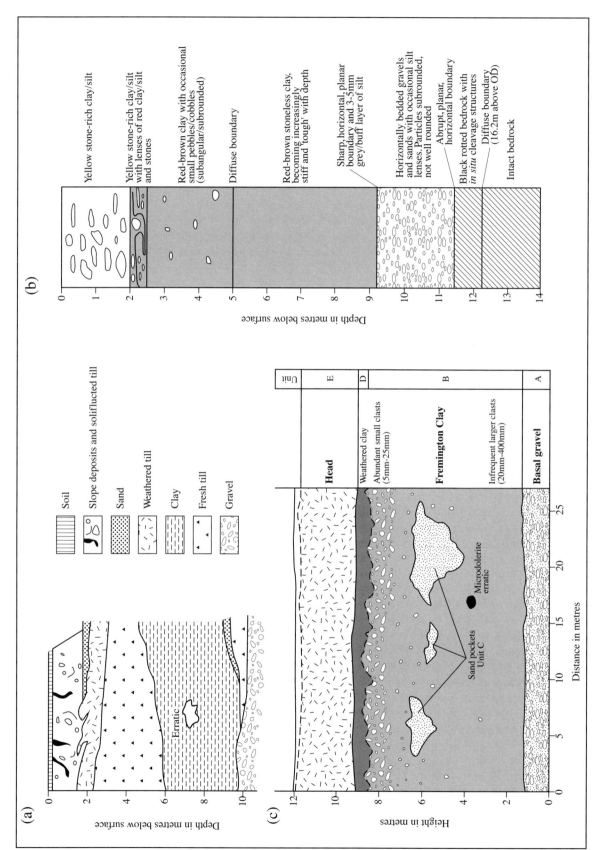

Figure 7.3 The Quaternary sequence at Brannam's Clay Pit, Fremington. (a) Composite section of the former eastern and southern working faces, adapted from Stephens (1966a, 1966b, 1970a). (b) The succession recorded by Croot in 1987. (c) The sequence recorded by Croot *et al.* (1996).

Kidson and Heyworth (1977) recorded laminae from stoneless clays in unit B. Croot *et al.* (1996), however, suggest that original sedimentary structures are only rarely visible even at a microscopic scale.

(Numerous descriptions of the sequence at Brannam's Clay Pit have referred to the 'Fremington Clay' or 'Till' without reference to specific beds within the sequence. This confusion was recognized by Wood (1970, 1974) and Croot *et al.* (1996) who adopted the term 'Fremington Clay Series' to describe the sequence of clays and stony clays at the site. The term 'Fremington Clay' is retained here to describe the clay and diamicton sequence (units B–D; Croot *et al.*, 1996) without implying specific origins.)

Interpretation

Maw (1864) correlated the gravel beneath the Fremington Clay in the Bickington area with shingle deposits exposed on the coast at Fremington. These had previously been interpreted by De la Beche (1839) as raised beach deposits (Figure 7.2). Maw noted the close correspondence in height of the gravel beds at some 10.6 m above HWM. He demonstrated the relationship of the clay and gravel deposits over a wide area (between Bickington and exposures on the south side of the Taw) – a practice followed by later workers including Mitchell (1960) and Stephens (1966a). However, he contrasted the uncemented and unstratified gravels at Bickington and Fremington with the layered and cemented raised beach sands and gravels at Hope's Nose (Torquay) and Croyde Bay which, in addition, contained abundant marine shell fragments. He entertained the possibility, therefore, that the gravels at Bickington were related more appropriately to the overlying clays which he considered, albeit tentatively, were glacial in origin. He argued that erratics at the base of the Croyde raised beach deposit and towards the top of the Fremington Clay were probably of the same age, and that the Fremington gravels were therefore older than the raised beach at Croyde, being ' ... separated by at least the interval in which most, if not all, of the Fremington clay-bed was deposited.' (Maw, 1864; p. 450). Ussher (1878) regarded the Fremington gravels and the local raised beaches as the same age (unspecified) and of estuarine origin.

Many subsequent authors (e.g. Prestwich, 1892; Dewey, 1910, 1913; Taylor, 1956, 1958; Vachell, 1963; M. Arber, 1964) established the petrology and sources of erratics associated with the Fremington Clay. Most of the boulders appear to have been derived from relatively local sources in Devon and Cornwall (Dewey, 1910, 1913; Taylor, 1956; M. Arber, 1964). Some, however, had more distant origins, including the hypersthene andesite recorded by Dewey (1910) which was believed to have originated from the west coast of Scotland. The precise stratigraphic context of many of the early finds, however, is doubtful, although Maw (1864), Dewey (1910) and M. Arber (1964) presented evidence to suggest that some of the larger erratics had been derived from within the Fremington Clay (unit B) itself. Although such evidence has been used to support a glacial origin for the bed (M. Arber, 1964), others have been more cautious. Taylor (1956) suggested that the erratics may have been transported by floating ice, and Dewey (1913) suggested that the clay had been deposited ' ... under such conditions as would permit large erratic boulders to be dropped in it.' (Dewey, 1913; p. 155).

The Fremington Clay was also briefly referred to by Balchin (1952) who regarded it as an alluvial infilling of reworked Keuper Marl.

The Fremington Clay was next referred to in regional stratigraphic correlations by Zeuner (1959) and Mitchell (1960). Zeuner regarded the clay as a 'bottom-moraine' of an ice sheet from the Irish Sea, which had penetrated inland on the southern shore of the Bristol Channel. Mitchell (1960) argued that the raised beaches of the area (at Fremington and Saunton) lay stratigraphically beneath the Fremington Clay, which he too regarded as a till, but of Gipping (Saalian) age (Figure 7.2). He correlated the 'Fremington Till' with the Ballycroneen Till of Ireland, and assigned the raised beaches of the area to the Hoxnian Stage, and not to the last (Ipswichian) interglacial as suggested by Zeuner (1959).

Mitchell's (1960) proposal for a Pleistocene chronology in north Devon was also upheld by Stephens (1961a, 1961b, 1966a, 1966b, 1970a, 1973). From evidence at Brannam's Clay Pit, Stephens argued that the pebbly gravels (his bed 1; unit A) could be correlated with raised beach deposits exposed at the coast; according to him, they contained well-rounded clasts and occurred at approximately the same height. These proposed Hoxnian-age marine sediments were overlain by a sequence of tills (his beds 2, 5 and 6) and lake clays (bed 4), the diagnostic characteristics of which (including pebble lithology and orientation,

(a)

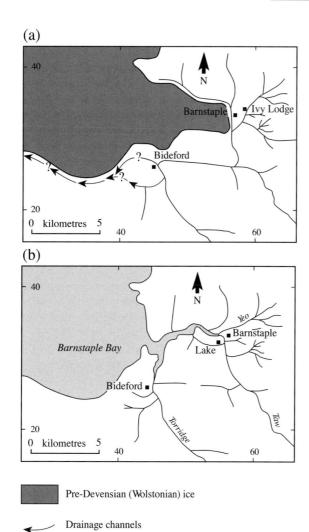

(b)

███ Pre-Devensian (Wolstonian) ice

←‿ Drainage channels

Figure 7.4 A reconstruction of the proposed Wolstonian (Saalian) glaciation of the Barnstaple Bay area, after Edmonds (1972), illustrating: (a) The development of ice-marginal drainage at the height of glaciation; (b) Present-day drainage.

texture, carbonate, shell and lignite content) indicated derivation from the Irish Sea Basin. Stephens correlated these beds with the Ballycroneen Till (Eastern General Till) of Eire of proposed Saalian age (cf. Mitchell, 1960). During the Ipswichian the surface of the till was chemically weathered (accounting for his bed 6), and during periglacial conditions in the Devensian the upper layers of the proposed till were solifucted and mixed with locally derived head (bed 8). Frost-cracks were also formed during this periglacial phase (Stephens, 1966a, 1970a).

In marked contrast, Edmonds (1972) argued that the proposed raised beach deposits exposed

around Fremington Quay were Ipswichian in age but were dissimilar to the gravels spasmodically exposed in the bottom of Brannam's Clay Pit (cf. Bowen, 1969); he thereby discounted Mitchell's and Stephens' stratigraphical correlations. He argued, however, that the overlying tills and lake clays were of Wolstonian (Saalian) age, suggesting that Wolstonian ice had moved south across the Irish Sea to the north Devon coastline, and in the process deposited erratics on Lundy (Mitchell, 1968). This ice was believed to have advanced across the Fremington area depositing till (bed 2 of Stephens), and then receded to an unknown position, but probably near Fremington (Figure 7.4). The ice front dammed surface waters, forming a lake basin to the east. The fine-grained sediments of the Fremington Clay (Stephens' bed 4; lower part of unit B) were believed to have been deposited in these relatively still lake waters (Edmonds, 1972). It was envisaged that the ice then readvanced perhaps as far as Barnstaple to deposit a further till (Stephens' beds 5 and 6) (Edmonds, 1972).

Wood (1970, 1974) and Kidson and Wood (1974) reinvestigated the Pleistocene deposits of the Barnstaple Bay area, and particularly those in the vicinity of Fremington, using boreholes and geophysical techniques. They demonstrated that the gravel (unit A) was distinctive in terms of clast lithology and roundness, and particle-size distribution, and that it was not therefore associated with the raised beach deposits exposed at the coast (e.g. at Fremington Quay). Instead, they suggested that the gravel at Brannam's Clay Pit was a glaciofluvial sediment – perhaps formed by the same ice lobe which deposited the overlying tills and clay (Kidson and Wood, 1974). They suggested that erratics in the proposed glacial beds at Brannam's Clay Pit could be correlated with the giant erratics at Croyde and Saunton – which rest on a shore platform and which are associated with raised beach and blown-sand deposits. They concluded that the raised beaches of the area were of Ipswichian age, and that the large erratics found along the Croyde–Saunton coast had been derived from glacial sediments of the same (Wolstonian/Saalian) age as the Fremington tills and clay. They did not discount the possibility, however, that large erratics farther south around the coast of South-West England (e.g. the Giant's Rock at Porthleven) had been ice-rafted into place during an earlier (Anglian) glacial phase (cf. Mitchell, 1960). During the Devensian, the upper layers of the proposed Saalian till were redistributed by solifluction and the upper beds were disrupted by frost-action (Kidson and Wood, 1974). A similar simple sequence of Middle to Late Pleistocene events was also followed in a series of papers by Kidson (1971, 1977) and Kidson and Heyworth (1977).

Bowen *et al.* (1985) and Bowen and Sykes (1988) applied amino-acid geochronological dating methods to the raised beach deposits at nearby Saunton and Croyde, establishing an Oxygen Isotope Stage 9 age for the oldest faunal elements within them. On the basis of local stratigraphic relationships, this placed the Fremington Clay, albeit very tentatively, in Stage 12 (Anglian). Later results on shell fragments recovered from the Fremington Clay itself indicated a range of ages from Early and Middle Pleistocene to Late Devensian (Bowen, 1994b). The latter provided some support for the suggestion by Eyles and McCabe (1989, 1991) and Campbell and Bowen (1989) that the Fremington Clay could have accumulated in a glaciomarine environment during the Late Devensian.

Excavations at Brannam's Clay Pit during 1986/1987 (Croot, 1987) and in 1994 (Croot *et al.*, 1996; Gilbert, in prep.) have provided significant new data regarding the character, age and depositional environments of the succession. Croot *et al.* interpret unit A (basal gravels) as a fluvial deposit, of uncertain age, that has undergone slight deformation since deposition. They cite the relative angularity and local lithology of clasts within the unit as evidence for its origin.

Interpretation of the overlying units B, C and D, however, is much more problematic. Croot *et al.* suggest that the lower levels of unit B (mainly clay) were deposited in a quiet-water environment. Clasts, which become more frequent towards the top of the unit, are regarded as dropstones; with a single exception, their origin is relatively local. Although micromorphological examinations of the clay show strong similarities with known glaciomarine deposits, the lack of a marine fauna and the local lithology of dropstones found within the clay are taken by Croot *et al.* to indicate deposition of units B and C in a glaciolacustrine setting. Although clasts in unit B are clearly dropstones, their means of transport to the quiet-water body remains unclear. With the single exception of the striated microdolerite cobble, the remaining locally derived clasts could have been introduced on or in floes of river ice: the single glacial cobble, however, must have been dropped from a mass of glacier ice. Whether this mass of ice was a single iceberg or a more continuous sheet of floating ice remains uncertain (Croot *et al.*, 1996).

Croot *et al.* also present data from micromorphological and engineering tests to demonstrate that units A, B and C were gently deformed and partially over-consolidated following deposition. This adds weight to the likelihood that units A, B and C were overridden by glacier ice, but does not accord with the interpretation (Croot *et al.*, 1996) of the overlying material (unit D) simply as a weathered variant of unit B. Although previous workers have interpreted unit D as a basal till (e.g. Stephens, 1966a, 1966b, 1970a; Edmonds, 1972), Croot *et al.* argue that the material demonstrates a weak fabric with clast lithologies identical to those in the underlying unit. Unit E, which caps the sequence, is interpreted as a typical head deposit formed by solifluction during periglacial conditions.

Dating of the sequence also remains highly controversial. Although recent work has shown that units B, C and D contain abundant pollen and spores, there is no material of uniquely Pleistocene age (Croot *et al.*, 1996). Indeed, some of the assemblage is directly comparable with that of the Bovey Basin clays and other Tertiary clays found in South-West England (Wilkinson and Boulter, 1980; Freshney *et al.*, 1982; Hunt, 1984; Croot *et al.*, 1996). The rest is derived from other sources to the north-west (Hunt, 1984). Equally, the nannofossil assemblage comprises a dominance of reworked Cretaceous forms and Croot *et al.* conclude that the fossil evidence alone would imply an early Tertiary age.

The application of Optically Stimulated Luminescence (OSL) dating techniques to sand samples from unit C, however, has provided potentially significant results (Croot *et al.*, 1996; Gilbert, in prep.). OSL measurements show that the sands are older than Oxygen Isotope Stage 2 (Late Devensian) and are more probably of Anglian (Stage 12) age. Large degrees of uncertainty associated with these dates, however, mean that deposition during any of the intervening evenly numbered (cold) Oxygen Isotope Stages 4, 6, 8 and 10 cannot be ruled out. On balance, however, these preliminary and unconfirmed results would point to the Fremington Clay (units B, C and D) being of Anglian (Stage 12) age.

Conclusion

Brannam's Clay Pit is undoubtedly one of the most important Pleistocene sites in Britain, and the

Figure 7.5 The Pleistocene sequence towards the western end of the Fremington Quay exposure. The vertical 'pipe' structures are infilled with lighter-coloured silt and clay, and penetrate beyond the base of the exposure: they may be frost or desiccation cracks (Wood, 1970). (Photo: S. Campbell.)

Fremington Clay has long been regarded as the ' … most significant glacial deposit of the peninsula' (Kidson, 1977; p. 294). The oldest deposits at the site are gravels of probable fluvial origin. The overlying Fremington Clay, in fact a complex sequence of clay, silt, sand and stony clay, has traditionally been regarded as the product of a Wolstonian (Saalian) Irish Sea ice sheet. However, recent work suggests that the deposits are older, possibly of Anglian (Oxygen Isotope Stage 12) age, that they are overwhelmingly of local derivation, and that they may have accumulated, at least in part, in a glaciolacustrine environment. The evidence is still insufficiently precise to rule out a glaciomarine origin. The site may well prove to be one of the most southerly points of Britain to have been overrun by glacier ice.

FREMINGTON QUAY
S. Campbell and D. G. Croot

Highlights

Exposures at Fremington Quay show a complex and controversial sequence of stony clay, gravel, sand and silt which has long figured in reconstructions of regional Pleistocene history. Some authorities claim that the site shows raised beach deposits overlain by glacigenic sediments, others that the sequence comprises soliflucted and fluvially sorted materials. Recent evidence shows that the site may demonstrate glacially dislocated and thrusted 'rafts' of bedrock overlain by glaciofluvial materials.

Introduction

Exposures at Fremington Quay have figured prominently in interpretations of the Pleistocene history of the Barnstaple Bay area, particularly in establishing the crucial relationship between the raised beaches and possible glacial deposits. A sequence of gravels and stony clays has been interpreted as a Hoxnian raised beach deposit overlain by Wolstonian till (e.g. Stephens, 1966a). An alternative view, however, is that the raised beach deposits date from the Ipswichian and that the overlying stony clays are head deposits (including reworked till) emplaced during the Devensian. The site was referred to in early studies by Maw (1864), Ussher (1878) and Dewey (1913), and in regional

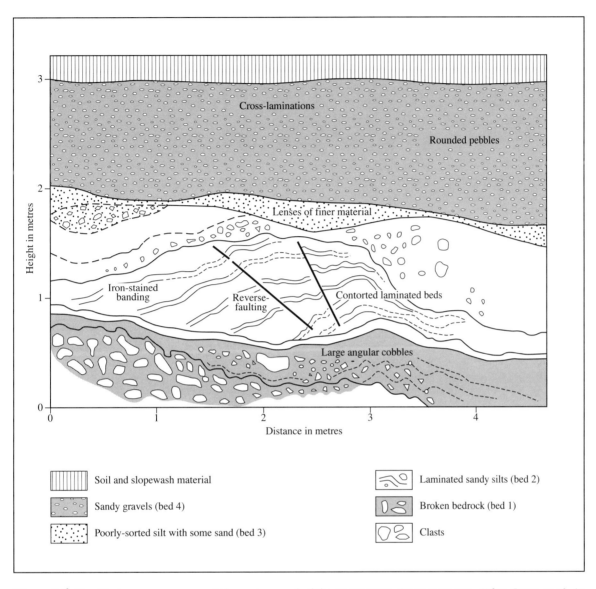

Figure 7.6 illustration with labels: Cross-laminations, Rounded pebbles, Lenses of finer material, Iron-stained banding, Reverse-faulting, Contorted laminated beds, Large angular cobbles. Axes: Height in metres (0–3), Distance in metres (0–4).

Legend:
- Soil and slopewash material
- Sandy gravels (bed 4)
- Poorly-sorted silt with some sand (bed 3)
- Laminated sandy silts (bed 2)
- Broken bedrock (bed 1)
- Clasts

Figure 7.6 The Quaternary sequence at the eastern end of the Fremington Quay exposure. (After Croot *et al.*, in prep.)

Pleistocene syntheses by Zeuner (1959), Mitchell (1960, 1972) and Everard *et al.* (1964). More detailed interpretations of the site were provided by Stephens (1966a, 1974), Kidson (1971), Edmonds (1972) and Kidson and Wood (1974). The site was also referred to by Wood (1970, 1974), Stephens (1973) and Kidson and Heyworth (1977), and is the subject of current reinterpretation (Croot *et al.*, in prep.).

Description

The Fremington Quay exposures run for 0.5 km in a low coastal cliff on the south edge of the Taw Estuary, between SS 514332 and SS 509331. Despite having featured in numerous Pleistocene stratigraphic correlations, the exposures and sediments have not been described in great detail.

Edmonds (1972) stressed the sedimentary variability of the Quaternary deposits around Fremington Quay, and referred to them as the 'pebbly drifts of the estuary'. At SS 511331 he recorded that the pebbly clay, sand and silt enclosed a large clay lens (5 m × 1 m), and that at other locations stratification was evident. Seams and lenses of clay, silt and gravelly sand are a common feature of the beds (Edmonds, 1972). He recorded that the pebbly drift rises to 18 m OD, and that it underlies much of Fremington Camp.

On the east side of the Pill, these sediments extend to form the promontory of Penhill Point.

At the eastern end of the Fremington Quay exposures, at Fremington Pill, about 4 m of Quaternary sediment overlies a small cliff some 4–5 m high cut in steeply dipping shale bedrock. Although this rock surface is uneven, it has been described as a 'wave-cut' platform (Stephens, 1966a). It falls in height westwards and the western end of the GCR site comprises a low, 1–2 m-high, cliff of Quaternary sediment disrupted by vertical 'pipes' or cracks which are infilled with lighter-coloured silt and clay (Figure 7.5). Stephens (1966a) records a generalized sequence of:

4. Weathered, pebbly sandy clay with striated stones and erratics
3. Gravels and sands
2. Silts with pebbles
1. Shale, distorted and brecciated

Croot *et al.* (in prep.) records the following section at the eastern end of the Fremington Quay outcrop (SS 513332; Figure 7.6) (maximum bed thicknesses in parentheses):

4. Sandy gravels, partly cross-laminated (2.0 m)
3. Poorly sorted silt with some sand (0.2 m)
2. Laminated sandy silts with inclusions of grey-blue silt. The whole unit is contorted and reverse-faulted, with planes dipping north-west (1.4 m)
1. Broken bedrock: large cobble- to small gravel-sized blocks of bedrock dislocated south-eastwards from source by up to 0.5 m (1.5 m)

Interpretation

Raised beach deposits were first described in the Fremington area by De la Beche (1839), and a gravel bed was recorded at Fremington Quay by Maw (1864). Maw traced this shingle bed inland, via open-sections in the railway cutting west of Fremington Pill, through Bickington and Lake to Combrew, where he believed it underlay a considerable depth of clay – the Fremington Clay (Figure 7.2). Although he regarded the shingle as a raised beach deposit, he noted the possibility that the bed could also be related to the overlying clays which he regarded as glacial in origin. Maw's paper raises a critical issue which has been central to most subsequent studies of Pleistocene sediments in the

area; namely, do the sands and gravels at Fremington Quay form a raised beach and can they be correlated with the gravels underlying the Fremington Clay (e.g. at Brannam's Clay Pit)?

Ussher (1878) believed that the gravels in the Fremington area were estuarine deposits of the Taw. Dewey (1913) regarded the gravels as marine and described them as overlain by head and stony clay. The latter contained deeply striated clasts and led Dewey to correlate the stony clay at Fremington Quay with the Fremington Clay (and till) inland.

Mitchell (1960) proposed a stratigraphical model for the Pleistocene history of the Irish Sea using sections from the Barnstaple Bay area, including Fremington Quay and Brannam's Clay Pit, as critical evidence for his arguments. He suggested that the erratic-bearing raised beach gravels at Fremington Quay overlay a raised shore platform and that the gravels could be traced inland where they underlay the Fremington Clay (Figure 7.2). He argued that Maw's (1864) section

' ... shows clearly and correctly that there is no justification for pretending that the beaches at Fremington and at Saunton are stratigraphically above the Fremington boulder clay ... and if the boulder clay is of Gipping age, the Fremington and Saunton beaches cannot lie in the last inter-glacial period' (Mitchell, 1960; p. 319).

Mitchell envisaged that the shore platform had been cut probably in Cromerian times and that the overlying raised beach gravels could be ascribed to the Hoxnian. The Fremington Clay or Till was regarded as a glacial deposit of Gipping/Wolstonian (Saalian) age.

Stephens (1961a, 1966a, 1966b) examined the sections at Fremington Quay and at Brannam's Clay Pit in more detail, and independently came to the same conclusions as Mitchell (1960). However, his analysis showed that, unlike the Fremington Clay at Brannam's Clay Pit, the stony clay at Fremington Quay contained no shells and was non-calcareous; a glacial origin was nonetheless favoured. He proposed that the beds were subsequently weathered during the Ipswichian and then deeply cryoturbated in the Devensian when, he believed, glacier ice did not reach north Devon (Stephens, 1966a, 1966b).

Subsequent workers, however, have tended to follow Zeuner's (1959) interpretation of regional Pleistocene stratigraphy. Zeuner argued that there was no evidence for a raised beach anywhere in the region having been overridden by ice. He concluded that the glacial deposits of the Barnstaple

Bay area pre-dated the local raised beaches; the latter he therefore regarded as having formed during the Ipswichian Stage.

This view has been reiterated by others including Bowen (1969), Kidson (1971), Edmonds (1972) and Kidson and Wood (1974). Edmonds did not differentiate between raised beach and other deposits at Fremington Quay and referred to the beds simply as the 'pebbly drifts'. He argued that this gravelly drift could be traced a short distance inland, but certainly not to the point of demonstrating the bed's equivalence with the gravelly material spasmodically exposed in the bottom of Brannam's Clay Pit. The 'pebbly drift' rested on a shore platform at Fremington Quay and had been derived, at least in part, from till, some sorting having occurred as the glacial material was soliflucted downslope and reworked by fluvial and, finally, estuarine processes. The 'pebbly drift' was therefore regarded as reworked (Saalian) till, soliflucted and fluvially sorted, and emplaced on to the shore platform in an estuarine environment during the Ipswichian (Edmonds, 1972). He argued that this pebbly sediment graded inland into a river terrace (Terrace 1 in his classification), and that this terrace, the pebbly drift and the local raised beaches (at Saunton and Croyde) were all Ipswichian in age.

Wood (1970, 1974), Kidson (1971) and Kidson and Wood (1974) in part followed earlier interpretations of the sequence, recognizing a raised beach deposit overlain by more poorly sorted sediments. However, they disputed Mitchell's and Stephens' chronostratigraphic interpretations and argued that the raised beach had formed during the Ipswichian and that the overlying sediments were head deposits, soliflucted into their present position during cold conditions in the Devensian. They emphasized the importance of the stratigraphic relationship of the coastal Pleistocene sediments with those inland, particularly at Brannam's Clay Pit; a detailed analysis of gravel samples from both sites was undertaken (Wood, 1970). At Brannam's Clay Pit the gravels were composed predominantly of Culm grits and sandstones. No erratics were recovered, and the deposit was both more poorly sorted and contained more angular clasts than the gravel exposed at the coast (Wood, 1970; Kidson and Wood, 1974). This, they suggested, discounted Mitchell's and Stephens' assertion that the raised beach was traceable inland where it could be seen to underlie the Fremington Clay. They also discounted Stephens' correlation of the stony clay at Fremington Quay with the Fremington Clay at Brannam's Clay Pit. Instead, the stony clay at Fremington Quay was interpreted as a solifluction deposit. The upturned beach materials, which Stephens (1966a) argued had been disrupted by Wolstonian glacier ice, were believed to have resulted from periglacial activity. 'They are inadequate testimony to the powerful machinery of an advancing ice-front' (Kidson and Wood, 1974; p. 233). No glacial deposits *in situ* were recognized above raised beach material in the region, and the raised beach deposits were therefore demonstrably younger and accordingly ascribed to the Ipswichian.

However, recent studies have shown that the exposures at Fremington Quay may in fact show evidence for glacial activity (Campbell and Scourse, 1996; Croot *et al.*, in prep.). During the 1996 Annual Field Meeting of the Quaternary Research Association, members were shown dislocated, possibly thrusted, blocks of bedrock at the base of the Pleistocene succession (Figure 7.7). Croot *et al.* (in prep.) argue that the distinctive reverse-faulting of bed 2 at the eastern end of Fremington Quay, related to the localized transport of bedrock blocks, is characteristic of small-scale glaciotectonism associated with a thin-ice margin. He suggests that bed 2 is therefore a basal/subglacial unit. Although he does not record any erratics in bed 2, and finds no evidence of overconsolidation, other authors have recorded striated erratics at this locality (Dewey, 1913; Stephens, 1966a). The more widely developed overlying unit (bed 4) is interpreted by Croot *et al.* as glaciofluvial material deposited in an outwash fan.

There is some degree of equivalence, therefore, between Croot *et al.*'s record of Fremington Quay (east) and Stephens' record for Penhill. However, there is clear evidence that the gravels exposed at Penhill and Fremington Quay are quite different from the basal gravels at Brannam's Clay Pit. There is, therefore, no basis for establishing a common lithostratigraphy for these sites. However, the proposed glacial event responsible for the glaciotectonic structures at Fremington Quay may also have caused the overconsolidation of the Brannam's Clay Pit sequence. This event remains imprecisely dated (see Brannam's Clay Pit).

Conclusion

Fremington Quay is an important site for interpreting Pleistocene stratigraphy in South-West England. The origin of the sediments here has caused

Figure 7.7 Members of the Quaternary Research Association discuss possible evidence for glaciotectonism at the base of the Pleistocene succession towards the eastern end of the Fremington Quay exposures. (Photo: S. Campbell.)

considerable debate and has led to widely disparate opinions as to the sequence and timing of Pleistocene events in the region. Fremington Quay is an essential reference site for resolving two principal and crucial stratigraphic questions: firstly, whether the clay overlying the postulated raised beach deposits is an *in situ* till as claimed by Stephens (1966a); and whether the shingly gravel extends inland to underlie the Fremington Clay. Some workers have maintained that the sequence at Fremington Quay shows a Hoxnian raised beach deposit overlain by a Saalian-age till. Others have argued that the raised beach deposit is Ipswichian in age and that the overlying stony clay is a Devensian solifluction deposit. A recent proposal is that the site shows evidence of glacially thrusted (tectonized) bedrock, formed at a thin-ice margin, overlain by glaciofluvial sediments. However, there is still no firm agreement as to the age or origin of these controversial sediments. The interpretation of Pleistocene stratigraphy, and therefore of events and conditions in the Barnstaple Bay area, relies very heavily on exposed sequences at Brannam's Clay Pit and Fremington Quay. Together with exposures at Croyde and Saunton and at Westward Ho!, these stratigraphic reference sites are central to any reinterpretation of Pleistocene events in the region.

THE CROYDE–SAUNTON COAST
S. Campbell and A. Gilbert

Highlights

One of South-West England's most famous Pleistocene localities, the Croyde–Saunton Coast exhibits one of the finest compound shore platforms in Britain, a series of spectacular, possibly ice-rafted, erratics and a thick sequence of raised beach, blown sand and head deposits. Although the sections have been studied for over 150 years, the deposits still present major problems of interpretation. Optically Stimulated Luminescence (OSL) dates indicate an Early Devensian age (Stage 4) for the aeolian 'sandrock'.

Introduction

The extensive exposures of Pleistocene sediments between Saunton (SS 445378) and Baggy Point (SS

214

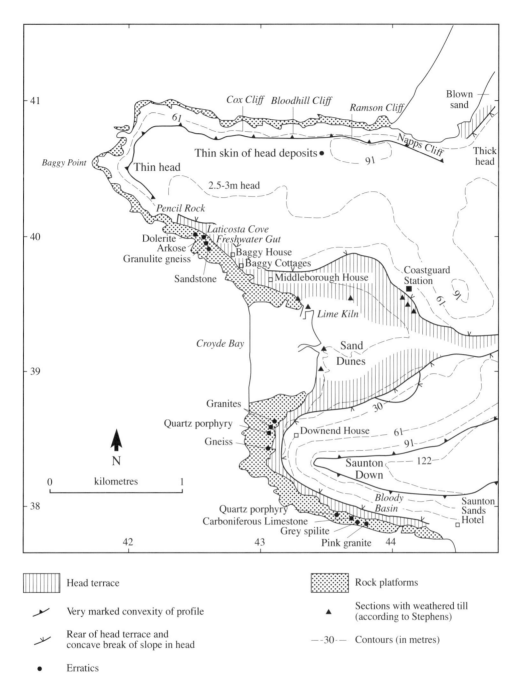

Figure 7.8 The Quaternary deposits and coastal morphology of the Croyde–Saunton Coast. (Adapted from Stephens, 1970a.)

419406), here referred to as the Croyde–Saunton Coast, are some of the best-studied in Britain, and they demonstrate significant evidence for interpreting Pleistocene events in South-West England. The raised beach deposits and erratics at the site first stimulated scientific interest in the early nineteenth century, and discussions have continued until the present day (e.g. Williams, 1837; De la Beche, 1839; Sedgwick and Murchison, 1840;

Maw, 1864; Bate, 1866; Pengelly, 1867, 1873a; Ussher, 1878; Hughes, 1887; Dewey, 1910, 1913; Evans, 1912; Baden-Powell, 1927, 1955). Detailed descriptions of the erratics were later given by Taylor (1956, 1958), Madgett and Inglis (1987) and Sims (1996). Flint artefacts found in the Croyde Bay and Baggy Point areas were described by Whitley (1866). The site has been set in a wider context by Zeuner (1945, 1959), Mitchell (1960, 1972) and

Cullingford (1982). Detailed accounts of the Pleistocene sediments were provided by Stephens (1966a, 1966b, 1970a, 1974), Wood (1970), Greenwood (1972), Kidson and Wood (1974) and Kidson (1977). Amino-acid ratios were presented and discussed by Andrews *et al.* (1979), Davies (1983) and Bowen *et al.* (1985). The site has also been widely referred to elsewhere (e.g. Green, 1943; Arkell, 1945; M. Arber, 1960; Stephens, 1961a, 1961b, 1973; Linton and Waters, 1966; Waters, 1966b; Macfadyen, 1970; Edmonds, 1972; Kidson, 1974; Kidson and Heyworth, 1977; Edmonds *et al.*, 1979, 1985; Keene and Cornford, 1995). Studies on 'pipe' structures (palaeokarst) in the 'sandrock' were carried out by West (1973) and Morawiecka (1993, 1994). The most comprehensive analysis of the sequence is that of Gilbert (1996, in prep.), who provides sedimentological evidence and OSL dates.

Description

Sections through Pleistocene sediments occur around Saunton Down, in parts of Croyde Bay, and from Middleborough to Morte Bay and Putsborough Sand. Two main areas of Pleistocene interest can be identified: sections around Saunton Down from Saunton Sands Hotel (SS 445378) to just beyond Cock Rock (SS 436392); and sections from Middleborough House (SS 432396) to just beyond Pencil Rock (SS 423402). Possible weathered till deposits east of Middleborough Hotel were described by Stephens (1970a, 1974) but are now obscured by a sea wall. Extensive, but thinly developed, head is exposed around the remainder of the Baggy Point promontory. The principal Quaternary features of this coastline are shown in Figure 7.8. Schematic sections through the Pleistocene sediments at Pencil Rock and Croyde Bay are shown in Figure 7.9 and selected cross-sections of the marine-cut platforms, which occur widely between Saunton and Baggy Point, are shown in Figure 7.10. The generalized sequence for this part of the north Devon coast can be summarized thus:

5. Head
4. Cemented sand
3. Raised beach conglomerate
2. Erratic boulders
1. Shore platform(s)

However, considerable lateral and vertical variations occur, and the site's features are best described under the following headings (Stephens, 1970a, 1974):

Raised shore platforms

Three raised shore platforms have been recognized between Saunton and Baggy Point (Stephens, 1970a, 1974), all having been exhumed from beneath Pleistocene sediments by Holocene marine erosion (Figure 7.10). These platforms range in height from 0–6 m OD, through 5.5–7.6 m OD to 10.7–15.0 m OD (Stephens, 1970a, 1974). Although the lowest of these passes are below present HWM, at Bloody Basin (SS 438378) and Middleborough the thick sequence of Pleistocene deposits cemented to the platform, and currently being eroded by the sea, attests to the considerable age of even the lowest platform (Stephens, 1974). The higher platforms are best seen between Middleborough and Pencil Rock and at Freshwater Gut (SS 427400) where a 50 ton block of granulite gneiss is located near to the upper limit of the middle platform at a point where it notches the upper platform (Figures 7.8 and 7.10; Stephens, 1974).

Erratics

Large erratic boulders have long been known from the site (e.g. Williams, 1837; Maw, 1864; Bate, 1866; Pengelly, 1867, 1873a; Hall, 1879b; Hughes, 1887; Prestwich, 1892; Dewey, 1910, 1913; Hamling and Rogers, 1910; Evans, 1912), and the distribution of the principal examples is shown in Figure 7.8 (Taylor, 1956, 1958; Stephens, 1966a, 1966b). These include the famous 'pink granite' erratic weighing some 10–12 tons (Taylor, 1956) at Bloody Basin near Saunton, and the granulite gneiss (weighing an estimated 50 tons) at Freshwater Gut; the former is firmly trapped between the shore platform and the raised beach deposits (Figure 7.11). Some of these erratics compare in size with the Giant's Rock at Porthleven (Chapter 6) in southern Cornwall (Flett and Hill, 1912; Stephens and Synge, 1966; Stephens, 1970a, 1974) and with others elsewhere around the Devon and Cornwall coast (Prestwich, 1892; Worth, 1898; Ussher, 1904; Reid and Scrivenor, 1906; Reid, 1907).

Raised beach deposits

At this site, the raised beach sequence has frequently been described as consisting of two elements; a lower raised beach conglomerate and an overlying cemented shelly sand. The latter has

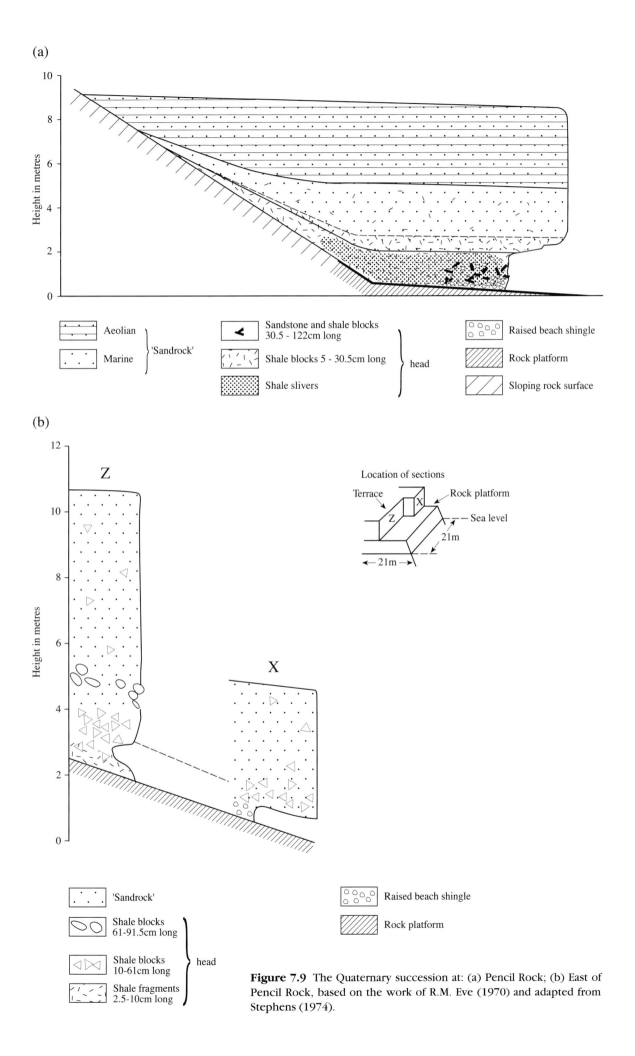

(a)

Height in metres

Aeolian ⎫
Marine ⎬ 'Sandrock'

Sandstone and shale blocks 30.5 - 122cm long ⎫
Shale blocks 5 - 30.5cm long ⎬ head
Shale slivers ⎭

Raised beach shingle
Rock platform
Sloping rock surface

(b)

Height in metres

Z

X

Location of sections

Terrace   Rock platform
X
Z   Sea level
              21m
21m

'Sandrock'

Shale blocks 61-91.5cm long ⎫
Shale blocks 10-61cm long ⎬ head
Shale fragments 2.5-10cm long ⎭

Raised beach shingle
Rock platform

Figure 7.9 The Quaternary succession at: (a) Pencil Rock; (b) East of Pencil Rock, based on the work of R.M. Eve (1970) and adapted from Stephens (1974).

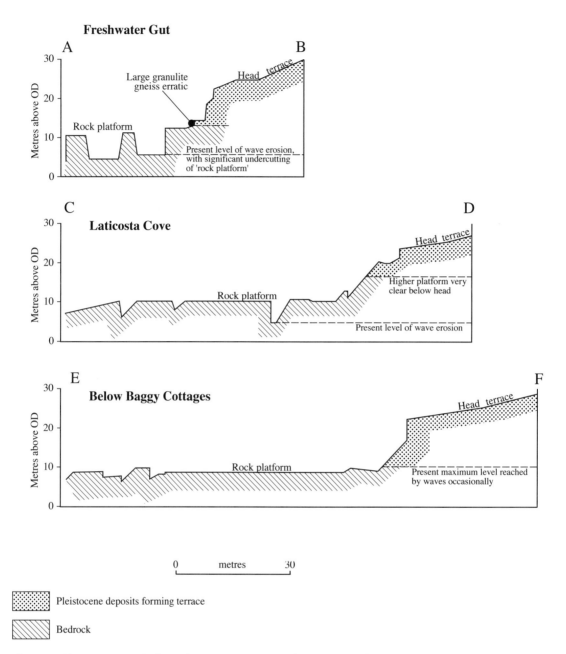

Figure 7.10 Marine-cut platforms between Saunton and Baggy Point, based on the work of R.M. Eve (1970) and adapted from Stephens (1974). (See Figure 7.8 for locations of cross-sections.)

been shown to be of marine origin in its lowest layers, but aeolian towards the top of the beds (Greenwood, 1972; Gilbert, 1996). This sandstone ('sandrock' or 'aeolianite' as it is commonly described) reaches up to 9 m in thickness and sometimes overlies the raised beach conglomerate (2 m maximum thickness), but frequently overlies the raised shore platform(s) and giant erratics directly. It is almost completely cemented and, near Saunton, the horizontal bedding in the lower 1.8–3 m of the deposit is striking (see Figure 7.11). This bedding, however, gradually changes upwards to sloping beds of sand with small dispersed slate fragments. The high shell content of the upper sand beds (Hughes, 1887) and the bedding structures suggest that it is a wind-blown deposit, probably a series of fossil dunes (Greenwood, 1972; Stephens, 1974; Gilbert, 1996, in prep.). Layers and lenses of head are found in the sandrock at all levels.

Figure 7.11 Saunton's famous 'pink granite' erratic, sealed beneath cemented sand and seen during the 1996 Annual Field Meeting of the Quaternary Research Association. (Photo: S. Campbell.)

The raised beach shingle is composed of well-rounded pebbles and cobbles with occasional larger boulders and erratic blocks. The matrix is of coarse sand with shale slivers, small pebbles, shells and shell fragments (Stephens, 1970a, 1974). Prestwich listed 26 species of marine mollusc from the shingle, all indicating temperate conditions. In some places, for example at Pencil Rock, the shingle is mixed with more angular head material, including a number of large blocks (Figure 7.9). The raised beach conglomerate is frequently interbedded with sand.

Head deposits and till

Along much of the coast, the sandrock is succeeded by considerable thicknesses of head (up to 21 m) which forms a large terrace or apron at the foot of the coastal slope. This head is thickest in the east, being particularly well exposed near Saunton Sands Hotel. To the west, the cliff of Pleistocene sediments becomes progressively dominated by the sandrock. The relationship of the head terrace to the shore platforms is shown in Figure 7.10. Kidson (1974) saw little reason to subdivide the head deposits, but Stephens (1970a,

1974) recognized a thick 'Lower' or 'Main' Head and a thin 'Upper' Head of finer material, including slopewash sediments. Stephens noted that the Main Head was well weathered and cryoturbated, with frost-wedges extending into the deposit from the top of the bed some 1.5–1.8 m, and being sealed at their tops by the thin Upper Head. The latter was also disturbed by frost structures but was less weathered than the underlying deposits. The Upper Head can be seen in roadside cuts near Saunton Sands Hotel where angular shattered bedrock overlies *in situ* strata.

Stephens described a number of localities within this large coastal site where he believed till deposits to occur (Figure 7.8). These included sections near the old lime kiln east of Middleborough House, and temporary sections farther east. A low cliff at the head of Croyde Bay, at the junction with the Croyde Brook, was also thought to show till. The stratigraphic relationship of this deposit to others exposed in the coastal cliffs, however, is difficult to determine (Stephens, 1974). Soil horizons and slopewash deposits over the head contain many small flint flakes (some Mesolithic microliths?), indicating human activity during the Holocene (Stephens, pers. comm.).

Interpretation

Interest in the Croyde–Saunton Coast was stimulated, in particular, by the large erratics which lie on the shore platforms and which are sometimes overlain by raised beach, sand and head deposits (Figure 7.11). Williams (1837) was the first to mention the large granite erratic at Saunton and, by the turn of the century, this, and other erratics strewn along the coast, had attracted considerable comment and debate (e.g. Bate, 1866; Pengelly, 1867, 1873a; Hall, 1879b; Hughes, 1887; Prestwich, 1892). Some of these early studies discussed possible sources for the boulders and mechanisms for their transport and emplacement. Dewey's detailed petrological work established that some of the large erratics were foreign to the region, a number having probable sources in north-west Scotland. This, and additional work (Hamling and Rogers, 1910; Evans, 1912), established something of a consensus that the erratics had probably been emplaced on ice floes, a mechanism also to find favour with later workers (e.g. Stephens, 1966a, 1966b, 1970a, 1974).

However, whether these large erratics were moved into their present position by a regional ice sheet (Mitchell, 1960; Kidson, 1971, 1977) or by floating icebergs in the Early Pleistocene (Stephens, 1966a, 1974) has not been satisfactorily resolved. Stephens (1974) nonetheless suggested that the widespread occurrence of such large erratics throughout the Bristol Channel coastlands and even as far south as the northern French coast, supported the latter hypothesis (Mitchell, 1965; Stephens, 1966b), particularly since very large erratics like the pink granite at Saunton and the Giant's Rock at Porthleven are confined to very narrow zones along the coast, below 9 m OD and within the reach of present-day storm waves. Had they been emplaced by a regional ice sheet, their expected distribution might be much less selective (Stephens, 1974). Madgett and Madgett (1974) refuted this argument, citing the occurrence of a large erratic of epidiorite (apparently of Scottish origin) on Baggy Point at some 80 m OD. It is possible, however, that this boulder was dragged up from Croyde Bay to act as a boundary marker. Comprehensive reviews of erratics in the Barnstaple Bay area are given by Taylor (1956), Madgett and Inglis (1987) and Sims (1996).

Early studies by Williams (1837), De la Beche (1839), Sedgwick and Murchison (1840) and Bate (1866), among others, established that the shingle and sand beds in section around the Croyde–Saunton Coast were similar to those found around the present shore. They interpreted the beds as ancient beach deposits and explained their present position by changes in the relative level of the land and sea. Hughes (1887), however, disputed this and maintained that although the beds comprised marine sediments overlain by blown sand and capped by talus, the marine material lay well within the reach of present-day waves; and thus that the deposits in section did not necessarily reflect a former higher sea level.

Prestwich (1892) summarized much of the earlier work and provided a comprehensive stratigraphic analysis of the sections, comparing them with others in southern England and Wales. He argued that three main types of deposit overlay the shore platform:

3. The 'usual local angular rubble', composed of large and small fragments of slaty Devonian rocks in a brown earth without apparent bedding (3–15 m) – 'head'.

2. Blown sands (1.5–9 m), horizontally bedded with frequent oblique laminations and partly or wholly concreted. The sands include large numbers of land snails with occasional weathered valves of *Mytilus* and '*Cardium*'. He interpreted these beds as old dunes and correlated them with the Fremington Clay – a deposit he regarded not as till, but as a lake clay.

1. Raised beach deposits consisting of 'hard grey and micaceous sandstones, chalk flints, and pebbles of white quartz and reddish quartzite in a matrix of sand, with a large proportion of comminuted shells' and frequently cemented. He noted 26 species of marine mollusc from these beds, all of a 'temperate' character and indicating 'interglacial' conditions.

This simple stratigraphy has formed the basis for subsequent interpretations of the sequence; the origin of the beds as raised beach conglomerate and associated marine sand, blown sand, head and hillwash, is not generally disputed. Two very different schools of thought, however, have pertained regarding a chronology of events at the site, based principally on assumptions of the age of the raised beach deposits.

Workers including Mitchell (1960, 1972) and Stephens (1966a, 1966b, 1970a, 1974) have argued that the raised beach sediments accumulated in the Hoxnian, and that the overlying beds can be subdivided to represent the main remaining stages of

Figure 7.12 Extensively developed rock platforms at the western end of Saunton Down, looking north across Croyde Bay. (Photo: S. Campbell.)

Pleistocene time. Alternatively, Zeuner (1945, 1959), Edmonds (1972), Kidson (1974, 1977), Kidson and Wood (1974), Kidson and Heyworth (1977) and Edmonds *et al.* (1979), among others, have ascribed the raised beach deposits to the Ipswichian Stage, and invoked a different sequence of events.

Mitchell (1960) attempted to correlate the drifts of South-West England with those in Ireland. He suggested that the rock platform at Croyde and Saunton (Figure 7.12) had been cut in the Early Pleistocene, probably in the Cromerian, and he believed that the large erratics had been placed directly on to the platform by an ice sheet of Anglian age. He later revised this view on the extent of the Anglian ice sheet (Mitchell, 1972) and acknowledged that the erratics could have arrived on ice floes, becoming incorporated and buried by raised beach sediments during high sea levels in the Hoxnian.

Stephens (1966a, 1966b, 1970a, 1974) re-examined the sections around Barnstaple Bay including those along the Croyde–Saunton Coast. Like Mitchell, he concluded that the raised beach conglomerate and associated marine sands had been deposited during temperate, high sea-level condi-

tions in the Hoxnian. He argued that where the raised beach sediments were mixed with more angular material, a reworking of an ancient head deposit had taken place, this head perhaps having formed in the same cold period in which the giant erratics were emplaced by ice floes (?Anglian). Alternatively, and more simply, he suggested that the head may have accumulated contemporaneously as cliff fall material during the period of raised beach formation. As sea level fell towards the end of the Hoxnian, substantial areas of sea floor were exposed and sand was blown inland and banked up against the old cliffline. Thin layers of head found interbedded with the sandrock at all levels probably attest to intermittent falls of cliff material. However, as environmental conditions deteriorated into the Saalian, periglacial conditions pertained and head formation became the dominant process, and a massive terrace of head was formed seawards of the old cliffline (Stephens, 1974).

The dating of this thick (Lower or Main) head as 'Wolstonian' (Saalian) by Stephens rests very largely on analogies with head sequences (and thicknesses) in Ireland. There, he argued, no great thickness of solifluction deposits was associated with deposits of the last glaciation (Devensian).

This was in contrast to the thick head associated, stratigraphically, with the pre-Devensian (suggested Saalian Stage) Ballycronneen Till. To reinforce this dating, Stephens argued that the head at Croyde and Saunton showed evidence for a considerable length of weathering (in the ensuing Ipswichian). He noted that the head was also disturbed by cryoturbation, fossil ice-wedge casts and festoon structures, and he interpreted this as indicating that head material had ceased downslope movement by the time renewed periglacial activity had churned and disturbed the upper layers of the deposit. This phase of freeze-thaw activity could have taken place in either of the Saalian or Weichselian (Devensian) cold stages (Stephens, 1974). During the Devensian, less severe periglacial conditions returned and a further, but thinner (Upper), head accumulated together with hillwash sediments.

The sections showing possible till deposits in Croyde Bay and near Middleborough (Stephens, 1974) are difficult to interpret, and are not easily related to the coastal stratigraphy elsewhere within the site. Stephens (1974) nonetheless was convinced of the presence of till in Croyde Bay, since the deposit contained erratics and striated clasts. It is not clear, however, if this weathered deposit rests *in situ*. It may, for example, be mixed with head and, near the base of the bed, even with beach shingle (Stephens, 1974). In the absence of absolute dates and detailed sedimentological data, little more can be gleaned from these limited exposures which lie away from the main and extensive coastal sections.

The second main school of thought regarding a chronology of events at this site stemmed from Zeuner's claim that there was no evidence in the region for glacier ice having overridden any of the raised beaches; the latter could thus be assigned to the last (Ipswichian) high sea-level event. Following this premise, Kidson (1971, 1974), Kidson and Wood (1974) and Kidson and Heyworth (1977), among others, proposed the following scheme of Pleistocene events. A shore platform, being the oldest of the Pleistocene features, was likely to be composite in age, formed during high sea levels probably in the Hoxnian and earlier 'interglacial' events. The large erratics lying on the surface of this composite platform were believed to have been emplaced in the same (Wolstonian/Saalian) glacial event which deposited the nearby Fremington Clay (also containing some large erratics from comparable sources). The raised beach conglomerate and sands (including the sand-

rock) which directly overlie some of the erratics were believed to be of Ipswichian (not Hoxnian) age. It followed that the succeeding head deposits which, according to Kidson (1974), showed little sign of differentiation, were periglacial deposits formed during the following Devensian Stage. Such a chronological interpretation was also followed by Edmonds (1972) and Edmonds *et al.* (1979), although these workers favoured that the large erratics had been emplaced during the Anglian rather than the Saalian.

More recently, fossil marine molluscs from the site have been subjected to analysis by both radiocarbon and amino-acid dating techniques. An infinite radiocarbon age determination on *Balanus balanoides* (Linné), taken from the surface of a shore platform, and a clearly unrealistic date of 33 200 + 2800/− 1800 BP (I–2981) (Kidson, 1974) from shells in the raised beach conglomerate, failed to determine the age of the raised beach sequence.

On the other hand, amino-acid ratios obtained from shells in the raised beach conglomerate and overlying sand (Andrews *et al.*, 1979; Davies, 1983; Bowen *et al.*, 1985) have provided significant results. Andrews *et al.* (1979) showed that most of the shells subjected to the technique gave ratios that were higher than the two principal groups of ratios obtained from shells at other raised beach sites in South-West England and Wales. The latter were believed to have lived and then been deposited during Oxygen Isotope Stage 5e (Ipswichian); such ratios were calibrated by Uranium-series dating to a high sea-level event at *c.* 125 ka BP (Keen *et al.*, 1981). The significantly higher amino-acid ratios from Saunton were interpreted as indicating a greater age (or significantly different temperature history) for the raised beach there (Andrews *et al.*, 1979). However, a single shell yielding a typical Stage 5e ratio introduced the possibility that the raised beach could be the product of two distinct (but similar in height) sea-level (interglacial) events. Davies (1983) provided additional amino-acid ratios from Saunton which were also significantly higher than most of those derived from sites elsewhere. This led to the correlation of the Saunton raised beach with the Inner Beach at Minchin Hole Cave, Gower: both were tentatively ascribed to Oxygen Isotope Stage 7 (*c.* 210 ka BP) (Davies, 1983).

Bowen *et al.* (1985) published amino-acid ratios on shells taken from the raised beach sequence both at Pencil Rock (near Baggy Point) and at Saunton. Shell samples were taken from both the raised beach conglomerate and overlying sands at

Figure 7.13 Thick cemented sand (marine and aeolian) and overlying head deposits near Saunton Sands Hotel. (Photo: S. Campbell.)

these sites. This work confirmed the earlier ascription of the beds (Davies, 1983) to Stage 7. However, younger shells were also present in the samples. It was suggested that these had been banked up against the older Stage 7 beach deposits during a high sea-level event in Stage 5e (Ipswichian) (Bowen *et al.*, 1985). In addition, some ratios from the overlying sandrock were grouped with those from the 'unnamed' D/L stage at Minchin Hole Cave, but these were also ascribed to some part of Stage 7 (Bowen *et al.*, 1985), although other possible correlations were discussed.

These data have significant repercussions for interpreting the succession at Croyde and Saunton. An Oxygen Isotope Stage 7 age for the raised beach and associated sands allows the possibility that the overlying head deposits date from a variety of Pleistocene cold phases as Stephens (1974) originally suggested.

Gilbert (1996, in prep.) argues that previous interpretations of the raised beach-sandrock sequence (Figure 7.13) have been over-simplified. Instead, he proposes that five facies, widespread along the coast here, can be recognized in the deposits. These show the progression from: an initial marine transgression (facies 1 – the

well-cemented raised beach conglomerate described by numerous previous workers); to a foreshore environment dominated by nearshore intertidal activity (facies 2); to a deeper-water environment on the flank of a wave-/tide-dominated river-fed embayment (facies 3); to a backshore environment with palaeosol development (facies 4); finally to a backshore dune environment (facies 5). He presents OSL dates which place the lowest marine deposits in Stage 5e (Ipswichian) and the aeolian sediments in Stage 4 (Early Devensian; *c.* 70 ka BP). These data conflict with previously published amino-acid ratios (Andrews *et al.*, 1979; Davies, 1983; Bowen *et al.*, 1985) which would point to a Stage 7 age for Gilbert's facies 1 and 2 (Campbell and Scourse, 1996).

Conclusion

The exposures at Croyde and Saunton provide a key stratigraphic record and exhibit a number of features crucial to the reconstruction of Pleistocene history in South-West England. First, the shore platform between Croyde and Saunton is one of the finest examples anywhere in Britain; the extensive development at this site graphically demonstrates its compound nature and age, with at least three main platform surfaces (of unknown age) having been recognized.

Second, the site is also particularly important for the profusion of large erratics which are found overlying the shore platforms and below raised beach, blown sand and head deposits. Although sporadic examples (such as the Giant's Rock in Cornwall) occur elsewhere, those between Croyde and Saunton have proved especially important in reconstructions of earlier Pleistocene conditions in the region. Their lithological diversity, confined coastal location and limited altitudinal range, and their stratigraphic context have stimulated debate concerning their origins and mode(s) of emplacement. Whether these large erratics were introduced to this coast by a regional ice sheet or on ice floes (ice-rafted) has not been satisfactorily established; these examples will undoubtedly play an important role in the resolution of this debate.

Third, the raised beach sequence at the site shows a transition from fully marine conditions (raised beach conglomerate and marine sand) to terrestrial conditions (blown sand and fossil dunes). Amino-acid ratios reveal a complex age for the raised beach sequence (probably reflecting high sea-level conditions in both warm Oxygen Isotope Stages 7 and 5e). The overlying beds of blown sand are some of the best and most extensive examples in Britain, preserving a rare and detailed record of terrestrial conditions. Recent OSL dates have shown that most of the blown sand accumulated during the Early Devensian (Stage 4; *c*. 70 ka BP).

Finally, the stratigraphic importance of the Croyde–Saunton Coast cannot be seen in isolation. Although providing a crucial record of Pleistocene conditions including evidence for glacial, interglacial and periglacial climatic cycles, the true value of the sections lies in their relationship to others in the Barnstaple Bay area, particularly critical exposures at Brannam's Clay Pit, Fremington Quay and Westward Ho! These form a core of stratigraphic sites indispensable to any reconstruction of Pleistocene history in the region.

WESTWARD HO!
S. Campbell

Highlights

Westward Ho! is a classic site for studies of the Quaternary in South-West England. It provides both an important Pleistocene stratigraphic record and detailed evidence, in the form of submerged forest and associated beds, for the Holocene evolution of the Barnstaple Bay area.

Introduction

The submerged forest beds at Westward Ho! have featured in numerous studies (e.g. De la Beche, 1839; Ellis, 1866, 1867; Pengelly, 1867, 1868a; Hall, 1870, 1879a; Rogers, 1908; Worth, 1934; Rogers, 1946; Churchill, 1965; Churchill and Wymer, 1965; Stephens, 1970a, 1973, 1974; Jacobi, 1975, 1979; Kidson and Heyworth, 1977). The definitive account of the Holocene sequence, however, is that of Balaam *et al.* (1987) which integrates detailed archaeological and palaeoenvironmental evidence. The Pleistocene features, including a spectacular cobble raised beach, have been referred to by Ussher (1878), Dewey (1913), Rogers (1946), Everard *et al.* (1964), Stephens (1966a, 1970a, 1973), Kidson (1974), Kidson and Heyworth (1977) and Campbell *et al.* (in prep.).

Description

The Pleistocene sequence

Raised beach deposits were first described in the area by De la Beche (1839) and Sedgwick and Murchison (1840), who noted that they extended along much of the Appledore coastline and Taw Estuary. They are particularly well developed between SS 420291 and SS 422291. The Pleistocene sediments (mainly raised beach and head deposits) overlie a cliffed and elevated rock shore platform (at 8-9 m OD) cut in near-vertically bedded Carboniferous sediments, and with a surface lying approximately 6 m above the present tidal platform (*c*. 2-3 m OD) (Stephens, 1970a; Kidson and Heyworth, 1977). The old wave-cut notch at the back of the elevated platform lies at an estimated 12.2-13.7 m OD.

The most detailed description of the sequence was given by Stephens (1966a, 1970a) who recorded:

9. Soil
8. Sandy clay with stones, including flint and granite erratics
7. Angular head in sandy matrix
6. Head

(a) Pleistocene sequence

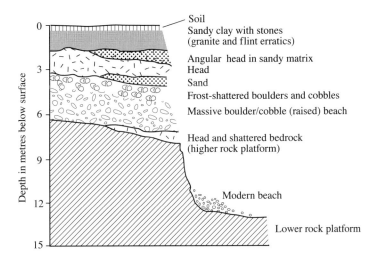

(b) Holocene sequence

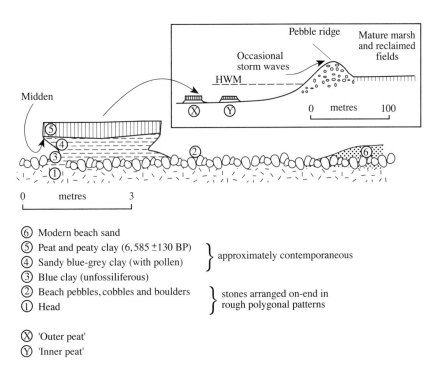

Figure 7.14 Quaternary landforms and deposits at Westward Ho! (Adapted from Stephens 1970a.)

5. Sand
4. Frost-shattered boulders and cobbles
3. Massive boulder/cobble beach
2. Head and shattered bedrock
1. Rock platform

The Pleistocene sequence averages 6–7 m in thickness and forms a terrace which slopes gently seawards and buries the fossil coastline. At the eastern end of the sections, the sediments comprise mostly head (beds 6–7) up to 4 m in thickness, but to the west, the raised beach deposits thicken to *c.* 2.5–3 m. The latter comprise mostly well-rounded boulders and cobbles of Carboniferous grit, with some clasts up to 0.3 m in diameter; apparently no erratic lithologies are present (Kidson and Heyworth, 1977).

Most workers have recognized a sequence of head deposits overlying the raised beach. Stephens (1970a) divided the head deposits into a number of beds, some with erratics (Figure 7.14). He also recognized a thin development of head in places beneath the raised beach cobbles. The head deposits are highly variable and include beds of fine shale head, blocky head, head with more rounded clasts and erratics, layers of sand and gravel, and silt and clay lenses. The sequence is capped in places by a sandy silt, possibly loess, from which microliths and flints have been recorded (Rogers, 1946).

The Holocene sequence

The submerged forest and associated beds in Barnstaple Bay were recognized over 300 years ago by Ridson (Rogers, 1946), who described a 9 m-long oak trunk embedded in them at Braunton Burrows. Those at Westward Ho! are justly the most famous and have been described by numerous workers (e.g. Hall, 1879a; Rogers, 1946; Churchill, 1965; Churchill and Wymer, 1965). Considerable problems exist, however, in relating the earlier descriptions to more recent surveys; the seaward exposures are now much more limited in extent due to coastal erosion, and the precise locations of earlier finds and descriptions are often extremely unclear. Two descriptions of the Holocene sequence are given here: Stephens' (1966a, 1970a) representation of the sequence is useful in showing the relationship of the main Holocene deposits both to the Pleistocene succession and to more recent landforms (Figure 7.14); the description given by Balaam *et al.* (1987) is by far the most comprehensive and up-to-date.

Stephens follows Rogers' (1946) account by illustrating two main exposures of Holocene sediment, the so-called 'inner' and 'outer' peats (Figure 7.14), and describes the following succession from the outer peat:

6. Modern beach sand
5. Peat and peaty clay
4. Sandy blue-grey clay (with pollen)
3. Blue clay (unfossiliferous)
2. Beach pebbles, cobbles and boulders
1. Head
   Intertidal rock platform

Stephens noted that clasts in the head and cobbles in the beach were frequently arranged with their long-axes vertical; in plan, some sorting of the deposits into polygonal patterns had also occurred. A kitchen midden found between beds 4 and 5 was noted by Rogers (1946) and was analysed in some detail by Churchill and Wymer (1965), who recorded worked flints, vertebrate remains (including hedgehog, fallow deer, red deer, pig and wild boar), molluscs, pollen and seeds. Numerous similar finds were made by earlier workers although their stratigraphic context is not always clear. A sample from the top of the peat (bed 5) was radiocarbon dated to 6585 ± 130 BP (Q–672) (Churchill and Wymer, 1965), and a date from the same bed of 4995 ± 105 BP was subsequently recorded by Kidson (1977).

Balaam *et al.* (1987) surveyed the site during 1983 and 1984, and confirmed the presence of the inner and outer peats. In addition, they mapped an extensive area of estuarine silt (Figure 7.15; Area 4) running north-east from the inner peat. The distribution of the main Holocene sediments is shown in Figure 7.15 and two cross-sections of the midden, which is found on the west edge of the outer peat, are shown in Figure 7.16. The full succession is not superposed in any single section and must be pieced together from the separate outliers.

Areas 2 and 3

Area 3 corresponds approximately with the outer peat described by earlier workers, but in fact comprises a fourfold sequence (maximum bed thicknesses in parentheses):

4. Upper clay (0.7 m)
3. Outer peat (0.8 m)
2. Mesolithic midden (*c.* 0.2 m)
1. Lower blue clay (1.1 m)

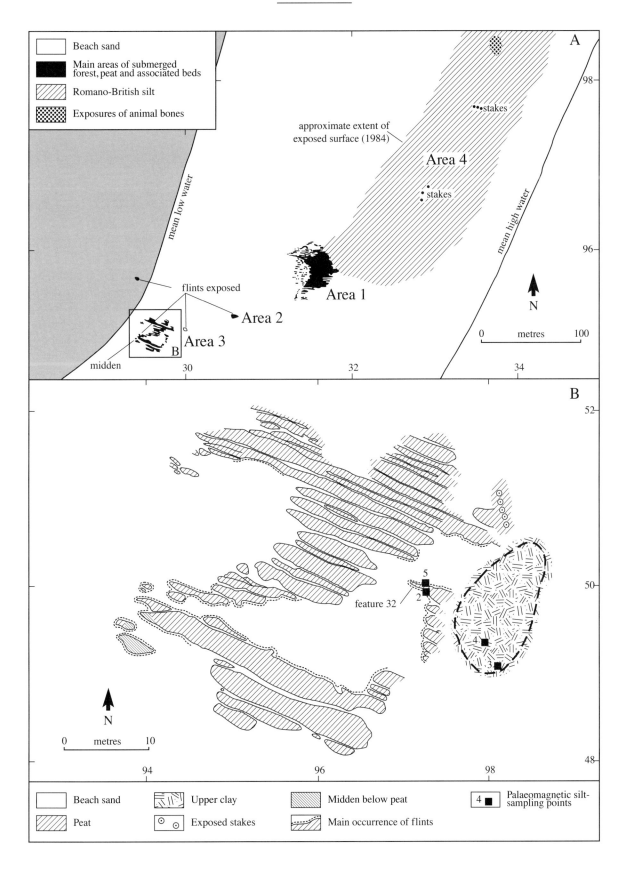

Figure 7.15 The distribution of Holocene deposits at Westward Ho! (Adapted from Balaam *et al.*, 1987.)

According to Balaam *et al.* (1987), the lower blue clay (bed 1) rests on 'drift', pebbles and rock – presumably the head, 'raised beach' deposits and wave-cut platform described by Stephens (1970a; Figure 7.14). It contains a small amount of organic matter, mainly roots and stems of monocotyledonous plants. Flint artefacts and charcoal fragments are evenly distributed throughout its upper layers (Figure 7.16) which lie at *c.* − 2.5 m OD, but there is no evidence of stratification. Charcoal from the upper layers of the deposit has yielded radiocarbon dates of 6250 ± 120 BP (HAR-6215), 6770 ± 120 BP (HAR-5644) and 8180 ± 150 BP (HAR-5643), while archaeomagnetic dating suggests an age of around 8400 to 7800 cal BP. The deposit contains no molluscs, forams, diatoms or ostracods (Balaam *et al.*, 1987).

The midden (bed 2) is restricted to Area 3 and survives only as an isolated remnant, just west of the outer peat (Figure 7.15). It occupies a gentle hollow in the surface of the clay and comprises fragmented shell material (mostly of mussel, *Mytilus* sp., and peppery furrow shell, *Scrobicularia plana* (da Costa)), mineral matter (mainly silt), humified organic matter and, occasionally, substantial roots, presumably the remnants of trees and shrubs which grew in the overlying peat and which rooted in a more stable substrate (Balaam *et al.*, 1987). The bed contains pollen, insect remains, charcoal, flint artefacts and bone and is much bioturbated (Figure 7.16). Charcoal from the midden has yielded radiocarbon dates of 6100 ± 200 BP (HAR-5632) and 6320 ± 90 BP (HAR-5645) (Balaam *et al.*, 1987).

The outer peat (bed 3) overlies both the midden and, elsewhere, the lower blue clay (Figures 7.15 and 7.16). It comprises humified monocotyledonous material with wood, occasional hazelnuts and leaves of common sallow. Substantial remains of trunks and stools of oak and willow survive locally on the peat, contradicting Rogers' (1946) assertion that the last of these remains had been washed out of the peat in 1935; their roots are found throughout the peat, which also contains fine laminations of both inorganic and organic material, abundant pollen and insect remains. The outer peat is only a few centimetres thick where it overlies the midden (bed 2), but thickens a few metres away where its upper surface is sealed by the upper clay (bed 4). Balaam *et al.* (1987) provided six radiocarbon dates from the outer peat in Areas 2 and 3 (Figure 7.15; Table 7.1).

These dates complement those provided earlier from materials in the same bed by Churchill and Wymer (1965) (6585 ± 130 BP; Q-672), Jacobi

TABLE 7.1 Radiocarbon dates from the outer peat

Material	Lab. Ref.	Result (years BP)	Height (metres OD)
wood	HAR-5642	4840 ± 70	−1.0
peat	HAR-6363	5190 ± 80	—
peat	HAR-5640	5200 ± 120	−2.1
wood	HAR-5631	6100 ± 100	−2.0
wood	HAR-5630	5630 ± 80	−2.0
peat	HAR-5641	5740 ± 100	−2.2

(1975) (6680 ± 120 BP; Q-1249) and Welin *et al.* (1971) (5004 ± 105 BP; IGS-42).

The upper clay (bed 4) is present only within an isolated hollow in Area 3 (Figure 7.15) where it seals part of the outer peat (Balaam *et al.*, 1987). It is generally similar to the lower blue clay (bed 1), but contains more sand and less silt; its organic content is low. The material was not analysed in detail by Balaam *et al.* (1987) and no faunal or botanical data are available.

Area 1

The Holocene sequence in this area can be summarized as follows:

7. grey silt and silt-filled channels
6. inner peat (0.4 m)
5. silt

The inner peat (bed 6) is overlain in places by grey silt (bed 7) which is extensively exposed in Areas 1 and 4, and which lies roughly at OD (Figure 7.15). These silts are incised by a complex network of silt-filled channels. The deposits have yielded a variety of wooden structures (stakes and brushwood) and are locally rich in animal bone, marine molluscs and pollen. Bone from one of the channel-fills has yielded a radiocarbon date of 1560 ± 80 BP (HAR-6513), and one of the wooden stakes (embedded in silt but overlain by channel-fill silt) is dated to 1600 ± 80 BP (HAR-6440) (Balaam *et al.*, 1987).

Interpretation

The Pleistocene sequence

The Pleistocene sequence at Westward Ho! was described by numerous early workers (e.g. De la

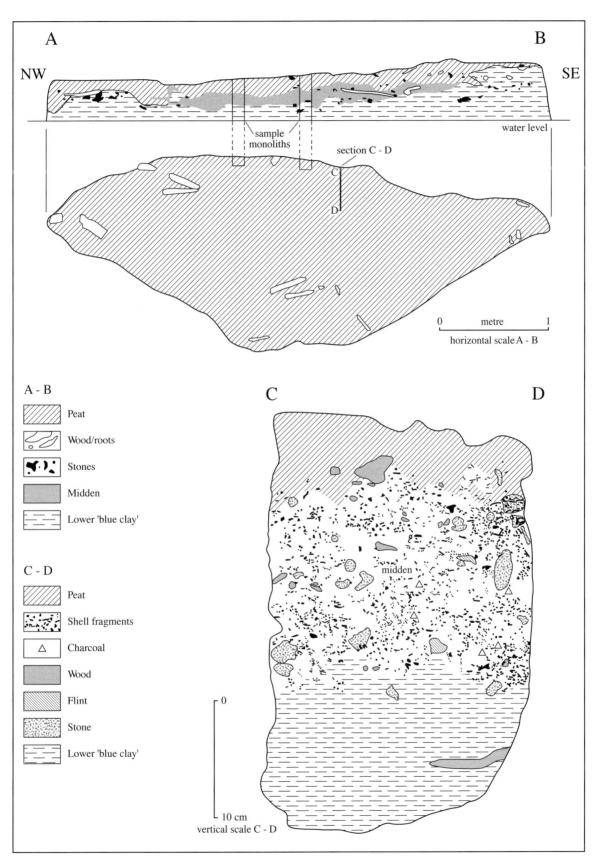

Figure 7.16 Plan and section of the midden 'island' at Westward Ho! (Adapted from Balaam *et al.*, 1987.)

Figure 7.17 The Pleistocene sequence at Westward Ho!, showing the higher marine-cut platform overlain by raised 'cobble' beach and head deposits, with a lower platform extending into the distance. (Photo: S. Campbell.)

Beche, 1839; Sedgwick and Murchison, 1840; Pengelly, 1867; Ussher, 1878; Dewey, 1913; Rogers, 1946) who established that it comprised a raised beach deposit overlain by head (Figure 7.17). Mitchell's (1960) study of Pleistocene sequences throughout the Irish Sea Basin and in South-West England provided a considerable stimulus for further work. A detailed interpretation of the sequence at Westward Ho! was provided by Stephens (1966a, 1970a, 1974) who upheld Mitchell's arguments and suggested the following sequence of Pleistocene events.

The elevated platform was believed to have been planed during the Cromerian (cf. Mitchell, 1960), and its surface to have been shattered during an ensuing cold phase (?Anglian) when large erratics were believed to have been ice-rafted into position around the coast of South-West England. During this stage, cold-climate head (bed 2) accumulated on the platform. The arrangement of the raised beach sediments (beds 3 and 4), deposited during the Hoxnian, suggested a period of sea level higher than at present. This was followed by climatic conditions sufficiently severe to disturb and crack the upper layers of the beach cobbles (bed 4). During this proposed Wolstonian (Saalian) event, head (bed 6) and blown sand (bed 5) accumulated on the raised beach sediments. At this time, an ice sheet was believed to have impinged on the north Devon coast, depositing tills and associated sediments in the Fremington area. The lower head (bed 6) was weathered during warmer conditions in the Ipswichian. During cold conditions in the Devensian, when glacier ice did not reach the north Devon coast, an upper head (beds 7 and 8) containing erratics reworked from glacial deposits equivalent to the Fremington Clay, was deposited under periglacial conditions. Such an interpretation was founded on the belief that the raised beach deposits found widely around the coastlands of the Irish Sea Basin and South-West England were Hoxnian in age. This chronostratigraphic interpretation was reinforced by the perceived relationship of deposits at Fremington Quay and at Brannam's Clay Pit. The proposed raised beach (Hoxnian) at Fremington Quay was correlated on the basis of altitude and sedimentary characteristics with a gravel sporadically exposed beneath till and lacustrine sediments at Brannam's Clay Pit; the latter were therefore believed to be of Wolstonian age (e.g. Mitchell, 1960; Stephens, 1966a, 1966b, 1970a).

Other workers have argued against this correlation, suggesting that nowhere in the region is there evidence for a raised beach deposit having been overridden by ice (e.g. Zeuner, 1959; Kidson, 1971, 1974; Edmonds, 1972; Kidson and Wood, 1974; Kidson and Heyworth, 1977). These workers have therefore assigned raised beach deposits in the region, such as those at Westward Ho!, to the Ipswichian, and have considered the overlying head deposits to have accumulated during the Devensian. In this scheme, the giant erratics around the coast and the Fremington Clay were assigned to the same, Wolstonian (Saalian), glaciation – although such an event has now been thrown into considerable doubt (Chapter 2; Anglian and Saalian events).

In addition to the present-day shore platform and the elevated platform at 8–9 m OD, two further platform remnants at Westward Ho!, at − 1.5 m and + 5.0 m OD, have been recognized (Kidson, 1977; Kidson and Heyworth, 1977), the latter being represented by isolated stacks. It is therefore likely that there was more than one phase of platform formation (Kidson and Heyworth, 1977), although Everard *et al.* (1964) have argued that the detailed form of the platforms is, above all, controlled by structure and lithology.

Stephens (1970a, 1974) alluded to the possibility that beach cobbles (Figure 7.14; bed 2) beneath the Holocene submerged forest, peat and clay sequence at the site, were Ipswichian in age, and cited their lower altitudinal position with regard to the cliffed raised cobble beach deposits farther west, as support for such a dating. The polygonal sorting and vertical-stone structures observed in this bed were believed to have formed under periglacial conditions in the Devensian Stage. The underlying head (Figure 7.14; bed 1) was correlated tentatively with bed 6 (Pleistocene site description) in the cliff sections, and was believed to be of Saalian age (Stephens, 1970a). Wood (1970), however, observed that locally the beach cobbles (bed 2) are intermixed with the blue clay (bed 3), implying a Devensian late-glacial age for both.

The Holocene sequence

The Holocene sediments have attracted much attention. Hall (1879a) recorded 70–80 tree stumps in the peat. He demonstrated that some of their roots penetrated up to 1.2 m into the deposits, thereby establishing that the forest had grown *in situ*. Bate (1866) described the stratigraphy of the site and ascertained the relationship between the

deposits and those at Braunton Burrows. Ellis (1866, 1867) recorded bones, teeth, flint flakes and cores, marine shell fragments (largely *Ostrea edulis* Linné and *Cerastoderma edule* (Linné)) from the beds, and although he established that the deposit of oyster shells (up to 0.6 m thick) was mixed up with the bones, their stratigraphic context was not given. It seems likely, however, that such a bed was probably part of the once more extensive kitchen midden later described by Churchill and Wymer (1965) and Balaam *et al.* (1987).

Pengelly (1868a) suggested that the relationship between the forest beds and the large pebble/cobble ridge on their landward side (Figure 7.14), was important for interpreting the relative levels of the land and sea in the region. He argued that the cobble beach indicated an upheaval of the land surface subsequent to a period of subsidence, during which the coastal forest had been swamped by the sea.

Rogers' (1908) observations on the forest beds added considerable data on the flora and fauna but, however, added little to the interpretation of the sequence; the stratigraphic context of most of the finds was not recorded although the mammalian remains were found *in situ* in a tough blue clay which contained abundant remains of the snail *Hydrobia ulvae* (Pennant) (Figure 7.15; possibly bed 7 of Balaam *et al.*, 1987).

A detailed account of the stratigraphy was provided by E.H. Rogers (1946) who excavated at the site, and provided additional floral, faunal and archaeological finds. He noted that the seaward outlier of the peat (outer peat) overlay a shelly calcareous mud containing numerous split bones, teeth, flint flakes and cores. The peat also yielded a microlith, flakes and a core.

Churchill and Wymer (1965) provided a detailed account of the kitchen midden establishing its 'Mesolithic' character, and arguing that it had been formed in the zone between neap and spring high tides. Seeds and fruits extracted from the overlying peat (presumably the outer peat of Balaam *et al.* (1987)) showed a plant succession from a dry fen with *Quercus* and a ground flora of *Ajuga reptans*, *Carex*, *Ranunculus* and *Rubus fruticosus*, to an even drier fen with *Corylus avellana*, *Cretaegus monogyna*, *Populus*, *Prunus spinosa*, *Thelycrania*, *Sanguinea* and *Solanum dulcamara*. Such an assemblage was considered characteristic of present-day fen woods, with no traces of salt-marsh plants or deposits as recorded by Rogers (1946). An early Atlantic age was suggested for the peat on the basis of a radiocarbon date of 6585

± 130 BP (Q–672) derived from a sample near the top of the peat bed. The presence of *Plantago lanceolata* pollen was taken to be an indicator of progressive forest clearance by Neolithic humans (Churchill and Wymer, 1965), although flint artefacts from the midden, the peat and the nearby clay surface are Mesolithic in character.

Churchill and Wymer (1965) and Churchill (1965) argued, on the basis of estimates of mean sea level derived from radiocarbon-dated submerged forest and associated marine beds around the coast of Britain, that there has been no measureable tectonic displacement at Westward Ho! (and much of South-West England) since *c.* 6500 BP. This contrasted, they suggested, with sites to the north-west in the Irish Sea Basin (such as Ynyslas and Borth), which revealed an upward vertical displacement of *c.* 3 m, since that time, and sites in eastern England and The Netherlands which they argued had fallen by up to 6 m since then. This has since been strongly disputed by Kidson (1977) and Heyworth *et al.* (1985), among others, who have shown there to have been no significant differential movement of the west Wales and South-West England land surfaces during the Holocene.

Balaam *et al.* (1987) concluded, despite a lack of fossils, that the lower blue clay (their bed 1) found in Area 3 (Figure 7.15) was an estuarine deposit. They argued that the artefacts and charcoal found in its upper layers were not necessarily the same age as the clay. The age of the latter was enigmatic, and it could not be presumed that it immediately pre-dates the overlying midden material. However, they favoured an age for the clay of at least *c.* 8000 BP (based on archaeomagnetic dating) and believed that a lack of evidence for weathering and soil development indicated that the clay and midden materials were not greatly separated in time (Balaam *et al.*, 1987).

These workers also confirmed that bed 2 was a Mesolithic shell midden, pollen, molluscan and insect remains providing significant evidence of environmental change during its deposition. They argued that the midden had started to accumulate amidst a fairly dense fen carr closely surrounded by mixed oak woodland with a few herbs, of which grasses and sedges were dominant. The mollusc and insect remains are taken to indicate that the midden accumulated as domestic rubbish in a stagnant (but not brackish) pool: the insect remains show strong evidence for decaying wood and other vegetable matter.

Pollen from the upper layers of the midden showed evidence for a change in local vegetation with higher levels of willow, birch and ivy. There was no evidence, however, for the modification of the local vegetation by humans. An environment of relatively closed, damp woodland was further suggested by the remains of *Anguis fragilis* Linné (slow worm), the scales and bones of *Rana* sp. (frog) and bones of *Clethrionomys glareolus* (Schreber) (bank vole). Insect remains indicate that open country and sand dunes lay beyond the woodland and midden at no great distance (Balaam *et al.*, 1987), contradicting Churchill's and Wymer's (1965) and Rogers' (1946) assertion that the midden lay at or near the strandline.

The human origins of the midden were confirmed, however, by the accumulation of marine molluscs (including *Scrobicularia*), charcoal and flint artefacts. The bones of red deer, roe deer and fish were also probably introduced by humans. Balaam *et al.* (1987) noted that 1074 flints had been recovered from Area 3 indicating that Mesolithic people had knapped local flint (?beach cobbles) to make microliths for hunting and various blades and scrapers for more domestic activities.

Radiocarbon dates show that the overlying peat sequence (outer peat; bed 3) accumulated during a roughly 500- to 800-year timespan. Both radiocarbon and pollen evidence confirm its Atlantic (Godwin Zone VIIa) age (Balaam *et al.*, 1987). The pollen and plant evidence indicates a willow-dominated fen carr amidst a vegetation of deciduous woodland, with oak, elm, ash, hazel and willow strongly represented. Both plant macrofossil and insect evidence confirms the presence of deciduous woodland with only localized waterlogging and small streams; laminae within the peat probably attest to minor flooding episodes. According to Balaam *et al.*, the outer peat shows surprisingly little evidence for human activity. However, although hazelnuts found in the bed appear to have accumulated naturally, pollen in the upper layers of the peat show a decline of elm and ash and the appearance of goosefoot and ribwort plantain. Such changes may be associated with human activity within the catchment, reducing evapotranspiration and increasing surface flows (Balaam *et al.*, 1987). The reduction in the levels of ash and elm could also be associated with increased wetness (climatic) and waterlogging caused by a rising groundwater table which would have heralded the marine transgression responsible for the overlying upper clay (bed 4). The evidence afforded by bones of Bovid and Cervid animals, fish bones and artefacts found in the outer peat is also equivocal: these may have been derived from the underlying midden by bioturbation (Balaam *et al.*, 1987). The overlying clay (bed 4) denotes a dramatic change in conditions and demonstrates marine/estuarine inundation of the outer peat by *c.* 5200 BP (Balaam *et al.*, 1987).

Importantly, Balaam *et al.* demonstrated clearly that the outer and inner peats at Westward Ho! are of different ages. (It appears likely that some of the earlier accounts have referred to material from both the Mesolithic midden and later Romano-British deposits without apparent differentiation.) Archaeomagnetic measurements from silt (bed 5) underlying the inner peat (bed 6), show the latter to date from the Romano-British period; the pollen spectra also corroborate a much later date for the inner peat (Balaam *et al.*, 1987). The latter appears to have accumulated in a more open environment than the outer peat: similar floral elements are present but in different abundances. The vegetation of the inner peat is of a sedge/grass fen community, not the *Salicetum* of the outer peat. Pollen of trees and shrubs are also substantially fewer. Although there is no direct evidence of human activity, the upper levels of the inner peat show significant increases in the amount of *Plantago lanceolata* and cereal pollen (Balaam *et al.*, 1987).

The overlying silts (bed 7) are discontinuous and difficult to interpret. Molluscan assemblages, which include *S. plana* and *H. ulvae*, however, suggest an estuarine origin for the deposits and this is supported by the pollen evidence which also indicates a saltmarsh and seashore environment; radiocarbon evidence points to a Romano-British age. The presence of cattle, sheep/goat and dog bones in these deposits, and the stake and brushwood structures, provides clear evidence of human activity. However, it remains unclear if the stakes were driven into the silts or whether the silts accumulated around them. Balaam *et al.* suggested, tentatively, that some of the stakes may have been used for mooring while others may have formed some part of a fish trap. No other artefacts are recorded by Balaam *et al.* from these deposits, although numerous flints have been recovered *ad hoc* from this general area over the years.

Conclusion

The interpretation of the classic raised beach and head sequence at Westward Ho! has proved controversial, with claims being made for either Ipswichian or Hoxnian ages for the raised beach deposits. Unfortunately, the latter are unfossiliferous and their relative age cannot therefore be resolved by amino-acid geochronology.

The site also provides one of the finest examples of a compound shore platform in western Britain. The extensive development of two principal platforms which lie at significantly different heights, and possibly two more which can be identified from more isolated remnants (Kidson and Heyworth, 1977), clearly disproves the earlier concept of a single platform planed during the Cromerian. It is clear from this evidence that there was more than one phase of platform formation, the ages of which are unknown, and that lithology and structure may have been controlling factors. The beach deposits overlying the highest of these platforms are one of the best examples of a raised cobble beach in Britain. The age and stratigraphic relationships of beach material (possibly soliflucted) underlying the submerged forest, are also a subject of debate and enhance the scientific value of the site.

Westward Ho! also provides perhaps the most famous and best studied of South-West England's coastal Holocene deposits, interest in the site having been stimulated by its evidence for Mesolithic occupation. The site reveals an important record of changing terrestrial and coastal conditions in the Holocene, demonstrating clear evidence for the transition from the very low sea-level conditions of the Late Devensian, through the initially rapid rise of the Holocene sea and the swamping of the coastal forest that had developed by about 6000 BP, to a late stage in which the rate of sea-level rise slowed and the present-day coastal configuration, including the landwards migration of the massive cobble ridge, was established. Its radiocarbon-dated peat and forest bed is important for establishing the pattern of relative land (isostatic) and sea (eustatic) movements around the coast of Britain.

THE VALLEY OF ROCKS
S. Campbell

Highlights

The Valley of Rocks is one of Devon's most spectacular and controversial landforms. Some authorities maintain that it was cut by glacial meltwater, others that it was formed by marine capture of a formerly more extensive East Lyn River.

(a)

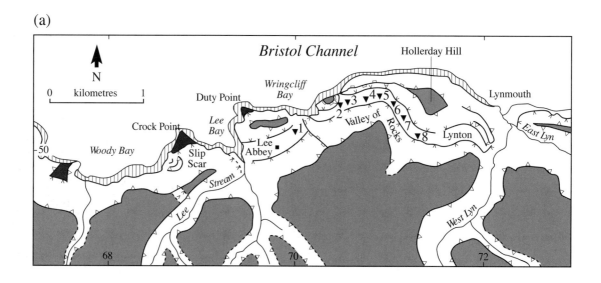

(b)

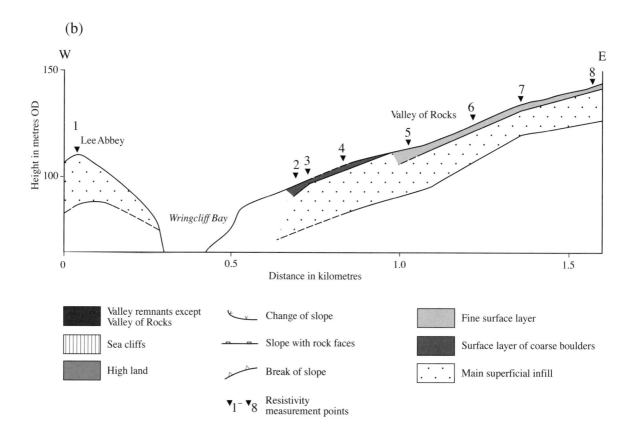

Figure 7.18 (a) Landforms and Pleistocene deposits between Lynmouth and Woody Bay. (b) Profile of Pleistocene deposits within the Valley of Rocks and at Lee Abbey. (Adapted from Dalzell and Durrance, 1980.)

The Valley of Rocks

Figure 7.19 The Valley of Rocks, looking east from Wringcliff Bay. (Photo: S. Campbell.)

Introduction

The Valley of Rocks is noted for a large dry valley and a series of periglacial features. The origin of the valley is much disputed, but has a major bearing on coastal and drainage evolution in north Devon. The site has been referred to by E. Arber (1911), Steers (1946), Simpson (1953), Mottershead (1964, 1967, 1977c), Stephens (1966a, 1966b, 1970a, 1974, 1990), Gregory (1969), Pearce (1972, 1982), M. Arber (1974) and Cullingford (1982). A detailed description and reinterpretation of the landforms was given by Dalzell and Durrance (1980).

Description

The Valley of Rocks or 'the Danes' (SS 700495) extends some 2 km from Wringcliff Bay in the west to the western side of Lynton, and lies roughly parallel with the east-west-trending north Devon coastline (Figures 7.18 and 7.19). It is cut in the sandstones and slates of the Devonian Lynton Beds. On its southern margin it is backed by the high ground of Exmoor which locally rises from *c.* 260 m to 318 m OD. To the north, the valley is separated from the sea by the mass of Hollerday Hill and a narrow westward-running ridge, precipitous on its seaward side, capped by tor-like buttresses and mantled with scree. Although the Valley of Rocks terminates at Wringcliff Bay, its perceived course continues west to Lee Bay through a col in which Lee Abbey is situated. The principal 'tors' crop out on the valley's northern margin and are of both the crestal and valley-side types. These include the castellated turrets of rock known as Castle Rock, Rugged Jack and Chimney Rock. Valley-side tors also crop out on the south-facing slopes of the valley (e.g. the Devil's Cheesewring).

Considerable thicknesses of head are exposed in the coastal cliffs. Mottershead (1967, 1977c) recorded up to 25 m of such deposits comprising angular slate and sandstone fragments in a poorly sorted matrix of fines. At Lee Bay, a thick sequence of Pleistocene sediments is exposed in the coastal cliff at the head of the bay (Dalzell and Durrance, 1980). This sequence overlies a rock platform at *c.* 7 m OD and comprises (maximum bed thicknesses in parentheses):

4. Coarse angular rock fragments with some sub-rounded pebbles (*c.* 18 m) (head)

(a)   (b)

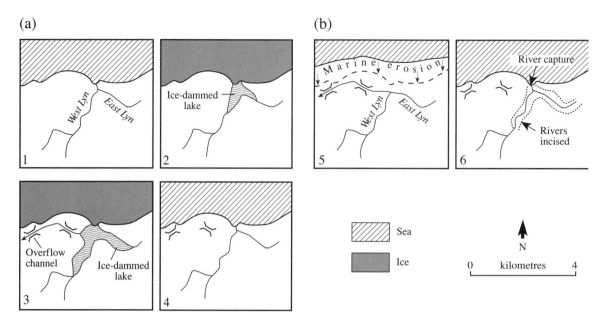

Figure 7.20 The evolution of the Valley of Rocks by: (a) Pre-Devensian glacial meltwaters; (b) Marine erosion and river capture. (Adapted from Mottershead, 1967, 1977c.)

3. Subrounded pebbles (*c.* 3.5 m)
2. Mixture of well-rounded and subrounded pebbles with sand layers (*c.* 3.0 m)
1. Well-rounded pebbles in a sandy matrix (*c.* 3.5 m)

Beds 1–3 were described by Dalzell and Durrance (1980) as waterlain, comprising a mixture of fluvial and marine materials. They also showed that the Valley of Rocks and the smaller valley near Lee Bay (the Lee Abbey gap) have a substantial infill of Pleistocene scree and solifluction deposits.

Interpretation

Following Balchin's (1952) work on the erosion surfaces of Exmoor, Simpson (1953) put forward the view that the dry valley remnants could be explained by marine erosion and capture of a formerly more extensive River Lyn, which then flowed west. He concluded that the East Lyn river originally flowed from Lynmouth through the Valley of Rocks, the Lee Abbey gap, Crock Point and Martinhoe Manor to Heddon's Mouth. Such an interpretation, based entirely on the present form and location of the remnant dry valley floors, was followed by Mottershead (1964, 1967, 1977c) (Figure 7.20) who stressed the similarity between the channel form and direction of the present East

Lyn river east of Lynmouth, and the Valley of Rocks. Thus, he suggested that the Valley of Rocks represents the former course of the East Lyn before it cut down to its present level, and before its outlet at Lynmouth existed. As a result of the capture of the East Lyn and the abandonment of the Valley of Rocks, the course of the East Lyn to the sea was dramatically shortened. Initially, it probably reached the sea via large waterfalls, but with continued erosion upstream, the course became graded and more subdued to its present, but still sharply incised, form (Mottershead, 1977c).

In marked contrast, Stephens (1966a, 1966b) suggested that the valley remnants had formed as ice-marginal drainage channels cut in Wolstonian times (Saalian Stage) when glacier ice was believed to have reached Barnstaple Bay – and consequently may have impinged upon the north Devon coast for substantial portions of its length. Stephens' model implies that the pre-Wolstonian drainage pattern must have been substantially similar to that of today. As ice advanced and fringed the Exmoor coast, a lake formed in the Lyn Valley behind present-day Lynmouth (Figure 7.20). The Valley of Rocks was believed to have been cut by water spilling westwards from this impounded lake. As the ice wasted, the rivers reverted to their original courses and the Valley of Rocks was left dry. A meltwater origin was also considered plausible by Gregory (1969), but Mottershead (1967, 1977c) reviewed this mechanism and although he, and

Figure 7.21 Tor-like buttresses and precipitous rock slopes on the northern margin of the Valley of Rocks, looking west. (Photo: S. Campbell.)

subsequent workers, have found no positive evidence against it, he argued that the concept of marine capture was probably more straightforward. In support of his model, Stephens noted that erratics and striated pebbles had been found in the area (including an example of the renowned Ailsa Craig microgranite), and that a number of often abrupt and now dry channels in the area had probably been formed in the same manner.

Dalzell and Durrance (1980) used electrical resistivity techniques to establish the origin of the dry valley system at the Valley of Rocks and Lee Bay (Figure 7.18). Their results showed that considerable, but variable, depths of Pleistocene deposits capped the Devonian strata. For the most part, the infill was interpreted as solifluction material derived from the valley sides, distinguished by a high proportion of fine sediment. A coarse surface layer of boulders and blocks, averaging between 2–5 m thickness, was discerned to the west end of the Valley of Rocks (Figure 7.18). At the eastern end a similar layer, but of finer material, was also noted. The thickness of superficial material in the Valley of Rocks increases west from 27 m at the highest point in the valley floor to 35 m at its lowest point. The pattern at the west end, however, is complicated by mass movement caused by marine erosion; here the rock-head profile is obscured (Dalzell and Durrance, 1980).

These results led Dalzell and Durrance (1980) to suggest that the rock-floor profile of the Valley of Rocks shows a gradation to a level lower than that of the Lee Abbey gap. Instead, the Valley of Rocks grades more readily to the heights and erosional remnants at Duty Point and Crock Point (Figure 7.18), which were therefore regarded as fluvially rather than marine-eroded surfaces. At the same time, the present Lee stream, with its gently profiled upper section, appears to grade more naturally with the rock-floor profile of the Lee Abbey gap (Figure 7.18).

Thus, Dalzell and Durrance rejected Simpson's (1953) and Mottershead's (1964) argument that the East Lyn had once flowed westwards through the Lee Abbey gap, and proposed instead that it had flowed in a course through the present Wringcliff Bay around Duty Point to Crock Point. The Lee stream would then have flowed east through the Lee Abbey gap to join the East Lyn as a tributary. A first stage of river capture by marine erosion in the Lynmouth area left the Valley of Rocks dry, save for the Lee tributary. Subsequent marine erosion, which formed Lee Bay, then captured the Lee to its present course. Since the platform of Devonian rocks in Lee Bay is overlain by raised beach and fluvial deposits (of presumed Ipswichian age) and then by periglacial head, Dalzell and Durrance argued that the coastal dissection had happened, at latest, in Ipswichian times; aggradation of head deposits in both the Valley of Rocks and the Lee Abbey gap could not have occurred if these valleys had been fluvially active throughout the Devensian.

Conclusion

The Valley of Rocks is a spectacular landform, the age and origin of which have been much disputed. The site has nonetheless played a focal role in the development of ideas concerning coastal and drainage evolution in north Devon. Two main theories have been put forward to account for the dry valley system. One explanation is that the Valley of Rocks was formed by marine capture of a formerly more extensive East Lyn River; another is that the feature is a marginal glacial drainage channel cut as water overspilled from an ice-impounded lake. Recent work graphically shows the problems of interpreting landforms such as this from their present surface morphology. The Valley of Rocks, and also that at Lee Abbey, are in fact underlain by considerable thicknesses of Pleistocene solifluction and head deposits which mask the true profiles and gradients of the rock floors. The application of electrical resistivity techniques shows that the Valley of Rocks and its perceived extension into Lee Bay (the Lee Abbey gap) could in fact be the result of a more complex sequence of marine erosion and river captures, with the Lee Abbey gap being the abandoned channel of a tributary of the East Lyn.

No detailed evidence to reject the glacial drainage hypothesis has, however, been put forward. Nonetheless, in favour of the river capture theory, M. Arber (1974) has argued that many streams flowing northward off Exmoor fall to the sea via waterfalls. She has cited this as evidence for rapid coastal retreat in the Pleistocene, the rivers still not having adjusted significantly to the most recent change in base level (but see Chapter 7; Introduction).

The conservation value of this site is enhanced by the well-developed tor-like buttresses and scree slopes on the margins of the dry valley system (Figure 7.21), and by the head and solifluction deposits which infill the valleys. The head deposits are well exposed in the sections at Lee Bay, and their association with raised beach deposits of proposed Ipswichian age there provides rare stratigraphic evidence for the relative dating of such landforms. Much of the head which infills the dry valleys is believed to have accumulated under periglacial conditions in the Devensian, when glacier ice is not thought to have reached this part of the Peninsula. Although the tor-like features have not yet been studied in detail, they are believed to have been significantly modified at this time, although in common with granitic tors, it is likely that they evolved in response to processes operating over more protracted timescales.

DONIFORD
S. Campbell

Highlights

Doniford displays the finest sections through Quaternary periglacial river and associated mass movement deposits in South-West England. Its sequence of loams, sands and gravels has yielded artefacts and mammal bones, and exhibits well-developed periglacial structures.

Introduction

This site shows Quaternary sediments, including fluvial deposits and head, formed in a periglacial environment. The deposits were first described by Ussher (1908) and mapped by Thomas (1940). Accounts of archaeological and mammalian remains from the site were given by Wedlake (1950) and Wedlake and Wedlake (1963). A detailed stratigraphic interpretation was provided by Gilbertson and Mottershead (1975) and the site has also been referred to by Norman (1975, 1977), Kidson (1977), Mottershead (1977a), Edmonds and Williams (1985) and Campbell *et al.* (in prep.).

Description

Quaternary sediments crop out more or less continuously for 2 km along the north Somerset coast. They are particularly well exposed between Helwell Bay (ST 078434) and the Swill (ST 091433), and occupy a valley which runs eastwards from Watchet Station (ST 072433), through Helwell Bay to the Swill (Figure 7.22). To the east, gravels are exposed continuously on the foreshore as far as grid line 100 and, beyond there, discontinuously (Mottershead, 1977a).

In Helwell Bay, the gravels rest on a platform of Liassic shale and limestone beds (Figure 7.23). In the western part of the bay, the rock surface reaches 10–12 m OD, and the overlying Quaternary deposits are only about 2 m in thickness. This rock platform falls in height eastwards to *c.* 5 m near the concrete jetties (ST 082431) and to around 3 m OD at the Swill where the junction between the platform and gravels is obscured by beach and recently tipped materials. Here, the entire cliffline was formerly cut in Quaternary sediments some 5 m in thickness (Gilbertson and Mottershead, 1975). *In situ* Quaternary sediments are also present along the now overgrown western bank of the Swill (Figure 7.24). The Doniford exposures are constantly affected by marine erosion, and several stretches of the coastline have been obscured in recent years by coast protection works. Although the present sections are very different to those decribed by Gilbertson and Mottershead (1975), the same basic sequence can be seen (maximum bed thicknesses in parentheses):

6. Brown loam - generally structureless with occasional pebbles and sometimes thinly bedded (0.6 m)
5. Red clay-silt (red loam) - with marked prismatic fracture and frequent angular pebbles of Liassic material (0.6 m)
4. Buff silty loam (buff loam) - with less well-defined prismatic structure and with layers and lenses containing angular stones (0.6 m)
3. Red sandy silt - thinly laminated with fine gravel throughout (0.3 m)
2. Cobbly gravels - partially rounded to angular cobbles and boulders, often split, in a matrix of platy pebbles and fine silt; generally poorly bedded and sorted (2.9 m)
1. Rock platform (Lias)

Although the stratigraphy is complex, several more or less continuous beds can be traced along the coast, the cobbly gravels (bed 2) being the most distinctive and extensive. While they usually show little bedding, some stones lie with their long-axes in the horizontal plane (Gilbertson and Mottershead, 1975). In addition, lenses of well-sorted, cross-stratified sand and gravel occur in the otherwise massive beds; near the Swill, particularly in the river sections, channels up to 8 m across contain cross-stratified deposits (Gilbertson and Mottershead, 1975; Figure 7.24).

These coarse gravels (2–60 mm), with common cobbles (60–200 mm) and occasionally boulders (> 200 mm), comprise mainly purple and green slates, sandstones and cherts derived from Devonian strata in north-west Somerset (Gilbertson and Mottershead, 1975). Blocks of Lias clay have also been recorded from bed 2 (one measuring 2 m × 0.3 m). Most clasts are subangular to subrounded, although fresh fractures on otherwise rounded cobbles suggest post-depositional breakage. Clast long-axis measurements show a dominant orientation in the range between *c.* 320–360°, but there is no clear-cut pattern of clast orientation in relation to the directions of local slopes, and clast dip values appear random (Gilbertson and Mottershead, 1975). Many clasts, however, give the impression of lying with their long-axes vertical, and the lower *c.* 2 m of the bed is sometimes affected by involutions. At the west end of Helwell Bay, large involutions (1–2 m in wavelength and amplitude) affect the surface of the Lias platform and the cobbly gravels above, and also disturb the overlying loams (Gilbertson and Mottershead, 1975). At one locality, a 'raft' of Lias clay has been suspended in a large involution.

The cobbly gravels (bed 2) have yielded the remains of tusks and molars of *M. primigenius* (woolly mammoth) (Wedlake, 1950). Over the last 60 years or so, this bed has also yielded a substantial number of Palaeolithic artefacts, collected mainly between high and low water marks from the surface of the gravels exposed to the east of the Swill (Wedlake, 1950). These include over 20 hand-axes of Acheulian Culture made mostly from Greensand chert, and some large flakes. All the implements are heavily rolled. More recently, Norman (1975, 1977) discovered flint artefacts *in situ* in gravel (bed 2) exposed in cliff sections near

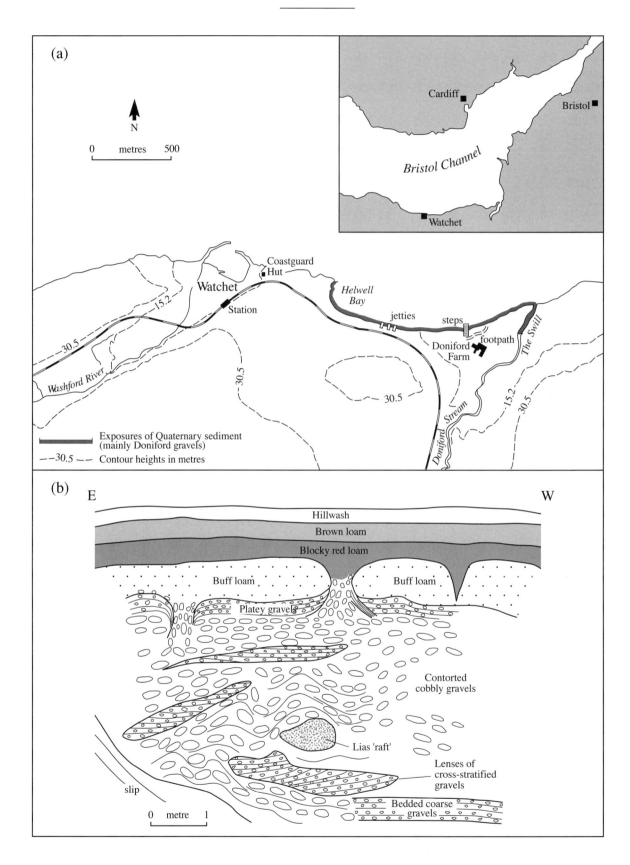

Figure 7.22 (a) The location of the Doniford gravels. (b) Typical section through the Doniford gravels west of the footpath. (Adapted from Gilbertson and Mottershead, 1975.)

Figure 7.23 The Doniford gravels overlying Liassic bedrock at the eastern end of Helwell Bay. (Photo: S. Campbell.)

the Swill. The largest was an incomplete blade (10.5 cm in length) found in a lens of fine, well-sorted gravel. The smallest was another segment of a large blade found at the junction of the same lens with the more poorly sorted overlying gravels.

The succeeding beds (3–6) comprise sandy silt and loam and can be distinguished by colour, structure and stone content. These beds are discontinuous but are best seen to the west of the old lime kiln (ST 087431). The red sandy silts (bed 3) are often thinly laminated, and contain fine gravel throughout, including many strongly weathered clasts of Jurassic sediment (Gilbertson and Mottershead, 1975). The overlying buff silty loam

(bed 4) is highly calcareous (nearly 32% calcium carbonate) and contains scattered lenses and seams of angular chips and fragments of Lias limestone. The deposit has a prismatic structure, although not so pronounced as the succeeding red loam (bed 5). The latter contains up to 27% clay in a matrix which supports sand and gravel, apparently derived from Devonian strata, and which itself appears to have been derived from more local Triassic rocks. Clasts in this bed show a marked preferred orientation downslope; mean values change progressively as the beds are traced around the variable local slopes. The orientation values are stronger in this bed than in the cobbly gravels (bed 2) (Gilbertson and Mottershead, 1975).

Finally, a brown loam (bed 6) lies with marked unconformity on the lower deposits. It incorporates fragments of weathered local limestones, and has yielded profuse Mesolithic artefacts. These include microliths, scrapers, burins and a crude tranchet axe (Wedlake, 1950; Norman, 1975, 1977).

Beds 3–6 are frequently affected by involutions and, in several places, the loams are let down into the cobbly gravels (bed 2) in pipes and V-shaped wedges; dislocated boulders and gravel clasts sometimes occur within these pipes (Figure 7.22; Gilbertson and Mottershead, 1975).

Interpretation

Although mapped, described and interpreted by Thomas (1940) as river gravels, the Quaternary deposits at Doniford were first studied in detail by Gilbertson and Mottershead (1975); a précis of this work was given by Mottershead (1977a) with an accompanying account of the archaeological finds by Norman (1977).

Since far-travelled erratics found in northern Somerset have been used as evidence for ice sheets moving eastwards up the Bristol Channel, and penetrating well inland near Weston-super-Mare (Hawkins and Kellaway, 1971), Gilbertson and Mottershead (1975) reinvestigated the Doniford sections to ascertain if glacier ice had played any part in their formation. They found that no clasts in either the gravel (bed 2) or overlying loams (beds 3–6) were foreign to the area. Apart from local Jurassic and Triassic rocks, only those of Devonian age were found, even these implying a not too distant source. A glacial origin was therefore ruled out for any part of the sequence, and analyses of the sediments and structures indicated that the beds were most likely periglacial in origin, consisting of a series of alluvial gravel, head and slopewash sediments (Gilbertson and Mottershead, 1975; Mottershead, 1977a).

Gilbertson and Mottershead (1975) argued that the gravels occupied a valley running from the west beyond Watchet Station, through the Memorial Ground and Helwell Bay and then parallel with the Doniford coast to the Swill. The gravels are believed to occupy an ancestral valley of the Washford River which once had a confluence in the area of the present Swill, this valley system having now been dismembered by marine erosion

causing the Washford River to flow directly into the sea at Watchet (cf. Mottershead, 1967). The gravels of the former river system are therefore exposed in cross-section at the west end of Helwell Bay and in the Swill river sections, and in long-profile along the Doniford coast as far as the Swill (Gilbertson and Mottershead, 1975; Mottershead, 1977a).

The cobbly gravels, however, cannot simply be regarded as fluviatile sediments since they also display characteristics of periglacial mass movement. Fluvial characteristics are clearly shown by the degree of bedding, including cross-stratification and sorting in parts of the bed. The gravel clasts are also significantly more rounded than typical periglacial head deposits in the region (Mottershead, 1976). Similarly, their non-local (but not distant) provenance also mitigates against an origin solely as periglacial head. Alternatively, Gilbertson and Mottershead suggested that parts of the gravel bed showed little evidence for sorting, having undergone mixing and reworking in a periglacial environment. In places, clast orientation measurements showed a preferred orientation downslope, indicating possible mass movement of gravels from higher ground located to the south of the sections. The evidence for fluvial activity is strongest near the Swill, where well-sorted, cross-stratified sands pick out channels presumably cut and filled by surface streams. In this area, deposition was effected by streams trending generally north, perhaps running off local hillslopes, but primarily by a forerunner of the present Doniford stream (Gilbertson and Mottershead, 1975). The presence of Lias 'rafts' incorporated into the overlying gravels, the intimate contortion of the Lias platform with the gravels, cryoturbation and vertical stone structures found at all levels elsewhere in the beds and the presence of many cracked clasts (Mottershead, 1977a) were taken to show that periglacial conditions had predominated throughout the accumulation of the sequence.

Gilbertson and Mottershead (1975) noted that loams (beds 3–6) which overlie the gravels, although differing in composition and structure, are also poorly sorted and show a consistent downslope orientation of clasts, characteristic of solifluction deposits. These sediments form an 'apron' sloping off adjacent hillsides, the red loam (bed 5) probably having been derived by erosion and subsequent redeposition of adjacent red Triassic rocks, and the buff loam (bed 4) consisting largely of redeposited (highly calcareous) Lias clays and Jurassic and Rhaetic limestone fragments. Thin

Figure 7.24 Quaternary deposits (mainly fluvial cobbly gravels; bed 2) exposed on the western bank of the Swill in 1980. (Photo: S. Campbell.)

bedding in some of these sediments suggests local deposition by slopewash, probably in a sparsely vegetated, periglacial environment.

Dating of the beds is more problematical. The capping brown loam (bed 6), interpreted as hill-wash sediment (Gilbertson and Mottershead, 1975), is the stratum from which the Mesolithic artefacts were recovered. Norman (1975, 1977) has argued that these artefacts are probably divisible on typological grounds, and may represent both ear-lier (*c.* 10 000–8500 BP) and later (*c.* 8500–6000 BP) Mesolithic industries in the area. There is every likelihood, therefore, that the brown loam, upon

which the modern soil has developed, is Holocene in age.

The artefacts and mammalian remains from the cobbly gravels (bed 2), however, only give a vague maximum age for these deposits. Lower Palaeolithic material of Acheulian Culture has tra-ditionally been regarded as of Hoxnian and early 'Wolstonian' (Saalian) age. (Recent evidence from Waverley Wood, Warwickshire, shows that the Acheulian Culture extends back to pre-Anglian times (Shotton *et al.*, 1993).) The highly abraded and rolled condition of the artefacts shows that they are derived and renders them of limited use for dating the gravels, which must simply be 'no older'. Similarly, woolly mammoth remains elsewhere are first recorded in deposits assigned to the Wolstonian Stage (*sensu* West, 1968). The species was also present in the latest Ipswichian and became more common in the Devensian (West, 1968). It is impossible, in the absence of reliable absolute dates, to assign the remains at Doniford with certainty to any of these stages. There are, however, other indirect clues as to the age of the sediments (beds 2–5). If the upper loam (bed 6) is ascribed to the Holocene, as seems likely, and the underlying sediments are accepted as representing a cold environment, then a pre-Devensian age (?Saalian) would mean that considerable breaks in sedimentation occur in the sequence. The sedimentological data show no evidence for such a protracted hiatus (Gilbertson and Mottershead, 1975). Under such circumstances, it is most likely that beds 2–5 (the gravels and most of the loam sequence) at Doniford were formed during cold conditions in the Devensian when sea levels were lower (Gilbertson and Mottershead, 1975; Mottershead, 1977a).

Conclusion

The Doniford sections show a classic example of deposits and structures formed by fluvial and mass movement processes in a periglacial environment, probably during the Devensian. The sediments lie along the former course of the Washford River and consist largely of reworked river gravels including Devonian material probably brought from the Brendon Hills by an ancient Washford river (Gilbertson and Mottershead, 1975). The sediments, now exposed by marine erosion, show a complex interaction between fluvial, mass movement and freeze-thaw processes. There is clear evidence that much of the sediment has been subject to solifluction, although the lenses and channel-fills of better-sorted sediment, particularly near the Swill, demonstrate reworking of deposits by small, temporary streams running over the aggrading sediment surface (Gilbertson and Mottershead, 1975). As such, Doniford is an excellent fossil analogue of modern periglacial environments, where spring meltwater release plays an important role in reworking deposits of previous seasons (e.g. McCann *et al.*, 1972). The sediments at Doniford also confirm that glacier ice did not reach the north Somerset coast during the Devensian.

Finally, the site is notable for the discoveries of mammal remains and artefacts. The remains of *M. primigenius* (woolly mammoth) are an important palaeontological record, but are of little use for dating the sequence. Similarly, the Lower Palaeolithic artefacts recovered provide important evidence for the Acheulian Culture in this part of Somerset. They too, however, provide little additional evidence regarding the age of the sediments. The later Palaeolithic artefacts, however, raise the interesting possibility of human activity in the Doniford Valley in latest Devensian times. Discoveries of Mesolithic artefacts in the sections at Doniford are significant, showing two distinct typological assemblages and fixing a Holocene age for the uppermost bed of the sequence.

THE CHAINS
S. Campbell and R. Cottle

Highlights

The Chains GCR site provides a detailed pollen record, calibrated by radiocarbon dating, of mid- to late Holocene vegetational and environmental changes on Exmoor. It permits comparisons with other upland sites in South-West England and demonstrates the impact of humans on the landscape from Neolithic times onward.

Introduction

The peat mires of the Chains provide an important record of the changing vegetation cover on Exmoor during the last 5000 years or so. Radiocarbon dating of the deposits has allowed the pollen record to be correlated with periods of human activity in the area, thereby throwing light on the problem of anthropogenic activity in the initiation of blanket peat. The site was studied in detail by Merryfield and Moore (1974), and its evidence has been reviewed by Moore (1973), Crabtree and Maltby (1975), Bell *et al.* (1984) and Moore *et al.* (1984).

Description

The Chains is an upland plateau ridge running approximately 10 km from Radworthy (SS 700435) in the west to Raven's Nest (SS 780406) in the east, at an average height of 475 m. The ridge acts as the

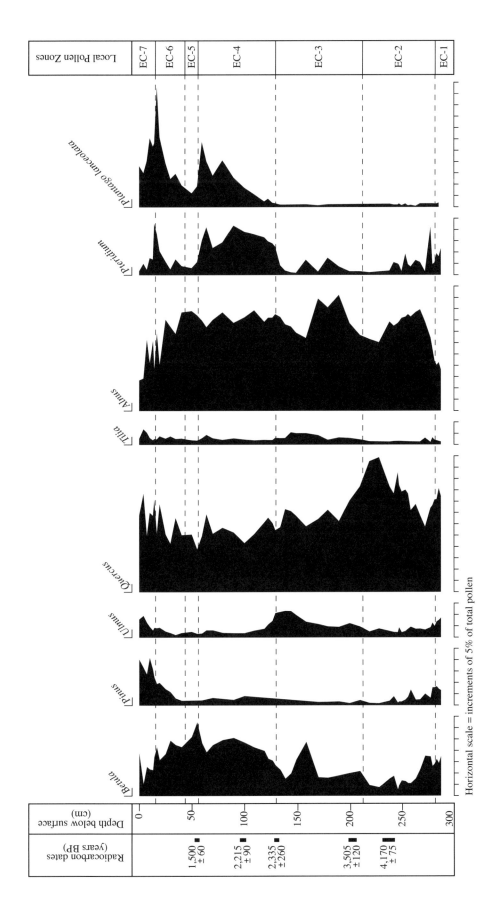

Figure 7.25 Selected pollen data and radiocarbon dates from a peat profile at The Chains, Exmoor. (Adapted from Merryfield and Moore 1974.)

Horizontal scale = increments of 5% of total pollen

Local Pollen Zones

EC-7
EC-6
EC-5
EC-4
EC-3
EC-2
EC-1

Plantago lanceolata

Pteridium

Alnus

Tilia

Quercus

Ulmus

Pinus

Betula

Depth below surface (cm)

0
50
100
150
200
250
300

Radiocarbon dates (years BP)

1,500 ±60
2,215 ±90
2,335 ±260
3,505 ±120
4,170 ±75

major watershed for north Exmoor, with streams running off the highest area (487 m) to form the River Barle in the south and the West Lyn in the north.

The Chains GCR site (SS 732424) comprises an area of rough moorland and peat bog some 200 × 150 m. The site lies *c.* 600 m north-west of Chains Barrow and 1 km east of Pinkworthy Reservoir on a gentle north-west-facing slope. The GCR site coincides with the deepest blanket mires (up to 3 m thick) so far found on Exmoor. Merryfield and Moore (1974) cut a monolith from the mire for pollen analysis and radiocarbon dating. Selected results of these analyses are shown in Figure 7.25 which also illustrates local pollen assemblage biozones. However, no lithological details of the succession are given by Merryfield and Moore (1974).

Interpretation

Local pollen assemblage zone EC1, identified from the basal layers of the peat, is characterized by relatively high values of *Ulmus* (elm) and *Pinus* (pine) pollen. Merryfield and Moore (1974) take this to indicate the undisturbed state of local woodland prior to the arrival of farming cultures. The succeeding zone (EC2) demonstrates a decline of pine and elm and an increase in *Alnus* (alder) at the expense of *Quercus* (oak). It also shows an increase in *Pteridium* and the consistent presence of *Plantago lanceolata* (ribwort plantain). Merryfield and Moore (1974) suggest that these changes denote the arrival of farming cultures and the response of local woodlands to their activities. Based on the radiocarbon date of 4170 ± 75 BP (UB–821) and the assumption of a constant rate of peat formation, Merryfield and Moore have suggested that this biozone covers the period between 5000 to 3800 BP (Neolithic).

Zone EC3 demonstrates a gradual recovery of elm to its maximum representation at the end of the zone. In contrast, there is no recovery in pine and both *Pteridium* and *P. lanceolata* appear sporadically. According to Merryfield and Moore (1974), the pollen evidence indicates an unsettled period between *c.* 3800 and 2300 BP, with no permanent human settlements being established (Bronze Age).

A sudden decline in the values of elm pollen and an increase in *Betula* (birch), *Pteridium* and *P. lanceolata* characterize Zone EC4, and are taken to indicate increased farming activity, woodland clearance and settlement between *c.* 2300 and 1500 BP (Iron Age to Roman) (Merryfield and Moore, 1974). An abrupt decline in *Pteridium* and *P. lanceolata* and a further increase in birch characterize Zone EC5, and indicate reduced human activity from about 1500 BP (end of Roman times) onward. A gradual recovery of *Pteridium* and *P. lanceolata* to their maximum values are key features of Zone EC6, which also shows a decrease in alder and a rise in elm and pine. Merryfield and Moore have argued that these changes indicate increased settlement and deforestation, especially of the lowlands and valleys. The duration of the biozone is uncertain, but may have culminated in the Napoleonic Wars about 180 years ago. The final biozone identified (EC7) shows a decrease in *Pteridium* and *P. lanceolata* with pine reaching its maximum value, perhaps reflecting the reduced intensity of farming during the last 180 years or so (Merryfield and Moore, 1974).

The pollen record for The Chains is similar to others derived from blanket peat deposits in mid-Wales (Moore, 1968) and the southern Pennines (Tallis and Switsur, 1973). In particular, the dating of the basal peat layer to approximately the time of arrival of farming cultures lends support to previous suggestions that the introduction of grazing animals or the ploughing of marginal land may have assisted in the initiation of some upland peat deposits (Merryfield and Moore, 1974).

A shallower peat profile at Hoar Tor on Exmoor was correlated by Merryfield and Moore, using pollen biostratigraphy, with the Chains' profile. Both showed marked evidence for vegetation clearances around Iron Age to Roman times, events which may have triggered the further spread of peat. Merryfield and Moore (1974) have emphasized that both periods of blanket peat formation on Exmoor (Neolithic and Iron Age/Roman) coincide with times of deteriorating climate. They strongly favoured, however, that ' … Man, or his domesticated animals, provided a final stress on an ecosystem already under climatic strain' (Merryfield and Moore, 1974; p. 441).

Conclusion

The Chains' peat deposits provide an important record of changing climatic, vegetational and environmental conditions on Exmoor during the last 5000 years. The radiocarbon-dated profile provides strong evidence for human modification of the local vegetation and landscape during both Neolithic and Iron Age to Roman times, and adds weight to the argument that the spread of upland blanket peat was caused by a combination of climatic and anthropogenic factors. Evidence from The Chains provides a stark reminder of human effects on an ecosystem already under climatic stress.

Chapter 8

The Quaternary history of the Isles of Scilly

Introduction

INTRODUCTION
J. D. Scourse

The Isles of Scilly contain a wealth of exceptionally important and interesting Quaternary sediments and landforms quite out of proportion to their relatively small area. Situated 45 km west-south-west of Land's End (Figure 8.1), the group consists of some 100 islands, most of which are devoid of soil and terrestrial vegetation. Most major appraisals of the Quaternary stratigraphy and geomorphology of the islands (Barrow, 1906; Mitchell and Orme, 1967; Scourse, 1991) have been based on evidence from the five largest and permanently inhabited islands of St Mary's, St Martin's, St Agnes, Tresco and Bryher, and the larger uninhabited islands of Samson, St Helen's, Northwethel, Tean, Nornour, Great Ganilly, Great Arthur, Little Arthur and Annet (Figure 8.1).

The Quaternary significance of the islands lies in the evidence they contain for: 1. changes in relative sea level, in the form of raised beaches; 2. former periglacial conditions elucidated from sedimentary, geomorphological and palaeobotanical sites including the first radiometric dates for the coastal 'head' deposits of South-West England; 3. glaciation of the northern islands with a well-defined glacial limit straddling the archipelago, indicated by a range of glacial, glaciofluvial and associated aeolian sediments; 4. some of the finest granite landforms in the British Isles whose evolution and form can be demonstrated to be intimately linked to the periglacial and glacial history of the islands; and 5. the successful establishment of forest tree taxa on the islands during the Holocene post-dating the severance from mainland Cornwall by sea-level rise. Of particular significance and importance are the palaeobotanical records from periglacial sequences which provide evidence of the vegetation of the late Middle and early Late Devensian, a phase very poorly covered by sites elsewhere in Britain, and the glacial sequences, which, together with glacigenic sediments from the adjoining continental shelf, provide evidence of the most southerly advance of ice during the Quaternary in north-west Europe.

The solid geology of the islands is almost exclusively granite (Figure 2.2), Scilly representing the highest parts of an almost completely submerged elliptical cupola which is the westernmost extension of the Variscan batholith of South-West England. Barrow (1906) divided the granite into coarse- and fine-grained facies and believed the only existing exposures of country rock to be the highly tourmalinized slates ('killas') of White Island, St Martin's. Although this material has been re-interpreted as either sheared or greisened granite (Hawkes, 1991), one small exposure of phyllitic country rock containing sporadic perthite megacrysts does occur on Shipman Head, Bryher (Hawkes, 1991). Although no *in situ* Mesozoic sediments occur on the islands, Barrow (1906) interpreted gravel, consisting largely of flint and greensand, on the summit of Chapel Down, St Martin's, to be of Eocene (?) age. This he correlated with the Eocene fluvial gravel capping the Haldon Hills in east Devon.

The homogeneity of the bedrock assisted in the early identification of pebbles of 'foreign' lithology on the northern islands. Smith (1858) recorded chalk-flints and greensand from Castle Down, Tresco. These were later interpreted by Whitley (1882) as glacial in origin. Barrow exhibited a striated boulder from the islands to the Geological Society in 1904 (Barrow, 1904), and he was later (1906) able to place the occurrence of the erratics within a stratigraphic framework. He took Chad Girt (see below), on White Island, St Martin's as his type-site for the Pleistocene of the islands. At Chad Girt, and at many other sites, Barrow observed that in cliff section the erratics were usually set within a silty matrix often cemented by iron oxides, and that this 'glacial deposit' was both underlain and overlain by head deposits, the whole resting on a raised beach. Barrow accepted Whitley's (1882) glacial hypothesis, but preferred deposition from 'floe-ice'.

In 1957 Dollar attributed the St Martin's gravels to the Pliocene, and in 1960 Mitchell further revised their probable age to Lower Pleistocene, citing a fluvial origin. Mitchell and Orme (1965, 1967) undertook a major re-examination of the Pleistocene stratigraphy of the islands and identified a more complex sequence than Barrow:

6. Upper Head
   Weichselian (= Devensian)
5. Raised Beach (Porth Seal)
   Eemian (= Ipswichian)
4. Glacial Deposit
   Gipping (= Wolstonian)
3. Lower Head
   Gipping (= Wolstonian)
2. Raised Beach (Chad Girt)
   Hoxnian
1. Shore platform

(revised stage names from Mitchell *et al.*, 1973b)

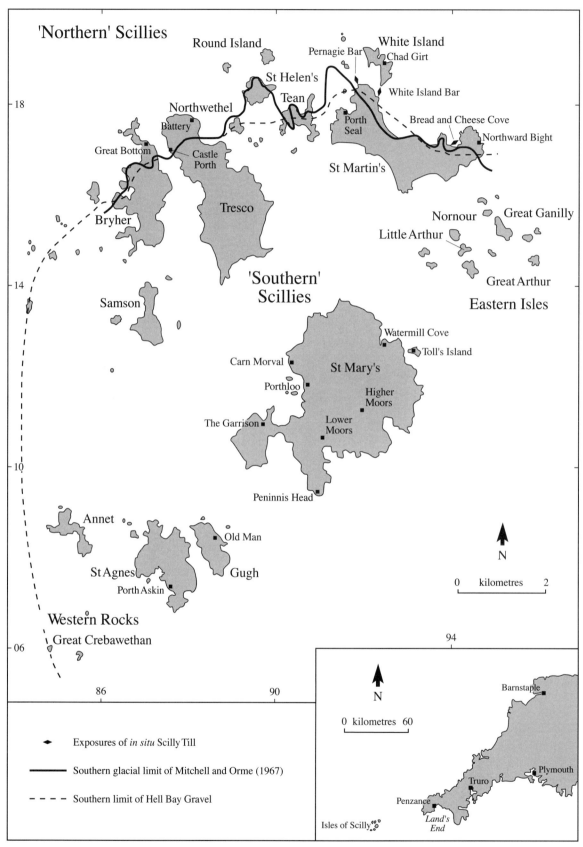

Figure 8.1 The Isles of Scilly: critical sites, exposures of the Scilly Till, the southern limit of the Hell Bay Gravel and Mitchell and Orme's (1967) glacial limit. (Adapted from Scourse, 1991.)

Figure 8.2 Carn Morval on St Mary's (Figure 8.1) is one of five sites on the Isles of Scilly where organic sediments, which have yielded pollen evidence and radiocarbon dates, lie beneath or interbedded with periglacial head. Although yielding the most detailed of these palaeoenvironmental records, Carn Morval was not selected as a GCR site because coastal erosion has since removed much of the critical organic sequence. (Photo: J.D. Scourse.)

They divided Barrow's (1906) 'glacial deposit' into two facies, till and outwash gravel, and identified an ice limit running across the northern islands based on the distribution of these sediments (Figure 8.1). By comparing their stratigraphy with similar sequences in mainland South-West England and in Ireland, Mitchell and Orme suggested a Gipping (= Wolstonian) age for the glacial event. This interpretation was largely based on the suggestion that the erratic-free Chad Girt raised beach is Hoxnian in age by correlation with other raised beaches of supposed Hoxnian age at similar elevations elsewhere, for example, the Courtmacsherry raised beach in southern Ireland. Direct correlation of lithostratigraphic units with chronostratigraphic stages on a one-to-one basis then dictated that the Porth Seal raised beach should be Eemian (= Ipswichian) in age. The interpretation of a Wolstonian glacial limit on the islands as a result became firmly established in the literature (cf. Catt, 1981; Lowe and Walker, 1984), despite some speculation that the glacial material on the Isles of Scilly might be younger in age (John, 1971; Synge, 1977, 1985).

Bowen (1969, 1973b) questioned Mitchell's and Orme's interpretations, proposing that the lenticular form of the glacial material and its geomorphic association with coastal valleys was more consistent with an origin as soliflucted till. He later (1981) suggested that the critical stratigraphy identified by Mitchell and Orme (1967) at the Porth Seal site was 'inferred and superposed' and argued that granite corestones had been misinterpreted as a marine deposit. Bowen (1973b) regarded the *single* raised beach as Ipswichian in age, the soliflucted glacial deposits having been originally emplaced in the Wolstonian.

Coque-Delhuille and Veyret (1984, 1989) proposed a much more southerly ice limit on the islands. They concurred with Mitchell and Orme (1967) and Bowen (1973b) that the glaciation was Wolstonian in age.

The dating of the glacial material on the Isles of Scilly by Mitchell and Orme (1967) and Bowen (1973b) was therefore heavily dependent on the number and age of stratigraphically juxtaposed raised beach units, and lithostratigraphic correlation with neighbouring regions.

The Quaternary history of the Isles of Scilly

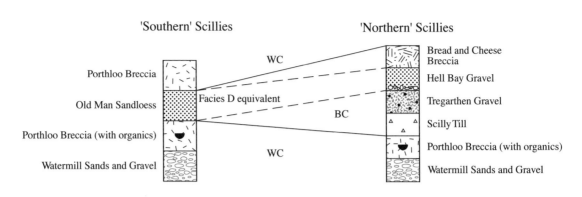

WC - Watermill Formation   BC - Bread and Cheese Formation

Figure 8.3 A lithostratigraphic model for the Isles of Scilly. (Adapted from Scourse, 1991.)

Scourse (1991), in the most recent re-examination of the islands, proposed a revised local stratigraphy independent of the stratigraphies erected in neighbouring regions. He reported organic sediments (Figure 8.2) from a number of sites which provided a chronology through multiple radiocarbon determinations, and which yielded palaeobotanical evidence critical in palaeoenvironmental reconstruction. Thermoluminescence (TL) dates (Wintle, 1981) and optical dates (Smith *et al.*, 1990) further assisted in the establishment of a radiometric chronology for the sequence.

Scourse (1991) defined eight lithostratigraphic units of member status which he incorporated into two lithostratigraphic models for the 'southern' (extra-glacial) and 'northern' (glaciated) Isles of Scilly (Figure 8.3). The southern limit of the Hell Bay Gravel defines the boundary between these two areas, and corresponds closely to the ice limit identified by Mitchell and Orme (1967; Figure 8.1). The same lithostratigraphical units are used, albeit in slightly modified form, in the Geological Society's recently revised correlation of British Quaternary deposits (Campbell *et al.*, in prep.).

Overlying the raised beach sediments of the Watermill Sands and Gravel in the southern Isles of Scilly is the Porthloo Breccia, a variable unit of soliflucted material derived entirely from the weathering of the granite bedrock (Scourse, 1987). The organic deposits were found towards the base of this unit at five sites (Figure 8.1); Carn Morval (SV 905118), Watermill Cove (SV 925123), Toll's Island (SV 931119), Porth Askin (SV 882074) and Porth Seal (SV 918166). Scourse interpreted these organic sequences as the infillings of small ponds or lakes impounded by active solifluction sheets or lobes.

Radiocarbon dates from these organic sediments are critical since they pre-date the units associated with the glacial advance (Figure 8.3), the Scilly Till, the Tregarthen Gravel, the Hell Bay Gravel and the Old Man Sandloess. The radiocarbon determinations indicate deposition of the organic material between 34 500 + 885/− 800 (Q-2410) and 21 500 + 890/− 800 (Q-2358) BP. They provide a maximum age for the glacial event and the first radiometric dates for the coastal 'head' deposits of South-West England.

All samples of organic sediment used for these radiocarbon determinations were taken from permanent open coastal sections which posed considerable problems of contamination by younger carbon derived from rootlet and groundwater sources. However, multiple dates of both humic and humin fractions of samples from different locations within the deposits enabled Scourse (1991) to identify the extent and sources of contamination and therefore to assess the reliability of the resultant determinations.

The pollen sequences from the Pleistocene organic deposits are all very similar in recording open grassland vegetation. Carn Morval on St Mary's (Figure 8.1) yielded the most detailed pollen profile, but has not been selected as a GCR site because its organic material has now been largely removed by coastal erosion. The organic beds at the site were situated towards the base of the Porthloo Breccia, and the pollen assemblages (Scourse, 1991; Fig. 10) are dominated by Gramineae (grasses), Cyperaceae (sedges) and other herb taxa. *Pinus* is the most important tree taxon, but Scourse (1991) interprets this as a long-distance component rather than indicating local presence; the increase in *Pinus* towards the top of the diagram is probably the result of the relative decline in the pollen productivity of the local herbaceous flora set against a relatively constant supply of long-distance *Pinus*.

These pollen sequences represent the earliest Quaternary vegetational record for South-West England west of east Devon. The spectra are broadly similar to others of the same age from elsewhere in north-west Europe (Bell *et al.*, 1972; Morgan, 1973; West, 1977b).

In the southern Isles of Scilly, the Porthloo Breccia is overlain by the Old Man Sandloess, a coarse aeolian silt with subdominant fine sand and minor amounts of clay (Catt and Staines, 1982). This has yielded two TL dates, both of 18 600 ± 3700 BP (QTL1d and QTL1f) (Wintle, 1981) and two optical dates, 20 ka ± 7 ka and 26 ka + 10/− 9 ka BP (738al and 741al; Smith *et al.*, 1990). This material occurs in a variety of facies related to different modes of reworking.

In the northern Isles of Scilly, the Porthloo Breccia is overlain by three units that are all related to a single glacial event. The Scilly Till, a massive, poorly sorted, clay-rich, pale-brown diamicton containing abundant striated and faceted erratics of northern derivation, occurs at Bread and Cheese Cove, at Pernagie Bar and White Island Bar (Figure 8.1). The precise depositional environment of this material is uncertain but Scourse (1991) argues, on the basis of sedimentological, structural, lithological and fabric (eigenvalue) data, that it is likely to be of lodgement origin. This conflicts with the suggestion by Eyles and McCabe (1989), echoing the 'floe-ice' hypothesis of Barrow (1906), that this material is of glaciomarine origin. At Bread and Cheese Cove the Scilly Till occurs in association with a matrix-supported sandy gravel, the Tregarthen Gravel, which has a similar erratic assemblage. The Tregarthen Gravel is best displayed at its type-site, Battery (Castle Down) on Tresco (see below – Battery; Figure 8.1).

Aeolian loessic processes in association with the glacial advance resulted in the deposition of the Old Man Sandloess in the southern Isles of Scilly. The relative coarseness of this material is interpreted by Scourse (1991) as a function of its proximity to glacially derived source material. The mineralogy of the Scilly Till is sufficiently similar to that of the Old Man Sandloess (Catt, 1986) to suggest a genetic link between the two units.

Overlying the Scilly Till and the Tregarthen Gravel in the northern Isles of Scilly is the Hell Bay Gravel, an extremely widespread matrix-supported gravel containing a similar assemblage of striated and faceted erratics to the underlying till, but alongside a considerable proportion of locally derived granitic material. The lithology and mineralogy of the Hell Bay Gravel is identical to colluvially reworked facies (facies D) of the Old Man Sandloess (Figure 8.3). This material represents an initial phase of solifluction, post-dating the glacial event, in which the Scilly Till, Tregarthen Gravel and Old Man Sandloess were mixed and transported downslope. In situations where these sediments were stripped from the land surface, weathered granite once again became the dominant raw material for solifluction, this subsequent phase being represented by the Bread and Cheese Breccia in the northern islands and the upper Porthloo Breccia in the south. The Hell Bay Gravel is synonymous with Barrow's (1906) iron-cemented 'glacial deposit'.

There is a strong spatial relationship between the distribution of the glacigenic and glacial-derived sediments and marine bars, tombolos and granite landforms in the Scillies. Many marine bars occur within the limit of the Hell Bay Gravel, and two of these, Pernagie Bar and White Island Bar (Figure 8.1), are directly underlain by Scilly Till. Although most of the boulders comprising these bars are granite, many are erratics, dominantly flint, red siltstones and sandstones. Whereas contemporary marine processes clearly control their detailed morphology, the internal structure of the bars and their distribution in relation to the sedimentary units discussed above suggests a possible morainic origin (Scourse, 1991).

Scourse (1987) defined four tor forms from the Isles of Scilly, and there appears to be a good correlation between highly smoothed, eroded tors and the southern limit of the Hell Bay Gravel, suggesting that the glacial advance, though unable to penetrate far into the Scilly massif, was nevertheless capable of eroding the solid granite. Tor forms in the southern islands are, by contrast, 'mammilated' or 'castellated' (see below – Peninnis Head; Figure 8.6).

The evidence presented by Scourse (1991) therefore indicates that ice advanced as far as the northern Isles of Scilly during the Dimlington Stadial of the Late Devensian around 18 600 ± 3700 BP. However, this may not have been the first glacial event to affect the islands because erratics are widespread in some exposures of the Watermill Sands and Gravel, the basal lithostratigraphic member of the Pleistocene sequence. The age of this earlier event remains uncertain.

Scourse's (1991) interpretation clearly conflicts

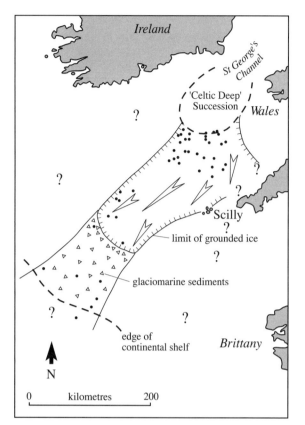

Figure 8.4 A reconstruction of the Celtic Sea ice lobe and glaciomarine terminus at 19 ka BP, adapted from Scourse *et al.* (1991). Dots represent vibrocoring sites which have yielded glacigenic sediment.

in a number of important respects with Mitchell and Orme's (1967) sequence of events. The main differences include: 1. recognition of only one raised beach unit stratified with other sediments; 2. recognition of widespread loessic sediments, and the interpretation of a sedimentary suite of till, outwash gravel and loess related to a single glacial event; and 3. independent radiometric dating rather than relative dating based on the inferred ages of raised beach units.

The interpretation of the Late Devensian glaciation of the Isles of Scilly, and the evidence on which it is based, have proved controversial. Scourse (1991) identifies the main points of contention as: 1. the reliability of the radiometric dates; 2. the validity of the lithostratigraphic correlations between sites and islands; and 3. the *in situ* status of the Scilly Till. Scourse (1991) discusses these points in turn and concludes that the evidence in support of the Late Devensian model is much more substantial than that proposed by Mitchell and

Orme (1967) to support the hypothesis of Wolstonian glaciation.

Glacial and glaciomarine sediments have been discovered on the continental shelf adjoining the Isles of Scilly as far south as 49° (Scourse *et al.*, 1990, 1991). These sediments have been grouped into a northern facies containing sparse reworked microfaunas, interpreted as either overconsolidated lodgement tills or proximal glaciomarine sediment, and a southern facies containing abundant cold-water microfaunas, interpreted as distal glaciomarine silty clays. Scourse *et al.* (1990, 1991) correlate both facies with the Scilly Till, thereby enabling a quantitative reconstruction of ice thicknesses, grounding-line, sea level and shoreline elevations in the Celtic Sea at *c.* 19 ka BP.

The Late Devensian ice advance responsible for the glaciation of the northern Isles of Scilly and the adjoining shelf is thought to have terminated in marine waters towards the shelf edge break and is likely to have constituted a thin lobate surge over deformable marine sediments at the southern terminus of the Irish Sea ice stream (Scourse *et al.*, 1990, 1991; Figure 8.4).

Many of the GCR sites on the Isles of Scilly have been selected because they provide critical evidence relating to this glacial event and its dating; these sites include Bread and Cheese Cove (St Martin's), Watermill Cove (St Mary's), Porth Seal (St Martin's), Battery (Castle Down, Tresco), Castle Porth (Tresco) and Old Man (Gugh, St Agnes). Other sites have been chosen for their historical significance or because they contain evidence which has proved controversial in the evolving ideas on the Quaternary history of the islands; these include Chad Girt (White Island, St Martin's), Northward Bight (St Martin's) and Porthloo (St Mary's). Peninnis Head (St Mary's) has been selected because it contains spectacular granite landforms typical of features which lie outside the limit of the Hell Bay Gravel, and Higher Moors (St Mary's) because it contains the palaeobotanical evidence for Holocene forest development (Scaife, 1984, 1986).

PORTHLOO, ST MARY'S
J. D. Scourse

Highlights

Fine exposures of head at this easily accessible site have been chosen as the stratotype for this material on Scilly since the early twentieth century. The stratigraphical relations of this material to underlying raised beach sediments are also clearly demonstrated.

Introduction

Barrow (1906) defined the section of granitic head at Porthloo (Figure 8.1) as the type-site for the 'Main Head' on Scilly. He commented that occasional exposures of raised beach could be observed at the base of the section. Mitchell and Orme (1967), while confirming the quality of the head outcrops at the site, commented that the raised beach exposures here were inferior to those at other sites, and did not correlate the raised beach here with either their 'Chad Girt' or 'Porth Seal' raised beaches. Scourse (1991), following Barrow (1906), took Porthloo as the type-site for the Porthloo Breccia (Figure 8.3), and correlated the raised beach with the Watermill Sands and Gravel.

Description

Two sedimentary units are exposed at Porthloo (SV 908115). Up to 5 m of coarse granitic head overlies occasional large rounded granite cobbles at the base of the section. The clasts within the head are all extremely angular, vary in size from pebbles to boulders, and are exclusively of granite. Barrow (1906) noted that the deposit locally varies in texture. The unit is mostly clast-supported and the matrix is extremely poorly sorted. In places it is stratified and occasionally displays lobate structures with clast concentrations along the margins of the lobes. The clast fabric is consistently oriented parallel with the local slope, but with a predominant dip into the flow direction.

Interpretation

The head at this site has been consistently interpreted as a solifluction deposit, and the underlying rounded cobbles as the remnants of a raised beach. In explaining the variation in texture of the deposit, Barrow (1906) commented that this ' … is clearly due to the fact that the steep rock-face behind the Head alternately recedes from and approaches the present low cliff-face; in the former case, the Head is finer, in the latter coarser' (Barrow, 1906; p. 19). In terms of the model put forward by Scourse (1987) to illustrate the stratigraphic and sequential development of such

deposits, this variation is between facies B (coarse blockfield/felsenmeer facies) and facies D (finer solifluction facies).

Barrow further noted that there had been little lateral transport of these deposits parallel to the shoreline as shown by the disposition of fragments of the Porthloo elvan or quartz-porphyry; these do not extend more than a few feet on either side of the dyke. He contrasted this with the occurrence of transported raised beach cobbles within the deposit testifying to considerable forward and downward movement. This downslope movement has been confirmed through clast fabric studies in deposits of the Penwith Breccia at other sites (Scourse 1987, 1991). The characteristic dip of clasts into the section at angles between 5° and 45° from the horizontal have been explained by Scourse (1987) in terms of penecontemporaneous upfreezing accompanied by mass movement of material under gravity processes in the seasonally thawed layer characteristic of periglacial environments (French, 1976).

The source material for the head was overwhelmingly weathered local granite, though in places local lenses of silt occur within it which are probably of loessic origin, aeolian deposits derived from a wider source area.

Conclusion

This site affords excellent and accessible exposures of sediments deposited downslope as a result of the seasonal thawing of ground in cold Arctic-type environments. These sediments are quite widespread on the Isles of Scilly, but the exposures at Porthloo have long been regarded as the best available.

WATERMILL COVE, ST MARY'S
J. D. Scourse

Highlights

This is the most important Quaternary site south of the glacial limit on the Isles of Scilly. Exposures here contain the finest raised beach sequence, and the most impressive Middle–Late Devensian organic sequence, on the islands. The raised beach is unique because it is the only example on Scilly to contain a distinct bed of unconsolidated sand above the beach shingle, cobbles and boulders. The organic sequence has yielded pollen assemblages indicative of open grassland vegetation thought to be contemporary with periglacial conditions. Radiocarbon dates from the organic material are critical in showing that it pre-dates the glacial event which impinged on the northern islands.

Introduction

Mitchell and Orme (1967) recorded raised beach sediments at Watermill Cove, noting that the beach of small cobbles and boulders is overlain by *c.* 2 m of 'coarse sand stained black and red by manganese and iron', in turn covered by undifferentiated head. Though they did not explicitly correlate the raised beach here with either their upper 'Porth Seal' or lower 'Chad Girt' raised beaches, Page (1972), in a paper concerned with the radiocarbon dating of organic interglacial deposits in Britain, argued that both the 'Upper Head' and 'Lower' or 'Main Head' of Mitchell and Orme's scheme (1967) could be identified at the site. He therefore implied that raised beach sediments underlying the Lower Head were 'pre-Gipping or Hoxnian' in age, thereby effecting a correlation of these deposits with the earlier Chad Girt raised beach. Page (1972) was the first to identify organic beds lying stratigraphically between the raised beach and the head at Watermill Cove. He interpreted this material as Hoxnian in age in view of its stratigraphic position in relation to the units identified by Mitchell and Orme, and reported radiocarbon dates of around 22 ka BP from bulk organic samples from the site. He used these dates, along with others from sites of Hoxnian or supposed Hoxnian age elsewhere, to argue that the Hoxnian occurred between about 25 500 and 21 100 BP. Page's (1972) controversial paper was criticized by Shotton (1973) on the basis that there is no evidence to justify Page's assumption that the organic material at Watermill Cove is Hoxnian in age.

Scourse (1991) described the sections at Watermill Cove and, on the basis of further radiocarbon dating, pollen analysis and correlation with other sedimentary units on Scilly, proposed that the entire sequence above the raised beach is Middle or Late Devensian in age, much younger than envisaged by either Mitchell and Orme (1967) or Page (1972).

Description

The section containing organic material lies at SV 925123, to the south-east of Watermill Cove proper on the north-east coast of St Mary's (Figure 8.1). Lying below 1–2 m of coarse granitic head (Figure 8.5; bed 7) there is a relatively undisturbed unit of sand and silt containing organic material (Figure 8.5; beds 3–6). The unit averages between 1 and 1.5 m thick, and extends laterally for about 20–25 m along the base of the section at around 4.4 m OD. This unit can be subdivided into four beds (Figure 8.5; beds 3–6). The lowest (bed 3) is a brown to black organic silt, highly humified, containing some quartz granules. It is overlain by a fawn to light brown sand (bed 4), less organic than bed 3, again containing quartz granules; the contact between beds 3 and 4 is gradational. Bed 5 is a black, richly organic, highly humified silt, while bed 6 is very similar to the upper parts of bed 4.

Pits dug at the base of the section (Scourse, 1991) revealed a unit (bed 1) of homogeneous coarse to medium light brown sand, separated from the overlying organic material by a layer of coarse cemented granitic head (bed 2) just a few centimetres thick (Figure 8.5). The sands are uncemented and appear to rest on solid granite. The contact between beds 1 and 2 is sharp and erosional.

Farther to the north-west, on the southern shore of Watermill Cove itself (SV 924123), a raised beach consisting of both rounded granite cobbles and sand is exposed at the same elevation. No raised beach cobbles are visible within the main exposure, but it is clear that the sand (bed 1) there forms part of the same exposure of raised beach.

The coarse granite head (bed 7) overlies the entire sequence. Most of the Devensian organic sites on Scilly (e.g. Carn Morval, Porth Askin, Porth Seal; Figure 8.1) have been severely disrupted by deposition of the overlying head. At Watermill Cove, however, only the uppermost parts of bed 6 have been deformed.

Interpretation

The sand unit (bed 1), along with the raised beach cobbles along the southern shore of Watermill Cove, form the stratotype of the Watermill Sands and Gravel (Scourse, 1991). This is the basal stratigraphic unit of the Quaternary succession on the islands, and is clearly of littoral origin. The sand has a dominant mode of 1.0ϕ on the coarse/medium sand boundary with moderate to good sorting. Though it resembles dune sand it lacks the high kurtosis characteristic of such sediments and may therefore be of backshore origin.

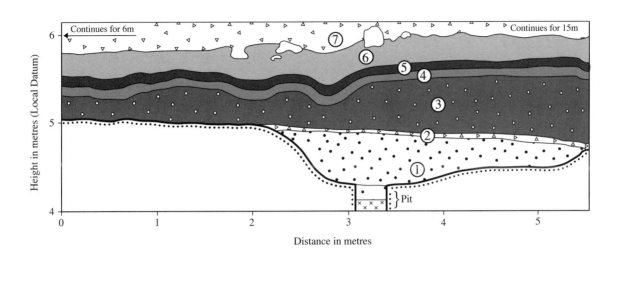

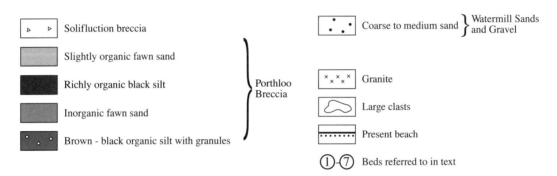

Figure 8.5 The Pleistocene sequence at Watermill Cove. (Adapted from Scourse 1991.)

The thin head (bed 2), like the more massive head (bed 7) above the organic sequence, is interpreted as a solifluction deposit which must have spread across the surface of the exposed beach. To the north-west of the organic sequence a prominent quartz-porphyry dyke (elvan) crosses the shore platform from south-west to north-east. Clasts derived from this dyke can be found in the main body of head (bed 7), indicating that the dominant direction of solifluction was from the east, north-east or north in response to the local slope morphology.

Solifluction is interpreted by Scourse (1991) to be the cause of the accumulation of the organic sequence at the site. Solifluction down the slopes of the neighbouring headlands would have ponded any small stream flowing down the valley into the Cove, forming a small lake or large pond into which the sediments accumulated. The organic content of beds 3–6 varies conversely with coarse minerogenic content; organic content reaches 10% in beds 3 and 5 coincident with minima in the

coarse minerogenic fraction. The varying sediments of these beds are therefore interpreted to represent inwashings of coarse minerogenic sediment, probably associated with active solifluction within the catchment, contrasting with phases of quiescent organic sedimentation.

The lacustrine origin of the organic sequence is further supported by the occurrence of obligate aquatic taxa in the pollen spectra. These include *Sparganium* type, *Potamogeton*, cf. *Sagittaria*, and the algae *Botryococcus* and *Pediastrum*. These are all represented by quite high frequencies in beds 3 and 5.

The pollen diagram (Scourse, 1991; Fig. 14, p. 420) from beds 3–6 can be divided into two assemblage zones on the basis of changes in the frequency of *Pinus*. Zone WC1 is dominated by herb taxa, particularly Cyperaceae, Gramineae, *Solidago* type and Rubiaceae, and occurs within bed 3. *Pinus* values remain below 1% throughout zone WC1, but rise to a peak of 51% in WC2 with a concomitant decrease in the frequencies, though not the diversity, of the herb taxa. Zone WC2 coincides with beds 5 and 6. Intact *Pinus* grains consistently only contribute around 10% of total *Pinus*. As in the case of the Carn Morval pollen diagram (see above), which also contains a distinctive *Pinus* peak, this is not interpreted as an increase in the local representation of this tree. Rather, it is explained as a combination of long-distance transport with differential weathering and a decline in the pollen productivity of the local vegetation probably resulting from climatic deterioration.

The importance of the obligate aquatic taxa, and the behaviour of Cyperaceae, lends support to the evidence of the sediments themselves, and the geomorphological context of the site, that this sequence formed in a small lake, and that the high humic content of the organic material represents allochthonous material from within the catchment. The very high values for Cyperaceae and *Sparganium* type in the basal level suggest ponded conditions, the subsequent decline in both representing the infilling of the basin with minerogenic sediment (bed 4); this is supported by the lack of pollen in this bed. The renewed importance of the obligate aquatic taxa, and in particular cf. *Sagittaria*, which reaches a peak of 12%, in beds 5 and 6 suggests a second phase of ponding. The stratigraphy and pollen record therefore indicate two ponding episodes at Watermill Cove.

Page (1972) obtained two radiocarbon dates from the organic material corresponding to Scourse's (1991) beds 3–6; these were 21 200 + 900/− 600 BP (GaK-2471) and 22 200 + 400/− 400 BP (T-833). Scourse (1991) obtained a further eight dates from humic and humin fractions of beds 3 and 5; these were from samples taken up to 3 m back from the face of the section ('second series' samples) in addition to samples from only 0.75 m from the face ('first series' samples). Apart from the bed 5 second series determinations, the humic extract samples are younger than the humin residues, and the second series dates older than the first series. This pattern is very similar to the radiocarbon results from other similar organic sites on Scilly; the differences in radiocarbon content indicate that contamination by modern humus, probably transported by percolating groundwater from recent/modern soil, is a problem at the face and the base of the profile, but that this contamination is less of a problem higher in the sequence. It is probable that the thin head (bed 2) acts as an impermeable layer,

and that groundwater charged with modern carbon has percolated, and continues to percolate, through bed 3, which acts as an aquifer, along the contact with the head, modern humus being deposited in the process. The older second series bed 5 humic extract date can be explained in terms of the inwashing of older mor humus into the basin; the likelihood of such allochthonous humus being present is supported by the pollen diagram.

On the basis of this analysis of the radiocarbon content of the materials in beds 3–6, Scourse (1991) suggests that the most reliable determination for bed 3 is 33 050 + 960/− 860 BP (Q-2408) and for bed 5, 26 680 + 1410/− 1200 BP (Q-2407). This demonstrates that the organic sequence accumulated during the latter part of the Middle Devensian, and provides maximum and minimum ages for the underlying and overlying units. The overlying head (bed 7) must post-date 26 ka BP, and the underlying head (bed 2) and the Watermill Sands and Gravel must pre-date 33 ka BP. The age of the Watermill Sands and Gravel (bed 1) is uncertain everywhere on Scilly, but by correlation with similar raised beaches on mainland Britain it is probably of interglacial age.

Scourse (1991) correlates all the sedimentary units at Watermill Cove, other than the basal raised beach, with the Porthloo Breccia (Figure 8.3). In the absence of the Old Man Sandloess at the site it is not possible to correlate beds 2–7 definitively with either the upper or lower Porthloo Breccia. However, the radiocarbon dates and pollen spectra are very similar to analogous sequences at other organic sites on Scilly where stratigraphic relationships can be established with the Old Man Sandloess (Carn Morval, Porth Askin) indicating that probably at least bed 2 and beds 3–6 can be correlated with the lower Porthloo Breccia.

Conclusion

Watermill Cove contains the best exposures on Scilly of raised beach sediments deposited in an interglacial temperate stage, before a glacier advanced as far as the northern islands around 19 ka BP. The site also contains excellent exposures of organic sediments deposited under cold tundra-like conditions between 33 ka and 26 ka BP. Pollen analysis of these sediments reveals an open grassland type vegetation with no trees.

OLD MAN, GUGH, ST AGNES
J. D. Scourse

Highlights

This site affords the finest exposures of aeolian sandloess on the Isles of Scilly. This material, dated by thermoluminescence (TL) and optical techniques to the Late Devensian, is widespread on the islands but the quality of the exposures here have led to the selection of Old Man as the stratotype for the Old Man Sandloess. This unit lies between layers of granitic head at the site, thus demonstrating its stratigraphic context. The Old Man Sandloess is of significance because it has been demonstrated that there is a genetic link between this material and the Scilly Till; the absolute dates on the Sandloess therefore relate directly to the age of the glacial event which affected Scilly.

Introduction

Barrow (1906) reported the widespread occurrence of 'iron-cement', a ferruginous and micaceous sandy silt, on Scilly. He noted that this material often forms the highest part of the Head, but that its contact with the Head is ' ... distinctly sharp, suggesting a somewhat different origin from the latter' (Barrow, 1906; p. 20). He commented that 'The origin of this curious deposit is no means clear; it is distinctly micaceous and far finer in texture than normal blown sand, as well as of a totally different colour and composition' (Barrow, 1906; p. 21). He speculated that the deposit might be of aeolian origin derived from material released from frost action; 'The idea of invoking the aid of frost in the production of this material is based on its substantial identity both in composition, mode of occurrence, and geological age, with the matrix of an undoubted glacial deposit ... ' (Barrow, 1906; p. 21). Barrow therefore hinted at a possible genetic relationship between the 'iron-cement' and the glacial material found on the northern islands.

Arkell (1943) commented that ' ... in the Scilly Isles, where the main head is covered by loess identical with the *limon* of Brittany ... ' (Arkell, 1943; p. 159) and later quotes W.B.R. King in a personal communication linking the loess of Brittany, Normandy, the Channel Islands and the Isles of Scilly. This was the first recognition that the 'iron-cement' described by Barrow was in fact loessic.

However, Mitchell and Orme (1967) did not recognize loessic sediments on the islands. They were critical of Barrow's interpretation of the 'iron-cement' and proposed a different hypothesis for its origin:

'In post-glacial time, soil-forming processes have developed a podsol profile on the Upper Head. The B-horizon is deeply coloured by iron and humus, and is sometimes cemented by silica and iron oxide. This material Barrow described as iron-cement, and he regarded it as being essentially the same as the cemented outwash gravel ... This confusion of two deposits of entirely different origin vitiates much of Barrow's description of the glacial deposits' (Mitchell and Orme, 1967; p. 68). This hypothesis was rejected by Scourse (1991).

The existence of loessic sediments on the islands was supported by Catt and Staines (1982). On the basis of particle-size analysis and heavy mineralogy they recognized the significance of coarse loess as a soil parent material on Scilly. Wintle (1981) dated two samples of this material from St Mary's and St Agnes to 18 600 ± 3700 BP (QTL–1f and QTL–1d). Further absolute dates on this material were published by Smith *et al.* (1990) using the optical technique.

Scourse (1991) identified coarse loess as a significant sedimentary unit within the Pleistocene sequence on the islands (Figure 8.3). He was able to demonstrate a genetic relationship between this material and the Scilly Till on the basis of particle-size distributions, mineralogy, patterns of sedimentation, stratigraphy and absolute age. He defined the unit as the Old Man Sandloess from this site on Gugh, St Agnes, and noted a limited distribution of the material south of the ice limit across the northern islands (Figure 8.1). North of the ice limit the Sandloess has been mixed with glaciofluvial (Tregarthen Gravel) and glacial (Scilly Till) sediment, and solifluted downslope as the Hell Bay Gravel (Figure 8.3).

The Old Man Sandloess is of stratigraphic significance because it provides evidence in support of the Late Devensian glaciation of the northern islands (Scourse, 1991). The absolute age of this unit, derived from the TL and optical methods, directly relates to the age of the glacial event because of the demonstrated genetic link between the Old Man Sandloess and the Scilly Till.

Description

The Old Man section is located on the north-east coast of Gugh (St Agnes; SV 893085); the name 'Old Man' is derived from a prominent standing stone positioned a few metres upslope. The section comprises:

3.   Granitic head (0.2 m)
2.   Light brown to ochre sandy silt (1.0 m)
1.   Coarse granitic head (1.5 m)

The lower head contains occasional very well-rounded granite boulders, probably incorporated into the head from an underlying raised beach. There are good exposures of cobble-rich raised beach deposits underlying the head at other sites along the eastern coast of Gugh indicating that this is the probable source.

The sandy silt is moderately well sorted with a dominant mode in the coarse silt fraction, and a subdominant mode in the medium sand fraction; the grain-size distribution has a high kurtosis. The unit possesses a columnar structure and small pin-hole voids are visible in the sediment fabric.

Interpretation

The grain-size and structural characteristics of the sandy silt (bed 2) are characteristic features of *in situ* loess (Mellors, 1977). This material at Old Man is typical of the Scilly loesses in containing consistently more sand than clay (Catt and Staines, 1982; Scourse, 1991). It is too coarse to be defined as true loess, but too fine to be defined as coversand. The Old Man material is typical in having a dominant modal class either in the coarse silt or fine sand fraction, with total sand usually more than 25% and total clay less than 10%. In The Netherlands and Belgium, sediment with these characteristics is commonly distributed in a transitional belt between true loess and coversand, and is known as 'sandloess'. Comparable material on Scilly was thus defined as the Old Man Sandloess by Scourse (1991) from the type-site at Old Man.

Scourse (1991) defined four facies of the Old Man Sandloess on the basis of structural and grain-size criteria. The stratotype of the Old Man Sandloess is typical of facies A. This is interpreted as *in situ* material because it contains the columnar structure and pin-hole voids characteristic of *in situ* loess (Mellors, 1977). The other facies are interpreted to have been deposited through water (facies B), or reworked by fluvial (facies C) or soliflual (facies D) processes. All the facies described by Scourse (1991) represent stages along a continuum from *in situ* material to sandloess intermixed to such an extent with other material that its identity is only

barely recognizable. Scourse (1991) further noted that the matrix of the soliflual Hell Bay Gravel is identical to the Old Man Sandloess, indicating that the Hell Bay Gravel comprises glacially derived material (Scilly Till, Tregarthen Gravel) thoroughly mixed with Old Man Sandloess and reworked downslope by solifluction. He noted that the Old Man Sandloess is confined to the area outside the ice limit (Figure 8.1).

Two samples of the Old Man Sandloess have been dated to 18 600 ± 3700 BP (QTL–1f, sample from St Mary's; QTL–1d, sample from St Agnes) using the TL method (Wintle, 1981). These dates therefore suggest that this material is of Late Devensian age. This interpretation has been partially supported by two optical dates of 20 ka ± 7 ka BP and 26 ka + 10 ka/− 9 ka BP (Smith *et al.*, 1990) from the same material. These dates are stratigraphically significant because Scourse (1991) has interpreted a genetic link between the glacigenic units of the northern Scillies (Scilly Till, Tregarthen Gravel) and the Old Man Sandloess. The relative coarseness of the Sandloess is interpreted as a function of proximity to glacially derived source material, and the mineralogy of the Scilly Till is sufficiently similar to the Sandloess to suggest a genetic relationship between the two units (Catt, 1986; Scourse, 1991). Aeolian loessic processes in association with the glacial advance therefore resulted in the contemporary deposition of the Old Man Sandloess in the southern Isles of Scilly with glacial deposition in the northern islands. The Late Devensian dates for deposition of the Old Man Sandloess can therefore be used to constrain the age of the glacial event.

At locations where the Old Man Sandloess is absent, it is not possible to discriminate between the upper and lower units of Porthloo Breccia (Figure 8.3). At sites such as Old Man, however, the Sandloess lies clearly between two units of granitic head. The dates on the Sandloess therefore also help to constrain phases of active solifluction in the the Isles of Scilly.

Conclusion

This site contains the best exposures on the Isles of Scilly of sandy silts deposited by wind action beyond the southern limit of an ice sheet which crossed the northern islands at about 19 ka BP. The silts were picked up from the outwash plain in front of the glacier by strong winds, and therefore contain the same minerals found in the glacial deposits in the northern islands. The silts have yielded absolute dates which help to constrain the age of the glaciation of the Isles of Scilly to the last major ice-sheet glaciation of the British Isles, the Late Devensian.

PENINNIS HEAD, ST MARY'S
J. D. Scourse

Highlights

This site contains some of the most spectacular granite landforms in the British Isles. These include the finest development on the Isles of Scilly of the 'castellated' or 'mammilated' tor forms which are characteristically found only outside the glacial limit which straddles the islands.

Introduction

Although the granite landforms of Peninnis Head (Figure 8.1) have been mentioned briefly by a number of authors (Barrow, 1906; Mitchell and Orme, 1967), they have never been the focus of a major individual study. Scourse (1986, 1987) discussed the general aspects of tor morphology on the islands and cited the examples at Peninnis Head as constituting some of the finest granite landforms in Britain. He (Scourse, 1986, 1987, 1991) further drew attention to the association between different tor forms and the glacial limit (Figures 8.1 and 8.6) which he, and others (Mitchell and Orme, 1967), were able to identify across the islands.

Description

Scourse (1986, 1987, 1991) mapped four tor forms on the Isles of Scilly (Figure 8.6). Peninnis Head (SV 911094) contains the largest concentration of forms A, B and C in the islands, and the finest individual examples of each form (Figure 8.7). Horizontal tors (form A) are characterized by extensive sub-horizontal discontinuities separating large granite slabs which typically touch at only a few points. These resemble 'pedestal' features locally called 'logan stones'. Elsewhere this particular tor form has been described as 'mammilated', 'castellated' or 'lamellar' (Waters, 1955). Vertical tors (form B) are characterized by vertical discontinuities with granitic rubble often filling the voids between granite slabs. Form C, hillslope tors, are a coastal variant of forms A and B.

Interpretation

Granite tors occur throughout the Isles of Scilly, but they are particularly well developed to the south of the ice limit which has been identified across the islands (Figure 8.1). South of this limit all tors can be classified as forms A, B or C, as described above, or resultant transitional or composite forms (Figure 8.6). These forms are well displayed on St Agnes, where ornate, individual, weathered granite slabs have been assigned popular local names, on Annet, the Western Rocks, St Mary's and the southern part of Tresco. However, the scale, concentration and extent of the features at Peninnis Head justify selection of this site to represent these landforms which typify the Scillonian landscape south of the ice limit.

Examples of tor form D (Figure 8.6) have only been identified north of the reconstructed ice limit. This form is smoothed and rounded, with all loose material removed. Alternatively, forms A, B and C are all ornate and contain a large proportion of loose or delicately shaped rock masses.

The distribution of the different tor varieties provides a key to understanding their evolution. The lower slopes of tor forms A, B and C at Peninnis Head, as elsewhere on the islands, are mantled by thick periglacial slope deposits or 'head' (Porthloo Breccia; Figure 8.3) which have been interpreted by most workers as being of soliflual origin (e.g. Scourse, 1987). This provides strong evidence that exhumation of the corestones of these tors took place under periglacial conditions during the Pleistocene. The radiometric dates on organic sediments from within the mantling Porthloo Breccia (Figure 8.3) provide valuable evidence on the timing and rate of tor exhumation during periglacial conditions of the late Middle and early Late Devensian.

Tors attributable to form D, however, are quite distinct in resembling roches moutonnées; a good example is provided by Round Island (Figures 8.1 and 8.6) in the northern Isles of Scilly. The clear association between these different landforms and the proposed ice limit suggests that the smoothed form D is a product of glacial erosion. Despite the fact that some authors have interpreted elaborate tor forms as occurring within former ice limits (Dahl, 1966; Sugden, 1968; Clapperton, 1970), Scourse (1987) points out that it is difficult to envisage how the more delicate tor forms, typified by those at Peninnis Head, could have survived being overridden by ice.

The origin of tors has long been a matter of

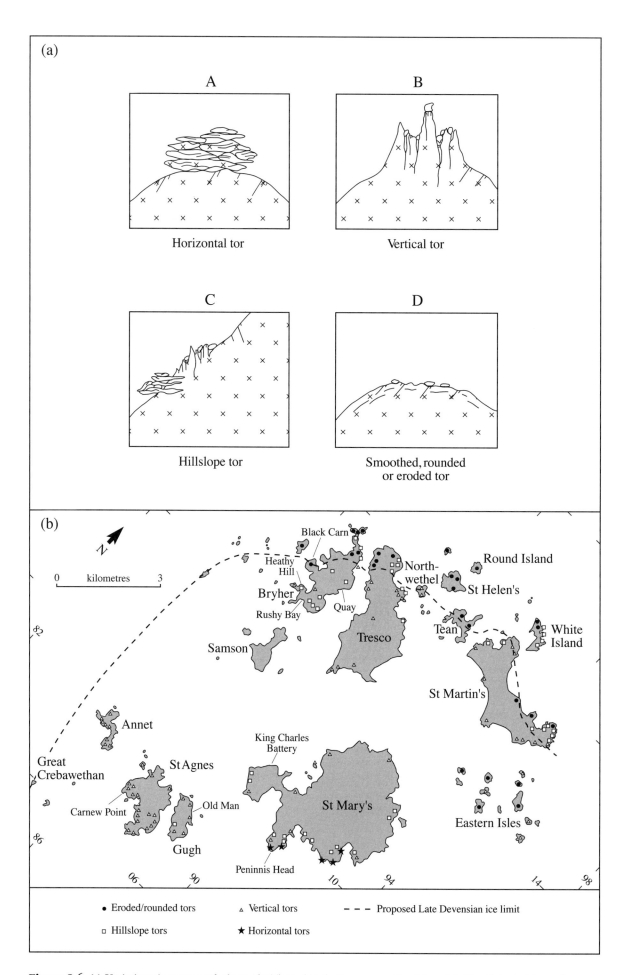

Figure 8.6 (a) Variations in tor morphology. (b) Their distribution across the Isles of Scilly. (Adapted from Scourse 1986.)

Figure 8.7 The spectacular development of horizontal tors (form (a); Figure 8.6) on the eastern side of Peninnis Head. (Photo: S. Campbell.)

considerable debate, and the various mechanisms and processes proposed are discussed more fully in Chapter 4. Some authors have invoked a two-stage process of tor formation involving deep chemical weathering during the Tertiary followed by exhumation of corestones by Pleistocene periglacial processes (Linton, 1955). Others have argued that Pleistocene periglacial activity alone was responsible for the tors (Palmer and Radley, 1961; Palmer and Nielson, 1962). The Scilly tors, and the Peninnis Head features in particular, therefore support the widely recognized implication of periglacial slope processes in tor exhumation. However, there is no evidence at Peninnis Head, or

at any other tor site on Scilly, which might point to an earlier phase of deep chemical weathering during the Tertiary.

Conclusion

Peninnis Head contains the largest concentration, and the finest examples, of granite tors in the Isles of Scilly. The association of these features with dated slope deposits formed under cold climates elsewhere on Scilly provides unique evidence of the timing and conditions of their formation. The intricate and delicate tor forms at Peninnis Head contrast with the much smoothed and apparently ice-eroded tors which occur in association with glacial sediments in the northern Isles of Scilly. They therefore provide supporting evidence that the southern islands remained unglaciated during the proposed Late Devensian (*c.* 19 ka BP) glaciation of the northern islands.

PORTH SEAL, ST MARTIN'S
J. D. Scourse

Highlights

Mitchell and Orme (1967) defined Porth Seal as the type-site for the younger of the two raised beach deposits they identified on Scilly. Organic beds found within the head at this site have yielded pollen consistent with deposition under periglacial conditions, and the beds have been radiocarbon dated to the late Middle and early Late Devensian. These absolute dates are critical for establishing a chronology of Pleistocene events on Scilly.

Introduction

Mitchell and Orme (1967) recognized two distinct raised beach deposits on the Isles of Scilly. An erratic-free deposit, the Chad Girt Raised Beach, was assigned a Hoxnian age, while an erratic-rich deposit, the Porth Seal Raised Beach, was believed to post-date the proposed Gipping (= Wolstonian/Saalian) glaciation of the northern islands, and was accordingly assigned to the Eemian (= Ipswichian). Mitchell and Orme were unable to identify a site where both beach deposits could be seen unambiguously in stratigraphic superposition, although they believed that two separate raised beach deposits were present at Porth Seal (Mitchell and Orme, 1967; p. 73). However, the sequence at Porth Seal was subsequently depicted by Stephens (1970a; Figure 8.8), based on an unpublished field sketch made in 1965 by F.M. Synge, as showing two raised beach deposits separated by periglacial head (Figure 8.8; Section 3). The lower raised beach deposit was correlated with the 'erratic-free' Chad Girt Raised Beach, the upper being the 'erratic-rich' Porth Seal Raised Beach.

Bowen (1981) suggested that the Porth Seal Raised Beach was not marine in origin, but that instead it comprised soliflucted granite corestones. Scourse (1991) re-interpreted the upper beach as a

soliflucted facies of the lower (= Chad Girt Raised Beach) (Figure 8.8). Similar stratigraphic relationships were also described at Northward Bight. He correlated this raised beach with the Watermill Sands and Gravel (Figure 8.3).

Description

Porth Seal (SV 918166) lies on the north-west coast of St Martin's (Figure 8.1), and the most complete sections occur on the south side of the bay. Scourse (1991) records the critical section described by Mitchell and Orme (1967) and Stephens (1970a), and demonstrates the stratigraphic relations between this sequence and organic materials at the site (Figures 8.8 and 8.9). Scourse's (1985a, 1991) composite sequence from Sections 1 and 2 (Figure 8.8) can be summarized thus:

9.   Coarse, granitic solifluction breccia
8.   Fawn granular sand
7.   Organic silty sands with granite clasts and quartz granules
6.   Coarse, white granular sand
5.   Black, richly organic fine silt
4.   Coarse, white granular sand
3.   Soliflucted raised beach deposits (= Porth Seal Raised Beach of Mitchell and Orme (1967))
2.   Solifluction breccia
1.   Raised beach deposits (= Chad Girt Raised Beach of Mitchell and Orme (1967))

Matrix-supported, rounded, raised beach cobbles (bed 3) are exposed towards the western end of the described section overlying a unit of granitic blocky head (bed 2). A complex unit (beds 4-8), including organic materials, overlies these beach cobbles laterally to the east. This metre-thick deposit consists of several distinct, internally stratified, beds. The lowest (bed 4) consists of coarse white granitic sand and granules, and rests directly on the underlying (soliflucted) beach cobbles. This sand lenses out to the west such that the overlying bed of black, richly organic, fine silt (bed 5) also rests directly on the beach cobbles at its western limit. In places the matrix of the cobble deposit (bed 3) comprises this organic silt. Bed 6 is similar to bed 4, and is overlain by bed 7, a relatively thick deposit of internally variable, dark brown, organic silty sand with granite clasts and quartz granules. Beds 4-8 are overlain by up to 5 m of coarse granitic head (bed 9).

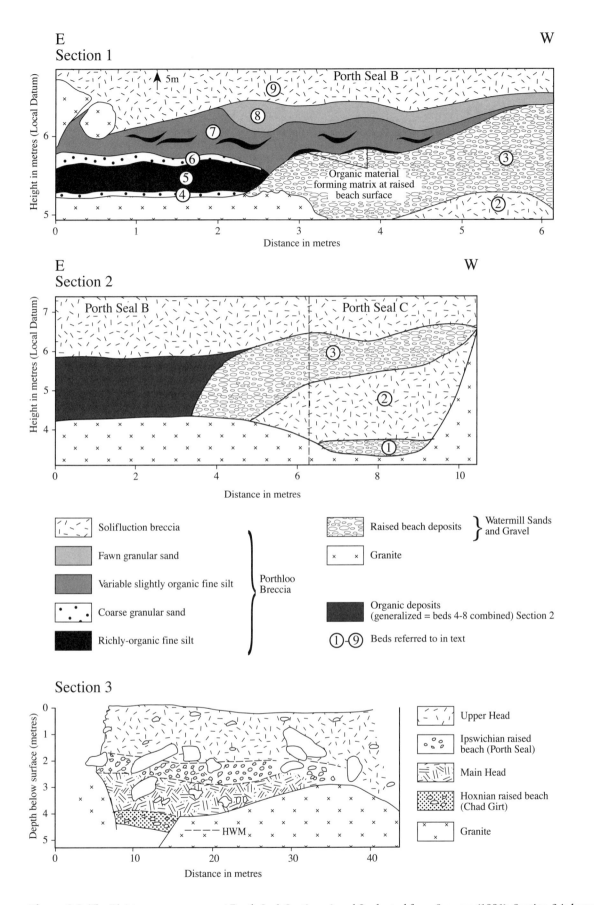

Figure 8.8 The Pleistocene sequence at Porth Seal: Sections 1 and 2 adapted from Scourse (1991); Section 3 is based on a field sketch made in 1965 by F.M. Synge and subsequently figured in Stephens (1970a) and Kidson (1977).

The granitic head (bed 2) can be traced to the west as the Gipping (= Wolstonian) 'head without erratics' of Mitchell and Orme (1967) (Figure 8.8; Section 3). At this point, in a small gully within the basal granite bedrock (Figure 8.8; Section 2 = Porth Seal C), the head is underlain by another unit of raised beach cobbles (bed 1). Mitchell and Orme (1967) describe this section as follows: ' … at one point on the south of the bay there is a channel about 15 yds long and 2 yds deep in the shore-platform. The bottom of the channel has 6 inches of beach shingle on its floor, and the channel itself is filled with coarse head' (Mitchell and Orme, 1967; p. 73).

Interpretation

Mitchell and Orme (1967) correlated the lowermost raised beach deposit (bed 1) at Porth Seal with the Chad Girt Raised Beach (Hoxnian). The overlying head was regarded as the Gippingian (= Wolstonian) Lower (or Main) Head. They inferred that a shore platform was then trimmed across both the Lower Head and the bedrock, and the main unit of cobbles (the Porth Seal Raised Beach; bed 3) deposited. They defined bed 3 as the stratotype for the Porth Seal Raised Beach, which they suggested was equivalent to the Upper Belcroute Bay Beach of Jersey (Mourant, 1933, 1935) and of Eemian (= Ipswichian) age. They identified the head covering the Porth Seal Raised Beach as Upper Head of Weichselian (= Devensian) age.

Bowen (1981) suggested that the Porth Seal Raised Beach might not be a raised beach at all, but rather a body of soliflucted granite corestones. Scourse (1991) alternatively interpreted the Porth Seal Raised Beach as a soliflucted facies of the underlying (= Chad Girt) raised beach. Rather than the direct litho-chronostratigraphical interpretation offered by Mitchell and Orme (1967), he identified only one *in situ* raised beach deposit at Porth Seal, represented by bed 1 or the 'Chad Girt Raised Beach'. He suggested that the upper parts of this deposit have become entrained within the overlying solifluction flow and appear in section as an upper, younger, raised beach deposit (bed 3). Such tongues of incorporated beach material occur at many sites in South-West England where beach sediments have been overridden by solifluction lobes and sheets. These often appear in two-dimensional cliff sections as stratigraphically distinct raised beach units. Sedimentological criteria invoked at

Porth Seal by Scourse (1991) to support this interpretation include the matrix- rather than clast-supported character of bed 3, and the clearly soliflual rather than littoral origin of this matrix.

Scourse (1991) correlates bed 1 (the Chad Girt Raised Beach of Mitchell and Orme) with the Watermill Sands and Gravel (Figure 8.3), and beds 4–9 with the Porthloo Breccia. Beds 2 and 3 are classified as facies Aa 'deformation breccia' (Scourse, 1987).

Pollen analysis of samples from organic materials in beds 4–8 have yielded spectra very similar to those described from comparable sequences at Carn Morval and Watermill Cove (Scourse, 1991; see above). Important elements include Gramineae, *Solidago* type and Rubiaceae, with Cyperaceae, *Achillea* type and *Plantago* spp. as minor contributors. A low but consistent presence of the obligate aquatic taxon *Sparganium* type, with *Potamogeton* and *Typha latifolia* in some samples, supports a lacustrine origin for this organic material. These spectra are consistent with open grassland vegetation. The organic sediments probably accumulated in a basin on the exposed shore platform during the onset of active solifluction (Figure 8.9).

The richly organic beds (5 and 7) were radiocarbon dated by Scourse (1991). The samples were taken from different distances from the cliff face and each sample split into humic and humin fractions to identify the extent and sources of contamination. Six radiocarbon determinations were obtained; the most reliable dates were interpreted by Scourse (1991) to be 34 500 + 885/− 800 BP (Q-2410) for bed 5 and 25 670 + 560/− 530 BP (Q-2409) for bed 7. These determinations therefore suggest that beds 4–8 were deposited during the late Middle and early Late Devensian, and that the underlying head and raised beach units (beds 1–3) pre-date the late Middle Devensian. It follows that the overlying head (bed 9) post-dates the early Late Devensian. These dates form part of a series of radiocarbon determinations from a network of organic sites within the Porthloo Breccia throughout Scilly which provide critical chronological data on the age of the identified sedimentary units (Figure 8.3).

Conclusion

Porth Seal is a critical site because it contains rare organic beds which can be dated by the radiocarbon method. Such dating has shown that these

Figure 8.9 Richly organic sediments lying beneath periglacial head at Porth Seal, St Martin's. (Photo: J.D. Scourse.)

sediments were deposited between 25 ka and 35 ka BP. These dates help to constrain the ages of the underlying and overlying sediments. The organic beds also contain abundant fossil pollen grains and spores which indicate that between these dates Scilly was characterized by open grassland vegetation in a cold Arctic climate. Porth Seal is also important because it shows controversial evidence of possibly two interglacial highstands of sea level.

BREAD AND CHEESE COVE, ST MARTIN'S
J. D. Scourse

Highlights

This is the most important Quaternary site on the Isles of Scilly. It is crucial to the arguments concerning the evidence for, and age of, the Late Devensian glaciation of the northern islands. The

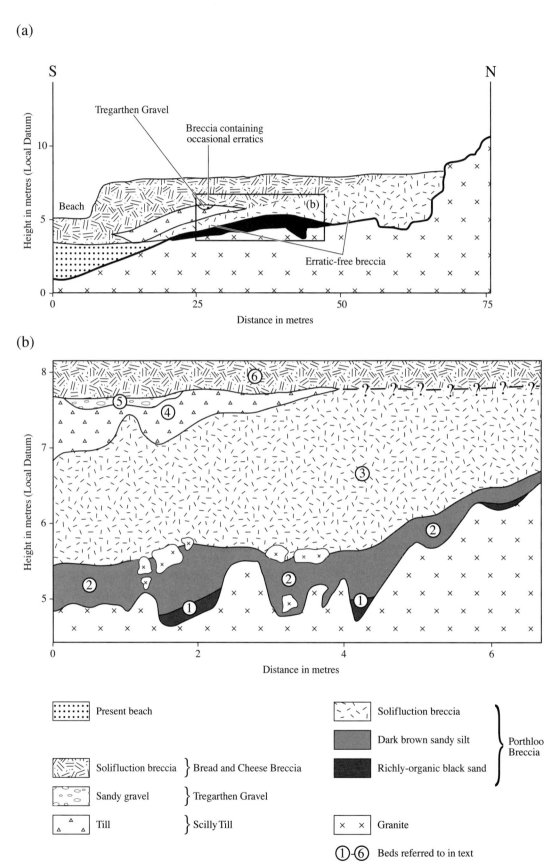

Figure 8.10 The Pleistocene sequence at Bread and Cheese Cove. (Adapted from Scourse, 1991.)

site contains the stratotypes of the Scilly Till and the Bread and Cheese Breccia.

Introduction

The significance of Bread and Cheese Cove was first recognized by Mitchell and Orme (1967) who described glacial sediments (presumed Wolstonian) cropping out in the low coastal cliffs at the cove. Scourse (1991) undertook a reinvestigation of the sedimentology of these 'glacial' deposits to establish their precise depositional origin and, in particular, whether post-depositional downslope movement had occurred. He also discovered brown humic organic beds underlying the supposed glacial sequence at the base of the cliffs. Pollen analyses of these organic deposits indicated open grassland vegetation consistent with deposition in a periglacial regime, and enabled correlation with other organic sequences in the islands – at Watermill Cove, Carn Morval, Porth Askin and Porth Seal (Figure 8.1) – all of which have been radiocarbon dated to the late Middle and early Late Devensian. Radiocarbon determinations on the organic sediments at Bread and Cheese Cove yielded Holocene ages (*c.* 10–7.5 ka BP) which are clearly incompatible with the lithostratigraphical and pollen evidence, and indicate sample contamination by younger carbon.

Description

Bread and Cheese Cove (SV 940159) lies on the northern coast of St Martin's (Figure 8.1). Mitchell and Orme (1967) recorded the presence of till and outwash gravel between the Lower (Main) Head and the Upper Head at this site. Their section is still readily observed, but Scourse (1991) was further able to identify the presence of organic sediments at the base of the section (Figure 8.10). A granite wave-cut platform rises towards the south-west and underlies a coarse granitic solifluction breccia (bed 3). At the base of the section there is a deposit of very coarse granite rubble and boulders, which extends outwards from the modern cliff across the granite platform. Forming the matrix between these boulders, and also a coherent unit towards the base of the section, is a humic horizon (bed 1), a dark brown sandy silt with quartz granules. Pits dug at the base of the section by Scourse (1991) revealed a richly organic black sand (bed 1) resting either on the granite boulders or the wave-cut plat-

form. Bed 3 is overlain gradationally by bed 4, a clay-rich light brown diamicton containing abundant erratic clasts. Bed 4 forms a lens some 22 m in length with a maximum thickness of 2 m. At one point (Figure 8.10) the diamicton is overlain by a small lens of iron-cemented sandy gravel (bed 5), which also contains abundant erratic clasts. Both the diamicton and the gravel are overlain by up to 4 m of coarse, dominantly granitic, breccia (bed 6) which contains occasional erratic clasts. The section is capped by a number of large granite boulders.

Interpretation

On the basis of lithological characteristics and pollen assemblage zones, Scourse (1991) correlated beds 1–3 with the Porthloo Breccia and bed 5 with the Tregarthen Gravel (Figures 8.3 and 8.10). He defined (Scourse, 1991) bed 4 as the stratotype of the Scilly Till, and bed 6 as the stratotype of the Bread and Cheese Breccia.

Pollen spectra from the basal organic deposits (bed 1) are dominated by herb taxa, in particular Gramineae, *Solidago* type, Rubiaceae and *Ranunculus repens* type. Tree taxa are almost completely absent, as are plant macrofossils. There are no significant changes in the pollen stratigraphy through the sequence. This assemblage is typical of open grassland vegetation, and is very similar to the spectra from other comparable organic sequences on the Scillies, all radiocarbon dated to the late Middle or early Late Devensian.

The coarse breccia deposits (beds 3 and 6) were interpreted by Scourse (1987, 1991) as solifluction deposits laid down under periglacial conditions. There is no evidence for breccia deposition, and therefore of solifluction, between the granite platform and the organic sediments. The organic sediments fill cracks in the surface of the platform, and the interstices between granite boulders associated with the platform. These boulders probably represent an immature beach deposit, organic sedimentation having occurred directly on the surface of the beach. Many of the sand grains and granules found within the organic sediment are extremely well rounded, suggesting a beach origin. This interpretation is supported by the presence of unambiguous raised beach deposits (lithostratigraphically assigned to the Watermill Sands and Gravel) in a similar stratigraphic position beneath the Porthloo Breccia at many sites on the Isles of Scilly. The organic deposits are probably lacustrine,

The Quaternary history of the Isles of Scilly

ponding having been effected by contemporaneous local solifluction. The sediments and pollen spectra indicate that beds 1–3 accumulated during periglacial climatic conditions.

The stratotype of the Scilly Till (bed 4) is dark yellowish-brown, drying to light yellowish-brown. It is largely non-calcareous, but mineralogical data (Catt, 1986) suggest that it is not heavily weathered, containing a number of easily weatherable minerals such as muscovite, glauconite, chlorite, biotite, augite, apatite, olivine and calcite. It contains abundant siliceous sponge spicules (Jenkins, 1986), but no calcareous microfossils or macrofossils. These data suggest a marine derivation with subsequent partial decalcification. The Scilly Till at this site is geochemically similar to other Late Devensian tills from the Irish Sea Basin (Burek and Cubitt, 1991).

The Scilly Till is extremely poorly sorted and crudely stratified with a number of sub-horizontal iron-stained sand partings up to 1 mm thick. The contained clasts are very freshly striated and faceted, and consist of a wide variety of lithologies, including Cretaceous flint, Variscan greywackes and quartzites, red sandstones and schistose metamorphic rocks, in addition to local granitic material (Hawkes, 1991). The till also contains small granules of a distinctive light green glauconitic micrite which is derived from the Miocene Jones Formation offshore to the north (Pantin and Evans, 1984).

Three features set the Scilly Till apart from the underlying Porthloo Breccia: the high clay content, low clast concentrations and the diverse erratic and mineralogical assemblage. The geochemistry, distinctive erratic and mineralogical assemblage, abundance of clay and silt and presence of sponge spicules are all consistent with derivation from the offshore area to the north of the Isles of Scilly.

A coarse lag of angular granite boulders occurs at the base of the Scilly Till at its contact with the Porthloo Breccia. The upper contact with the Bread and Cheese Breccia is clearly soliflucted.

Statistical (eigenvalue) analysis of clast macrofabric data shows that the central and upper parts of the Scilly Till are very similar to lodgement tills from both modern (Dowdeswell and Sharp, 1986) and fossil (Rose, 1974) contexts, and are quite different in fabric character from the overlying and underlying solifluction deposits (Porthloo Breccia and Bread and Cheese Breccia). However, the base of the deposit is more characteristic of remobilized or slumped till. Given that the Scilly Till may be largely of lodgement origin, the fabric data are consistent with ice flow from north-west to south-east (Scourse, 1991).

Scourse (1991) concludes that while it is impossible to interpret the precise depositional origin of the Scilly Till from this one small exposure, it is different in a number of fundamental characteristics from undoubted soliflucted till, as represented by the Hell Bay Gravel at other sites (Figure 8.3).

The grain-size and lithological content of the sandy gravel (bed 5) suggest clear affinities with the stratotype of the Tregarthen Gravel (Figure 8.3) at the Battery (Castle Down) site on Tresco (see below). This material is interpreted by Scourse (1991) as a glaciofluvial gravel deposited very close to the Scilly Till.

In its general characteristics the Bread and Cheese Breccia (bed 6) is very similar to the Porthloo Breccia, and is similarly interpreted as a periglacial solifluction deposit. However, it does contain occasional erratic clasts clearly derived from the underlying Scilly Till and Tregarthen Gravel. It cannot, therefore, be correlated with the Porthloo Breccia on lithological grounds, even though it was deposited in a similar environment. It has therefore been defined as a separate stratigraphic unit (Scourse, 1991).

The large granite boulders capping the section have been interpreted by Scourse (1987) as fossil 'ploughing blocks' (Tufnell, 1972). Large boulders lying on the surface of solifluction sheets and lobes have been extensively reported from contemporary periglacial environments where they have been termed 'ploughing blocks' because they move at velocities faster than the flow of the deposits on which they rest.

The Bread and Cheese Cove sequence is unique because it is the only site on the Isles of Scilly where dateable organic sediments occur beneath the glacigenically derived units of the Bread and Cheese Formation of the northern Scillies (Figure 8.3). Humic extract dates of 7880 ± 180 BP (Q-2369) and 7830 ± 110 BP (Q-2411), and a humin date of 9670 ± 65 BP (Q-2368), are however clearly aberrant, with values very much younger than ^{14}C and TL dates on related stratigraphical units elsewhere on the Scillies, and the occurrence of glacigenic and solifluction deposits overlying material yielding Holocene-equivalent radiocarbon determinations clearly suggests that the samples are wholly unreliable. Furthermore, these results are inconsistent with the pollen spectra from this same unit.

Scourse (1991) attributes these erroneous results to contamination by modern carbon derived from rootlet penetration and/or groundwater deposition of modern humus. In particular, the impermeable surface of the granite immediately beneath the organic deposit concentrates groundwater flows through the overlying permeable organic material. It is therefore probable that post-depositional concentration of groundwater-derived humus has occurred in this material. Scourse (1991) has demonstrated, through split humic/humin radiocarbon analyses of other similar organic sequences on the Isles of Scilly, that such hydrogeological contexts consistently produce erroneously young radiocarbon determinations.

Conclusion

Bread and Cheese Cove provides the most complete vertical succession of Pleistocene sediments on the Isles of Scilly. It also contains the most extensive exposure of glacial sediments on the Scillies, providing vital evidence for the glaciation of the northern islands. These sediments appear not to have moved since they were originally deposited, and provide evidence for the direction of movement of the glacier with which they are associated. The sequence also contains meltwater sediments associated with the ice, and slope deposits laid down in cold non-glacial climates both before and after the glacial event. The site is also unique because it is the only place on the Isles of Scilly where organic deposits rich in fossil pollen have been discovered beneath the glacial sediments. These organic deposits provide evidence of open grassland vegetation on the islands prior to the ice advance. From other sites on the Scillies it is known that this type of vegetation was present on the islands around 30 ka BP. This therefore provides evidence that the glaciation must have occurred after this time, during the Late Devensian.

CHAD GIRT, WHITE ISLAND, ST MARTIN'S
J. D. Scourse

Highlights

This site affords excellent exposures through a sequence of raised beach and overlying periglacial slope deposits containing reworked glacial sediments of Late Devensian age. The site is of historical importance because it was selected in the early twentieth century as the type-site for the Quaternary sequence of the Isles of Scilly.

Introduction

Barrow (1906) was the first worker to place the erratic pebbles found at a number of locations on the northern islands into a stratigraphic context. He used Chad Girt to exemplify and demonstrate this context, in addition to showing the stratigraphic relations between other Pleistocene deposits. The site was discussed by Mitchell and Orme (1967), who used it as the type-site for their 'Chad Girt' Raised Beach. The site was reinterpreted by Scourse (1991).

Description

Barrow (1906) defined four main stratigraphic units in the Isles of Scilly: raised beach, head, ironcement and a glacial deposit. Barrow recognized the conglomerate of an old beach, now raised above the level of the present beach, resting on a shore platform. He regarded the beach as once much more extensive but having been largely eroded away leaving the old platform exposed in many places. He applied the term 'head' to the ' ... accumulation of angular or subangular fragments of granite in an advanced state of decomposition' (Barrow, 1906; p. 17). This head had reoccupied the position of the eroded beach and, resting on the old platform, had a terrace-like contour imparted to it. He was able to divide the head in certain localities into two parts, an 'Upper' and a 'Lower' or 'Main' Head separated by a 'curious glacial deposit' (Barrow, 1906; p. 18).

Barrow cited the section at Chad Girt (SV 926174) on White Island, St Martin's, as fixing the stratigraphic position of these identified units. He described the Isles of Scilly as surrounding an 'interior-sea'; the northern coast of St Martin's, including White Island, forms part of this island rim. The north and east shore of White Island consists of a linking chain of granite tors forming a steeply cliffed coastline; the western slopes of these tors descend gradually to the beach-dune coastline on the sheltered western shore adjoining Porth Morran. The rugged cliffs of the east and north shore are bisected by some deep fissures in the solid granite and overlying Pleistocene sediments, trending north-west to south-east, typified by Underland Girt and Chad Girt. The deep embay-

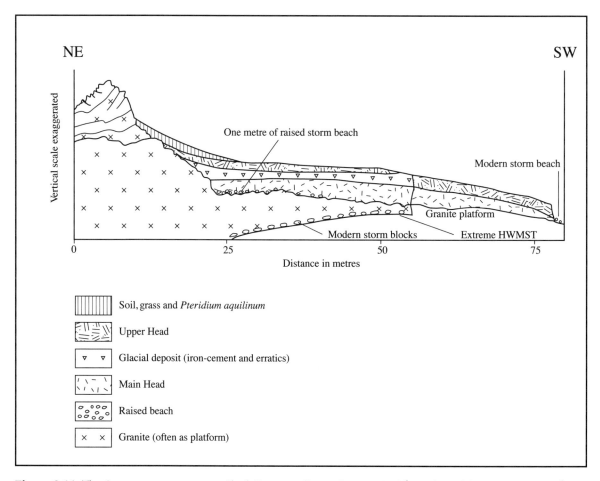

Figure 8.11 The Quaternary sequence at Chad Girt according to Barrow (1906). (Adapted from Scourse, 1986.)

ment of Chad Girt is separated by two parallel granite ridges.

The southern side of the Chad Girt fissure affords an excellent exposure of Pleistocene sediments resting on the solid granite. Here Barrow (1906; Figure 8.11) noted that the old beach rests on the bare granite and underlies the 'Main Head', ' … a glacial deposit in turn reposing on the latter' (Barrow, 1906; p. 16). Barrow's (1906) sequence is still clearly exposed, and is little disputed; there have, however, been differences of interpretation.

Interpretation

Mitchell and Orme (1967) provided an excellent interpretation of the geomorphological evolution of the site. They recognized that the Chad Girt embayment did not exist when the shore platform was cut, and that the east side of White Island then formed a continuous north-south rock ridge. Higher sea levels eroded a shore platform on both sides of this ridge, and beach deposits were laid down on the sheltered, western, Porth Morran side. Marine erosion in the 'post-glacial' then attacked the ridge from the east, cutting through the rock ridge and into the shore platform and overlying deposits on its west side.

Although Barrow (1906) was explicit in his interpretations of the basal cobble deposit at Chad Girt as a raised beach, and was able to recognize that the Main and Upper Heads were derived from weathered granite and showed evidence of having moved downslope, he was less sure about the 'glacial deposit'; 'The origin of this curious deposit is by no means clear … ' (Barrow, 1906; p. 21). He noted the high content of silica and iron oxide in the matrix, and the occurrence of lenticular patches of foreign stones within it. He argued that ' … a considerable portion of them [the foreign stones] must have been derived from an older deposit, as many of them are too well rounded to leave any doubt that they were derived from some gravel and not directly from the parent rock'

(Barrow, 1906; p. 23). Whatever their precise origin, Barrow was quite certain that they had been transported to the islands by ice; 'That these stones have been brought into their present position by ice admits of little doubt' (Barrow, 1906; p. 27). He believed their curious distribution to be unintelligible except by invoking some other means of transport than water. Further, 'It is quite clear that they [the foreign stones] must have been carried by floe-ice' (Barrow, 1906; p. 27).

Mitchell and Orme (1967) interpreted the head deposits as the products of solifluction under periglacial conditions, and the glacial deposit as of glaciofluvial (meltwater) rather than strictly glacial origin. Whereas Barrow only identified a single raised beach on the islands, Mitchell and Orme recognized what they believed to be two stratigraphically distinct raised beach deposits, the lower being erratic-free, the upper erratic-rich. They identified the raised beach at Chad Girt as the stratotype for the lower, erratic-free, beach. Mitchell and Orme placed the meltwater sediments in the Gipping (= Wolstonian) glacial stage, the raised beach in the Hoxnian, thus assigning the Lower Head to the Wolstonian and the Upper Head to the 'Last Cold Period' (= Devensian).

Scourse (1991) concurred broadly with previous sedimentological interpretations of the raised beach and solifluction deposits, assigning these units to the Watermill Sands and Gravel (raised beach), Porthloo Breccia (Lower or Main Head) and Bread and Cheese Breccia (Upper Head). However, he differed from Mitchell and Orme (1967) on the age of the various units, and from Barrow (1906) and Mitchell and Orme on the interpretation of the glacial deposit.

The grain-size, lithology and mineralogy of the 'glacial deposit' at Chad Girt is identical to the Hell Bay Gravel (Figure 8.3). This is a solifluction deposit derived from the Old Man Sandloess, the Scilly Till and the Tregarthen Gravel, in which all these sediments were mixed and redistributed downslope. The very distinctive silt matrix of this deposit is derived largely from the Old Man Sandloess, an aeolian sandy silt deposited penecontemporaneously with the glacial event. This material is not therefore an *in situ* glacial or glaciofluvial sediment, but rather a periglacial slope deposit derived from sediments associated with the glaciation.

The Old Man Sandloess has been dated by TL on the southern islands to 18 600 ± 3700 BP (Wintle, 1981; Scourse, 1991), and the Porthloo Breccia, which underlies the Hell Bay Gravel, contains organic beds at other sites on Scilly which have been dated by radiocarbon methods to the late Middle and early Late Devensian. The age of the raised beach remains uncertain. Scourse (1991) therefore interprets most of the sediments at Chad Girt as considerably younger than the ages envisaged by Mitchell and Orme (1967).

Conclusion

The fine exposures of Pleistocene sediments at Chad Girt have been used as a reference section by a number of workers, and are therefore of historical significance. The site demonstrates evidence for a high sea-level event pre-dating periglacial conditions in the Devensian in which Late Devensian glacial deposits were moved downslope.

NORTHWARD BIGHT, ST MARTIN'S
J. D. Scourse

Highlights

This site shows controversial evidence for two raised beach deposits separated by periglacial head. Different interpretations of this sequence have important implications for the age and sequence of Quaternary events on the Isles of Scilly.

Introduction

In 1906 Barrow identified a single raised beach deposit at the base of the Pleistocene sequence on the Isles of Scilly. Mitchell and Orme (1967) proposed a more complicated sequence consisting of two raised beaches separated by periglacial and glacial sediments. They argued that the lower beach was typified by the exposure at Chad Girt (see above; Figure 8.1) and was typically erratic-free, whereas the upper beach, with a type-site at Porth Seal (see above; Figure 8.1), contained abundant erratic material derived from the underlying glacial sediments. In their 1967 study Mitchell and Orme were unable to identify a site where the two beaches could be observed strictly in stratigraphic succession, but subsequently Synge (*in* Stephens, 1970a; p. 294) described such a complete sequence from Porth Seal. The upper beach was reinterpreted by Bowen (1981) as a body of granite corestones, and by Scourse (1985a) as soliflucted beach sediment derived from the lower beach.

Mitchell (1986) identified Northward Bight as a second site where the two beaches could be observed in stratigraphic succession.

Description

Northward Bight (SV 944159) is the northernmost of two deep gullies trending south-west to north-east on the north-eastern side of St Martin's Head. Mitchell (1986) identified the following Pleistocene sequence, overlying bedrock, at this site:

4. Head with erratics
   Devensian
3. Unconsolidated beach with large cobbles
   Ipswichian
2. Head without erratics but with shattered beach cobbles
   Wolstonian
1. Consolidated beach with small cobbles
   Hoxnian

Interpretation

Mitchell (1986) correlated the upper unconsolidated beach deposit (bed 3) with the Porth Seal Raised Beach and the lower consolidated beach (bed 1) with the Chad Girt Raised Beach (Mitchell and Orme, 1967). He thus reinforced his belief that two raised beach deposits, of different ages, are present on the Isles of Scilly, assigning the earlier to a 'pre-glacial' Hoxnian sea-level event, and the later to a 'post-glacial' Ipswichian event. Such an interpretation therefore constrained the glaciation of the Isles of Scilly to the Wolstonian (Mitchell and Orme, 1967).

However, Scourse (1986) does not accept this interpretation. As in the case of Porth Seal (Scourse, 1991), he interprets the upper of the two 'beaches' at Northward Bight as beach material reworked by solifluction from the lower beach. Therefore, he recognises only one *in situ* raised beach deposit at this site. In discussion of the number of raised beach deposits present in the Pleistocene succession of the Isles of Scilly and Cornwall, Scourse (1987, 1991) points out that at nearly all sites where raised beach sediments have been overridden by solifluction lobes and sheets, 'tongues' of incorporated beach material can be observed in the overlying solifluction deposits. These can often appear as two stratigraphically distinct 'raised beach' units in section; in a facies model of solifluction deposits from this

area he defines such incorporation of underlying materials as facies Aa 'deformation breccia' (Scourse, 1987).

Scourse (1986) assigns the basal raised beach deposit at Northward Bight (Mitchell's Chad Girt Raised Beach) to the Watermill Sands and Gravel Member of his lithostratigraphic classification (Figure 8.3), the 'head without foreign stones' and the 'Porth Seal beach' to the Porthloo Breccia, and the 'Head with foreign stones' to the Bread and Cheese Breccia. On the basis of radiocarbon dates from organic beds within the Porthloo Breccia at other sites on Scilly, he interprets the Porthloo Breccia as of late Middle or early Late Devensian age. The age of the Watermill Sands and Gravel he regards as uncertain.

Conclusion

Northward Bight contains a controversial sequence of deposits, the interpretation of which has considerable implications for the age of all the Pleistocene deposits of the Isles of Scilly. Mitchell (1986) identifies evidence of two high sea-level stands related to separate Pleistocene interglacials at this site, but Scourse (1986) denies that the upper beach is *in situ* and prefers a younger age for most of the succession.

BATTERY (CASTLE DOWN), TRESCO
J. D. Scourse

Highlights

This site contains the finest exposure on Scilly of Late Devensian meltwater sediments. Head deposits interbedded with the meltwater sediments demonstrate that solifluction was active on the slopes of the meltwater channels. Rounded granite tors on Castle Down have been interpreted as the products of glacial erosion.

Introduction

Pebbles of probable glacial derivation on the northern Isles of Scilly (Figure 8.1) have been known since the middle of the last century. Smith (1858) was the first to record these erratics, or 'foreign pebbles', making a collection of chalk-flints and greensand from Castle Down, Tresco. In reporting their occurrence to the Royal Geological Society of

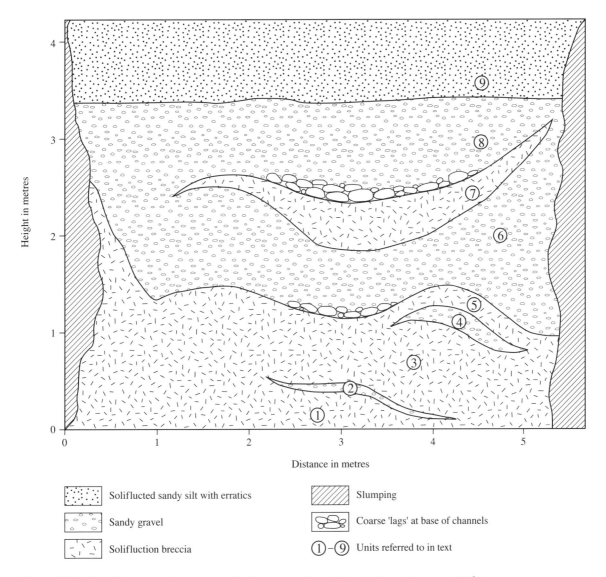

Figure 8.12 The Pleistocene Sequence at the Battery section. (Adapted from Scourse, 1986.)

Cornwall he commented that 'The flints and green-stones varied little in size ranging from that of a hen's egg to that of a blackbird – How they got to Scilly was a mystery which it was for gentlemen of more scientific knowledge than he professed to explain' (Bishop, 1967; p. 91). Such a gentleman proved to be Whitley (1882), who interpreted these foreign stones as glacial in origin.

More recently, Scourse (1991) described an important site of interbedded glaciofluvial and solifluction sediments from an embayment on the north-eastern side of Castle Down known as the Battery section.

Description

Castle Down is an expanse of undulating heath moorland forming the northern part of Tresco. Small eroded granite tors, locally called 'carns', in and around the moor rise to *c.* 40–45 m OD, and the coastal fringe of the moorland consists of granite headlands and small exposures of Pleistocene sediments in cliffs backing embayments and coves. The northern part of Castle Down lies to the north of the glacial limit identified across the northern islands by Mitchell and Orme (1967) and Scourse (1991) (Figure 8.1). The tors in this area are of the 'smoothed', 'rounded' or 'eroded' form (form D; Figure 8.6) defined by Scourse (1987), and contrast markedly with the elaborate 'castellated' or 'mammilated' forms found to the south of the proposed ice limit and typified by those at Peninnis Head (see above). Scourse (1987) regards this rounded form as the product of direct glacial erosion.

Erratics similar to those described by Smith (1858) are still abundant on Castle Down. They can be observed in small patches on the ground surface where the heath is being eroded. The stratigraphical context of the erratics, and the associated glaciofluvial and solifluction sediments, are exceptionally well exposed in the Battery section (SV 887165) which backs the small embayment on the north-eastern side of Castle Down immediately to the south of Piper's Hole. At the extreme northern end of this section a 5.8 m-thick complex of sandy gravels and head deposits is exposed (Figure 8.12). In all, nine sedimentary units have been identified in this section, units 1, 3, 5 and 7 comprising coarse granitic head, and units 2, 4, 6 and 8 sandy gravels. All the contacts in the centre of this section are erosional. The sandy gravels, especially units 6 and 8, occur as channel-fills with very coarse, clast-supported basal lags, and they fine and become matrix-supported upwards. The head units form lobate rather than channelized bodies, most being continuous with the more massive bodies of head on either side of the section. Unit 7, however, is a lens entirely enclosed by gravel units 6 and 8 (Figure 8.12). The sequence is capped by unit 9, a moderately well-sorted sandy silt containing an erratic clast assemblage similar to that in the underlying sandy gravels.

The head units are extremely poorly sorted and contrast with the well-sorted sandy gravels. However, the gravels are even more distinctive in containing a rich and diverse erratic assemblage that is strikingly similar to that obtained from the stratotype of the Scilly Till at Bread and Cheese Cove (Scourse, 1991). Unit 6 has also yielded a clast of green Miocene glauconitic micrite, a distinctive lithology which occurs in the Scilly Till and which is thought to be derived from the offshore Jones Formation (Pantin and Evans, 1984; Scourse, 1991). The head units contain more clasts than the sandy gravels.

Interpretation

The head units within the Battery section are interpreted as solifluction deposits comprising material derived largely from the breakdown of the local granite, whereas the gravels are interpreted as glaciofluvial outwash material. The erratic assemblage and stratigraphic context of these gravels suggest that they were deposited during the same (Late Devensian) glacial event responsible for the Scilly Till (Scourse, 1985a, 1987, 1991).

The sequence demonstrates that solifluction occurred penecontemporaneously with pulses of glaciofluvial deposition. The outwash palaeocurrent is estimated to have been in an easterly direction on the basis of the bed geometry, with solifluction lobes flowing normal to this from the south and north. Head unit 7, however, represents a solifluction lobe moving downslope parallel with the dominant outwash direction; it is perhaps significant that this particular head unit contains a relatively high sand content, for this sand must have been incorporated from the underlying gravel during longitudinal movement down-channel. Solifluction lobes crossing the channels transversely had less opportunity to rework underlying material. At the base of the sequence, solifluction was clearly the dominant process, but gradually the solifluction lobes were overwhelmed by outwash activity.

The sandy gravels at this site were selected by Scourse (1991) as the composite stratotype for the Tregarthen Gravel (Figure 8.3); he assigns the head units to the Bread and Cheese Breccia, and the uppermost pebbly silt to the Hell Bay Gravel members of his lithostratigraphic classification.

Conclusion

This site shows the best evidence on Scilly for the meltwater activity associated with the advance of a glacier to the northern islands at about 19 ka BP. Granite tors on Castle Down were eroded and smoothed by this glacial advance, and the Battery section demonstrates that the relatively rapid downslope movement of materials (solifluction), a common feature of Arctic environments, occurred at the same time as the meltwater deposits were accumulating. Castle Down is also of historical importance because pebbles transported to Scilly by glacier ice were first discovered here during the middle of the last century.

CASTLE PORTH, TRESCO
J. D. Scourse

Highlights

This site lies at the southernmost limit of a glacier which reached Scilly at about 19 ka BP, and exemplifies the evidence used to reconstruct the limit of the ice across the northern islands. Sediments of glacial derivation are abundant at the northern end of the section but absent at the southern end.

Introduction

Barrow (1906) was the first to comment on the distinctive distribution of foreign pebbles at Castle Porth, and the significance of this distribution was further elaborated by Mitchell and Orme (1967) and Scourse (1986).

Description

Castle Porth (SV 882160) lies on the north-west coast of Tresco immediately to the south of Cromwell's Castle, and to the south-west of Castle Down (see above; Figures 8.1 and 8.13). In describing the distribution of the 'curious glacial deposit' on Scilly, Barrow (1906) included a detailed report of its distribution on Tresco which he regarded as a model for the northern islands. He noted that the glacial deposit occurred in a small hollow at the northern end of Gimble Porth on the east side of Tresco, and that

' ... this hollow is continued up the hill inland with a comparatively gentle slope. Up the whole of this slope the pebbles in considerable numbers can be traced to the crest of the ridge [Castle Down] ... and down the corresponding slope on the opposite side of the headland. Moreover on the crest of the ridge they may be traced to some little distance to the south. On the opposite, or western, side, owing to the exposed nature of the coastline, practically no Head is met with north of Cromwell's Castle, but immediately south of this is a small and more protected bay [Castle Porth] in which another patch of Head has escaped denudation. On top of this patch the foreign stones are abundant in the northern portion only of the outcrop; further south they cease just as suddenly as they did on the opposite side of the island [Gimble Porth]. If now the two points of cessation of this deposit on the two shore-lines be joined up their line of junction is

seen to pass approximately along the southern limit of the stones on the crest of the hill. Their distribution at the north end of Bryher is on exactly similar lines' (Barrow, 1906; p. 26).

He went on to interpret this distribution of foreign stones to be ' ... unintelligible except by invoking some other means of transport than water ... It is quite clear that they [the foreign stones] must have been carried by floe-ice' (Barrow, 1906; p. 27). This constitutes the first recognition of an ice limit across the northern islands.

Mitchell and Orme (1967) reported the same distinctive distribution of the foreign pebbles on Tresco:

' ... north along the shore of Gimble Porth, foreign stones suddenly appear in great quantity before the end of the bay is reached. This part of the island is the northern end of a granite ridge 100 to 130ft high. The foreign stones can be followed up to the crest of the ridge at 100ft and down the other side into the north end of Castle Porth ... at the end of the retaining wall east of Cromwell's Castle, the base of the section was in coarse Lower Head. Above this was a finer Upper Head, and just along the junction there were layers and pockets of outwash gravel, together with thin lenses, some kneaded and disturbed, of silty till ... One hundred yards south of the wall the erratic gravels ended, and, as on the east side of the island, south of this point only a very few erratic pebbles were seen in the Upper Head. They were not seen anywhere south of a line joining New Grimsby on the west and Merchant's Point on the east' (Mitchell and Orme, 1967; pp. 75–77; Figure 8.1).

Interpretation

The limit of erratic material which can observed so clearly in the section at Castle Porth has been interpreted by Barrow (1906), Mitchell and Orme (1967) and Scourse (1991) as recording the southern limit of ice on Scilly (Figure 8.1). Unlike some ice limits which are marked by morainic landforms, the maximum extent of ice on Scilly is simply delimited by the distribution of glacial or glacially derived sediment. This characteristic is exemplified by the Castle Porth site.

In terms of the stratigraphy proposed by Scourse (1991), the foreign pebbles at the northern end of Castle Porth occur within a matrix of sandy silt interpreted as a solifluted mixture of sandloess, outwash gravel and till, formally defined as the Hell Bay Gravel. This conflicts with Mitchell and Orme's

Figure 8.13 Quaternary sediments exposed in coastal cliffs at Castle Porth, Tresco. The Hell Bay Gravel at the northern end of the exposure (left) is rich in erratics, whereas erratics are absent at the southern end (right) of the section. (Photo: S. Campbell.)

(1967) suggestion that the section contains 'silty till'.

The Hell Bay Gravel occurs above a granitic head devoid of erratic material, correlated by Scourse with the Porthloo Breccia. It is overlain by another unit of granitic head which contains a proportion of erratic material incorporated from the underlying Hell Bay Gravel; this is the Bread and Cheese Breccia. At the southern end of the section the Hell Bay Gravel is absent, and here the upper unit of granitic head (the 'Upper Head' of Mitchell and Orme) rests directly on the lower unit (their 'Lower Head'). Lithologically this upper unit is indistinguishable from the lower unit, hence Scourse's classification of the entire sequence here as Porthloo Breccia. This demonstrates two points: first, that the Bread and Cheese Breccia is confined to the area north of the ice limit and second, that at sites in the 'southern' Scillies where the Old Man Sandloess is absent, for example, the southern end of Castle Porth, the upper Porthloo Breccia rests directly on the lower Porthloo Breccia and is indistinguishable from it. All the sedimentary units at Castle Porth are therefore of soliflual origin and the

basis for their differentiation lies in the lithological composition of the source material.

Conclusion

Castle Porth helps to establish the southern limit of a glacier which reached Scilly at *c.* 19 ka BP. Although this limit is recorded at a number of other sites on Scilly, it is most clearly demonstrated here. Sediments deposited by the glacier can be observed at the northern end of this site but are absent at its southern end. Since deposition, these glacial sediments have been moved downslope by gravity flows in an Arctic climate.

HIGHER MOORS, ST MARY'S
J. D. Scourse

Highlights

This site provides the most complete record of Holocene vegetational history on the Isles of Scilly.

Pollen and radiocarbon evidence indicates the existence of Holocene forest prior to clearance associated with the spread of agricultural activity during the Neolithic and Bronze Age.

Introduction

The Isles of Scilly are currently devoid of indigenous woodland, but the extremely mild climate of the islands has encouraged the development of a diverse flora and led to the success of many introduced species, some arboreal and many subtropical in character (Lousley, 1971). There would therefore appear to be no natural physical constraint on the development of indigenous woodland, raising the possibility that the islands were wooded prior to their settlement by humans.

Early descriptions of the occurrence of oak trunks at submerged forest localities first suggested the former existence of woodland around the islands (Scaife, 1986). These finds were supported by numerous records of stumps, mostly of *Quercus*, having been removed from the ground on St Mary's, St Martin's and Tresco, those on the latter island having been recorded by Augustus Smith during the last century (Scaife, 1986).

The first pollen-analytical evidence that the vegetation of St Mary's had formerly been dominated by deciduous woodland was provided by Dimbleby (1977) from analyses of a soil profile at Innisidgen (SV 919128). Though it was not possible to date this profile directly, it demonstrated that *Quercus* and *Corylus* had been present in open-canopy woodland with a Gramineae and *Pteridium* ground flora.

Palynological data from soil profiles at a number of archaeological sites have provided clear evidence of post-forest clearance agriculture (Butcher, 1970, 1971, 1972, 1974; Dimbleby *et al.*, 1981; Evans, 1984). These profiles, of various ages postdating the Neolithic, all provide evidence of very open vegetation with some indicating cereal cultivation.

In order to provide a more continuous record of Holocene vegetational change, Scaife (1980, 1981, 1984) investigated the only two remaining areas of peat accumulation on the islands, Higher Moors and Lower Moors on St Mary's. The cores from Higher Moors have yielded pollen and radiocarbon data which provide a reference stratigraphy against which the more isolated soil pollen profiles from archaeological sites can be compared; these data also support the hypothesis of former Holocene deciduous woodland cleared as a result of prehistoric agricultural activity.

Description

Higher Moors is a topogenous mire forming part of Porth Hellick Nature Reserve which extends inland from Porth Hellick (SV 925107) to Holy Vale (SV 921115). The present vegetation consists of a mosaic of wetland species ranging from *Salix caprea* and *S. viminilis* carr associated with the sedge *Carex paniculata*, to mesotrophic sedge and *Phragmites* fen (Scaife, 1984); however, the mire has been extensively disturbed by peat cutting (Lousley, 1971), land drainage and, more recently, by extraction of groundwater for public supply, and these activities have significantly altered the hydrology of the site with resultant impacts on the vegetation (Scaife, 1984). Extensive augering through the length of the mire indicated the thickest peats in the area close to Holy Vale. Here, black highly humified detritus and monocotyledonous peat reaches a maximum thickness of 76 cm, abruptly overlying white bleached sand of very low organic content. Samples for pollen analysis and radiocarbon dating were taken from open sections cut through the peat (Scaife, 1984). Five pollen zones (HM1-5) were recognized by Scaife (1984; Figure 8.14).

HM1: 76-70 cm. Arboreal and shrub pollen dominate this zone with *Quercus*, *Betula*, *Corylus* and *Salix* dominant, and *Pinus*, *Ulmus* and *Alnus* subdominant. Herbaceous frequencies are low and mainly represented by Cyperaceae and other mire taxa.

HM2: 70-62 cm. This zone is characterized by a reduction in arboreal pollen and increases in herbaceous taxa. Although *Corylus* and *Salix* remain important, *Betula* and *Quercus* decline, and *Calluna* increases. Herb taxa, chiefly Gramineae and Cyperaceae, dominate the zone. There are significant records of cereals, weeds and ruderals.

HM3: 62-50 cm. Significant increases in *Betula* and later *Quercus* dominate this zone with concomitant decreases in herbaceous taxa.

HM4: 50-4 cm. An abrupt decrease in arboreal and shrub frequencies occurs in this zone coinciding with increases in Gramineae, Cyperaceae and ruderal taxa. Pollen of Umbelliferae, *Pteridium* and ericaceous taxa also increase in this zone.

HM5: 4-0 cm. This zone reflects the uppermost level of the peat profile. It contains slightly higher frequencies of arboreal pollen than HM4 alongside

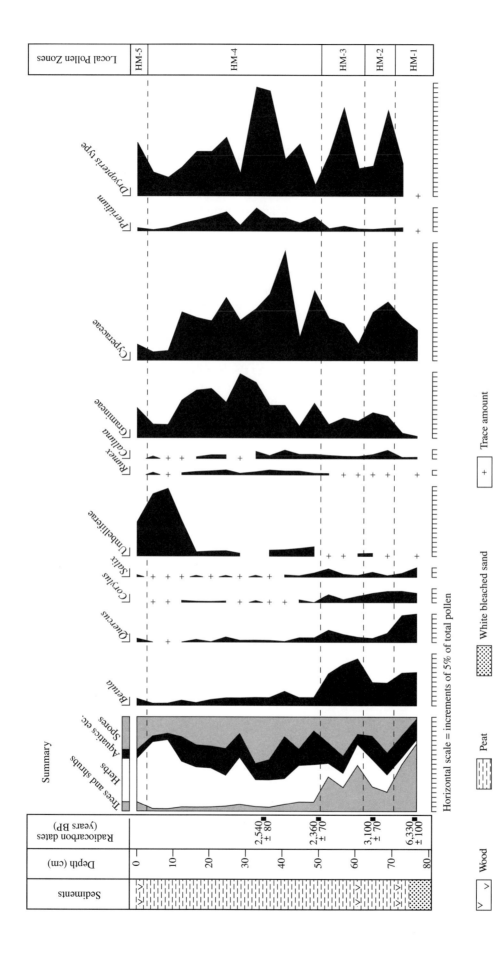

Figure 8.14 Selected pollen data and radiocarbon dates for a peat profile at Higher Moors, St Mary's. (Adapted from Scaife, 1984.)

Pittosporum and some other introduced exotic taxa.

Four 2 cm-thick samples of the peat were submitted for radiocarbon assay and the uncalibrated results are shown in Figure 8.14.

Interpretation

Although there is clearly a problem with the radiocarbon determinations (Figure 8.14), with an inversion towards the top of the profile, Scaife (1984) attributes this to humification effects, and accepts the lowermost date, from the base of the profile at 75 cm, as reliable. This therefore indicates the presence of forest, at least in this part of St Mary's, during the middle Holocene. Scaife (1984) interprets the pollen record as indicating open *Quercus* woodland allowing a *Corylus* understorey to flower freely. He correlates this forested phase with the undated soil profile from Innisidgen (Dimbleby, 1977) which provided the first pollen evidence for the former presence of woodland during the Holocene on the islands.

A decline in arboreal pollen and increases in herb taxa occur above 68 cm (HM2), including evidence of cereal cultivation, followed above 62 cm (HM3) by forest regeneration with increases in *Betula* and *Corylus*. Problems with the radiocarbon assays in this section of the profile hinder attempts to correlate this evidence of clearance and subsequent forest regeneration with specific archaeological events or periods. However, Scaife (1986) suggests that the clearance phase may be Neolithic with regeneration occurring in the late Neolithic or early Bronze Age. The major phase of forest clearance at 50 cm (HM4) probably dates to the middle Bronze Age and is consistent with the widespread archaeological evidence (Scaife, 1986).

Conclusion

The peat bog at Higher Moors has provided evidence that the Isles of Scilly were at least partly covered by deciduous forest at around 6 ka BP. This indigenous forest was then partly cleared by Neolithic humans, and finally disappeared as a result of ground clearance associated with agricultural activity during the Bronze Age. Higher Moors is the most important site providing evidence of the changing vegetation on Scilly since the end of the last ice age.

Chapter 9

The Quaternary history of the Somerset lowland, Mendip Hills and adjacent areas

Introduction

INTRODUCTION
C. O. Hunt

The sites described in this chapter were selected to document the extra-glacial development of the Somerset lowland, Mendip Hills and adjacent areas (Figure 9.1). This region has considerable potential significance, because it contains important and unique evidence for extra-glacial Quaternary environments. Especially important are the interglacial marine deposits of the Burtle Formation, the interstadial marginal marine Low Ham Member and the massive cold-stage gravel aggradations, possibly interstadial deposits and archaeological material of the Broom Gravel Pits in the Axe Valley. The northern and western parts of the region also contain extremely important examples of cold-stage periglacial, colluvial, aeolian and fan-gravel sedimentation and warm-stage palaeosol development.

The Pleistocene record of Somerset has great potential importance, but has been relatively neglected in comparison with regions such as East Anglia. Some of the leading nineteenth century earth scientists worked on aspects of the Pleistocene of the region. For instance, Buckland (1823) recorded mammal remains from a number of cave sites in his *Reliquiae Diluvianae* and Buckland and Conybeare (1824) identified the marine origin of the interglacial Burtle Beds of King's Sedgemoor. Later in the nineteenth century, raised beaches near Weston-super-Mare were described by Sanders (1841) and Ravis (1869). The first detailed synthetic work on the Quaternary deposits of the region was the Geological Survey Memoir of Woodward (1876). Prestwich included a number of sites in Somerset and Avon in his monumental reviews, providing faunal lists and proposing an interglacial age for the raised beaches and Burtle Beds (Prestwich, 1892).

Head, alluvial fan gravels and 'cold-stage' terrace gravels are an extremely important component of the Pleistocene deposits in this region, but until modern geochronometric methods are applied to them they will remain virtually impossible to date with any confidence. Occasional records of gravel and head deposits were made during the nineteenth century, and many of these are summarized in the early Geological Survey Memoirs. Thus, Woodward (1876) described a number of terrace-gravel and fan-gravel sites around Mendip. Ussher (1906) mapped and described 'old washes and talus fans' and 'terrace gravels' in the Isle Valley and Chard Gap and later mapped 'loamy gravel and head' in the Quantocks and Tone Valley (Ussher, 1908). At several localities, these deposits were associated with reindeer remains and, at Doniford, with mammoth (Ussher, 1908). The great hand-axe sites of the Axe Valley were first noted by D'Urban (1878) and these terrace sites were briefly described by J.F.N. Green (1943). Similar deposits were later mapped as 'spreads of head and gravel' in the upper Parrett, Yeo, and Cam valleys by Wilson *et al.* (1958). Palmer (1934) conducted studies of a number of cold-climate breccia and blown-sand sites, including Holly Lane (Chapter 10) and the important section at Brean Down, and demonstrated a southerly origin for the sands on mineralogical grounds. The first detailed account of the periglacial deposits of Brean Down was not made, however, until the work of ApSimon *et al.* (1961).

After the work of Woodward (1876), Prestwich (1892) and Ussher (1908), interest in the marine interglacial deposits of the Somerset lowlands waned. Research resumed with Bulleid and Jackson's (1937, 1941) detailed accounts of the Burtle Beds of King's Sedgemoor. This was followed by the work of ApSimon and Donovan (1956), who described marine Pleistocene deposits in the Vale of Gordano. The marine interglacial deposits became a source of major controversy, as morphometric work and then the first radiocarbon dates on the raised beach at Middle Hope suggested correlation with the Upton Warren Interstadial (Donovan, 1962; Wood *in* Callow and Hassall, 1969), though this was later rejected by Kidson (1970).

The influential reviews by Mitchell (1960, 1972) provoked much debate concerning the limits and timing of glaciation, the possible existence and age of proglacial lakes and the occurrence, nature and stratigraphic position of the interglacial marine deposits. Thus Kellaway (1971), Hawkins and Kellaway (1971, 1973), and Kellaway *et al.* (1975) suggested that most of South-West England had been overrun by ice sheets during an early glaciation. Convincing evidence of glaciation in the Bristol area was provided by Hawkins and Kellaway (1971) and these authors contended that the Burtle Beds of lowland Somerset were glacial outwash deposits. Stephens (1970a, 1970b, 1973) suggested an alternative, with limited glaciation in the coastlands of Somerset damming an enormous proglacial lake in lowland Somerset, which eventually discharged into the Axe Valley through the Chard Gap. These suggestions were vigorously disputed

Somerset lowland, Mendip Hills and adjacent areas

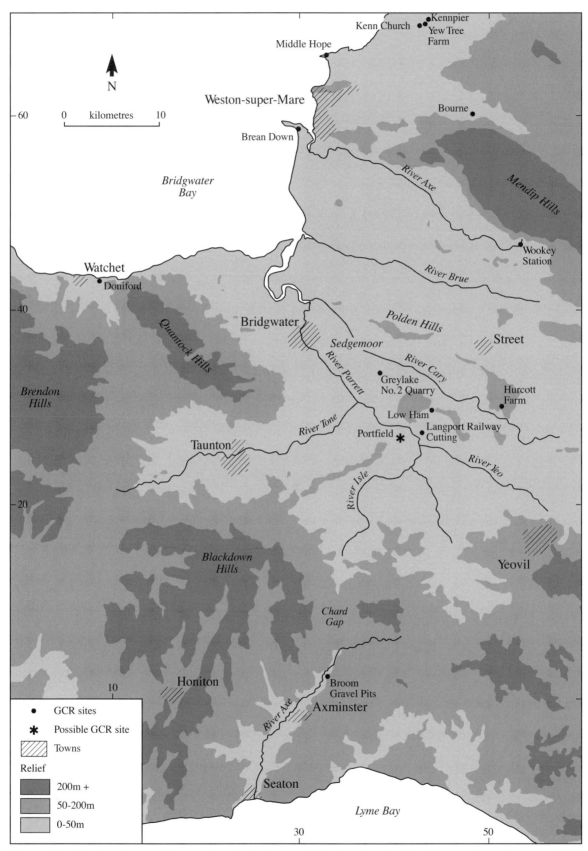

Figure 9.1 The Mendips and Somerset lowland, showing GCR sites described in this chapter, and selected GCR sites described in Chapters 7 and 10.

Introduction

by Kidson (1971), Kidson and Haynes (1972), Kidson *et al.* (1974) and Kidson (1977) who produced compelling evidence to show that the Burtle Beds were of interglacial marine origin, and C.P. Green (1974b) who showed that the gravels of the Axe Valley were locally derived. More recently, Hunt *et al.* (1984) showed that no unequivocally erratic material could be found in southern Somerset.

Eventually, the emergence of robust stratigraphical and palaeoenvironmental evidence led to a broad consensus: that the Burtle Beds of King's Sedgemoor are estuarine interglacial deposits with freshwater intercalations (Kidson *et al.*, 1978; Gilbertson, 1979; Hunt and Clarke, 1983), with the balance of evidence pointing toward an Ipswichian age. The raised beach at Swallow Cliff, Middle Hope (Gilbertson, 1974; Gilbertson and Hawkins, 1977; Briggs *et al.*, 1991), periglacial deposits at Holly Lane and elsewhere in Avon and north Somerset (Gilbertson, 1974; Gilbertson and Hawkins, 1974, 1983), the valley-fill deposits at Doniford (Gilbertson and Mottershead, 1975) and fan sediments in Mendip (Findlay, 1977; Pounder and Macklin, 1985; Macklin and Hunt, 1988) were also re-described.

Recently, the development and application of aminostratigraphy have led to the recognition of the extreme complexity of the Pleistocene sequence in Avon and Somerset and have prompted the re-examination of a number of key sites. The first application of the method in the region was by Andrews *et al.* (1979) who determined ratios from the Burtle Beds at Greylake and the Middle Hope raised beach, together with other sites in South-West England and South Wales. The Greylake No. 2 Quarry deposits were shown to be only marginally older than the Middle Hope raised beach deposits, although *Patella* shells from both produced ratios in the region of 0.11. The work of Davies (1983) suggested that the Middle Hope raised beach deposits were substantially older, and she attributed them to Group 3 of her classification (mean ratios of 0.2). This problem has not yet been fully resolved, though the latest views (Bowen, pers. comm., 1996) lean towards a last interglacial (Oxygen Isotope Stage 5e) age. Hunt *et al.* (1984) provided ratios on *Corbicula* of 0.18 from the Chadbrick Gravels and 0.18 and 0.26 from *in situ* assemblages from the Burtle Beds at Greylake No. 1 Quarry (Gilbertson, 1979). These ratios were interpreted as suggesting an Ipswichian age for the Chadbrick Gravels and some components of the Burtle Beds.

The important reassessment of the aminostratigraphy of British non-marine deposits by Bowen *et al.* (1989) renders many of the early aminostratigraphic interpretations in Somerset and Avon obsolete (Hunt, 1990a). Given that very few assays have been made and that correlations based on small numbers of ratios may not be trustworthy, it would nevertheless appear that an extremely complex picture of sea-level change is preserved in the sequences in Avon and Somerset. These changes can be related, albeit tentatively, to the global oxygen isotope stratigraphy (Campbell *et al.*, in prep.; Figure 9.2).

Important themes in the Pleistocene of the Somerset lowland, Mendip Hills and adjoining areas

Several important themes emerge from the scientific background outlined above. These have guided site selection, which was carried out in consultation with other relevant specialists. In several cases, there is only one site where a certain facet or feature of Quaternary history can be demonstrated clearly; where several closely comparable sites were available, secondary factors, such as other features of earth-science interest, vulnerability to development and ease of access were considered. The themes are:

1. Evidence for high Pleistocene sea levels

Somerset and Avon offer an exceptional sequence of marine interglacial and interstadial deposits. GCR sites were chosen to provide the best evidence for these high sea-level events, wherever possible. The sites are Swallow Cliff (Oxygen Isotope Stage 5e or 7), Portfield (Stage 7), Greylake No. 2 Quarry (Stages 7 and 5e) and Low Ham (Stage 5a). Low Ham is particularly important as the only site in the UK where high relative sea levels during a Devensian (i.e. post-Stage 5e) interstadial can be demonstrated. Complementary sites in the Bristol area (Chapter 10) include Kennpier (? Stage 15), Kenn Church (Stage 7) and Weston-in-Gordano (undated but with three marine interglacial sequences interbedded with ?till).

2. Terrace stratigraphy

Excluding the glacial and marine sequences, the fundamental framework for establishing a Pleistocene stratigraphy in Somerset and Avon is

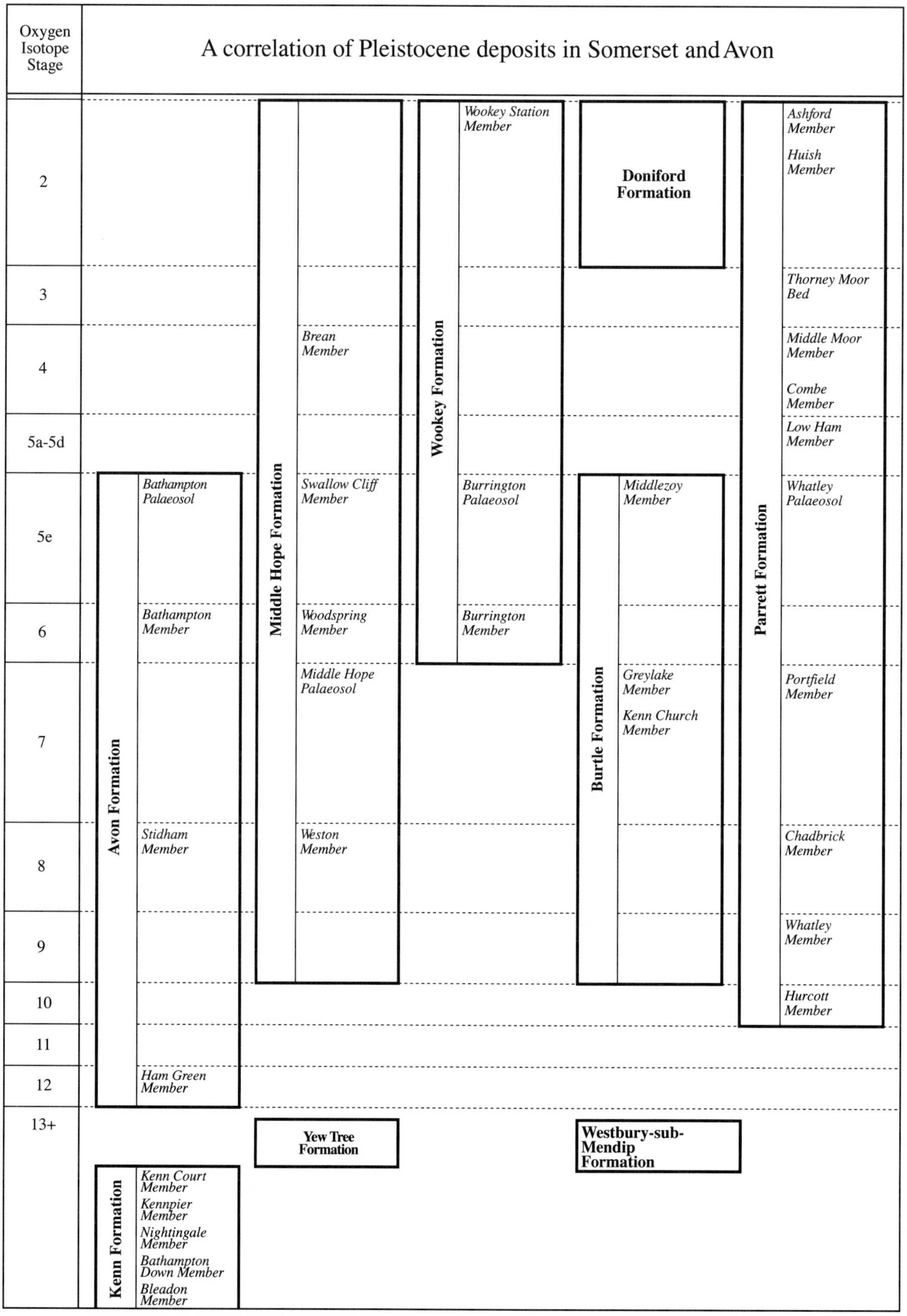

A correlation of Pleistocene deposits in Somerset and Avon

Oxygen Isotope Stage	Avon Formation / Kenn Formation	Middle Hope Formation	Wookey Formation	Doniford Formation / Burtle Formation	Parrett Formation
2			Wookey Station Member	**Doniford Formation**	Ashford Member / Huish Member
3					Thorney Moor Bed
4		Brean Member			Middle Moor Member / Combe Member
5a–5d					Low Ham Member
5e	Bathampton Palaeosol	Swallow Cliff Member	Burrington Palaeosol	Middlezoy Member	Whatley Palaeosol
6	Bathampton Member	Woodspring Member	Burrington Member		
7		Middle Hope Palaeosol		Greylake Member / Kenn Church Member	Portfield Member
8	Stidham Member	Weston Member			Chadbrick Member
9					Whatley Member
10					Hurcott Member
11					
12	Ham Green Member				
13+	Kenn Formation: Kenn Court Member, Kennpier Member, Nightingale Member, Bathampton Down Member, Bleadon Member	Yew Tree Formation		Westbury-sub-Mendip Formation	

Figure 9.2 A correlation of Pleistocene deposits in the Somerset lowland, Mendips, Bristol district and Avon Valley. (Adapted from Campbell *et al.*, in prep.)

290

provided by river terrace gravels. Key terrace sites were selected for the GCR to demonstrate the main elements and regional variations in terrace stratigraphy. In southern Somerset, key sites in the Parrett-Cary catchment are the interglacial (Stages 9 and 7) sites at Hurcott Farm and Portfield, the cold-stage (Stage 8) site at Langport Railway Cutting and the interstadial site (Stage 5a) at Low Ham.

3. Temperate-stage palaeobiology

Somerset and Avon have one of the most complete and richly fossiliferous sequences of marine interglacial and interstadial deposits in the British Isles. There are also some important non-marine palaeobiological sites. One site is of particular significance. The Stage 5a marginal marine interstadial site at Low Ham, with its mollusc and ostracod faunas, plant macrofossils and pollen is unique in north-western Europe. Important regional marine mollusc sites include Greylake No. 1 Quarry (Stage 9 and late Stage 7), Swallow Cliff (Stage 5e or 7) and Greylake No. 2 Quarry (Stage 5e). The Greylake sites have yielded regionally important mammal faunas and freshwater mollusc assemblages. The Stage 9 site at Hurcott Farm and the Stage 7 site at Portfield contain regionally important freshwater mollusc fossils and some pollen.

4. Cold-stage sedimentation and palaeobiology

Subsequent to the Kenn glaciation, Avon and Somerset lay beyond the glacial limits. Cold-stage sedimentation was ubiquitous, but good examples, particularly pre-Devensian ones, are very rare. Subgroups of sites were chosen to demonstrate this aspect of Pleistocene history. Two sites represent fan sedimentation around Mendip. Wookey Station provides an excellent example of fan sedimentation, a rare mollusc fauna and pollen dating from the coldest part of the Late Devensian. The Bourne sequence demonstrates at least two earlier periods of fan aggradation, separated by an interglacial palaeosol which is virtually unique in the region. Brean Down is important for periglacial aeolian and slope sediments and Early to Mid-Devensian vertebrates and land molluscs. At Swallow Cliff, Middle Hope, the raised beach deposits overlie rare Stage 6 or 8 slope deposits which contain land molluscs. Cold-stage river terrace gravels make up a further important group of sites. Outstanding among these is the cliff section at Doniford (Chapter 7), which exposes a considerable proportion, in cross-section, of a cold-stage valley-fill. Langport Railway Cutting provides an excellent sequence, of proposed Stage 8 age, with land mollusc faunas.

(A) SITES RELATING TO THE EXTRA-GLACIAL DEVELOPMENT OF THE SOMERSET LOWLAND AND ADJACENT AREAS

In this section, sites are documented which illustrate the Quaternary development of the Somerset lowland and adjacent areas. Two main themes are apparent: marine incursions into the Somerset Levels, and landscape development through the evolution of drainage nets and incision of valleys around the margins of the lowland, particularly in the Parrett catchment to the south of Sedgemoor.

There is still considerable scope for investigations of the marine record in the Somerset lowland, since the complexity of the sequences at localities such as Greylake may have been underestimated by previous workers. Nevertheless, marine and marginal marine sediments attributed to Oxygen Isotope Stage 7 (the Greylake and Kenn Church members at Greylake No. 2 Quarry and Kenn Church, respectively), Stage 5e (the Middlezoy Member at Greylake No. 2 Quarry), and Stage 5a (the Low Ham Member at Low Ham) have been identified and are documented here. The altitude reached by these marine transgressions can be determined; for Stage 7 at Portfield, for Stage 5e at Greylake No. 2 Quarry, and for Stage 5a at Low Ham. These sites are all richly fossiliferous and those at Greylake and Kenn Church provide important evidence for interglacial shallow-marine mollusc communities in South-West England.

A further set of sites documents landscape development around the margins of the Somerset lowland. Key stratigraphic sites in the terrace 'staircase' of the Parrett catchment include the ?Stage 9 site at Hurcott Farm, the Stage 8 site at Langport Railway Cutting, the Stage 7 site at Portfield and the Stage 5a site at Low Ham. Hurcott Farm, Portfield and Low Ham have been dated by aminostratigraphy and thus provide important chronological control in an otherwise undateable sequence. The vastly important site at Broom is perhaps the richest Palaeolithic site in the South-West, and provides important information on the evolution of the Axe Valley in east Devon. The selected GCR sites are all important for the light they throw on the palaeobiology of inland South-West England during the Middle and Late Pleistocene.

LANGPORT RAILWAY CUTTING
C. O. Hunt

Highlights

Langport Railway Cutting provides a good example of 'cold-stage' fluvial sedimentation, correlated with Oxygen Isotope Stage 8, and interglacial soil development seen in a rare permanent exposure. The site is located at the Langport Gap, a key locality in southern Somerset where the presence within a small area of several important fluvial and marine formations enables morphostratigraphical relationships to be elucidated. This is the type-site of the Chadbrick Member.

Introduction

Langport Railway Cutting contains well-exposed 'cold-stage' river gravels overlain by a well-developed palaeosol. The site is representative of the oldest gravel aggradation in southern Somerset, which can be traced upstream into the headwaters of the rivers Isle and Parrett. The gravels are probably best correlated with Oxygen Isotope Stage 8.

The site was first recorded by Woodward (1905) in his survey of railway cuttings between Langport and Castle Cary, but was not described. Subsequently, Hughes (1980) gave a very brief description of the site and attributed it to a 'fluvio-estuarine' facies of the interglacial marine Burtle Beds. More recently, Hunt (1987; in press) redescribed the site, carrying out geomorphological, clast lithological, sedimentological and palaeontological studies. He established that deposition was from a multi-channel river flowing through a sparsely vegetated landscape. The site evidence was attributed to an unspecified cold stage prior to the last interglacial. Subseqently, Campbell *et al.* (in prep.) used Langport Railway Cutting as the type-section for the Chadbrick Member which is correlated with Oxygen Isotope Stage 8.

Description

Langport Railway Cutting (ST 426272) is excavated in a broad terrace-like feature to the north-east of Langport. The feature is separated from the present course of the River Parrett by a low hill known as Whatley and lies some 14 m above the level of the present floodplain. The terrace surface is underlain by gravels which lie on a gently undulating erosion surface cut in Rhaetic shales. The gravels can be divided into three units (maximum bed thicknesses in parentheses):

3. Gravels, once probably sandy but now significantly clay-enriched and lacking in limestone clasts. The original trough cross-bedding is disrupted and many large clasts are orientated vertically. The upper part of the unit is generally bleached while the lower part is slightly reddened, but this pattern has been somewhat disturbed by involutions. A few very decalcified specimens of *Cepaea* sp. were present in the lower part of this unit. (1.4 m)

2. Gravels and sands infilling a channel, 0.9 m deep and over 10 m wide, incised into bed 1. The infill comprises a basal, imbricated clast-supported gravel passing into trough cross-bedded yellow-buff sands and sandy gravels. Clast orientation in the imbricated unit suggests a palaeocurrent direction of 332°. Most gravel clasts are lightly point-contact cemented with calcite, but some sand beds, particularly towards the top of the unit, are heavily cemented ('plugged'). A restricted molluscan fauna, including *Pisidium* sp., *Succinea* cf. *oblonga*, *Trichia* cf. *hispida*, *Pupilla muscorum* Linné, Limacidae and helicid fragments, was found in this unit, together with fragments of the case of the aquatic larvae of *Caddis* sp. and fragments of insect exoskeleton attributable to the Trichoptera. (0.9 m)

1. Pale brown, trough cross-bedded coarse sands and matrix-supported sandy gravels. The trough cross-beds are of the order of 0.1–0.3 m deep and 1.0–3.0 m across. A rather restricted molluscan fauna, including *Succinea* cf. *oblonga*, *P. muscorum*, *Vertigo* sp., Limacidae and helicid fragments, was found in this unit. Clast lithological analysis showed that these gravels comprise lithologies derived from the Mesozoic and Tertiary rocks upstream. The dominant lithotypes are grey micrites derived from the limestone beds in the Rhaetic and Lias, oolitic and bioclastic limestones from the Middle Jurassic formations, and flint and chert from the Chalk and Greensand. (1.2 m)

Interpretation

The sands and gravels of the Langport terrace and many of the other terrace deposits of Somerset can be exemplified by the site at Langport Railway Cutting. The trough cross-stratification of the lower unit (bed 1) is consistent with deposition from a multi-channel river with a highly peaked or even ephemeral discharge regime and an abundant sediment supply. Modern analogues are widely known from high altitudes, high latitudes and arid and semi-arid areas (Doeglas, 1962; Williams and Rust, 1969; Bull, 1972; Miall, 1977). The palaeontological evidence from bed 1 is consistent with this type of depositional regime. The absence of aquatic molluscs in this unit probably reflects the ephemeral nature of the river, particularly the short duration of flows. *P. muscorum* is a species typical of exposed arid environments (Kerney and Cameron, 1979), though it is thought to have been tolerant of damp places during the Pleistocene (Kerney, 1963; Kerney *et al.*, 1964). The other taxa were most probably living in damp, partially vegetated areas on the floodplain.

The imbricated gravels at the base of the middle unit (bed 2) are probably related to the scouring of the channel. The scour-and-fill structures in the sands and gravels overlying the imbricated gravels may be the result of the movement of sand waves in one channel, or may reflect deposition from a number of smaller braided channels. The fauna from bed 2 includes remains of the aquatic larvae of *Caddis* sp. and Trichoptera and the aquatic bivalve mollusc *Pisidium* sp.. It is therefore probable that flows were less ephemeral than they were during deposition of bed 1, or that standing water occupied a pool here after the recession of seasonal floods. The other molluscs are taxa of damp herbaceous vegetation (*Trichia* cf. *hispida*), wet terrestrial environments (*Succinea* cf. *oblonga*), generalists (Helicidae, Limacidae) or xerophiles (*P. muscorum*).

The remaining traces of trough cross-bedding in the upper unit (bed 3) are again consistent with deposition from a multi-channel stream. *Cepaea* is a thermophilous, generalist terrestrial mollusc genus, but other smaller taxa were most probably destroyed by decalcification. The leached and reddened upper horizons and especially the presence of plugged calcretes are consistent with weathering under a warm arid climate, probably during the last interglacial. However, most evidence shows that the last interglacial was characterized by a 'continental warm summer' rather than 'arid' climate (Jones and Keen, 1993). The disruption of soil horizons and bedding and the reorientation of stones probably took place under periglacial conditions during the Devensian.

Langport Railway Cutting is one of the key sites in the argument about the possible glaciation of southern Somerset, as suggested by Gilbertson and Hawkins (1978a, 1978b), or a more widespread glaciation of southern England, as argued by Kellaway (1971) and Kellaway *et al.* (1975), or the development of a proglacial lake in southern Somerset as hypothesized by Stephens (1966b, 1970b, 1973). A possible glacial erratic was recorded some 7 km north at Greylake (Kidson *et al.*, 1978; Gilbertson, pers. comm., 1982), but Langport Railway Cutting contains only clasts derived from within the Parrett catchment and therefore probably lay beyond any possible glacial or glaciolacustrine limits.

Conclusion

Langport Railway Cutting was selected as a good example of 'cold-stage' fluvial sedimentation seen in a rare permanent exposure. The site has well-preserved scour-and-fill stratification typical of deposition from a braided river and well-preserved, if sparse, molluscan assemblages. It is also unusual among British 'cold-stage' sites for having yielded remains of the larvae of aquatic insects. The upper part of the sequence shows good examples of post-depositional modification by periglacial and pedogenic processes including the rather unusual process of calcrete formation during the last interglacial. The site has stratigraphical significance: it is part of the oldest (and only pre-Ipswichian) terrace of the River Parrett and is located at the Langport Gap, a key locality in southern Somerset where the presence within a small area of several important fluvial and marine formations enables their morphostratigraphical relationships to be elucidated. The site also helps to establish the limits of glaciation in southern Somerset; it contains only rock types derived from within the Parrett catchment and therefore probably lay beyond any possible glacial or glaciolacustrine limits.

GREYLAKE (NO. 2 QUARRY)
C. O. Hunt

Highlights

This classic locality is the type-site for the Burtle Formation – a complex of shallow marine shelly sands, gravels and clays probably dating from more

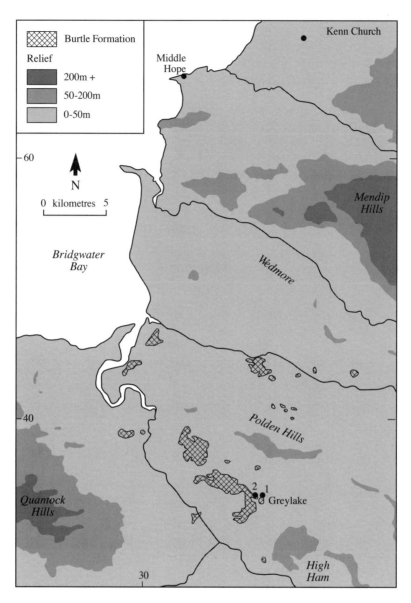

Figure 9.3 The Burtle Formation of the Somerset lowland. (Adapted from Kidson *et al.*, 1978 and Hunt, 1987.)

than one interglacial. The old Greylake No. 2 Quarry is the best known and most studied Burtle Beds locality, and has yielded rich fossil mollusc, foraminifer and ostracod assemblages as well as mammal bones. At least two phases of interglacial marine sedimentation can be demonstrated, with the marine beds separated by a well-developed calcreted palaeosol. The site may mark the southernmost limit of the Kenn glaciation in Somerset.

Introduction

The name 'Burtle Beds' was coined by Buckland and Conybeare (1824) for shelly sands and sandy

gravels which form 'batches' – the local name for areas raised above the alluvial deposits of the Somerset Levels (Figure 9.3). They described the distribution of the shelly sands around King's Sedgemoor and suggested a littoral marine origin. Later, in a number of rather inexact reports (e.g. Poole, 1864), shelly sands were described from the lower part of the Parrett Valley but no precise locality details were given. Ussher (*in* Woodward, 1876) noted shelly marine sand at a number of localities near Middlezoy and Westonzoyland. The presence of marine sand in the Greylake pits was first noted by Ussher (1914), but detailed description of the site awaited the work of Bulleid and Jackson (1937), who described shelly sands and gravels from the uppermost 5 m of the sequence, down to the water-table. They listed a fossil mammal fauna of *Dama* cf. *dama*, *Cervus elaphus* Linné and *Bos primigenius* Bojanus and a diverse marine mollusc fauna. Two valves of *Corbicula fluminalis* (Müller), a freshwater bivalve now extinct in Europe, were found in the marine deposits.

In the early 1970s, Kellaway and his co-workers (Kellaway, 1971; Kellaway *et al.*, 1975) suggested that the Burtle Beds were outwash deposits associated with a glacial incursion into southern England. Kidson and colleagues reinvestigated the Burtle Beds and vigorously restated their marine interglacial origin (Kidson and Haynes, 1972; Kidson *et al.*, 1974; Kidson and Heyworth, 1976; Kidson, 1977; Kidson *et al.*, 1978; Hughes, 1980). Kidson *et al.* (1978) gave a definitive description of the beds visible in Greylake No. 2 Quarry in the mid-1970s (Figure 9.4). These workers conducted an intensive investigation of the stratigraphy, sedimentology, molluscs, foraminifers, ostracods, plant macrofossils and pollen. They demonstrated a sequence of basal non-marine deposits overlain by a silt-clay intertidal mudflat unit in turn overlain by the marine Burtle sands and gravels, and interpreted the evidence as showing the progress of a marine transgression to a mean sea level about 12 m higher than today's.

Amino-acid racemization ratios derived from *P. vulgata* and *Macoma balthica* (Linné) from Greylake No. 2 Quarry were given by Andrews *et al.* (1979), who assigned the site to their amino-acid Group 2, which they regarded as of last interglacial age. More recently, Hunt *et al.* (1984) obtained ratios on *Corbicula* from Greylake No. 1 Quarry, less than 400 m from this site. These ratios, in the light of subsequent research, suggest that these shells, and thus probably the deposits at Greylake No. 1 Quarry, date from Oxygen Isotope Stages 9 and 7, whereas at least part of the sequence at Greylake No. 2 Quarry is more recent, dating from Stage 5e.

During reinvestigation of the site for the GCR, a series of boreholes was drilled to the west of the old quarry. These proved a sequence of fine silts overlying shelly sands and then a strongly calcreted gravelly palaeosol overlying further marine sands. Recently, Campbell *et al.* (in prep.) classified the lower sands as the Greylake Member of the Burtle Formation, and correlated them with Oxygen Isotope Stage 7; the upper marine sand unit is termed the Middlezoy Member and is correlated with Stage 5e.

Description

The deposits at Greylake No. 2 Quarry (ST 385336) extend to around 9 m OD and thus stand 2–3 m above the local Holocene deposits. The sequence in the old quarry (following Bulleid and Jackson (1937) and Kidson *et al.* (1978)), and confirmed by the GCR boreholes to the west, can be summarized thus (maximum bed thicknesses in parentheses):

16. Olive-grey silts, containing freshwater molluscs and large specimens of *Cepaea* and other land snails, on which is developed a modern soil profile. (1.4 m)
15. Reddish-brown sandy silts on which is developed a modern soil profile. Bed 15 passes gradually into bed 14. (1.0 m)
14. Pale yellow-brown fine-medium sand, cross-bedded, with heavily but irregularly carbonate-cemented horizons, and sometimes weathered and reddened towards the top. Contains occasional to abundant shells of *M. balthica* and *Hydrobia* spp. (2.0 m)
13b. Pale yellow-brown, sandy shelly gravel with abundant *H. ulvae*, some *Hydrobia ventrosa* (Montagu) and a few *Phytia myosotis* (Draparnaud), *Littorina* spp., *M. edulis*, *O. edulis*, *Retusa* sp. and *Patella* sp. *C. fluminalis* and bones of *B. primigenius*, *Dama* cf. *dama* and *C. elaphus* were reported from this horizon by Bulleid and Jackson (1937). This bed may pass laterally into bed 13a. (0.3 m)
13a. Strong brown, mottled dark red, very stony clay passing down into a coarse sandy gravel strongly cemented with pedogenic calcium carbonate (a plugged calcrete horizon). (0.3 m)
12. Pale yellow fine sand, plane-bedded, with *Hydrobia ?ulvae*. (0.4 m)
11. Pale yellow-brown gravel with rare *M. balthica*, *C. edule*, *M. edulis*, *Littorina neritoides* (Linné), *Helicella itala* (Linné), *Acroloxus lacustris* (Linné) and *Armiger crista* (Linné). (0.1 m)
10. Pale yellow fine sand, plane-bedded, with *H. ulvae*, some *H. ventrosa*, *M. balthica*, *Littorina* spp. and *Patella* sp. (0.9 m)
9. Blue-grey clay with sedge and grass stems in growth position, seeds of Chenopodiaceae and one seed of *Betula pendula*, and abundant *H. ulvae* and *H. ventrosa*. Foraminifer and ostracod biocoenoses suggestive of estuarine mudflats are present. (0.4 m)

8. Blue-grey, silty fine sand with a few *H. ulvae* and *H. ventrosa*. The foraminifers and ostracods are consistent with deposition on estuarine mudflats, with some inwashed outer estuarine and marine taxa. (0.4 m)

7. Blue-grey clay with sedge and grass stems in growth position, seeds of Chenopodiaceae, and extremely abundant *H. ventrosa*, very abundant *H. ulvae*, rare *P. myosotis* and *Cerastoderma* sp. Foraminifer and ostracod biocoenoses suggestive of estuarine mudflats are present. (0.1 m)

6. Blue-grey, silty fine sand with abundant *H. ventrosa* and some *H. ulvae*. The foraminifers and ostracods are consistent with deposition on estuarine mudflats, with some inwashed outer estuarine and marine taxa. (0.6 m)

5. Blue-grey silt with very abundant *H. ventrosa*, abundant *H. ulvae*, rare *Lymnaea peregra* (Müller), a planorbid and foraminifers and ostracods consistent with deposition on estuarine mudflats, with some inwashed outer estuarine and marine taxa. (0.5 m)

4. Grey-brown silt passing downwards into fine sandy silt, with wood fragments, probably roots, replaced by iron oxides. (0.6 m)

3. Grey-brown gravel with frost-cracked pebbles. (0.3 m)

2. Grey-brown sandy clay. (0.2 m)

1. Grey-brown diamicton with a clayey matrix. Clasts include chalk and frost-cracked pebbles. (> 0.4 m, base unseen)

The upper part of the sequence described here is very similar to the descriptions and photograph of Greylake No. 1 Quarry given by Bulleid and Jackson (1937). Of particular interest is their description of 'earthy gravel' (beds 13a and 13b above, and their units iii and vi at Greylake No. 1 Quarry). Their plate XXVII shows the 'earthy gravel' as irregular weathering horizons cutting across the marine beds. Bed 11 or 13 seems to have been associated with a sandy channel fill from which M.P. Kerney recovered large amounts of freshwater mollusc remains (Gilbertson, 1984).

Interpretation

At least six major sedimentary units are present at Greylake No. 2 Quarry. The interpretation of several of these units is equivocal, with the present state of knowledge, and there is considerable scope for further research. The oldest unit is most probably beds 1–4 in the old quarry, which are deposits of fluvial, periglacial or possibly glacial origin capped by what is probably a palaeosol which contains evidence of forest growth. The presence of a chalk clast within the basal diamict is possibly suggestive of glacial transport, since chalk clasts were quickly lost downstream from the Chalk outcrop in southernmost Somerset as the result of normal cold-stage and warm-stage fluvial transport (Hunt, 1987). If this is the case, then Greylake is the southernmost locality in Somerset where glacigenic deposits, most probably referrable to the Kenn Formation, have been identified.

The second unit at Greylake No. 2 Quarry is the body of silts, clays, fine silty sands, fine gravels and shelly sands (beds 5–12). Beds 5–9 contain molluscs, ostracods, foraminifers and plant macrofossils all of which indicate deposition on estuarine tidal mudflats and among reed communities near High Water Spring Tides with salinity values around 15‰. Sedimentation appears to have kept abreast of sea-level rise during deposition of these beds (Kidson *et al.*, 1978). The fine sands of beds 10 and 12 are dominated by the molluscs *H. ulvae* and *M. balthica* and were therefore most probably laid down below Mean High Water Neaps (Kidson *et al.*, 1978) and quite probably near the low tide mark (Hughes, 1980). The ostracod assemblages are a thanatocoenosis with a good proportion of fully marine species, but the foraminifera are consistent with deposition in shifting sand shoals near or below low tide mark (Kidson *et al.*, 1978). The gravel (bed 11) interbedded with these sands includes the freshwater taxa *A. lacustris* and *A. crista*, and also *H. itala*, which is a sand dune species in this country. Kidson *et al.* (1978) suggest that this assemblage is consistent with deposition in the vicinity of a small stream passing through sand dunes at the back of a beach. The channel itself may have been visible in the mid 1970s but was never adequately recorded. This unit is capped by the cemented palaeosol (bed 13a), which is regarded as partially reworked by tidal action in the area documented by Kidson *et al.* (1978). Here, it is listed above as bed 13b. The deposits of beds 5–12, the Greylake Member, can probably be referred to Oxygen Isotope Stage 7 (Campbell *et al.*, in prep.).

In this account, it is assumed that the marine sequence recorded in the old quarry reflects more than one marine transgressive cycle, contrary to the interpretation of Kidson *et al.* (1978). The third unit comprises the classic Burtle sands and gravels of the Middlezoy Member. The basal bed (13b) is a

Greylake (No. 2 Quarry)

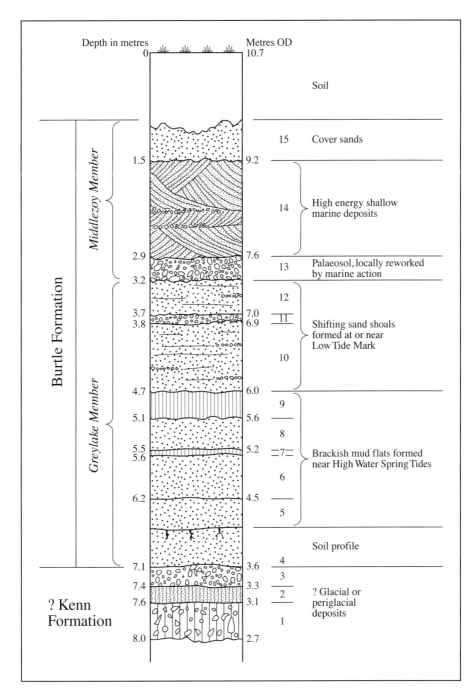

Figure 9.4 An interpretation of the Quaternary sequence at Greylake No. 2 Quarry, adapted from Hughes (1980). Beds 1–15 are described in more detail in the text.

lag deposit containing fossil molluscs, such as *C. fluminalis*, recycled from the Greylake Member. The mammal bones from this horizon – *C. elaphus*, *B. primigenius* and *Dama* cf. *dama* – are a warm-stage assemblage, but not ecologically very specific. The sands and gravels above this level contain mollusc, foraminifer and ostracod assemblages.

These and the mollusc assemblage in bed 13b are all consistent with deposition around Low Water Neap Tides. Kidson *et al.* (1978) computed a Mean Sea Level 12 m above the modern MSL and a high tide level of around 18 m OD for the peak of the marine incursion, which is equated here with Oxygen Isotope Stage 5e.

The fourth unit comprises the sandy silts and soil profile capping the quarry section (bed 15). Reddish sandy silts and silty sands are widespread around the Severn Estuary and seem to have been laid down as coversands during one or more of the cold stages (Gilbertson and Hawkins, 1978a).

The final unit is the olive silts (bed 16). The presence of thermophilous land snails like *Cepaea* suggests a Holocene age. It is suggested that these deposits occur as a veneer of overbank sediments which resulted from flooding in the Sedgemoor Levels.

Amino-acid racemization ratios derived from *P. vulgata* and *M. balthica* from the old quarry at this site were given by Andrews *et al.* (1979), who assigned the site to their amino-acid Group 2. The ratios were closely comparable with amino-acid ratios associated with a U/Th date of 121 ka + 14/− 12 ka BP from Belle Hougue Cave, Jersey (Keen *et al.*, 1981) and similar ratios and dates from the Gower coastal caves (Bowen *et al.*, 1985; Sutcliffe *et al.*, 1987). This correlation implies that the Middlezoy Member at Greylake No. 2 Quarry was laid down by the Oxygen Isotope Stage 5e marine transgression. A considerable scatter was evident in their data, however, with some ratios being 'old' enough to imply correlation with Stage 7. Unfortunately, the precise stratigraphic context of these shells was not recorded.

The strong palaeosol developed on the Greylake Member at Greylake No. 2 Quarry, together with limited aminostratigraphic evidence from Greylake No. 1 Quarry, suggest that the Greylake Member is considerably older than Stage 5 (Hunt *et al.*, 1984; Hunt, 1990a) since one ratio on *C. fluminalis* from Greylake No. 1 Quarry is consistent with an age equivalent to Oxygen Isotope Stage 7 and another with a yet older stage. (*C. fluminalis* is not known to occur in deposits of Stage 5e age in Britain (Keen, 1990; Bridgland, 1994).) The Greylake No. 2 Quarry is thus of considerable importance as a well-dated example of sedimentation by the Stage 5e transgression overlying an older transgressive unit.

Conclusion

Greylake No. 2 Quarry is important because it provides a well-dated, very fossiliferous sequence laid down during the Oxygen Isotope Stage 5e marine transgression. These deposits contain a classic interglacial estuarine fauna and microfauna. The interglacial deposits overlie a palaeosol and a second, older, interglacial marine sand unit and

then cold-stage deposits which may be of glacial origin. Further work is required to establish the precise origin and age of these deposits.

HURCOTT FARM
C. O. Hunt

Highlights

Hurcott Farm is a critical locality for the Pleistocene stratigraphy of the Cary and Yeo drainage basins in southern Somerset. Here, the Chadbrick Gravels provide a relatively unambiguous aminostratigraphic marker among a largely undateable set of terrace deposits. They also provide a *terminus post quem* for the capture of the upper Yeo catchment from the Cary by a tributary of the River Parrett. The site is proposed as the type-locality for the Whatley Member.

Introduction

Hurcott Farm demonstrates heavily cemented very shelly gravels of the fifth terrace of the River Cary. The fossil mollusc fauna and pollen are indicative of interglacial conditions and the mollusc assemblage includes the locally extinct species *C. fluminalis* and the extinct species *Pisidium clessini* Neumayr. An amino-acid ratio on *Corbicula* and the presence of *P. clessini* are consistent with an Oxygen Isotope Stage 9 age or older.

The site was 'lost' for some years and only recently rediscovered. In 1954, the Reverend J. Fowler presented a fragment of cemented shelly gravel to the British Museum (Natural History). The gravel was found 2.8 km north-east of Somerton in the Chadbrick Valley, but the exact position of the find was not reported. Gilbertson (1974) and Gilbertson and Beck (1975) suggested that the fragment had come from an area of terrace gravels of the River Cary mapped by the Geological Survey near Hurcott Farm. They identified a number of molluscs, including *Valvata piscinalis* (Müller), *L. peregra*, *Planorbis* sp., *Hygromia (Trichia)* cf. *hispida*, *C. fluminalis*, *Pisidium henslowanum* (Sheppard), *Pisidium nitidum* (Jenyns) and *Pisidium* sp. They interpreted the assemblage as indicating fluvial interglacial conditions.

The shelly gravels were relocated in 1982 (Hunt *et al.*, 1984; Hunt, 1987), lying on the valley side some 6 m lower than the terrace gravels mapped by the Geological Survey. The latter were referred (Hunt, 1987) to the sixth terrace of the Cary. Hunt *et al.* (1984) and Hunt (1987) named the fifth terrace deposits the Chadbrick Gravels and described their molluscan fauna, sedimentology and clast lithology. They also described the results of an amino-acid assay from a valve of *C. fluminalis* from the Chadbrick Gravels and two assays on *Corbicula* from the Burtle Formation at Greylake and, on the basis of comparisons with other published ratios, attributed the gravels to the Ipswichian. Hunt (1990a) described a pollen and algal microfossil assemblage from the Chadbrick Gravels and re-attributed the site to the 'Stanton Harcourt Interglacial' (Stage 7).

The presence of *P. clessini*, which became extinct after Oxygen Isotope Stage 9 (Keen, 1992), and a re-run of the amino-acid ratio to 0.225 (Bowen, pers. comm., 1996), make earlier interpretations untenable. Campbell *et al.* (in prep.) have therefore proposed correlation of the deposits with Oxygen Isotope Stage 9. They also designated the site as the type-locality of the Whatley Member.

Description

Six Pleistocene gravel units are present in the Cary Valley (Hunt *et al.*, 1984; Hunt, 1987). Near Hurcott Farm (ST 512296), coarse poorly sorted gravels with *P. muscorum* and *Trichia* cf. *hispida* underlie a well-developed terrace surface at 43 m OD (sixth terrace). The Chadbrick Gravels are cemented to an outcrop of Rhaetic limestone at 37 m OD (fifth terrace) and overlain by 0.5 m of silty sands with fragments of *P. muscorum*, succineids and hygromids and 0.5 m of stony colluvium. Farther downslope at 29 m OD are fragments of the fourth terrace, underlain by plane-bedded reddish sands and fine gravels. The third terrace is not found in this part of the Cary Valley, but is known both upstream and downstream. Near the present valley floor, the second terrace surface at 14 m OD is underlain by coarse, poorly sorted unfossiliferous gravels. Woodward (1905) described further coarse gravels underlying the Cary floodplain alluvium in foundation trenches for the Somerton railway viaduct. These can be traced upstream into the first terrace of the Cary.

The Chadbrick Gravels lie on a gently undulating surface cut into Rhaetic limestones. One unit of clast-supported, imbricated, epsilon cross-bedded shelly cemented gravel is present, truncated by erosion and overlain by uncemented silty sands and

stony silts. The gravels are well sorted, with an average clast size around 5.0 mm, and are well rounded. The cement is very uneven in development, consisting of irregular areas of grey-green micrite, which weathers rusty brown, interspersed with areas of sparry calcite. Voids are present under some large clasts (Hunt *et al.*, 1984).

The fossil mollusc fauna (Table 9.1) is dominated by *V. piscinalis* (35.5%), with lesser *L. peregra* (16.5%), *Bithynia tentaculata* (Linné) (9.6%), *C. fluminalis* (7.3%) and other species. The extinct bivalve *P. clessini* is present (Hunt *et al.*, 1984). The pollen assemblage is marked by abundant *Alnus* (24.7%) and Cyperaceae (20.5%), and a variety of broad-leaved tree and wetland species. The algal microfossils *Pediastrum* and *Spirogyra* are present (Hunt, 1990a).

An amino-acid assay on a valve of *C. fluminalis* gave a D-*alloisoleucine* to L-*isoleucine* ratio of 0.18 (Hunt *et al.*, 1984). This was re-run at 0.225 (Campbell *et al.*, in prep.).

Interpretation

The Chadbrick Gravels are epsilon cross-bedded, indicating deposition in a meandering river. They contain a diverse, temperate freshwater molluscan fauna, with a group of species typical of slow moving, calcareous, eutrophic mud-bottomed rivers and a group of species typical of faster moving water with sandy substrates and little aquatic vegetation. This is consistent with molluscan assemblages from rivers characterized by pool and riffle sequences (Hunt *et al.*, 1984; Hunt, 1987). The terrestrial molluscs are species typical of woodland and damp sheltered habitats. The pollen assemblage is of interglacial type, being dominated by broad-leaved arboreal taxa and wetland species. It probably reflects alder thickets and sedge marsh close to the river and mixed-oak woodland beyond. The presence of the planktonic *Pediastrum* and the benthonic *Spirogyra* is consistent with deposition in a fluvial environment (Hunt, 1990a).

The original amino-acid ratio is broadly comparable with ratios from the Stanton Harcourt Stage 7 interglacial deposits (Bowen *et al.*, 1989; Hunt, 1990a), but the re-run ratio is more compatible with ratios indicative of an Oxygen Isotope Stage 9 age (Bowen, pers. comm., 1996), as is the presence of *P. clessini*, which became extinct after Stage 9 (Keen, 1992). The high tree pollen percentages are also similar to those from other Stage 9 sites,

Table 9.1 Fossil molluscs from the Chadbrick Gravel

Species	Number	Percentage
Valvata cristata (Müller)	1	0.4
Valvata piscinalis (Müller)	88	35.5
Bithynia tentaculata (Linné)	24	9.6
Physa fontinalis (Linné)	1	0.4
Lymnaea stagnalis (Linné)	1	0.4
Lymnaea peregra (Müller)	41	16.5
Planorbis spp.	2	0.8
Gyraulus laevis (Alder)	7	2.8
Ancylus fluviatilis (Müller)	1	0.4
Unio sp.	3	1.2
Corbicula fluminalis (Müller)	18	7.3
?Sphaerium sp.	1	0.4
Pisidium amnicum (Müller)	9	3.6
Pisidium clessini Neumayr	5	2.0
Pisidium henslowanum (Sheppard)	5	2.0
Pisidium nitidum Jenyns	1	0.4
Pisidium subtruncatum Malm	4	1.6
Pisidium spp.	22	8.9
Oxyloma cf. *pfeifferi*	5	2.0
Cochlicopa cf. *lubrica*	1	0.4
Vallonia cf. *pulchella*	1	0.4
Discus rotundatus (Müller)	1	0.4
?Helicella sp.	1	0.4
Trichia cf. *hispida*	6	2.4
Total	**248**	

whereas most Stage 7 sites have relatively low tree pollen percentages. Capture of the former headwaters of the Cary by the Yeo, a tributary of the Parrett, occurred after aggradation of the second terrace of the Cary (Hunt, 1987).

Conclusion

The Chadbrick Gravels at Hurcott Farm have yielded important interglacial molluscan and pollen assemblages. A re-run of an amino-acid assay on a valve of *C. fluminalis* yielded a ratio of 0.225, comparable with ratios from sites attributed to Oxygen Isotope Stage 9. There is support for this ascription from the molluscan and palynological data. Hurcott Farm is a critical locality for the Pleistocene stratigraphy of the Cary and Yeo drainage basins in southern Somerset, where the Chadbrick Gravels provide a relatively unambiguous aminostratigraphic marker among a largely undateable set of terrace deposits. They help to determine when the upper Yeo catchment was captured from the Cary

by a tributary of the River Parrett. Non-marine interglacial deposits are unusual in South-West England, making Hurcott Farm's aminostratigraphically correlated interglacial molluscan and pollen assemblages of considerable importance to reconstructions of Pleistocene history.

PORTFIELD
C. O. Hunt

Highlights

Portfield is a key site in the terrace stratigraphy of the Parrett catchment because amino-acid ratios on molluscs from the site provide rare chronological control in a largely undated set of terraces. The site is also important for its interglacial, non-marine fossil mollusc fauna. It is the type-locality for the Portfield Member of the Parrett Formation.

Introduction

Portfield is the only surviving locality of the fifth terrace of the Parrett/Isle system. The site is complex. A basal weathered diamicton is overlain by silts which contain freshwater and terrestrial fossil molluscs of interglacial affinity. The silt unit extends to 12.5 m OD and is overlain by over 6 m of gravelly sands which contain a sparse and restricted mollusc fauna. These are succeeded by stony diamictons. Amino-acid ratios derived from shells in the interglacial silts suggest an Oxygen Isotope Stage 7 age.

Quaternary gravelly sands were briefly reported from Portfield by Hughes (1980), who equated them with the interglacial marine Burtle Formation of the Somerset Levels. The site was reinvestigated by Hunt (1987) and Hunt and Bowen (in prep.) who drilled a number of boreholes, conducted mollusc analyses and amino-acid geochronometric assays and attributed the interglacial deposits to Oxygen Isotope Stage 7. Campbell *et al.* (in prep.) accepted this correlation, and named the interglacial deposits the Portfield Member.

Description

Near Portfield, the former existence of deposits of the fifth terrace of the Parrett/Isle system can be seen from 'flats' in the landscape underlain by extensive spreads of shallow gravel-based soil

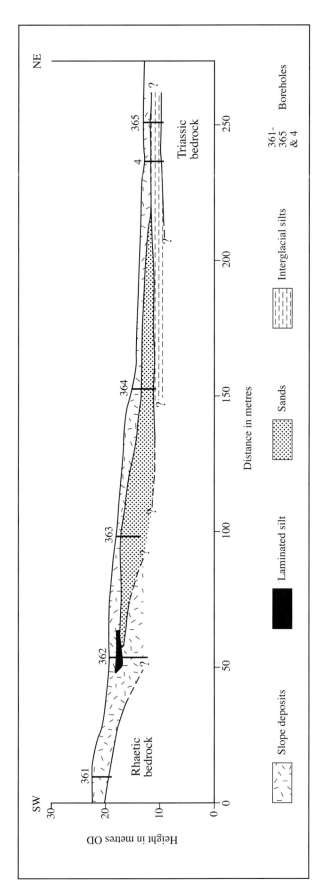

Figure 9.5 A cross-section of the Pleistocene deposits at Portfield. (Adapted from Hunt, 1987.)

between ST 409260 and ST 408274. Deep *in situ* Pleistocene deposits, however, survive only at Portfield. The sequence is shown in Figure 9.5 and can be summarized as follows (maximum bed thicknesses in parentheses):

6. Sandy silty clays with abundant angular limestone fragments, crudely stratified and with clast long-axes pointing downslope. These slope deposits interdigitate with and overlie beds 5 and 4. (2.0 m)
5. Sandy silts, indistinctly and irregularly laminated, with abundant calcareous root tubules and rare fossil molluscs; pale orange-yellow. (0.5 m)
4. Fine to medium sands, often silty with indistinct lamination and rare fossil molluscs, and sometimes gravelly; pale reddish-orange. (6.5 m)
3. Fine sands with fossil molluscs, interdigitating with bed 2, and having a transitional lower junction with it; pale yellow. (0.65 m)
2. Clayey silts, sometimes indistinctly laminated, with rare to abundant fossil molluscs; light green to blue-grey. (1.1 m)
1. Stiff clayey silt with angular clasts of Rhaetic limestone, with rare fossil molluscs; blue-grey, mottled strong brown. (0.7 m)

Fossil molluscs are present in all of these stratigraphical units (Hunt, 1987; Hunt and Bowen, in prep.; Figure 9.6) and a few plant macrofossils are also present in some units. The basal stony clayey silts (bed 1) contain restricted assemblages characterized by taxa such as *Trichia* cf. *hispida*, *Succinea* cf. *oblonga*, *Lymnaea truncatula* (Müller), *Helicella* sp., Limacidae and extremely corroded fragments of *Pisidium* sp. and *Bithynia*. Plant macrofossils include seeds of *Saxifraga* sp., Polygonaceae, Chenopodiaceae and *Stellaria* sp.

The laminated silts of bed 2 (the Portfield Member of Campbell *et al.*, in prep.) contain mollusc assemblages characterized by considerable numbers of aquatic thermophilous taxa, especially 'river' species such as *B. tentaculata*, *V. piscinalis*, *Sphaerium corneum* (Linné) and *Pisidium amnicum* (Linné) but also 'ditch' taxa such as *Valvata cristata* (Müller), *Planorbis planorbis* (Linné), *Gyraulus laevis* (Alder), generalist species like *L. peregra*, and 'slum' species including *Anisus leucostoma* (Millet), *L. truncatula* and *Pisidium obtusale* (Lamarck). The terrestrial taxa are rare, but include occasional specimens of the shaded habitat species *Discus rotundatus* (Müller), and a variety

of other mostly ecologically indeterminate species. Plant macrofossils are rare, but include the thermophilous aquatic species *Zannichellia palustris*. The mollusc assemblages from bed 3 are essentially similar (Hunt, 1987; Hunt and Bowen, in prep.).

Beds 4, 5 and 6 contain sparse assemblages characterized by rare 'slum' aquatic taxa, such as *P. obtusale* and *L. truncatula*, the marsh taxon *Succinea* cf. *oblonga*, grassland taxa such as *Trichia* cf. *hispida*, *Vallonia pulchella* (Müller), *Vallonia excentrica* Sterki, *Vertigo pygmaea* (Draparnaud) and sometimes large numbers of the exposed-ground species *P. muscorum*. Also present is a mixture of rolled thermophilous taxa such as *M. balthica*, *P. amnicum*, *Viviparus* sp., *D. rotundatus* and *Pomatias elegans*. Hunt and Bowen (in prep.) have argued that the grassland taxa may be recycled from soil profiles of Stage 7 age. Plant macrofossils include *Carex* nutlets, seeds of *Stellaria*, and a prophyll of *Salix* (Hunt, 1987; Hunt and Bowen, in prep.).

Interpretation

A complex sequence of Quaternary events can be identified at Portfield. During an interglacial prior to Oxygen Isotope Stage 7, interglacial deposits were laid down. The evidence for these deposits is the presence of corroded specimens of freshwater molluscs such as *B. tentaculata* and *Pisidium* spp. in bed 1. At some time after the deposition of this ancient interglacial deposit, it was destroyed and the interglacial molluscs were incorporated into bed 1, most probably by mudflow, solifluction and wash processes in cold-stage conditions. The other molluscs and seeds in this unit are a mixed assemblage, but mostly point to open exposed (*Helicella* sp., Chenopodiaceae) or damp grassy (*Trichia* cf. *hispida*, *Saxifraga* sp.), or marshy conditions (*Succinea* cf. *oblonga*, *L. truncatula*) or are ecologically indeterminate (Limacidae, Polygonaceae and *Stellaria* sp.). It is presumed that these fossils were incorporated into the deposit as an early-interglacial soil profile began to form and was then swamped by rising waters as the interglacial marine transgression progressed and caused rising water levels on the site.

With further sea-level rise, the Portfield Member (beds 2 and 3) was laid down by a perennial, slow-moving freshwater river, perhaps under weak tidal influence. The preponderance of species typical of larger water bodies such as *B. tentaculata*, *V. piscinalis*, *S. corneum* and *P. amnicum* points to this

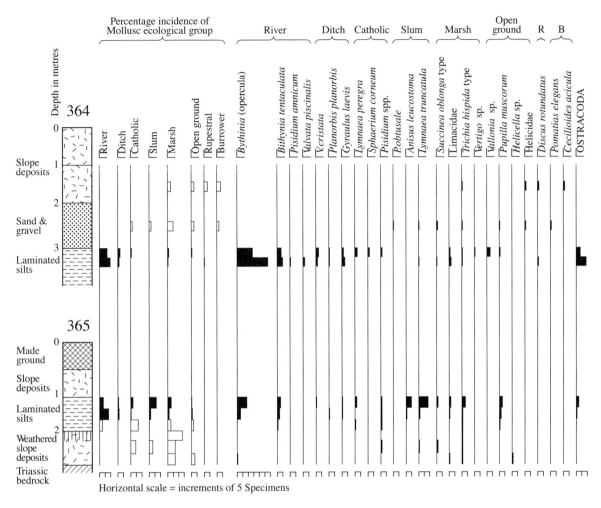

Figure 9.6 The molluscan biostratigraphy of Pleistocene deposits at Portfield, adapted from Hunt (1987). Numbers 364 and 365 refer to boreholes shown in Figure 9.5.

being more than a small stream, while the presence of 'ditch' taxa such as *V. cristata*, *P. planorbis*, *G. laevis*, generalist species like *L. peregra*, and 'slum' species including *A. leucostoma*, *L. truncatula* and *P. obtusale* indicates the presence of a variety of submerged habitats. The presence of *D. rotundatus* and *B. tentaculata* is consistent with the climate at this time not having been cooler than southern Scandinavia today (Hunt and Bowen, in prep.). The amino-acid ratios are compatible with the Portfield Member being of Oxygen Isotope Stage 7 age. Assessment of the sea-level relationships of the silts, after allowances for tidal funnelling, suggests a mean sea level around 4.5–7.5 m OD (Hunt and Bowen, in prep.), comparable with other Stage 7 sites in South-West England at Portland and Torbay (Chapter 6).

The deposition of beds 2 and 3 was followed by

climate deterioration and sea-level fall. The latter led to the exposure and deflation of substantial areas of interglacial marine sands of the Burtle Formation in Sedgemoor. The deflated sand then accumulated against the Langport-Curry Rivel escarpment, behind the Portfield site. At the same time, exposed soils on the escarpment were subject to erosion. The aeolian sands and eroded soil were then redeposited at the foot of the escarpment as bed 4. Sedimentary structures in this unit point to deposition in a multi-channel ephemeral stream. The molluscs and plant macrofossils in bed 4 may be partly recycled, especially the rolled thermophilous taxa such as *M. balthica*, *P. amnicum*, *Viviparus* sp., *D. rotundatus* and *P. elegans*. The 'slum' aquatic and marsh taxa, *P. obtusale*, *L. truncatula* and *Succinea* cf. *oblonga*, are the type of assemblage that might colonize pools and wet ground in the bed of an ephemeral stream. Exposed ground is indicated by the presence of *P. muscorum*, but *Trichia* cf. *hispida*, *V. pulchella*, *V. excentrica*, and *V. pygmaea* may have lived in grassy vegetation. Hunt and Bowen (in prep.) have argued that the grassland taxa may be recycled from soil profiles of Stage 7 age.

Deposition of bed 4 was followed locally by the accumulation of the wash deposits of bed 5. These laminated sandy silts were most probably laid down by shallow overland flow in a sparsely vege-tated landscape during 'cold-stage' conditions. Interdigitating with, and overlying beds 4 and 5, are the slope deposits of bed 6. These were most probably laid down by mudflow, solifluction and wash processes in cold-stage conditions.

Conclusion

Portfield provides vital evidence for the age of the upper part of the terrace sequence in the Parrett Valley. The site is also notable for its temperate-stage and cold-stage mollusc faunas and provides an important indication of sea levels during Stage 7. The sea level calculated for Portfield compares well with those for Stage 7 sites at Portland Bill and Torbay (Chapter 6), suggesting that South-West England has not been significantly affected by dif-ferential uplift since *c*. 200 ka BP.

LOW HAM
C. O. Hunt

Highlights

Low Ham is of national importance as the only locality where high sea levels during a Devensian interstadial can be demonstrated. The site contains the best Stage 5a interstadial pollen, mollusc and ostracod assemblages in South-West England and is the type-site of the Low Ham Member of the Parrett Formation.

Introduction

A thick drift sequence – the Low Ham Member and the Combe Member – can be found in the Leaze Moor Valley, which cuts through the Langport-Somerton escarpment in southern Somerset (Figure 9.7). The Low Ham Member consists of sands, silts, clays and peats, dated by aminostratigraphy to

Oxygen Isotope Stage 5a. These deposits contain abundant molluscs, ostracods, plant macrofossils, pollen and other microfossils, which together indi-cate interstadial conditions. Ostracods and rare dinocysts and diatoms show that the Low Ham Member accumulated in a back-estuarine situation in response to rising sea levels. The overlying Combe Member accumulated during a phase of cli-matic deterioration and falling sea level, through the recycling of marine sands by aeolian, wash and ephemeral fluvial processes.

The deposits of the Leaze Moor Valley were dis-covered only recently, though the type-site of the Combe Member, at Combe, near Langport, was first described by Ussher (1908). Drift-based soils were mapped at Low Ham by Avery (1955), and sands with a sparse mollusc fauna were first recorded near Low Ham by Hughes (1980). The Leaze Moor Valley was investigated in detail by Hunt (1987) and Hunt *et al.* (in prep.). Twenty-nine boreholes were drilled and twelve mollusc, seven plant macrofossil, two pollen and two organic-walled microfossil diagrams were con-structed. Mizzen (1984) analysed the ostracod content of one of these boreholes. Amino-acid ratios derived from mollusc shells taken from the Low Ham Member were described by Hunt *et al.* (in prep.), who attributed the unit to Stage 5a. The Low Ham and Combe members are major con-stituents of the Parrett Formation defined by Campbell *et al.* (in prep.).

Description

The Low Ham beds underlie 'terraces' standing up to 4 m above Holocene alluvium in the valleys between Langport and Somerton (Hunt, 1987). At Low Ham GCR site (ST 43902900), the terrace sur-face lies at 19 m OD. Underlying the terrace surfaces are thick stony diamictons, then sandy gravels and laminated silts of the Combe Member with a sparse mollusc fauna characterized by *P. muscorum*, *Succinea* cf. *oblonga* and occasional *Columella edentula* (Draparnaud) (Figure 9.8).

These deposits pass down conformably into the Low Ham Member, the first unit being dark brown laminated sedge peats and thin grey silts with *Succinea* cf. *oblonga* and *P. muscorum*. These overlie dark brown highly organic detritus muds, with abundant seeds of *Zannichellia palustris*, *Potamogeton* stones and *Hippuris* nodes and stones. The mollusc fauna of this unit includes *Succinea* cf. *oblonga*, *A. leucostoma*, *Pisidium*

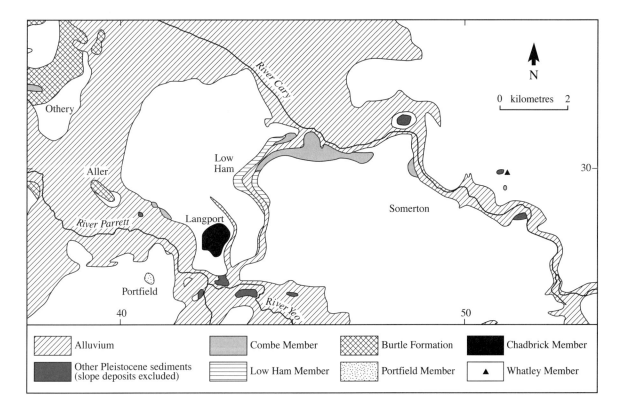

Figure 9.7 The distribution of Quaternary deposits near Langport, Somerset. (Adapted from Hunt 1987.)

spp. and *L. peregra*. The detritus muds overlie compacted mid-grey silts with *G. laevis*, *B. tentaculata*, *V. cristata*, *A. leucostoma*, *Pisidium* spp., *L. peregra* and *Succinea* cf. *oblonga*. These silts pass down into sparsely fossiliferous, pale blue-grey silty sands with *G. laevis*, *B. tentaculata*, *V. cristata*, *Pisidium* spp., and *L. peregra* and then unfossiliferous coarse red sands (Figure 9.8). Farther to the north along the Low Ham Valley, more diverse aquatic mollusc assemblages, including *Belgrandia marginata* (Michaud), were recorded (Hunt, 1987). Pollen assemblages from the detritus muds and grey silts are dominated by sedge, grass and herbs, with rare pine, birch, spruce, alder, willow and hazel. Assemblages from the laminated sedge peats and silts are of lower diversity and lack the tree and shrub species. Ostracod assemblages from the detritus muds are rich, often containing around 80 species. Occasional specimens of the salinity tolerant *Cyprideis torosa* (Jones) and a number of obligate halophilous taxa are present (Whatley *in* Hunt, 1987; Hunt *et al.*, in prep.). Also present are rare specimens of the marine dinoflagellate cysts *Operculodinium centrocarpum* and *Spiniferites* cf. *ramosus* and very rare marine diatoms. These

marine taxa are present in the borehole to 13.8 m OD.

Radiocarbon assays of the Low Ham Member yielded ages of >40 300 (SRR–2450) and > 41 ka BP (SRR–2451). Amino-acid racemization assays on molluscs from the Low Ham Member are technically very difficult because of the small size of most mollusc specimens, but ratios indicate that the most probable correlation is with Oxygen Isotope Stage 5a (Bowen, pers. comm., 1996; Campbell *et al.*, in prep.; Hunt *et al.*, in prep.). However, Keen (pers. comm., 1997) notes that *B. marginata* has not been recorded in deposits younger than Stage 5e.

Interpretation

The Combe Member (beds 1 and 2) contains *P. muscorum*, *Succinea* cf. *oblonga* and *C. edentula*, an assemblage typical of 'cold-stage' terrestrial sedimentation in open exposed landscapes. The diamictons, sandy gravels and silts of these deposits are consistent with deposition by a variety of mass movement and wash processes and thus with generally poorly vegetated stadial conditions.

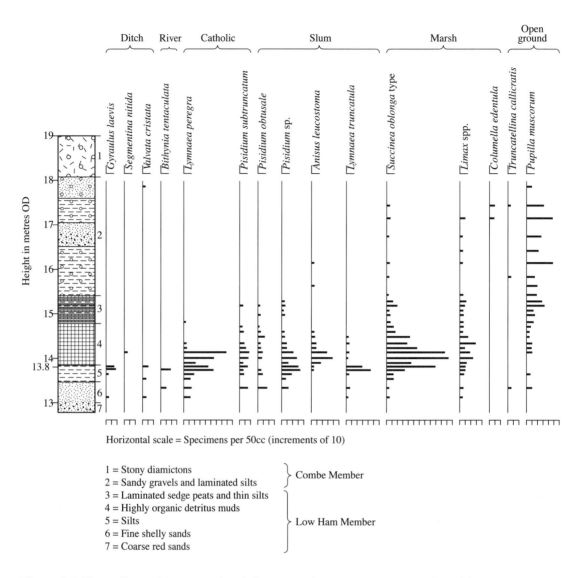

Horizontal scale = Specimens per 50cc (increments of 10)

1 = Stony diamictons   } Combe Member
2 = Sandy gravels and laminated silts

3 = Laminated sedge peats and thin silts
4 = Highly organic detritus muds
5 = Silts   } Low Ham Member
6 = Fine shelly sands
7 = Coarse red sands

Figure 9.8 The molluscan biostratigraphy of Pleistocene deposits at Low Ham. (Adapted from Hunt 1987.)

The Low Ham Member at this site has the characteristics of a channel-fill succession, with basal moving-water sands passing up into quiet-water detritus muds and then marsh peats. The succession of mollusc species also indicates a transition from a basal assemblage consisting largely of taxa typical of moving water with some weeds, such as *B. tentaculata* and *V. cristata*, to assemblages typical of rather poor quality stagnant water, typified by *A. leucostoma* and *L. peregra* in the detritus muds, and then into marshy conditions with occasional standing water characterized by *Succinea* cf. *oblonga*, occasional *Pisidium* spp. and terrestrial molluscs.

Although no brackish-water molluscs are present, the ostracods, dinoflagellate cysts and diatoms point to marine influence up to 13.8 m OD. The tidal range in the Bristol Channel is very large as the result of tidal funnelling and, assuming similar tidal ranges in the past, a mean sea level of 2–5 m OD is probable.

The age of the Low Ham Member is still open to question. The facies of the pollen and terrestrial mollusc assemblages are consistent with interstadial rather than interglacial conditions, though the presence of *B. marginata* at some localities is incompatible with later Devensian interstadials. The most compelling biostratigraphic comparisons are with the Wretton Interstadial of Norfolk (Hunt, 1987). The radiocarbon assays indicate only an age greater than 41 ka BP, while the amino-acid ratios are broadly consistent with an age late in Stage 5 (Campbell *et al.*, in prep.; Hunt *et al.*, in prep.). The Low Ham Member thus provides evidence of a marine incursion into the Somerset Levels after the Ipswichian (Stage 5e) transgression but before the Mid-Devensian, and probably during Stage 5a.

Conclusion

Low Ham GCR site exhibits a suite of sediments, the Low Ham Member, which contains ostracods, molluscs, pollen, plant macrofossils, dinoflagellate cysts and diatoms which together indicate a back-estuarine environment, a temperate climate and an open landscape of interstadial aspect. The elevation of the Low Ham beds is consistent with a maximum level of marine influence at 13.8 m OD, and after allowing for tidal funnelling, a mean sea level of 2–5 m OD. Amino-acid ratios are consistent with an age late in Stage 5 of the Oxygen Isotope scale and geomorphologically the Low Ham beds post-date the interglacial marine Burtle Beds of the Somerset Levels, some of which are Ipswichian (although some are certainly older). The site is therefore of national importance as the only location where high sea levels subsequent to the Ipswichian Interglacial (Oxygen Isotope Stage 5e) and prior to the Holocene marine transgression can be demonstrated clearly.

BROOM GRAVEL PITS
S. Campbell, N. Stephens, C. P. Green and R. A. Shakesby

Highlights

A site of exceptional interest to geomorphologists and archaeologists alike, Broom Gravel Pits expose a thick sequence of terrace deposits once attributed to waters spilling from glacially impounded 'Lake Maw'. However, the terrace sequence is now widely believed to have been deposited by braided streams in a periglacial environment. Broom is also notable for being the richest source of Lower Palaeolithic artefacts yet known from South-West England.

Introduction

The Axe Valley terrace gravels have a protracted history of investigation. They first attracted interest

in an archaeological context (D'Urban, 1878; Evans, 1897), and over the years have yielded a profusion of Lower Palaeolithic implements. The age and origin of the gravels have proved controversial. Stephens (1970b, 1973, 1974, 1977) suggested that the deposits were laid down when a large, glacially impounded, Saalian-age lake 'overspilled' south through the 'Chard Gap'. However, most other workers have found no evidence to support the existence of such a lake, and have proposed that the terrace gravels were deposited by braided streams in a periglacial environment (C.P. Green, 1974b, 1988; Campbell, 1984; Shakesby and Stephens, 1984). The site has also been widely referred to elsewhere (Ussher, 1878, 1906; Reid, 1898; Salter, 1899; Reid Moir, 1936; Hawkes, 1943; J.F.N. Green, 1947; Calkin and Green, 1949; Waters, 1960d; Lewis, 1970; Macfadyen, 1970; Stephens, 1970a; Edmonds *et al.*, 1975; Stephens and Green, 1978; Todd, 1987; Campbell *et al.*, in prep.). Preliminary SEM work (Campbell, 1984) and pollen analyses (Scourse, 1984) have also been conducted on deposits from the site. A comprehensive reappraisal of the deposits awaits publication (C.P. Green *et al.*, in prep.).

Description

Regional setting

The relationship of Broom GCR site to the 'Chard Gap' and other important topographic features is shown in Figure 9.9. Stephens (1977) has asserted that the Chard Gap is the largest and lowest (at 83–90 m OD) of a number of major breaks in the watershed between the Somerset Levels (and Bristol Channel) and the English Channel. Today, the headwaters of the north-flowing Isle are located within the gap; the Axe rises in the high ground to the north of Beaminster and flows past the southern end of the gap in a south-west direction. In east Devon and south Somerset, Cretaceous rocks (Chalk and Greensand) form multiple escarpments of varying height; less resistant Jurassic and Keuper clays and marls are exposed in the bottom of incised valleys (Ussher, 1906; J.F.N. Green, 1941; Gregory, 1969; Waters, 1971; Shakesby and Stephens, 1984). Structural and lithological variations have controlled dissection and erosion, giving rise to a 'cuesta' landscape (Stephens, 1977) with quite prominent west- and north-west-facing escarpments. Stephens (1970b) has also argued that the trench-like form of the Chard Gap is unlike

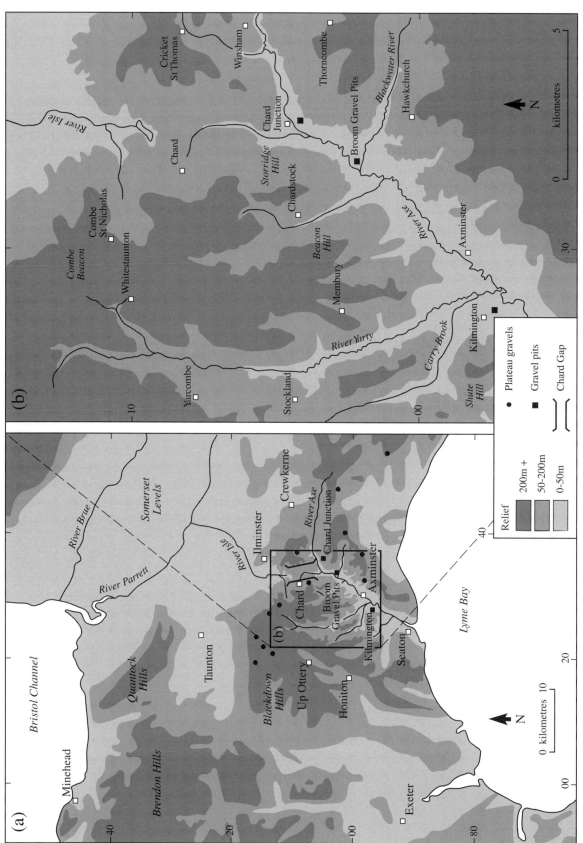

Figure 9.9 (a) The topographic setting of the Axe Valley and the distribution of plateau-gravel sites. (b) The principal exposures of the Axe Valley terrace gravels. (Adapted from Stephens, 1977 and Green *et al.*, in prep.)

any nearby watershed col except for that near Crewkerne, the Hewlish Gap, with a floor at *c.* 100–120 m OD. Local topography is therefore characterized by a series of comparatively flat-topped hills and low plateaux, although most of the Axe Valley lacks steep slopes. Between Chard Junction and Kilmington, the modern River Axe meanders across a broad floodplain; its main tributary, the Yarty, occupies a narrower, more confined valley (Shakesby and Stephens, 1984). Stephens (1974) identified five main types of superficial deposit in the local area:

1. Alluvium: mostly confined to existing river floodplains.

2. Valley gravels (described in more detail below): forming extensive outcrops on low ground to the north of the Chard Gap and also present, in small patches, within the gap and forming a major terrace in the lower Axe Valley (Figure 9.10). In places, a fine-grained 'brickearth' crops out on the surface of the terrace. The terrace surface itself declines in height seawards from *c.* 70 m OD at Chard Junction, to *c.* 65 m OD at Broom and finally to *c.* 30 m OD at Seaton on the coast (Stephens, 1973; Figure 9.10).

3. Head: this consists generally of locally derived material from various geological outcrops.

4. 'Clay-with-flints' and '-chert': this is composed largely of argillaceous material believed to have been derived from Tertiary strata, and mixed with flints and chert from Cretaceous beds. The material forms a discontinuous capping of irregular thickness on the plateau-like interfluves east and west of Chard and the lower Axe Valley (see Beer Quarry; Chapter 3).

5. Interfluve and plateau gravels: these are found in patches on both sides of the Axe Valley north of Axminster (Figure 9.9). Both Ussher (1906) and Waters (1960d) described the gravels as occurring on widely separated interfluves between *c.* 200 and 315 m; Reid (1898) interpreted them as river gravels, although Waters claimed the presence of ' … unmistakable beach cobbles of flint' (Waters, 1960d; p. 92). The material varies from sub-angular to well-rounded pebbles in a sandy matrix, with chert, flint, quartz, tourmalinized rocks, greywacke, a miscellany of Palaeozoic rock types, and a few chatter-marked flint cobbles (Stephens, 1974, 1977). C.P. Green (1974b) analysed the lithological composition

of these gravels which he regarded as Tertiary in age.

The Axe Valley gravels

Extensive Pleistocene terrace gravels border the River Axe and its tributaries, including the Blackwater and Yarty, lying above their modern floodplains (Figure 9.10). They form a major lithostratigraphic unit (the Axe Valley Formation) divisible into the Broom, Pratt's Pit, Chard Junction and Kilmington members (Campbell *et al.*, in prep.). The terrace 'gravels' are generally poorly sorted, comprising clasts from gravel to boulder grade. They are crudely bedded, occasionally exhibit imbrication and contain lenses of laminated sands; more continuous beds of clay, silt and sand are present locally. The gravels comprise mostly flint and chert (69–96%), but quartz (0–35%) and other pebbles derived from Palaeozoic rocks (0–12%) are present (C.P. Green, 1974b). Occasionally, the gravels contain distinctive 'blocks' or 'rafts' of laminated sand. Their upper layers are frequently disturbed by involutions.

The principal descriptions of the terrace gravels have been based on three sites: Chard Junction (ST 342044); Kilmington 'New' Pit (ST 277976) and Broom Gravel Pits (ST 326020; see below). At Chard Junction, over 11 m of terrace gravels, comprising mainly clasts of flint and chert, are exposed (Figure 9.11). The gravels are overlain by a loamy silt or 'brickearth' and the upper 1.5–2.0 m of the gravel are cryoturbated (Shakesby and Stephens, 1984). Kilmington New Pit shows at least 3 m of poorly sorted, crudely bedded terrace gravels with discontinuous lenses of silt and sand (Shakesby and Stephens, 1984). Chard Junction is currently the most substantial working, sections at many of the other pits having become somewhat degraded and overgrown. Nonetheless, the most recent stratigraphic descriptions of the terrace gravels are based on an extensive series of excavations carried out at Broom Gravel Pits between 1978 and 1981 by Stephens and colleagues (Campbell, 1984; Scourse, 1984; Shakesby and Stephens, 1984; C.P. Green *et al.*, in prep.). The Broom Gravel Pits have proved not only the most prolific source of Palaeolithic implements from the Axe Valley, but are, to date, stratigraphically the most informative – hence their selection for the GCR.

Broom Gravel Pits

The GCR site comprises two disused gravel pits on the east side of the River Axe between Wadbrook

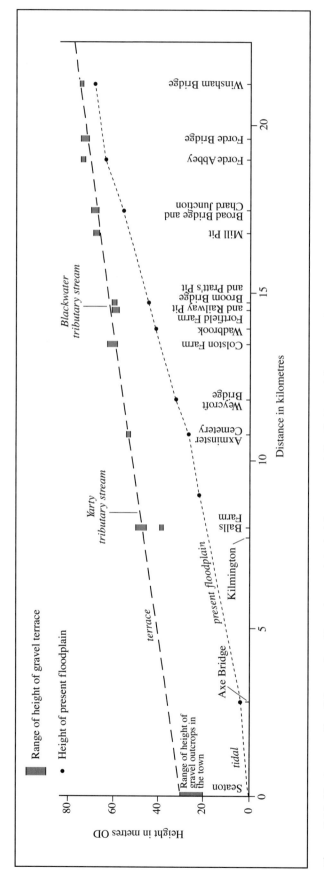

Figure 9.10 The long-profile of the modern River Axe, and the height-range and distribution of the principal terrace gravel outcrops. (Adapted from Green *et al.*, in prep.)

Figure 9.11 Extensive exposures at Chard Junction through the Axe Valley terrace gravels in 1985. (Photo: S. Campbell.)

Cross and Broom Crossing (Figure 9.12): 1. Pratt's New Pit (ST 328023); and 2. the Ballast or Railway Pit (ST 326020). A further disused pit, Pratt's Old Pit (ST 328024), occurs to the north of Holditch Lane (Figure 9.12; C.P. Green, 1988). These pits occur in a marked terrace which rises locally to *c.* 60 m OD.

Reid Moir (1936) gave the following composite stratigraphy for Broom (maximum bed thicknesses in parentheses):

4. Surface soil
3. Tumbled coarse gravels with partings of sandy clay and clayey matrix (derived implements) (7.6 m)
2. Stratified gravel with clayey and sandy seams, some black bands (fresh, unrolled implements) (2.4 m)
1. Unstratified sand and gravel (5.2 m)

The abandoned faces of the nineteenth century workings in the Ballast Pit at Broom reveal *c.* 13–15 m of chert-rich gravels disturbed by cryoturbation structures in their uppermost two metres. These gravels are overlain by a discontinuous stony

silt ('brickearth'). Recent excavations at the base of the disused faces have shown that the 'chert' gravels overlie a laterally persistent layer of pollen-bearing (Scourse, 1984), manganese- and iron-stained clay, silt and sand up to 0.5 m thick (Figure 9.12d; Shakesby and Stephens, 1984; Campbell *et al.*, in prep.). These in turn overlie flint-rich gravel which extends below *c.* 45 m OD to an unknown depth (Scourse, 1984; Stephens and Shakesby, 1984; C.P. Green, 1988).

A fresh cut in Pratt's New Pit, made in 1975, revealed a similar succession comprising 6–10 m of crudely stratified gravels overlying laminated sands and clays (Stephens, 1977). The Pleistocene sequence here extends below *c.* 46 m OD to an unknown depth.

Pratt's Old Pit, north of Holditch Lane, has been reclaimed and grassed over, and is not therefore included within the GCR site. The best descriptions of the sequence here are those gleaned by C.P. Green (1988) from the notebooks of the late Charles Bean. The sequence consisted of *c.* 16 m of terrace gravels separated by up to 2 m of clay, sandy clay and loam (the Upper Gravel, Middle Beds and Lower Gravel of Green (1988)). The red

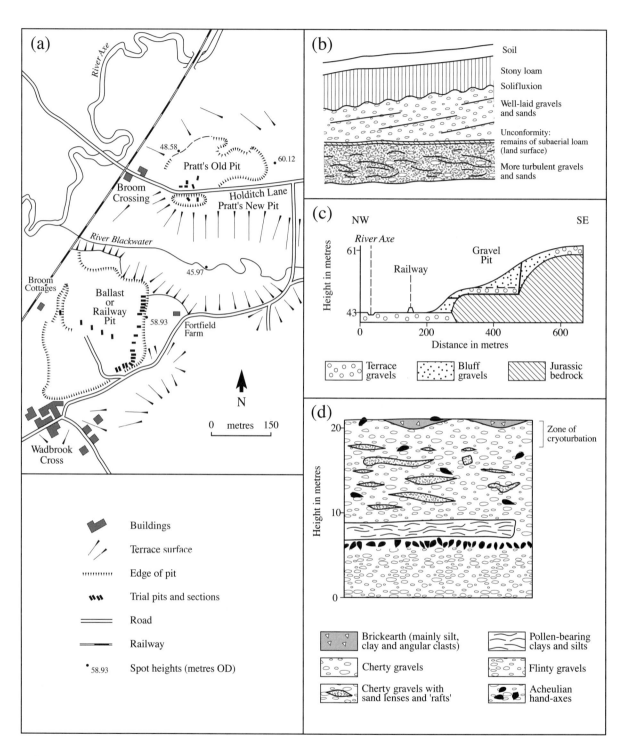

Figure 9.12 (a) Location of the Broom Gravel Pits, adapted from Green *et al.* (in prep.). (b) Schematic section of the Broom gravels, adapted from Reid Moir (1936) and Hawkes (1943). (c) An interpretation of a section of the Broom gravels, adapted from J.F.N. Green (1947) and Calkin and Green (1949). (d) A schematic composite section of the Broom gravels, adapted from Shakesby and Stephens (1984).

Upper Gravel appears to have reached a thickness of *c.* 9 m and contained lenses of sand and loam. The Middle Beds were described as brown and containing scattered stones and, notably, red- and black-stained gravel. They lay at levels between *c.* 45–50 m OD and their boundary with the Lower Gravel was sharp. The pale-grey to white Lower Gravel (> 5 m) contained smaller clasts than the Upper Gravel and was better stratified, exhibiting shallow cross-beds (C.P. Green, 1988).

Although full results of recent excavations at Broom Gravel Pits (C.P. Green *et al.*, in prep.) are not yet available, it appears that the broad three-fold sequence described at each of the major disused pits is part of a laterally continuous succession (Campbell *et al.*, in prep.). Whether this pattern holds more widely elsewhere in the Axe Valley is unknown.

Palaeolithic artefacts

Numerous Lower Palaeolithic artefacts have been recovered from a wide range of locations in the Axe Valley, from Chard to Seaton. The main concentrations of material have been found at Chard Junction, Kilmington and Broom. Of these, Broom has undoubtedly been the most prolific: over a century of gravel working has yielded some 1800 hand-axes (Stephens, 1977; C.P. Green, 1988), although virtually none of these has a clearly defined stratigraphical or archaeological context (Todd, 1987; C.P. Green, 1988). Many, including the 900 or so collected by Charles Bean, probably originated from Pratt's Old Pit, north of Holditch Lane, and there are strong indications that most originated from the 'Middle Beds' at Broom in particular. Many fine hand-axes are preserved in the Exeter, Salisbury, Brighton and British museums (Macfadyen, 1970; Todd, 1987). Less impressive implements and flakes have fared less well, frequently being ignored or discarded (Rosenfeld, 1969; Todd, 1987). Most of the implements from Broom are made of green-brown Upper Greensand chert, although a few are made from chalk-flint (Hawkes, 1943; Macfadyen, 1970). Some are sharp-edged (unrolled), others water-worn (rolled), but in both cases an ovate type (more properly termed chordate, namely asymmetrical ovate) predominates (D'Urban, 1878; Evans, 1897; Stephens, 1974; C.P. Green, 1988). They vary considerably in size, from *c.* 6–23 cm in length (Macfadyen, 1970). Roe (1968a, 1968b, 1981) described the industry as 'finely worked', and confirmed the view of earlier workers that

ovates, including twisted forms, predominate (60%); nearly 40%, however, are pointed forms, the remainder being narrow cleavers, several of which appear to have been sharpened by the tranchet blow (Roe, 1981). This confirms Charles Bean's analysis which shows that ovate forms are dominant (C.P. Green, 1988). The industry as a whole is classified as part of Roe's 'Intermediate Group IV' of British hand-axe industries. Whereas 34 out of the 37 other sites used in this classification fall very clearly into either 'pointed' or 'ovate' hand-axe traditions, the mixed assemblage found at Broom does not. The unknown stratigraphic provenance of most of the Broom finds leaves the interesting possibility that several industries are present. Nonetheless, most authorities seem to agree that the hand-axes are of Early–Middle Acheulian Culture (e.g. Roe, 1968a, 1968b, 1981; Wymer, 1968, 1970, 1977). Stephens (1974) added that since the industry included triangular hand-axes and twisted ovates, it was therefore probably comparable to the Acheulian at Swanscombe (but see Roe, 1981; Bridgland, 1994). On this basis, the manufacture of artefacts at Broom has been seen as broadly contemporaneous with the Hoxnian and Wolstonian (Saalian) stages (Stephens, 1970b, 1973, 1974, 1977).

Interpretation

At the outset, it should be pointed out that no first-hand detailed account of the Broom Gravel Pits has ever been published. Many of the early accounts, particularly those of the 1930s and 1940s, gleaned their evidence from meagre earlier-published accounts rather than from detailed field observation. Some had a tendency to 'force' the field evidence into preconceived stratigraphic schemes based on localities far from Broom and probably wrong even for those localities. Many of these accounts are either very slight, second-hand, or both, and tend to reflect interpretative traditions that have now been largely discarded. The best account is undoubtedly that of Charles Bean, recorded in his notebooks and summarized by C.P. Green (1988). His observations have been fully borne out by excavations undertaken in the late 1970s/early 1980s and reported in preliminary form by Shakesby and Stephens (1984).

D'Urban (1878) recorded that many chert palae-oliths had been found in the Ballast Pit during 1878, and that some had even been picked up from the gravel spread along the adjacent railway line;

none, however, had a known stratigraphic provenance. The hand-axes were considered to resemble closely those illustrated by Evans (1872) from Hoxne in East Anglia. D'Urban's remarks about the deposits from which the palaeoliths had come were brief. He noted that sections, up to *c.* 12–15 m high, through cherty gravel and clay occurred in the Ballast Pit; this material was believed to have been derived from the Greensand which caps the local hills (D'Urban, 1878). The earliest finds and descriptions of implements from the site were subsequently documented by Evans (1897).

The Ballast Pit was also referred to briefly by Salter (1899) who noted that the location and composition of the Axe Valley gravels showed that they had been emplaced 'by a strong current or stream from the north', and he regarded the Ballast Pit deposits as being made up of the debris from the 'high- and low-level plateau drifts' (cf. C.P. Green, 1974b). He observed that roughly shaped chert implements were abundant, being found largely in the 'bottom layers' of the pit.

Likewise, Jukes-Browne (1904a) regarded the valley gravels as having been formed by the 'action of rain and rivers during the excavation of the valleys to their present depth in the Pleistocene'. He thus considered them to have been derived from flint- and chert-rich clays (Eocene) found on the higher slopes and plateaux of the neighbourhood (Jukes-Browne, 1904a).

Between 1932 and 1941, Palaeolithic artefacts, including over 900 hand-axes, were recovered from the terrace deposits at Broom (the Holditch Lane Pits) by amateur archaeologist, Charles Bean. His detailed records, presented for the first time by C.P. Green (1988), show that most of the archaeological material originated from a complex bed (the Middle Beds), about 2 m in thickness, lying between two major gravel units (C.P. Green, 1988). Also from this early period of investigation comes an account of the site's stratigraphy by Reid Moir (1936) (see site description), and two further reviews of the archaeological material that had, by that time, been recovered (Gray, 1927; Smith, 1931).

Reid Moir (1936) described three gravel layers which he interpreted as a single aggradation. The mostly unstratified gravels (bed 3) were believed to have been formed by a mixture of periglacial solifluction and river processes. The stratified gravels (bed 2), on the other hand, showed dark bands which he interpreted as former interglacial land surfaces. The sharp, unrolled, implements found in

this bed led Reid Moir to assign the Broom Palaeolithic industry to the 'Third Interglacial', in today's terminology, the Ipswichian.

Also from around this time comes a sketch of the Broom deposits showing an unconformity, representing a subaerial weathering surface, between lower unstratified and overlying stratified gravels (beds 1 and 2) (Paterson *in* Hawkes, 1943; Figure 9.12(b)). Hawkes (1943) commented that the lowest gravels and sands were probably deposited during the 'Third Glacial' (= Riss or Saalian), with the unconformity (land surface) representing the Mousterian (= Ipswichian). The overlying gravels were attributed to the 'Fourth Glacial' (= Devensian). These workers stated explicitly that rolled implements had been recovered from the basal gravels, while sharp, fresh implements had been obtained from the ancient land surface and from the bedded gravels; the different artefacts were therefore considered to represent separate stages of the Acheulian and Clactonian industries (Hawkes, 1943).

J.F.N. Green (1947) undertook reconnaissance mapping of the terrace deposits of the Otter, Dart and Axe valleys, and confirmed that a series of 'flats' could be traced down valley towards the sea; these were classified on the basis of the height of the gravels and correlated with other known terrace remnants in the lower Thames Valley and in the Sleight district of Dorset (J.F.N. Green, 1947; Calkin and Green, 1949).

In referring back to Reid Moir's composite stratigraphy for Broom, Calkin and Green concluded that the Broom section presented a complex sequence with two platforms cut by successive erosional stages of the River Axe (Figure 9.12(c)), together with associated aggradational and reworked ('bluff') gravels. This schematic diagram, which illustrates the disposition of these various erosional and depositional elements, shows how the interpretation of two separate gravel deposits from two different terrace accumulations may have become complicated by the slumping and redistribution of gravels (Figure 9.12(c)). The three terraces illustrated were believed to have been formed by a proto-Axe river flowing at successive heights equivalent to the rivers which deposited the Sleight, Boyn Hill and Iver terraces (respectively descending in height and age). Calkin and Green argued that after the middle terrace (= Boyn Hill Terrace) accumulated at Broom and its river had formed a cliff at the edge of its floodplain, it was likely that gravels and other sediments from an older and higher terrace (= Sleight

Terrace) had slumped or been soliflucted down-slope as 'bluff' gravels (Reid Moir's bed 3) to overlie younger, *in situ*, well-stratified gravels (= Reid Moir's bed 2). Subsequently, erosion at a lower level led to cliffing and deposition of the lowest (= Iver) terrace; 'bluff' gravel was later deposited over these sediments in the same manner as described above (Figure 9.12(c)) (J.F.N. Green, 1947; Calkin and Green, 1949). On the basis that the principal source of artefacts had been the stratified gravels (bed 2), Calkin and Green assigned a Boyn Hill age to the industry (= Hoxnian).

Two main recent schools of thought pertain regarding the origin of the Axe Valley gravels. First, Stephens (1970b, 1973, 1974, 1977) speculated that a combination of Irish Sea and Welsh ice had blocked the Bristol Channel and pressed against the north Devon coast (see Brannam's Clay Pit; Chapter 7) in Saalian times, damming natural drainage and forming a large lake – 'Lake Maw' (Maw, 1864; Mitchell, 1960). He argued that such a lake would have overflowed at the lowest point of outlet to the south; the striking dry gap at Chard would have provided an ideal low-level routeway between the Somerset lowland to the north and the Axe Valley to the south (Stephens, 1970b). As this 'outwash' overflowed through the Chard Gap, it picked up Palaeolithic artefacts and incorporated them into a sizeable gravel terrace running all the way from Chard to Seaton. This consisted not only of locally reworked gravels (from plateau and inter-fluve areas) but, in addition, a variety of non-local rock types. The overspill from the lake was believed to have inundated a number of Palaeolithic working floors as attested by the vast numbers of artefacts found at localities such as Broom.

On the basis of the archaeological evidence for an Early–Middle Acheulian industry in the area, Stephens has argued that the Axe Valley terrace, with its incorporated artefacts, can be no earlier than the Hoxnian and no later than the Saalian (Stephens, 1970b). This fits neatly with the suggestion that proglacial Lake Maw built up during the Saalian Stage (= Wolstonian), approximately at the same time as the Fremington Clay was believed to have been deposited in north Devon (Stephens, 1970a, 1970b). Cryoturbation of the Axe Valley terrace gravels and deposition of brickearth (a silty loessic or colluvial deposit capping many local sequences) were attributed to periglacial conditions during the later part of the Saalian or in the Devensian cold stages (Stephens, 1974).

Second, an alternative explanation for the Axe Valley gravels, including those at Broom, was provided by C.P. Green (1974b) who determined the lithological composition of gravels both within the Axe Valley and in adjacent plateau and interfluve areas. He suggested that the Axe Valley terrace was unrelated to the Chard Gap. The source of erratic pebbles in the terrace gravels was thought to reflect the distribution and composition of adjacent Tertiary plateau gravels. Green has suggested that the incorporation of hand-axes into the terrace gravels at many sites in the Axe Valley was effected by extensive systems of braided streams, choked with chert and flint gravel, which occupied the valley floor where the Palaeolithic working sites had existed. Solifluction on valley sides and small tributary streams probably contributed material from plateau and valley-side sources to the valley floor, where shifting stream channels accomplished only limited sorting of material. All of these processes were believed to have been operative in a periglacial regime (C.P. Green, 1974b).

According to C.P. Green, erratics within the terrace gravels had been derived from Tertiary beds located on adjacent plateau and interfluve areas. A very strong argument in favour of this hypothesis is that deposits of similar composition to the Axe Valley gravels occur in other nearby valleys such as the Yarty and Otter – valleys which could not have been supplied via the Chard Gap. A glacial origin for the far-travelled material was thus rejected (C.P. Green, 1974b).

Campbell (1984) studied microtextural characteristics of sand from the Axe Valley terrace using Scanning Electron Microscopy (SEM). He noted that quartz-grain microtextural assemblages found in samples of the Axe Valley deposits are consistent with a marine origin, indicating reworking of material from Tertiary sources. At the same time, these preliminary SEM data indicated that probably very little reworking or abrasion of quartz grains had occurred since the material was removed from its plateau sources. This is consistent with the view put forward by C.P. Green (1974b) and Shakesby and Stephens (1984).

Shakesby and Stephens (1984) provided a preliminary account of recent excavations in the Ballast Pit, and attached considerable significance to the stained clays, silts and sands which occur within the gravel sequence (Figure 9.12(d)). Analysis of pollen extracted from this bed (Scourse, 1984) shows that the prevailing regional vegetation probably consisted of a boreal forest dominated by pine *Pinus*, spruce *Picea* and birch *Betula*, but also

Figure 9.13 'Cherty' gravels with sand lenses, seen in the south-east faces of the disused Railway Pit at Broom in 1985. (Photo: S. Campbell.)

with silver fir *Abies*. Scourse has argued that these trees were probably restricted to small stands interspersed within large expanses of open country, with ericaceous heath on the higher ground; a depositional environment at the end of a Middle Pleistocene (possibly Hoxnian) interglacial is suggested, although an interstadial origin, perhaps within the Saalian, cannot be ruled out (Scourse, 1984). C.P. Green (1988) has suggested that the apparent diversity of the sediment association reflects generally low energy deposition in a complex of pools and channels on a floodplain surface; in this case, one formed near the confluence of the proto-Axe and tributary Blackwater rivers.

The recent stratigraphic and pollen evidence somewhat complicates the simple depositional model originally put forward by C.P. Green (1974b); an apparently temperate floodplain deposit within a series of cold-climate gravels shows that periglacial braided stream deposition was interrupted by the accumulation of the pollen-bearing clays, silts and sands during a period of more temperate conditions. At present, no firm dating for this sequence is possible; on the basis of the archaeological evidence, however, it is likely that the gravels and associated deposits span the

temperate Hoxnian and cold Saalian stages (Shakesby and Stephens, 1984). There is no indication that any part of the terrace sequence accumulated during the Ipswichian; irregular cappings of brickearth, caused by solifluction and rain-wash, and cryoturbation structures in the upper 1–2 m of the terrace gravels, may have formed during the ensuing Devensian (Shakesby and Stephens, 1984).

In conservation terms, Broom Gravel Pits show key exposures in the controversial Axe Valley terrace gravels. In addition to providing representative examples of the gravels themselves, recent site excavations have shown laterally persistent silt and clay bands to be present within the terrace gravels; pollen sampled from these beds (Scourse, 1984) offer, for the first time, the opportunity to begin to reconstruct the palaeo-environment of the Axe Valley prior to the main formation phase of the large gravel terrace, perhaps at a time when the valley was extensively inhabited by Lower Palaeolithic Man. At present, the only possible clues as to the age of the terrace gravels come from the included Acheulian implements; even then, the age of the deposits is only loosely confined to between approximately the

Figure 9.14 Acheulian hand-axes from Broom, seen during the 1977 INQUA visit to South-West England. (Photo: N. Stephens.)

late Hoxnian and Saalian stages. Broom Gravel Pits have been, without doubt, the most important source of Lower Palaeolithic material from within the Axe Valley. Nonetheless, problems of interpreting the Palaeolithic assemblage from Broom still persist; it is not easily related to Acheulian assemblages found elsewhere in Britain (Rosenfeld, 1969), and it is quite likely that a mixture of industries is represented (Roe, 1968a, 1968b, 1981). Roe has suggested some affinities with the pointed hand-axe industries from Furze-Platt, Cuxton and Stoke Newington, but Broom differs markedly in showing a clear clustering of ovates within a narrow shape range (Rosenfeld, 1969). The unusual profusion of Palaeolithic material, particularly from the Ballast Pit, also presents a problem of interpretation; it may provide unique evidence in the South-West for a working floor. In any case, extremely rapid incorporation and burial of the hand-axes within the gravels is suggested (Shakesby and Stephens, 1984); the generally unrolled condition of the artefacts may suggest that they were originally discarded on the surface of the temperate floodplain deposits, and that they were only displaced over a short distance during low energy reworking on the floodplain (C.P. Green, 1988).

Conclusion

One of the classic geomorphological localities of South-West England, Broom Gravel Pits have long been central to arguments regarding the origin of the controversial Axe Valley terrace gravels. Most workers now hold that the gravels here were derived from Tertiary plateau deposits by Pleistocene solifluction and fluvial activity and then reworked, within the Axe Valley, by periglacial braided streams. Another more ambitious theory has tied the origin of the Axe Valley gravels to the Chard Gap, suggesting that the gravels were deposited (or at least their final terrace form created) as water spilled south from a large proglacial lake – 'Lake Maw' – dammed by Saalian-Stage ice in the Bristol Channel. Recent evidence from Broom shows that gravel accumulation was interrupted by a period of temperate climatic conditions when pollen-bearing clays, silts and sands were deposited. Whether this temperate event was part of a full interglacial within the Pleistocene, or merely a brief interstadial phase within one of the main cold phases, is as yet unresolved.

Broom also provides important evidence, in the form of a profusion of Acheulian implements (over 1800 hand-axes have so far been found at the site), for the activities of Lower Palaeolithic humans in the South-West. The large number of implements found here suggests that the site may have been a 'working floor' used for the manufacture of such tools; it is tempting to speculate that the relatively warm period detected within the sequence was coincident with a major occupation of the immediate area by Palaeolithic hunters who produced hand-axes in the Acheulian tradition. At present, the archaeological evidence is the only means of estimating the age of the sediment sequence here which, on that basis, has been assigned to between the late Hoxnian and Saalian stages.

(B) COLLUVIAL AND FAN-GRAVEL SITES IN MENDIP AND ADJACENT AREAS

This section documents sites which preserve evidence for terrestrial sedimentation patterns, mostly during cold stages, and palaeosol development, during temperate interstadial and interglacial conditions. The sites are chosen as prime examples of a variety of depositional settings and, where possible, because palaeobiological and dating evidence is available. Late Middle Pleistocene slope deposits, palaeosols and terrestrial mollusc faunas are located under the Swallow Cliff raised beach at Middle Hope. The Brean Member, an extensive Late Devensian sequence of rockfall breccias, aeolian sands, silts and palaeosols, spanning Oxygen Isotope Stages 4–2, is preserved at Brean Down. This sequence contains important fossil mammal and mollusc faunas. Alluvial fan gravels containing fossil molluscs and pollen, dating from the coldest part of the Devensian, are represented by the Wookey Station Member at Wookey Station. Older fan gravels, heavily affected by pedogenesis and overlain by aeolian sediments, can be demonstrated at Bourne.

MIDDLE HOPE
C. O. Hunt

Highlights

Middle Hope GCR site is significant because it contains a highly fossiliferous raised beach deposit of interglacial age overlying a complex of cold-stage slope deposits, some of which are affected by pedogenesis and some of which contain terrestrial mollusc fossils. Slope deposits and mollusc faunas

of this antiquity are extremely rare in South-West England and provide important evidence for the nature of Pleistocene cold-stage environments in the region. The site is the type-locality for the Middle Hope Formation.

Introduction

Middle Hope is a Carboniferous Limestone horst, bounded by alluvial lowlands to the south and east, and by the Bristol Channel to the north and west. Much of the Middle Hope massif is mantled by Quaternary deposits. These are mostly slope deposits, but at two localities they are interbedded with marine sediments. At Swallow Cliff, Middle Hope, a shore platform is overlain by the Middle Hope Palaeosol (pedogenically altered slope deposits), the Woodspring Member (silty slope deposits with fossil land snails), the Swallow Cliff Member (a raised beach deposit) and then by further slope deposits attributed to the Brean Member (Campbell *et al.*, in prep.).

The raised beach deposits here were first described by Sanders (1841) and Ravis (1869). They were briefly re-described by Woodward (1876) who, with Prestwich (1892), provided faunal lists. These early authors recognized a fossil fauna of *Tellina* (*Macoma*), *Littorina*, *Nassa*, *Cerastoderma*, *Murex* (*?Ocenebra*), *Purpurea* and *Ostrea*. They also recognized the presence of land molluscs.

Palmer (1931) assigned the raised beach to his '10 foot level'. Donovan (1962) recognised a series of erosion features around the Bristol Channel and correlated them with the Main Terrace of the Severn and the Swallow Cliff raised beach. He attributed them to an episode of sea level not far below OD during the Upton Warren Interstadial. Wood (*in* Callow and Hassall, 1969) obtained radiocarbon dates of 33 240 + 760/− 700 BP (NPL–126a) and 38 990 + 1690/− 1390 BP (NPL–126b) which seemed to support this hypothesis, but Kidson (1970, 1977) rejected them as 'almost valueless' since similar dates had been obtained from a number of sites of known interglacial status.

The site was definitively re-studied by Gilbertson (1974), Gilbertson and Hawkins (1977) and Briggs *et al.* (1991). The interglacial nature of the marine fauna was demonstrated, and the raised beach deposits were assigned to the Ipswichian by Gilbertson (1974) and Gilbertson and Hawkins (1977). These authors described in detail the stratigraphy, lithology, palaeontology, micropalaeontology and mineralogy of the sequence of shore platform, cold-climate slope deposits with land molluscs, wind-blown foraminifers and occasional recycled marine shells, raised (storm) beach and upper cold-climate slope deposits.

Andrews *et al.* (1979) provided an amino-acid ratio which was taken to support an Ipswichian (Oxygen Isotope Stage 5e) age for the raised beach deposit (Andrews *et al.*, 1979, 1984). Davies (1983) later provided a higher ratio which she interpreted as indicating a Stage 7 age (*c.* 210 ka BP). More recent reassessment (Campbell *et al.*, in prep.) suggests that attribution of the raised beach deposits to Oxygen Isotope Stage 5e (Ipswichian) is most appropriate.

Description

The deposits at Swallow Cliff, Middle Hope (ST 325661), lie on a shore platform and against a fossil cliff cut in Carboniferous Limestone, ashes and spilites. The shore platform lies between 12.5 and 11 m OD (Gilbertson, 1974; Gilbertson and Hawkins, 1977). It is overlain by *c.* 2.6 m of Quaternary deposits, in the following sequence (Gilbertson and Hawkins, 1977; Briggs *et al.*, 1991; Gilbertson, pers. comm., 1993; Figure 9.15) (maximum bed thicknesses in parentheses):

10. Modern topsoil. (0.25 m)

9. Brown, sandy very stony loam, with angular clasts of Carboniferous Limestone and volcanic rocks. The material has a weak, angular blocky structure and firm consistency, and exhibits abundant fine to medium pores. It has a sharp boundary with bed 8. (0.33 m)

8. Coarse, matrix- and clast-supported cobbly gravel of rounded Carboniferous Limestone clasts in a matrix of fragmented marine shell, porous and cemented at point contact. The fossil assemblage is dominated by *M. balthica*, with some *Littorina littoralis* (Linné) and *Cerastoderma* sp., a few *P. vulgata* and *Littorina littorea* Linné. *Littorina saxatalis* (Olivi), *Buccinum undatum* Linné, *Nucella lapillus* (Linné), *Nassarius reticulatus* Linné, *Lora* sp., *Trophonopsis truncatus* (Ström), *Ocenebra erinacea* (Linné) and *Ostrea* sp. are rare. Aminostratigraphical assays on *Patella* gave a combined D-*alloisoleucine* : L-*isoleucine* ratio of 0.101 ± 0.005 (AAL–771)

(Andrews *et al.*, 1979) and a mean ratio of 0.203 ± 0.016 (Group 3 of Davies, 1983). The deposits have a sharp uneven boundary with bed 7. (1.0 m)

7. Dark brown, sandy, slightly stony loam, with rare clasts of weathered subangular ?Triassic sandstone and Carboniferous Limestone and pockets, 3 mm deep, of silty clay loam with platy structure, medium pores and clay skins on ped faces. The bed contains numerous foraminifers, abundant marine shell fragments and some terrestrial molluscs – *Vallonia* cf. *pulchella*, *Trichia hispida* (Linné), *Agrolimax* cf. *agrestis* and Helicids. The material has a moderate, medium, subangular to blocky platy structure, is friable and exhibits abundant fine to medium pores in upper 50 mm; common fine pores are present below this level. Bed 7 has a sharp boundary with bed 6. (0.08 m)

6. Brownish-black silty clay loam with a fine, subangular blocky structure. The deposit is friable and exhibits abundant fine pores. It has a sharp uneven boundary with bed 5. (0.02 m)

5. Brown, extremely stony clay loam, with medium to coarse angular clasts of Carboniferous Limestone and flint, subrounded to subangular clasts of weathered Triassic sandstone and marl. The material has a subangular blocky structure, is friable and exhibits occasional clay skins on ped faces. It shows abundant fine pores and frequent medium to coarse pores and has a clear but uneven boundary with bed 4. (0.15 m)

4. Brown, silty, slightly stony clay loam, with angular to subangular clasts of ?Triassic marl. Abundant foraminifers are present. The deposits demonstrate a moderate, medium, angular, blocky platy structure and exhibit occasional clay skins on peds. They contain abundant fine and medium pores, and have a clear but uneven boundary with bed 3. (0.15 m)

3. Brown, extremely stony silty clay, with coarse subangular clasts of Carboniferous Limestone, weathered ?Triassic marl and sandstone and very infrequent clasts of Carboniferous volcanic rocks. It contains occasional pockets and individual specimens of land snails – *Vallonia* cf. *pulchella*, *T. hispida* and Helicids – and lenses (50–100 mm across) of fragmentary marine molluscs – *N. lapillus* and *L. littoralis*. It is slightly calcareous, cemented at point contact and demonstrates moderate,

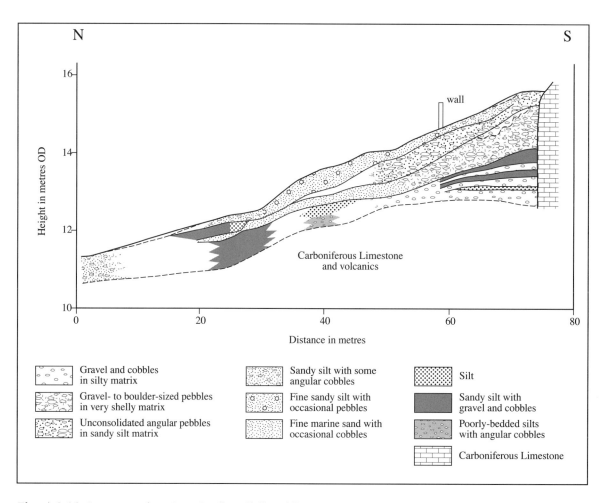

Figure 9.15 Quaternary deposits at Swallow Cliff, Middle Hope, simplified from Gilbertson and Hawkins (1977).

angular blocky structure and firm consistency. The material exhibits abundant fine and common medium pores and has a clear but uneven boundary with bed 2. (0.15 m)

2. Moderately mottled, red-brown, silty, slightly sandy stony clay, with common small to medium subrounded clasts of ?Triassic marl. Some foraminifers are present. The deposit has a moderate, medium, angular structure and firm consistency, and exhibits clay skins on peds. Abundant fine and common medium pores are present and the material merges down into bed 1. (0.10 m)

1. Red, silty, very stony clay, with subangular to subrounded coarse clasts of Carboniferous volcanic rocks and subrounded to rounded clasts of Carboniferous Limestone and ?Triassic sandstone. The bed demonstrates a moderate, medium, angular blocky structure,

shows clay skins on peds and has abundant fine to medium pores. It occurs in hollows on the shore platform. (0.35 m)

Interpretation

The basal shore platform and buried cliff most probably reflect one or more episodes of marine erosion of unknown age. The lowest deposits (bed 1) rest on the platform, contain rounded clasts and have a heavy mineral assemblage derived from the platform rocks. They are thus likely to represent a mixture of local weathering products and the remains of a former beach deposit (Briggs *et al.*, 1991; Gilbertson, pers. comm., 1993). The clay coats on peds in beds 1 and 2 are consistent with pedogenic activity (Gilbertson, pers. comm., 1993). Clay content can be used to differentiate the deposits of the Middle Hope Palaeosol (beds 1–3), from the overlying Woodspring Member (beds 4–7): clay enrichment in the former thus reflects a phase of weathering, of at least interstadial status, affecting older slope deposits.

Beds 2–7 contain predominantly angular to sub-rounded clasts and appear to reflect an alternation of hillwash and aeolian processes (Briggs *et al.*, 1991). The presence of marine molluscs and foraminifera in these deposits is suggestive of the former presence of marine deposits upslope, while the foraminifera could have been emplaced by wind following the deflation of nearby marine deposits (Gilbertson and Hawkins, 1977). The fossil land mollusc assemblages are consistent with open wet grassland.

Bed 8, the Swallow Cliff Member, represents a storm beach probably related to a mean sea level perhaps 5 m higher than that of today. The dominant *Macoma* was derived from sand and mud flats in the Bristol Channel. Many of the other taxa are species typical of rocky shores and are thus consistent with the depositional locality. Gilbertson (1974) and Gilbertson and Hawkins (1977) have demonstrated the interglacial nature of the fauna. The major discrepancy beween the amino-acid ratios given by Andrews *et al.* (1979) and Davies (1983) has not yet been accounted for, and both Ipswichian (Oxygen Isotope Stage 5e) and earlier ages for the beach deposit are possible. If Davies' (1983) interpretation is accepted, a correlation with the marine deposits at Kenn Church (Andrews *et al.*, 1984) becomes probable, although this was rejected by Campbell *et al.* (in prep.). Further work is needed to resolve this problem.

Bed 9 reflects the resumption of cold-climate depositional activity after the interglacial phase. This thin unit can be assigned to the Brean Member.

Conclusion

The deposits at Swallow Cliff, Middle Hope, are important because they provide evidence for depositional and pedogenic environments during two later Middle Pleistocene cold stages and an intervening ?temperate episode. Subsequent temperate-stage beach sedimentation, probably correlated with Oxygen Isotope Stage 5e, is also represented. Further work is necessary to resolve problems with the dating of the sequence.

BREAN DOWN
C. O. Hunt

Highlights

Brean Down provides a spectacular and most unusual example of cold-stage aeolian and slope sedimentation. It has important interstadial fossil mollusc and mammal faunas and preserves a very detailed record of conditions during a considerable part of the Devensian. It is the type-section of the Brean Member.

Introduction

At Brean Down, a thick late Quaternary sequence of sands, silts and breccias rests on a shore platform and against an ancient cliff. The sequence contains abundant mammal bones and fossil molluscs. During Pleistocene cold stages, aeolian and rockfall depositional processes were dominant. Interstadials within the sequence are marked by evidence of pedogenic activity, rich mollusc and mammal faunas, and appear to be characterized by slower rates of deposition.

The Quaternary deposits at Brean Down have been of interest for over a century. Ravis (1869) provided the first short account. Reindeer bones from Brean were described by Knight (1902). Ussher (1914) dealt with the site in a general account of the deposits of the Somerset Levels. Palmer (1930, 1931, 1934) gave brief accounts of a breccia containing reindeer antlers and bones, overlain by a thick sand and an upper breccia. A mineralogical analysis of the sand unit was compared with analyses from Clevedon, Bleadon and the Barnwood terrace of the Severn and from a number of possible sources; an aeolian origin, mostly from the Tertiary deposits of Devon and Cornwall, was suggested. The breccias were ascribed to 'alternations of abnormal cold and excessive moisture' and compared and correlated with the 'combe rock' of the chalklands of southern England. Balch (1937) listed a Pleistocene and Holocene fauna including Neolithic humans, horse, red deer, reindeer and ?northern vole from Brean, but supplied no stratigraphical details. The Pleistocene deposits were then described in passing in a description of archaeological remains from the Holocene sequence (Taylor and Taylor, 1949).

The first detailed account of the site was given by ApSimon *et al.* (1961). They described a

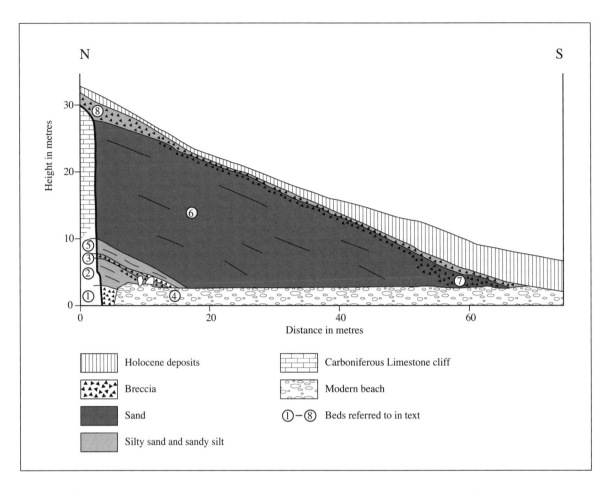

Figure 9.16 The Quaternary sequence at Brean Down, simplified from ApSimon *et al.* (1961).

complex Pleistocene stratigraphy, with a lower breccia, stony silt, middle breccia and bone bed, silty sand, main sand and upper breccia, overlain by an extensive Holocene sequence with Beaker, Bronze Age, Iron Age and post-Mediaeval artefacts. The lower breccia contained bones of vole, arctic fox, reindeer and bison and the stony silt contained reindeer bone and antler, and vole and bison bones. These assemblages were thought to indicate a tundra landscape. The middle breccia showed signs of soil development and contained a mammal fauna with remains of lemming, hare, Arctic fox, elephant, horse and reindeer together with indeterminate bird bones. Some bones showed signs of human workmanship. This horizon also contained both land and marine molluscs. The presence of horse was taken as evidence for a grassland environment but the land mollusc fauna was regarded as indicating a very exposed, poorly vegetated landscape. The marine molluscs were fragmentary and probably blown from nearby marine sediments. The Pleistocene sequence was attributed to the

Devensian late-glacial, but ApSimon (1977) later suggested that the lower and middle breccias might be of Early or Middle Devensian age.

The finding of two gold bracelets in 1983 and the necessity for sea defence works (McKirdy, 1990) led to a rescue excavation of the Holocene sequence, which unearthed a Bronze Age village (Bell, 1990, 1992a, 1992b) and a re-study and intensive sampling of the Pleistocene sequence (Hunt, in prep.). The Pleistocene stratigraphy and depositional environments described by ApSimon *et al.* (1961) were substantially confirmed by this recent work, but a more detailed picture of the mollusc fauna of the site has now emerged. The site was proposed as the type-section of the Brean Member by Campbell *et al.* (in prep.).

Description

The Pleistocene deposits of Sand Cliff, Brean Down (ST 295588), lie against the precipitous south face of the Carboniferous Limestone massif of Brean Down and extend *c.* 70 m southwards to where they pass below Holocene deposits and the modern beach. They lie on a platform cut in the Carboniferous Limestone which lies between OD and *c.* − 6 m OD (ApSimon *et al.*, 1961). The deposits dip south at 20–25°. The westerly portion of the deposits has been removed by marine erosion and the remaining deposits are well-exposed in a coastal cliff (Figures 9.16 and 9.17).

The stratigraphical terminology of ApSimon *et al.* (1961) is used in this account, but the bed descriptions, mollusc data and measurements are those from the 1986 re-study (Hunt, in prep.). These differ slightly from the descriptions given by ApSimon *et al.* (1961) because of lateral variability and cliff retreat over the intervening period. The Holocene deposits described by ApSimon *et al.* (1961) are not the central part of the GCR site interest and are not therefore repeated here. The sequence is as follows (maximum bed thicknesses in parentheses):

8. Holocene deposits.

7b. 'Earthy Breccia'. Reddish-brown breccia with gravel- to boulder-sized limestone clasts in a silty sand matrix. This horizon thickens downslope. (6.0 m)

7a. 'Sandy Breccia'. Reddish-brown to strong brown sandy breccias and sands. The layer becomes thicker and sandier downslope. (2.5 m)

6. 'Main Sand'. Orange-brown, fine to medium, sometimes silty sand with large-scale (0.5–1.5 m) cross-bedding. Occasional stone lines and silt lenses are present. *P. muscorum* and fragments of marine molluscs are occasionally present. Sheets and tubular structures of calcite occur in the uppermost 2 m. The horizon from which these structures originated has since been removed by erosion. (27.4 m)

5b. Angular limestone fragments. Clast-supported breccia of gravel-sized, very angular limestone fragments in an orange-brown sandy matrix. (0.2 m)

5a. 'Silty Sand'. Yellow-brown silts and sands becoming redder and sandier upward. The bed comprises centimetre-thick silt/sand couplets, with very occasional stones. Charcoal flecks occur at the base of the unit. *P. muscorum* is locally abundant and *Trichia* cf. *hispida*, *Cepaea* sp. and marine mollusc fragments are present. (2.7 m)

4. Openwork breccia comprising very angular

clasts with occasional *Rangifer* bone and antler and rare *Trichia* cf. *hispida*, *Catinella arenaria* (Bouchard-Chantereaux) type and marine mollusc fragments. This bed was not reported by ApSimon *et al.* (1961). (0.8 m)

3b. 'Bone Bed'. Red-brown stony silts in four thin units passing upwards into yellow-brown, stoneless sandy clay loam with occasional *Rangifer tarandus* Linné bone and antler. This is probably the lateral equivalent of the 'bone bed' which was rich in *R. tarandus* bone and antler, some showing signs of possible human working. This layer also yielded *Dicrostonyx* spp., *Lepus timidus*, *Canis lupus*, *Alopex lagopus* (Linné), '*Elephas*' sp., *Megaloceros giganteus* (Blumenbach), *Equus* sp. and indeterminate bird bones (ApSimon *et al.*, 1961; ApSimon, 1977). The mollusc fauna contains abundant *P. muscorum* (up to 92.5%), some *Trichia* cf. *hispida* (up to 22.9%), rare Limacidae, *Cochlicopa* sp., *Cepaea* sp., *D. rotundatus*, *Oxychilus* sp., *Succinea* cf. *oblonga*, *L. truncatula* and marine and freshwater mollusc fragments. (0.7 m)

3a. 'Middle Breccia'. Yellow-brown, passing up into red-brown mottled, yellow-brown, silty sandy breccias and sandy silts with occasional limestone fragments. Bones and antler of *R. tarandus* are fairly common. The terrestrial mollusc assemblage is dominated by *P. muscorum* (85.8%), with some *Trichia* cf. *hispida* (7.5%), Limacidae (3.8%) and rare *Cepaea* sp., *Vallonia costata* (Müller), and *C. arenaria* type. Fragmentary marine and freshwater molluscs are present. (3.35 m)

2c. 'Stony Silt'. Brown to orange-brown sandy clay loam, becoming sandier upwards, and containing occasional angular limestone clasts and thin chocolate-brown silty clay layers. Occasional bones and antler of *R. tarandus*, bird bones, land molluscs, mostly *P. muscorum*, and marine mollusc fragments. (3.1 m)

2b. 'Clay Band' in Stony Silt. 'Gravel' composed of pellets of green-grey and red-brown silty clay. (0.15 m)

2a. 'Stony Silt'. Yellow-brown becoming strong brown sandy loam with angular limestone clasts becoming more frequent upwards. Reindeer antler was found at the base of this unit and occasional bone and antler fragments are distributed throughout, while *Microtus* aff. *nivalis* is present towards the top. *P. muscorum* and marine shell fragments are present. (1.5 m)

Figure 9.17 The Pleistocene sequence at Brean Down. (Photo: S. Campbell.)

1. 'Lower Breccia'. Red-brown breccia of boulder-size limestone clasts, fining upward and becoming clay-rich in the upper 0.3 m where it contains *A. lagopus*, *R. tarandus*, *Microtus* sp. and *Bos* sp. (> 2.2 m)

Interpretation

The platform and cliff on which the Pleistocene deposits rest are most probably of marine origin, though marine deposits have not been found upon them. Such platforms are most probably the result of repeated marine transgressions to approximately the same altitude over long periods (Kidson, 1977). The limestone clasts in the overlying sequence were derived by cliff fall from the limestone cliff to the north, but the overwhelming proportion of the silts and sands in the sequence must have been derived from other sources by the wind, as has been noted at many sites in Avon and Somerset (Gilbertson and Hawkins, 1978a, 1983).

The boulder pile of bed 1 reflects cliff collapse, possibly, but by no means certainly, under frost weathering. The fauna in the top of this bed is certainly characteristic of cold-stage conditions.

ApSimon *et al.* (1961) interpreted the transition to bed 2 as a minor climatic amelioration, but the mollusc and mammal faunas of bed 2 reflect cold, exposed landscapes and the sediments probably reflect a considerable period of coversand-style aeolian sedimentation, being derived at least in part from local marine sands, so such an amelioration is by no means certain. The clay-rich layer (bed 2b) was interpreted by ApSimon *et al.* (1961) as derived from local Triassic deposits during a still-stand in aeolian sedimentation. In the 1986 re-study it was interpreted as pellets of an eroded soil (Hunt, in prep.).

Bed 3 was recorded as substantially thicker in the 1986 study than in the work of ApSimon *et al.* (1961), most probably as the result of local facies variation. Bed 3a was interpreted by ApSimon *et al.* (1961) as reflecting colder conditions than bed 2 because of an increase in the number of limestone clasts. The mammal and mollusc faunas of this bed are, however, richer than those of bed 2 and there are signs of soil development which together might indicate a minor climatic amelioration. Bed 3b, with its comparatively diverse faunas including higher incidences of the vegetation-loving *Trichia* and thermophiles such as *D. rotundatus*, *Cepaea* and *Oxychilus*, probably reflects further climatic amelioration and a herbaceous ground cover (Hunt, in prep.). The mammal assemblage may also be taken as compatible with steppe conditions. The breccias of bed 4 and the silts and breccia of bed 5 probably reflect a slow return to cold-stage conditions since there are relatively few mammal remains and mollusc diversity is low.

The aeolian sands of bed 6 are most probably the product of a very exposed and at least episodically arid landscape, and sand sedimentation continued during the deposition of the sandy breccia (bed 7a). A soil profile must then have formed with tree-sized vegetation and considerable carbonate mobilization by soil acids to give rise to the calcite structures. This had been completely eroded away before the deposition of the final Pleistocene unit (bed 7b).

The development of a soil profile and arboreal vegetation before the development of bed 7a most probably reflects a major climatic amelioration, perhaps the Windermere Interstadial. If this is the case, then the original attribution of the 'bone bed' by ApSimon *et al.* (1961) to the Devensian late-glacial cannot be correct and ApSimon's later (1977) suggestion that this reflects an Early or Middle Devensian event is much more convincing. Sea level in Stage 5e and probably Stage 5a would have been high enough to have emplaced raised beach deposits at Brean, rather as happened at Swallow Cliff, Middle Hope (this chapter). The absence of raised beach deposits might therefore be used as evidence that the whole Brean Down sequence post-dates Stage 5. It is perhaps reasonable to suggest, therefore, that the interstadial reflected by bed 3 is equivalent to Oxygen Isotope Stage 3. Broadly comparable mollusc faunas of Stage 3 age are known from Pin Hole Cave, Creswell Crags (Hunt, 1989). On this basis, beds 4–7a at Brean Down may have formed during Oxygen Isotope Stage 2, and bed 7b during the Younger Dryas (= Loch Lomond Stadial).

Conclusion

The Brean Down sequence is nationally important because it provides a very detailed, though inevitably incomplete, history of changing terrestrial depositional environments, and especially aeolian sedimentation, through much of the last cold stage. It has important mammal and mollusc faunas. A dating programme is urgently needed to place the environmental history of the site in context and enable its full potential to be realized.

BOURNE
C. O. Hunt

Highlights

Bourne shows an excellent example of ancient interglacial soil formation, periglacial aeolian deposition and mass movement on the older alluvial fan deposits of Mendip. Well-developed palaeosols have been reported extremely rarely in South-West England. The site is the type-section for the Burrington Member and the Burrington Palaeosol.

Introduction

At Bourne, deposits of the alluvial fan at the mouth of Burrington Coombe were exposed in temporary sections. A basal gravel unit, the Havyat unit, contains in its upper part a 'weathered horizon' – a soil profile with characteristics indicating pedogenesis in interglacial conditions. This is overlain by a second gravel body, the Ashey unit, which probably dates from the Devensian.

Gravel deposits along the foot of the Mendips have been known since the work of Woodward (1876) and Morgan (1888). Modern work started with the work of Clayden and Findlay (1960), Green and Welch (1965) and Findlay (1965) and has recently included the morphological work of Pounder and Macklin (1985). Gravels were first described from Bourne by Woodward (1876), who described 5 feet of sandy angular to subangular gravel in a roadside exposure. Findlay (1977) described the section at Bourne in detail, and suggested it showed evidence for a major episode of fluvial deposition and two phases of aeolian sedimentation separated by a phase of pedogenesis of interglacial status. Pounder and Macklin (1985) described the morphology of the fan deposits. They recognized four phases of aggradation, the Langford (oldest), Havyat, Ashey, and Link Lane units (Figure 9.18). Campbell *et al.* (in prep.) proposed the site as the type-section of the Burrington Member (Oxygen Isotope Stage 6) and the Burrington Palaeosol (Stage 5).

Description

Bourne GCR site lies at the eastern end of the Churchill–Burrington–Rickford alluvial fan complex, the morphology of which was described by Pounder and Macklin (1985). In temporary sections

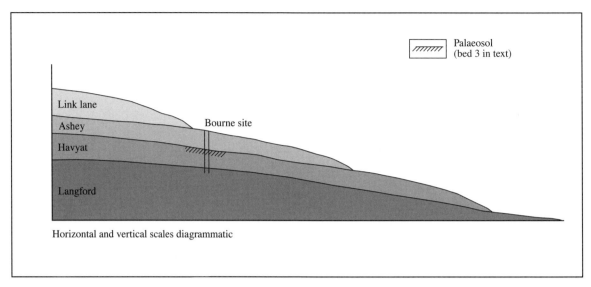

Palaeosol
(bed 3 in text)

Link lane

Ashey

Bourne site

Havyat

Langford

Horizontal and vertical scales diagrammatic

Figure 9.18 Schematic cross-section of the Burrington fan at Bourne, showing the aggradational components of the fan and their relationship to the Bourne section. (Adapted from Pounder and Macklin, 1985.)

at ST 483598, over 3 m of Quaternary deposits were exposed in the Ashey and Havyat units defined by Pounder and Macklin (1985). The following description is modified from that given by Findlay (1977) (maximum bed thicknesses in parentheses):

9. Dark brown, sandy silt loam topsoil. (0.15 m)
8. Brown and reddish-brown, slightly stony sandy silt. The clasts in this bed are mostly of sandstone with some of chert. Tongues or wedges containing material similar to the deposits in this bed extend from its base downwards into the underlying beds. (0.20 m)
7. Reddish-brown, slightly stony clayey silt passing down into a sandy clayey silt. The clasts are mostly sandstone with some chert. (0.30 m)
6b. Reddish-brown and red clay, with 20% black manganiferous mottle. The bed occurs as discontinuous lenses and merges with bed 5 when bed 6a is absent. (0.15 m)
6a. Yellowish-red to reddish-brown sandy silt. The bed occurs as discontinuous lenses. (0.20 m)
5. Reddish-brown and red very stony clay, with 20% manganiferous mottle on ped faces and stones. The clasts are mostly sandstone with some chert. When bed 5 is absent, this bed merges with bed 6 with decreasing black staining. (0.20 m)

4. Yellowish-red to reddish-brown sandy silt with layers of reddish-yellow silty sand, with occasional sandstone and chert cobbles and small boulders. The bed has sharp upper and lower boundaries and occurs as discontinuous lenses which attenuate downslope. (0.50 m)
3. Reddish-brown very stony clay. The clasts are predominantly sandstone with some chert and quartz. A wide range of clast sizes up to 250 mm is present and the clasts vary from round to subangular. In thin-section, the matrix of the bed can be seen to comprise almost entirely illuvial clay. There is a clear but highly undulating lower boundary to the bed. (0.75 m)
2. Reddish-brown to dark reddish-brown, very stony, sandy clayey silt. The clasts are predominantly of Carboniferous Limestone, with some sandstone and a little chert. In the top 0.3 m of this horizon, Carboniferous Limestone clasts are largely decalcified and surrounded by black clay. This bed passes gradually into bed 1. (0.55 m)
1. Loose, clast-supported cobbly gravel with gritty matrix. Base unseen. (> 0.20 m)

Interpretation

Beds 1 and 2 contain lithotypes derived from the Mendip Hills, including Carboniferous Limestone, Old Red Sandstone and Carboniferous chert, but contain no far-travelled material. The 'gravels' lie at the margins of the Churchill–Burrington–Rickford alluvial fan, which is well known from the work of Woodward (1876), Findlay (1965) and Pounder and Macklin (1985). There is no reason to interpret them as other than gravels of an alluvial fan. Clayden and Findlay (1960) and Macklin and Hunt (1988) have linked deposition of similar alluvial fan gravels around Mendip to short-lived fluvial flood events during stadial episodes in the Pleistocene.

Findlay (1977) suggested that bed 3 was a palaeosol which formed shortly after the deposition of the alluvial fan gravels of beds 1 and 2. The evidence for this includes the high incidence of illuvial clay in the bed, the 'clear but highly undulating' boundary between beds 2 and 3 ' ... so typical of weathering limestone materials' (Findlay, 1977; p. 24), and the decalcified higher part of bed 2. A weathering episode of considerable intensity, probably of interglacial status, would be needed to form a palaeosol of this type.

Silty sand and sandy silt deposits very similar to bed 4 have been widely reported around the Avon lowland as 'coversands' of relatively local aeolian origin laid down during periglacial phases (Gilbertson and Hawkins, 1978a). A similar origin can be suggested for this bed. Findlay's (1977) section appears to show that bed 4 had become disrupted by loading, with the underlying bed 3 being partially displaced. The most likely explanation is that these beds were deformed by mass-movement processes during a periglacial phase. Beds 5, 6a and 6b may represent material partially reworked from beds 3 and 4 at this time. They may, alternatively, reflect further phases of soil formation (bed 5), aeolian coversand deposition (bed 6a) and soil formation (bed 6b). Further research is needed to clarify this interpretation.

Findlay (1977) suggested that beds 7 and 8 were laid down during a final episode of aeolian silt sedimentation, and later altered by Holocene pedogenesis.

An important though enigmatic sequence can thus be seen at Bourne. The episode of alluvial fan sedimentation can be linked with flash-flood events originating from Mendip during a stadial phase. It was followed by a period of temperate weathering and soil development. This in turn was followed by aeolian coversand sedimentation during a stadial phase and a phase of periglacial mass movement. Other temperate soil-forming intervals and a further coversand depositional event may be represented, but this is by no means certain. Finally, a further

episode of aeolian silt sedimentation occurred, again most probably during a stadial episode.

Conclusion

The temporary section at Bourne showed a well-developed palaeosol formed on alluvial fan gravels. Aeolian 'coversand' was then laid down and was subsequently deformed by mass-movement processes before a final phase of aeolian silt sedimentation. The aeolian silt was affected by Holocene pedogenesis. Well-developed palaeosols have been reported extremely rarely in South-West England. Bourne is unusual in providing clear evidence of two phases of aeolian sedimentation.

WOOKEY STATION
C. O. Hunt

Highlights

Wookey Station provides an excellent example of alluvial fan sedimentation on the Mendip margin, dating from a stadial in the Devensian. It is of special importance because uniquely it contains fossil molluscs and pollen, critical to an understanding of the depositional environment of the Mendip alluvial fans. It is the type-section of the Wookey Formation and of the Wookey Station Member.

Introduction

Deposits of the Wookey alluvial fan are exposed in old railway cuttings at Wookey Station. Coarse, sometimes cryoturbated, gravels with occasional small palaeochannel features containing fossil molluscs, pollen and recycled palynomorphs are overlain by a broad silty palaeochannel-fill (Figure 9.17).

The deposits at Wookey Station were first mentioned by Woodward (1876), who noted that up to 10 feet of gravel was visible and that the long-axes of many stones were vertical. Green and Welch (1965) noted additionally that the gravels vary from round to angular and include clasts of Carboniferous Limestone and Old Red Sandstone. The site was described in detail by Macklin (1985, 1986) and Macklin and Hunt (1988), from whose work the following description is largely taken. Campbell *et al.* (in prep.) proposed the site as the type-section of the Wookey Formation and of the Wookey Station Member.

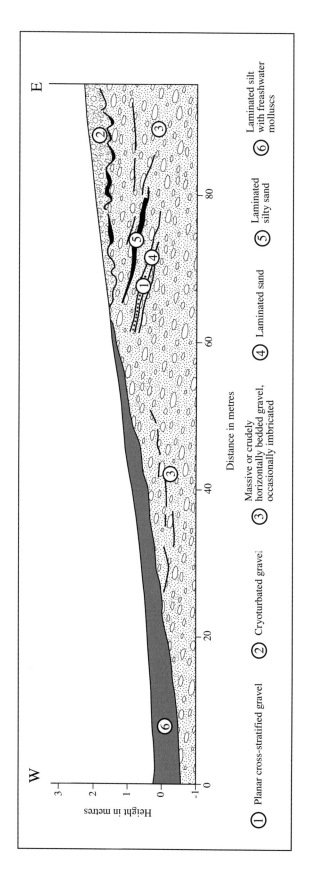

Figure 9.19 The Quaternary sequence exposed in the old railway cutting at Wookey Station. (Adapted from Macklin and Hunt, 1988.)

Wookey Station

Description

At Wookey, two gravel aggradations have been distinguished. The younger is a valley-fill of early Holocene age (Macklin and Hunt, 1988). This lies in a trench incised through the earlier aggradation, which has the morphological and sedimentological characteristics of an alluvial fan, with a convex upper surface and containing radiating palaeochannels (Macklin and Hunt, 1988). Cuttings at the old Wookey Station (ST 53154630; Figure 9.19), expose the following sequence (maximum bed thicknesses in parentheses):

3. Strong brown to dark red-brown slightly clayey silt with occasional stones, thickening towards the line of the modern River Axe and apparently filling a palaeochannel. (> 1.0 m)
2. Red-brown, massive- or crudely bedded, sandy cobbly gravel, involuted, with numerous vertical pebbles and cobbles and many split clasts. Disturbed gravel of this sort is present only in those parts of the section not overlain by bed 3. (1.0 m)
1. Red-brown, massive- or crudely bedded, clast-supported sandy cobbly gravel. Two graded units, separated by silty sand partings are present. In some places, the massive sandy gravel passes down-valley into planar cross-stratified openwork and clast-supported sandy gravel with occasional sand and silty sand beds. The gravels contain occasional scour channel-fills of normally graded plane-bedded sand or coarse silt. Some of these channel-fills contained shells of *P. muscorum* and *C. arenaria*, pollen of *Callitriche* and recycled palynomorphs derived from rocks of Carboniferous, Triassic, Jurassic and Pleistocene age. (> 2.0 m)

Interpretation

Beds 1 and 2 were interpreted by Macklin and Hunt (1988) as gravels of an alluvial fan laid down by a low-sinuosity single-channel stream. The fossil molluscs are taxa typical of British cold-stage deposits: *P. muscorum* is a xerophile tolerant of open exposed ground and *C. arenaria* today lives among sand dunes. Together, they indicate an exposed arid environment. The absence of pollen of terrestrial plants may also be taken as evidence for a largely unvegetated landscape or may be due to taphonomic problems. The *Callitriche* pollen probably reflects the vegetation of shallow, sun-warmed, relatively ephemeral pools. The recycled palynomorphs include taxa derived from rocks not present upstream from Wookey Station: the most probable explanation for their presence is that they were recycled by aeolian processes. The involutions of bed 2 post-date deposition of gravels in beds 1 and 2 and the colluvial deposits of bed 3. The latter is probably best interpreted as the colluvial-fill of a large channel, composed of sediments comparable with the aeolian coversands (Vink, 1949; Gilbertson and Hawkins, 1978a, 1983) of Avon and north Somerset. These sediments are thus probably at least partly of aeolian origin.

Conclusion

Wookey Station railway cutting exposes an excellent example of Devensian alluvial fan sedimentation on the margins of the Mendip Hills. The site is important because it contains fossil molluscs, pollen and recycled palynomorphs critical to an understanding of the depositional environment of the Mendip alluvial fans. The palaeobiological evidence suggests very open exposed landscapes. The recycled palynomorphs and sedimentary evidence points to considerable quantities of aeolian sediment being recycled by colluvial processes and to gravel deposition by streams on the alluvial fan.

Chapter 10

The Quaternary history of the Avon Valley and Bristol district

INTRODUCTION
C. O. Hunt

The sites described in this chapter were selected to document the glaciation and subsequent landscape development of the Bristol district and Avon Valley (Figure 10.1). This region contains important and unique evidence for a very early glaciation. Especially important are the glacial deposits of the Kenn Formation and associated landforms, distributed widely throughout the area, and the ?Stage 15 interglacial deposits of the Yew Tree Formation, which overlies the Kenn Formation of the Kenn lowlands. The Avon Valley contains a potentially important terrace sequence post-dating the glacial deposits, and important cold-stage aeolian and colluvial sediments are preserved at Holly Lane, Clevedon.

The Pleistocene record of the Bristol district and Avon Valley has considerable importance, but, with a few notable exceptions, has been relatively neglected in recent years. There is a rich history of research, spanning nearly two centuries. In the early years of the nineteenth century, many of the major elements of the Pleistocene geology of the region were described and interpreted by a variety of notable geologists. Much of this work still holds good today. Thus, Smith (1815) identified the alluvial origin of the Avon Levels and the presence of buried valleys under the alluvium, Conybeare and Phillips (1822) recorded erratic material on the hilltops around Bath, and Buckland (1823) recorded mammal remains from a number of cave sites in his *Reliquiae Diluvianae*.

Later, Weston (1850) described fossiliferous terrace gravels and erratic-rich plateau deposits near Bath, and Trimmer (1853) identified glacial erratics at Court Hill. Both argued that the erratics had been introduced during the 'deluge'. Considerable early attention was focussed on vertebrate localities in the Bath and Bristol districts (Dawkins, 1865; Moore, 1870). The first detailed synthetic work on the Quaternary deposits of the region was the Geological Survey Memoir of Woodward (1876). Prestwich (1890) later re-described the high-level gravels in the Bath area.

In the early years of the twentieth century, Harmer (1907) proposed that the river network of the Bristol district had resulted from glacial diversions of drainage. Although his suggestion was contested or ignored by authors such as Varney (1921), Davies and Fry (1929), Palmer (1931) and Trueman (1938), who favoured a solely fluvial origin for the network unhindered by glacial activity, recent work has tended to support his views. The terrace stratigraphy of the Bristol Avon was revised by Davies and Fry (1929) and Palmer (1931), who both proposed a tripartite terrace sequence with low, 50 foot and 100 foot terraces.

The investigation of periglacial deposits in Avon started with the discovery of the Clevedon bone cave in the Holly Lane 'gravel' quarry (Davies, 1907; Hinton, 1907a; Reynolds, 1907). Greenly (1922) recognized the cold-climate aeolian origin of the loamy sand units at Holly Lane, a conclusion supported by Palmer and Hinton (1929), Palmer (1934) and Vink (1949). Palmer (1934) conducted studies of a number of cold-climate breccia and blown-sand sites, including Holly Lane and the important section at Brean Down (Chapter 9), and demonstrated a southerly origin for the sands on mineralogical grounds.

Modern interest in the glacial geology of the region was stimulated by Mitchell's (1960) influential review, which provoked much debate concerning the limits and timing of glaciation, the possible existence and age of proglacial lakes and the occurrence, nature and stratigraphic position of the local interglacial marine deposits (e.g. Stephens, 1970a, 1970b, 1973; Hawkins and Kellaway, 1971, 1973; Kellaway, 1971; Kidson, 1971, 1977; Kidson and Haynes, 1972; Mitchell, 1972; Kidson *et al.*, 1974; Kellaway *et al.*, 1975).

Eventually, a growing body of stratigraphical and palaeoenvironmental research was to lead to a broad consensus on two major issues. First, that much of Avon had been glaciated during the Wolstonian (Gilbertson, 1974; Kidson, 1977; Gilbertson and Hawkins, 1978a, 1978b) and second that Somerset had not been glaciated (Kidson, 1977; Hunt *et al.*, 1984; Hunt, 1987). The Burtle Beds were shown to be estuarine interglacial deposits with freshwater intercalations (Kidson *et al.*, 1978; Gilbertson, 1979; Hunt and Clark, 1983), with the balance of evidence pointing toward an Ipswichian age. Similar sediments, post-dating the glacial deposits, were described from Kenn and were also thought to be Ipswichian in age (Gilbertson, 1974; Gilbertson and Hawkins, 1978a; Hunt, 1981). Post-Ipswichian periglacial deposits were described at Holly Lane and elsewhere in Avon and north Somerset (Gilbertson, 1974; Gilbertson and Hawkins, 1974, 1983). Some deposits, for instance the gravels on Bleadon Hill (Findlay *et al.*, 1972), remained more enigmatic, however.

The application of amino-acid geochronological techniques has since led to the reassessment of the

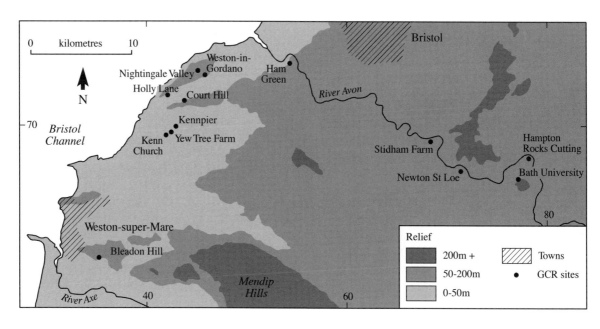

Figure 10.1 The Avon Valley and Bristol district, showing GCR sites described in this chapter.

Pleistocene sequence in the Bristol area and to the recognition that the glaciation of the region was of considerable antiquity. It has also become clear that marine interglacial deposits overlying the glacigenic sediments are of considerable complexity. Andrews *et al.* (1984) presented amino-acid ratios of *c.* 0.2 from estuarine interglacial deposits at Kenn Church and New Blind Yeo Drain, which they interpreted as Ipswichian; ratios of *c.* 0.38 for the upper estuarine deposits at Yew Tree Farm and Kennpier which overlay supposedly Wolstonian glacigenic deposits, were interpreted as equivalent in age to deposits at Purfleet in the Thames Estuary. Bowen *et al.* (1989) correlate these sites with later 'Cromer-complex' sites such as Waverley Wood and Oxygen Isotope Stage 15. The Kenn Church deposits have most recently been referred to Stage 7 (Campbell *et al.*, in prep.).

A number of important conclusions have emerged from the most recent work. First, many of the earlier ascriptions of sites to the Ipswichian Interglacial seem unfounded. Second, the great antiquity of the glaciation of Avon, pre-dating the Kennpier and Yew Tree Farm interglacial deposits and thus ?Stage 15, is also apparent. This glacial episode would appear to be substantially older than the Anglian glaciation of eastern England, which post-dates Stage 13.

Important themes in the Pleistocene of the Avon Valley and Bristol district

Several important themes emerge from the scientific framework outlined above, and were central to the process of site selection outlined in the introduction to Chapter 9. The themes are as follows.

1. Evidence for the age and limits of early glaciation

Sites in Somerset and Avon are of critical national importance since it is here that possible pre-Anglian glacial deposits and landforms are preserved in stratigraphic relationship with fossiliferous, and therefore datable, interglacial sediments. One group of sites was selected to demonstrate glacial deposits and landforms – the col-gully and glacial outwash at Court Hill, the till and glaciofluvial deposits at Nightingale Valley, and the tills and glaciofluvial gravels at Kennpier. A second group of sites provides additional evidence for glacial morphology and limits: glacial erratics contained in karstic fissures on the plateau of Bathampton Down at Bath University; recycled glacial erratics in fluvial gravels at Hampton Rocks Cutting in the Avon Valley, at Newton St Loe, Stidham Farm and Ham Green; ?glaciofluvial gravels at Bleadon Hill on Mendip; possible glacial deposits below the Burtle Beds at Greylake No. 2 Quarry in King's Sedgemoor (Chapter 9); and the erratic-free deposits at Langport Railway Cutting, 6 km farther south, which probably lay just beyond the glacial limit (Chapter 9). A further important group of sites provides dating evidence or the potential for dating Pleistocene events. At Kennpier, a channel incised into the Kennpier till contains interglacial deposits which have yielded amino-acid ratios indicative of an Oxygen Isotope ?Stage 15 age. At Weston-in-Gordano, till-like material lies stratified within interglacial marine deposits which may have formed during three separate high sea-level stands, thus providing further geochronological potential. The terrace stratigraphy of the Bristol Avon offers another potential dating tool since, in the nineteenth century, the Avon gravels were described as richly fossiliferous. Representatives of each of the main stratigraphic units were therefore selected – the plateau glacial deposits at Bath University, the 100' terrace at Ham Green, the 50' terrace at Stidham Farm, Saltford, and the low terraces at Newton St Loe and Hampton Rocks Cutting.

2. Evidence for high Pleistocene sea levels

Avon and Somerset offer an unparalleled sequence of marine interglacial and interstadial deposits and, wherever possible, GCR sites have been chosen to provide evidence for the high sea-level events. The sites are Kennpier (Stage 15), Kenn Church (Stage 7), and Weston-in-Gordano (undated but with three marine interglacial sequences interbedded with till-like material). Complementary sites are Swallow Cliff (Stage 5e or 7), Greylake No. 2 Quarry (Stages 7 and 5e) and Low Ham (Stage 5a) (Chapter 9).

3. Post-glaciation landscape development and river terrace stratigraphy

With the exception of the glacial and marine sequences, the fundamental evidence for establishing the Pleistocene stratigraphy of Somerset and Avon is provided by river terrace gravels. GCR sites were therefore selected to demonstrate the critical elements of this regional terrace stratigraphy. In the Bristol Avon Valley, Hampton Rocks Cutting, Newton St Loe, Stidham Farm and Ham Green were selected to represent the principal stratigraphic units. Complex deposits at Weston-in-Gordano show a long history of sea-level change and landscape development following the

glaciation of the area. In the same area, more recent landscape development under a periglacial regime is documented at Holly Lane.

4. Temperate-stage palaeobiology

Somerset and Avon have one of the most complete and richly fossiliferous sequences of marine interglacial and interstadial deposits in Britain. Two sites are of particular significance. The ?Stage 15 interglacial deposits at Kennpier and Yew Tree Farm are unique, with their rich fossil mollusc faunas, pollen and dinoflagellate cysts. Also important are the marine mollusc sites at Kenn Church (Stage 7) and Weston-in-Gordano (?Stage 7 and/or earlier stages).

5. Cold-stage sedimentation and palaeobiology

Subsequent to the Kenn glaciation, Avon and Somerset lay beyond the Pleistocene ice sheets. Although cold-stage sedimentation was widespread, good examples of pre-Devensian sediments are very rare. Fine examples of cold-stage river terrace gravels are seen in the Avon Valley at Hampton Rocks Cutting, Newton St Loe, Stidham Farm, Saltford, and Ham Green. Also important are the 'coversands' of the Avon coastlands, which pass laterally into thick colluvial and aeolian sequences found below steep slopes, for instance at Holly Lane, Clevedon.

(A) GLACIATION OF THE BRISTOL DISTRICT

This section describes sites selected to illustrate the pattern of the ancient glaciation of coastal Somerset and Avon and the Avon Valley. Till and erratic-rich outwash gravels are preserved in the Kenn lowlands at Kennpier and Yew Tree Farm. Sites such as Court Hill and Nightingale Valley show excellent examples of glaciofluvial deposits and flow tills, and show that the ice sheet downwasted against the Carboniferous Limestone massifs of the Failland Ridge and Clevedon Down. The Bleadon Hill site contains enigmatic deposits which may be glaciofluvial in origin and, if so, documents the incursion of a substantial ice sheet into Sedgemoor. Whereas the basal diamicton at Greylake (Chapter 9) may provide evidence for the maximum extent of glacial deposits in Sedgemoor, Langport Railway Cutting, 6 km to the south, is erratic-free (Chapter 9).

This section also documents sites which provide evidence for the age of the glacial episode, or which have significant potential for providing chronological control. This evidence derives from the fossil content of deposits at these sites. The palaeochannel-fills at Kennpier and Yew Tree Farm, which overlie the glacial deposits, have provided important aminostratigraphic evidence. The evidence at Weston-in-Gordano remains undated, but the complex stratigraphic sequence here, which includes evidence for three marine transgressions, offers the possibility of further geochronometric dating. These sites also offer important temperate-stage palaeobiological evidence, and evidence of former high sea levels. This is augmented by the Stage 7 marine deposits at Kenn Church. Morphostratigraphical evidence for dating the glacial episode is contained in the terrace sequence of the Avon Valley, which is described in the second section of this chapter.

COURT HILL
C. O. Hunt

Highlights

Although much of the Quaternary fill of the 'col-gully' at Court Hill was removed during construction of the M5 motorway, the site still provides secure and spectacular evidence for the glaciation of the Avon coastlands. It is the only well-documented example of this type of glacial landform in South-West England.

Introduction

At Court Hill, a col-gully cut in Carboniferous Limestone of the Failland Ridge contains up to 24 m of glacigenic sediments. These comprise unstratified tills, stratified boulder beds, gravels, sands and glaciolacustrine deltaic deposits.

Gravels with a component of quartzite and other erratic lithologies were first recorded at Court Hill by Trimmer (1853). Prestwich (1890) suggested that these gravels might be linked with the Westleton Beds of East Anglia, and argued that they might be a continuation of the drifts on the hills around Bath. In the early 1970s, a cutting for the M5 motorway was excavated through the deposits at Court Hill. The geotechnical investigations and the cutting itself revealed 24 m of sands, gravels and diamictons lying in a channel cut through the Failland Ridge (Hawkins and Kellaway, 1971; Gilbertson, 1974; Gilbertson and Hawkins, 1978b). The deposits included a number of clasts of lithologies erratic to the Failland Ridge, including Greensand chert, and some Cretaceous foraminifera. Gilbertson and Hawkins (1978b) interpreted the feature as a 'col-gully', of glacial origin. Campbell *et al.* (in prep.) assigned the glacigenic deposits to the Kenn Formation, which elsewhere in Avon and north Somerset pre-dates an interglacial of 'Cromer-complex', probably Oxygen Isotope Stage 15, age.

Description

A full geomorphological and stratigraphical description of the site was given by Gilbertson and Hawkins (1978b), from whose account the following is largely taken.

The deposits at Court Hill (ST 473723) lie in a channel some 24 m deep, excavated in the Carboniferous Black Rock Limestone of the Failland Ridge. The base of the channel falls and the channel widens from south to north. The fill of the feature varies laterally, with predominantly boulder-, cobble- and gravel-sized material to the south, passing into predominantly sand-sized material to the north (Figure 10.2). The following sedimentary facies can be distinguished.

5. The whole site is overlain unconformably by a veneer of red silty sand usually less than 0.5 m thick.

4. Unconformably overlying the main part of the sequence are lenticular red-brown diamicton and gravel bodies up to 3 m thick. The matrix of the diamicton is a sandy silt, with constituent boulders up to 0.5 m in diameter. Most of the large clasts are Carboniferous Limestone, but other lithologies, including Pennant Sandstone, Carboniferous chert, Triassic sandstone, Mercia Mudstone, Old Red Sandstone, Greensand chert and flint, are present.

3. On the south side, the main part of the fill comprises beds between 0.5 to 2.0 m thick of imbricated, well-sorted, often openwork well-rounded gravels, cobbles and boulders, which dip northwards at *c.* 37°. The beds exhibit both normal and inverse grading. The deposits contain a similar range of rock types to facies 4. The silty sand to gritty sand matrix of the clast-supported beds is indurated with

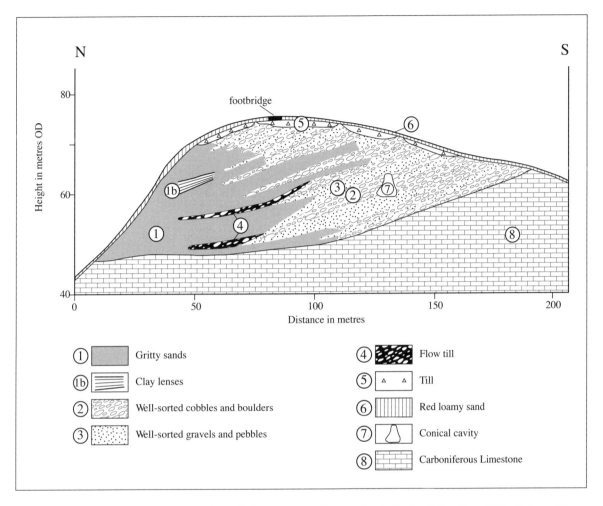

Figure 10.2 Schematic cross-section through Quaternary sediments in the 'col-gully' at Court Hill. (Adapted from Gilbertson and Hawkins, 1978b.)

calcium carbonate while openwork deposits are usually carbonate-cemented at point contact. A number of conical cavities, 2–3 m deep, 1 m wide at the top and 3–4 m wide at the base, occur within this facies.

2. The gravels and boulder beds interdigitate with uncemented, cross-bedded, coarse gritty sands with occasional very thin clay/silt partings. Gilbertson and Hawkins (1978b) suggest that this cross-bedding is of a deltaic type. Beds in the sands are 0.5–3.0 m thick and the cross-sets all dip northwards at 10–20°. A few Jurassic and Cretaceous foraminifera were found in the sands.

1. The sands are interbedded with occasional lenticular bodies of diamicton up to 0.75 m thick. The diamicton is poorly bedded and comprises cobbles and boulders in a sandy silt matrix. Most of the boulders are of

Carboniferous Limestone, but other lithologies, similar to those in facies 3 and 4, are also present.

Interpretation

A number of valley-fills of Triassic dolomitic conglomerate are known in the Bristol District (Kellaway and Welch, 1948). The presence of erratic materials, notably the Greensand chert and flint and the Cretaceous foraminifera, however, precludes the fill of the Court Hill channel being dolomitic conglomerate.

Gilbertson and Hawkins (1978b) regard the Court Hill channel as a glacial 'col-gully', cut by meltwaters of a downwasting ice sheet and infilled with glaciofluvial gravels (facies 3), rare flow tills (facies 1) and deltaic sands (facies 2). Further

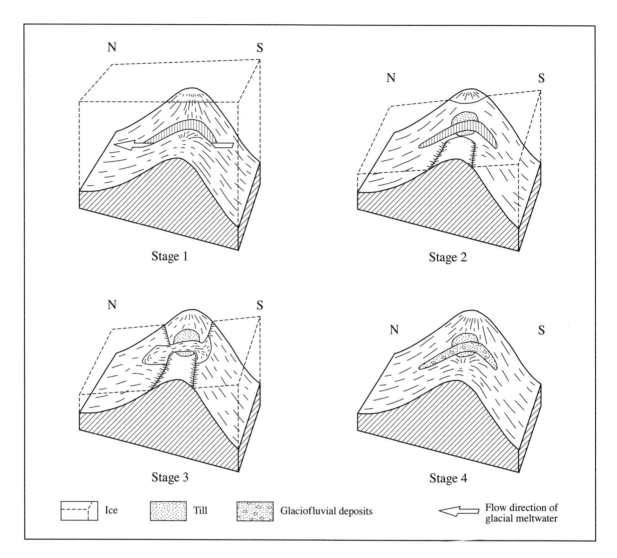

Figure 10.3 A four-stage model to explain the formation of the Court Hill 'col-gully'. (Adapted from Gilbertson and Hawkins, 1978b.)

patches of till (facies 4) overlie these deposits. The silty sand (facies 5) overlying the sequence was interpreted as an aeolian coversand. In their interpretation, the ice sheet, of proposed Wolstonian or Anglian age, is supposed to have been thicker in the Kenn lowlands, to the south of the Failland Ridge, than it was in the Vale of Gordano to the north, where an ice-marginal lake formed. The meltwaters which cut the 'col-gully' are believed to have flowed northwards into this lake (Figure 10.3).

Conclusion

Court Hill provides spectacular and unambiguous evidence for the glaciation of the Avon coastlands and, in particular, provides important evidence for the configuration of the ice masses located in the Vale of Gordano and the Kenn lowlands. The 'col-gully' contains an impressive infill of Quaternary sediments which includes till, glaciofluvial and glaciolacustrine deltaic deposits. The age of the glacial episode(s) responsible for the landforms and deposits is unproven, although the proposed pre-Anglian event responsible for glacigenic deposits of the Kenn Formation seems most likely at present. The M5 now runs through the middle of the Court Hill channel, making access difficult.

BATH UNIVERSITY
C. O. Hunt

Highlights

Bath University (Bathampton Down) is representative of sites at the far south-eastern extent of a spread of erratic material in the Bristol and Bath districts, and thus at the limits of a very early glaciation. The site has especial importance because it can be used to demonstrate that the original deposition of this 'exotic' material pre-dates the formation of most of the valleys in the Bath area. It is proposed as the type-site of the Bathampton Down Member.

Introduction

At Bathampton Down, fissures and karstic cavities contain gravels and diamictons rich in flint and other erratic lithologies. Good examples are exposed in the road cutting at the entrance to Bath University.

Buckland and Conybeare (1824) described transported chalk flints on the summits of the hills south and east of Bath. 'High-level' gravels in the Bath district were then described by Weston (1850), who attributed them to the 'deluge'. Prestwich (1890) attributed them to deposition by an eastward-flowing proto-Thames. Most of the subsequent writers (Varney, 1921; Davies and Fry, 1929; Palmer, 1931) regarded them as fluvial, deposited by a river system draining into the Solent. Later workers (Hawkins and Kellaway, 1971; Gilbertson and Hawkins, 1978a) have regarded them as of glacial origin. The gravels on Bathampton Down were described in general terms by Varney (1921), Davies and Fry (1929) and Palmer (1931), and in more detail by Hawkins and Kellaway (1971). The site is proposed as the type-locality of the Bathampton Down Member by Campbell *et al.* (in prep.), who correlated the deposits with the glacial deposits of the Kenn lowlands which are thought to pre-date Oxygen Isotope Stage 15.

Description

In the entrance-cutting to Bath University (ST 535759), solution cavities and fissures in the Great Oolite Limestone contain strong brown, clayey, clast- and matrix-supported fine gravels. These are sometimes overlain by pale brown, matrix-supported, crudely plane-bedded clayey and silty gravels. The gravel clasts are predominantly of Greensand chert, but flint, Carboniferous Limestone and chert, Oolitic Limestone, coal, shales, sandstone, 'bunter' quartzite and conglomerate are also present. Many of these rock types are not found upstream of Bathampton in the Bristol Avon catchment. Some of the fissure- and solution cavity-fills have been tilted and faulted by later cambering and landslips.

Interpretation

Both fluvial (Varney, 1921; Davies and Fry, 1929; Palmer, 1931) and glacial origins (Hawkins and Kellaway, 1971; Gilbertson and Hawkins, 1978a) have been suggested for these deposits. Hawkins and Kellaway (1971) suggested that the gravels may have been locally reworked from glacial deposits by fluvial or slope processes into the fissures and caves in the limestone. Most of these authors agree that deposition of these materials probably pre-dates the cutting of the valley system in the Bath area. The age of the glaciation which laid down these materials is not apparent from this site, but it has been argued to be the same as that which laid down the glacial deposits underlying interglacial sediments at Yew Tree Farm and Kennpier in the Avon Levels (Gilbertson and Hawkins, 1978a). Recent work (Andrews *et al.*, 1984; Bowen *et al.*, 1989) suggests that these interglacial deposits are of considerable antiquity and can be correlated with Oxygen Isotope Stage 15. Thus, the glacial deposits at Bathampton Down would appear to pre-date the Anglian and could date from before 600 ka BP.

Conclusion

The 'high-level' gravels of Bathampton Down are part of the evidence for a very early glaciation of the British Isles, although they have also been regarded as ancient fluvial deposits. This site lies towards the far south-eastern extent of a spread of 'exotic' material in the Bristol and Bath districts and

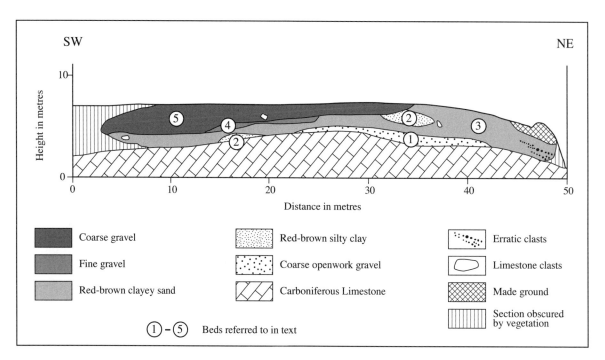

Figure 10.4 The Pleistocene sequence at Nightingale Valley, adapted from Hunt (in prep.).

is thus important for establishing the limits of a very early glaciation. Bath University GCR site is also important because it shows that most of the valleys in the Bath area were cut after this glacial event.

NIGHTINGALE VALLEY
C. O. Hunt

Highlights

Nightingale Valley contains unequivocal *in situ* glacial material and thus provides clear evidence for a glacial incursion in the Avon coastlands and the overriding of the Portishead–Clevedon ridge by ice from the Bristol Channel.

Introduction

Portishead Down is part of the Clevedon–Portishead ridge, a Carboniferous Limestone horst which separates the Vale of Gordano from the Bristol Channel. Near the summit of Portishead Down, overlooking the Vale of Gordano at above 85 m OD, up to 4 m of glacigenic deposits are exposed at the top of the old Black Rocks Quarry towards the head of Nightingale Valley. The

deposits include boulders, imbricated coarse gravels, sands and silty clays. Some of these deposits have a substantial erratic content.

The earliest mention of drift deposits on the coastal ridge between Clevedon and Portishead was the observation that on Walton Down ' ... the rabbits have thrown up a quantity of fine flint gravel. 250–270-ft. O.D.' (Davies and Fry, 1929; p. 164). These authors suggested a fluvial origin for the gravels as part of their Avon 'High Terrace'. Hawkins and Kellaway (1971) record that the site was visited by a British Association field excursion from the Bristol Meeting in 1955, when opinion on the 'cannon-shot' gravels was divided between glacial and marine Tertiary origins. Deposits at the site were mapped by Welch (1955) and considered to be of glacial origin by Hawkins and Kellaway (1971). The site has recently been re-described by Hunt (in prep.).

Description

Eroded remnants of glacial deposits cap the limestone plateau of Portishead Down. *In situ* deposits lie above *c.* 85 m at Nightingale Valley GCR site (ST 450752) and are well exposed at the south-eastern side of the site, at the top of the abandoned Black Rocks Quarry. The deposits overlie an erosion surface cut across sharply dipping Carboniferous Limestone. In general, this surface has a smoothly undulating relief of 2–3 m, though localized steep-sided depressions occur over fault-planes in the limestone below.

The deposits are up to 4 m thick and extremely variable in lithology. They show a complex stratigraphy (Figure 10.4), and can be summarized as follows (maximum bed thicknesses in parentheses).

5. The section is capped by extremely coarse, mostly angular, clast- and matrix-supported gravel. Clasts with B-axes between 0.05 and 0.1 m are very common. The largest limestone clast had a B-axis of 0.2 m and the largest erratic, a well-rounded brown quartzite clast, had a B-axis of 0.09 m. The matrix of the gravels varies between whitish-brown angular fine gravels and whitish-brown gritty sands. The gravels are imbricated, with A-axes typically orientated towards 140°. (2.6 m)
4. Fine gravel, imbricated and clast-supported, with a reddish-brown sandy clay matrix. About 61% of this gravel is made up of angular Carboniferous Limestone clasts; most of the rest are erratics including, in descending order of abundance, flint, sandstone, vein calcite, quartzite, vein quartz, siltstone, basalt, haematite (vein-fill) and coal. (0.6 m)
3. Reddish-brown slightly clayey sand with rare rounded quartzite and flint pebbles. At the north-east extremity of the exposure, occasional lines of well-rounded pebbles, mostly of quartzite, flint and brown sandstone lie towards the base of the unit. The largest clast exposed was of limestone and had an A-axis of 0.72 m and a B-axis of 0.3 m, but limestone clasts are comparatively rare in this bed. (3 m)
2. Reddish-brown silty clay with occasional quartzite and flint pebbles. A sample from near the base of this bed yielded a sparse assemblage of palynomorphs of Carboniferous, Mesozoic and Quaternary age. (0.6 m)
1. Extremely coarse openwork gravels. The typical clast has a B-axis of around 0.05–0.1 m and the gravels consist almost exclusively of angular fragments of the local Carboniferous Limestone. (1 m)

Interpretation

The clast size, erratic content and stratigraphy of the deposits are inconsistent with a marine or flu-vial origin, but are consistent with a glacial origin. The gravels and sands (beds 5, 4, 3 and 1) are most probably glaciofluvial in origin, though the reddish-brown silty clay (bed 2) may be a flow till. The presence of glaciofluvial deposits on the summit of Portishead Down and their south-eastward imbrication is not completely consistent with the suggestion (Hawkins, 1977) that glacial incursion was from the west. Neither is it fully consistent with Hawkins' (1977) suggestion that the Nightingale Valley to the south-east of the site originated as a glacial meltwater channel. Most of Nightingale Valley in fact drains south-westwards and therefore is unlikely to have been cut by melt-waters flowing south-eastwards. It is perhaps more probable that Nightingale Valley was cut by sub-aerial processes, probably in much the same way that the chalk combes were excavated under periglacial conditions (Kerney, 1963).

Some of the clast lithologies, such as the red sandstones, the brown sandstones and the coal, are erratic on Portishead Down but may be derived from nearby outcrops of Devonian, Triassic and Carboniferous age in the Bristol Coalfield, or possibly from the South Wales Coalfield. Other lithologies, principally the flint and the quartzites, are probably derived from farther afield, though the durability of these rocks and the roundness of the clasts is suggestive of an extremely long transportational history prior to their incorporation into the drift. A slightly disconcerting feature is the absence of Greensand chert, which is a common erratic lithology in glacial deposits elsewhere in Avon. The presence of erratic palynomorphs, including Rhaetic and Quaternary marine taxa, can be taken as evidence for the derivation of these sediments from the Bristol Channel.

Conclusion

The site is an important component in a network of sites which contains clear evidence for the glaciation of the Avon coastlands. Nightingale Valley is important because it contains evidence for the advance of ice inland from the Bristol Channel. The deposits are well preserved and rich in rocks that can only have been transported to the summit of Portishead Down by a glacier. The age of the glacial episode is unproven and the subject of considerable controversy. Nightingale Valley therefore has considerable potential for future research into the glacial history of the Avon coastlands.

BLEADON HILL
C. O. Hunt

Highlights

The enigmatic deposits on Bleadon Hill may be a Mesozoic sea beach deposit, Pleistocene shoreline materials, proglacial lake-shore sediments or glaciofluvial gravel. If either of the latter two possibilities is the case, this site provides evidence for a glacial invasion of at least part of Sedgemoor and is therefore of great significance for understanding the limits of Pleistocene glaciation in South-West England. Bleadon Hill has been proposed as the type-section of the Bleadon Member.

Introduction

Bleadon Hill lies on the southern flank of the Mendip Hills and its controversial deposits are unrelated to an obvious source, such as a valley or cave resurgence. Clasts are derived predominantly from the local Carboniferous Limestone, but the deposit contains rare Lower Jurassic foraminifera.

The site was found and described by Findlay *et al.* (1972) and re-described in the Geological Survey Memoir (Whittaker and Green, 1983). The following description is largely taken from their work. The site is proposed as the type-section of the Bleadon Member by Campbell *et al.* (in prep.), who accepted a glacial origin for the deposit.

Description

A body of sand and gravel lies at 82 m OD on the south side of Bleadon Hill at ST 350573. At its western end, the deposit lies upon a bench-like feature and against a near-vertical face cut in the Carboniferous Limestone, but most of the deposit lies upon siltstones of the Mercia Mudstone Formation.

At the south-east corner of the deposit, in an old gravel working, Findlay *et al.* (1972) recorded the following stratigraphy, with beds dipping at 35° to the north-east and all beds point-contact cemented. The base of the deposit was not seen. Not all bed maximum thicknesses were recorded by Findlay *et al.* (1972): those missing from their report were obtained during re-examination for the GCR, where possible, and are shown in parentheses.

5. Clast-supported, cobbly openwork gravel. The clasts are subrounded and up to 0.15 m in diameter. All clasts are of Carboniferous Limestone. (0.6 m)

4. Clast-supported, fine openwork gravel with occasional cobbles. The clasts are subrounded and most are between 5 and 20 mm, though the cobbles are up to 0.08 m. Most clasts are of Carboniferous Limestone, with some 'yellowish calcareous rock' and rare quartz and calcite. (1.2 m)

3. Clast-supported, cobbly openwork gravel. The clasts are subrounded and mostly 0.05–0.1 m in diameter, but with some up to 0.23 m in diameter. The transition to the underlying bed is irregular. (1.6 m)

2. Clast-supported, openwork very coarse gravel, cobbles and boulders. The clasts are up to 0.3 m in diameter. (*c.* 2 m)

1. Clast-supported fine and medium gravels. (> 2 m)

Findlay *et al.* (1972) recorded the following section at the eastern end of the site in an excavated pit:

4. Clast-supported, cobbly carbonate-cemented gravel. The clasts are subrounded and up to 0.15 m in diameter. They are all composed of Carboniferous Limestone. (0.6 m)

3. Reddish-brown, pebbly sandy silt. (1.2 m)

2. Pale brown carbonate-cemented sand.

1. Pale brown, 'laminated' and ripple-marked unconsolidated sand containing rare Liassic (probably Sinemurian) foraminifera. The bedding in the sands dips at 37° to the south. The base of the deposit was not seen.

Interpretation

Findlay *et al.* (1972) suggested a variety of origins for the deposits including a sea beach of either Mesozoic or Pleistocene age, a proglacial lacustrine beach deposit or a glaciofluvial gravel. Since it is now apparent that, in the Severn coastlands, sea levels have persistently returned only to levels close to or at most a few metres above present levels throughout the Middle and Upper Pleistocene (Andrews *et al.*, 1984), the presence of a Pleistocene shoreline deposit at 82 m OD at Bleadon Hill is considered unlikely. There is no evidence to disprove any of the other suggestions of Findlay *et al.* (1972), though the lack of demonstrably glacially transported erratic material could be taken as an indication that a glacial origin is unlikely. On the other hand, cementation of the deposit is never more than rather light point-contact; heavier cement might reasonably be expected from a deposit of Jurassic age.

Conclusion

The origin of the Bleadon Hill deposit is uncertain, but suggested possibilities include Mesozoic and Pleistocene sea beach deposits, a proglacial lakeshore deposit or a glaciofluvial gravel. This site thus potentially preserves evidence for a glacial invasion of at least part of Sedgemoor and therefore may be of great significance for the understanding of the limits of Pleistocene glaciation in South-West England. Its research potential is largely unrealized.

KENNPIER
C. O. Hunt

Highlights

Kennpier is of national importance because here the fossiliferous temperate-stage channel-fill and estuarine deposits of the Yew Tree Formation overlie the Kennpier and Nightingale members of the Kenn Formation. The latter comprises materials which have been interpreted as glacial outwash and till. Amino-acid ratios derived from fossil molluscs in the Yew Tree Formation suggest correlation with Oxygen Isotope Stage 15, and thus a pre-Anglian age for the underlying glacial deposits. Evidence for pre-Anglian glacial episodes is virtually unknown elsewhere in the British Isles. The site is the type-section of the Kenn Formation and of the Kennpier and Nightingale members.

Introduction

At Kennpier, exposures in deep drainage ditches and a borehole have shown a complex stratigraphy, with coversands of the Brean Member overlying estuarine sands of the Kenn Church Member and a richly fossiliferous channel-fill of the Yew Tree Formation. The latter can be correlated aminostratigraphically with Oxygen Isotope Stage 15. The channel-fill is incised into glacigenic diamictons and gravels of the Kenn Formation.

Pleistocene gravels have been known in the Kenn area since the work of Woodward (1876) and Greenly (1921). Gilbertson (1974) and Gilbertson and Hawkins (1978a) described the stratigraphy of Quaternary deposits at Kennpier Footbridge. These authors also carried out detailed studies on the freshwater, estuarine and terrestrial molluscs from the site, and Beck (*in* Gilbertson and Hawkins, 1978a) described six very small pollen assemblages. They regarded the palaeobiology of the site as indicative of the late stages of an interglacial, probably the Ipswichian. Hunt (1981) described the stratigraphy and palynology of a borehole through the channel-fill. The pollen was comparable with Ipswichian II–III assemblages. The organic-walled microplankton included both marine and freshwater forms, suggesting estuarine conditions. Andrews *et al.* (1984) presented amino-acid ratios on *Corbicula* shells from the Yew Tree Formation at Kennpier Footbridge. These are suggestive of an age of around 400 to 600 ka BP and may therefore indicate a pre-Anglian age for the underlying glacial deposits. This age was supported by Bowen *et al.* (1989) and accepted by Campbell *et al.* (in prep.). Andrews *et al.* (1984) also obtained an amino-acid ratio of 0.2 on *Macoma* from the Kenn Church Member at this site. Kennpier is designated the type-section of the Kenn Formation and of the Kennpier and Nightingale members (Campbell *et al.*, in prep.).

Description

The stratigraphic relationships at Kennpier are shown in Figure 10.5. Gilbertson (1974) recorded the following section at ST 427698, overlying Mercia Mudstones. Maximum bed thicknesses are shown in parentheses.

6. Grey estuarine silts of Holocene age. The bed has a sharp boundary with bed 5. (1.7 m)
5. Red silty fine sands (Brean Member). These overlie and are juxtaposed with beds 2, 3 and 4 in a series of involutions. (1.0 m)
4. Pockets of pebbly shelly sands (Kenn Church Member), for instance at ST 425708, containing marine molluscs and overlying beds 1 and 2. (0.5 m)
3. Greenish-grey, shelly silty sands (Yew Tree Formation), lying in a channel 30 m wide and of variable depth incised into bed 2. (2.0 m)
2. Pale brown to red silty and sandy diamictons (Kennpier Member), with striated boulders up to 1 m in diameter and weighing over 1 tonne. (1.7 m)
1. Grey-brown, poorly bedded cobbly gravels (Nightingale Member). (4.0 m)

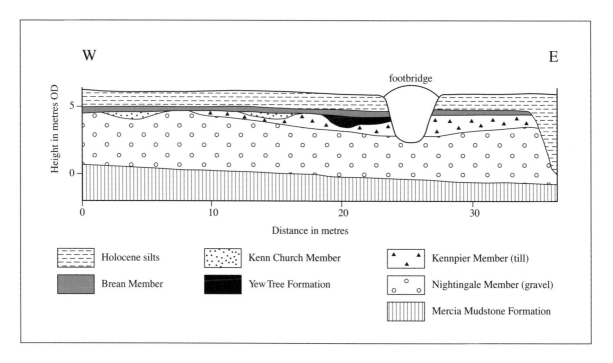

Figure 10.5 The Quaternary sequence at Kennpier. (Adapted from Andrews *et al.*, 1984.)

The Kenn gravels (bed 1) and Kennpier till (bed 2) contain numerous non-local rock types, and occasional very large striated clasts. The Yew Tree Formation channel-fill (bed 3) contains a diverse mollusc assemblage, with 30 taxa recorded (Gilbertson and Hawkins, 1978a; Table 10.1). The fauna is dominated by the opercula of *B. tentaculata*, with some *Agrolimax* spp. and *V. piscinalis*. Other taxa are comparatively rare. Estuarine molluscs are present but decrease in abundance up-section. The pollen assemblage (Hunt, 1981) contains abundant tree pollen, including *Pinus*, *Quercus*, *Alnus*, coryloid, *Carpinus*, *Betula*, *Picea* and *Tilia*, together with pollen of herbs, marsh plants and aquatics, and cryptogam spores. The algal microfossil assemblage (Hunt, 1981) is species-poor and includes the marine dinoflagellate cysts *Operculodinium centrocarpum*, *Achomosphaera andalousiense* (= *Spiniferites septentrionalis*; Figure 10.6), the prasinophyte *Cymatiosphaera* sp. and spores of the zygnemataceous alga *Spirogyra* sp. Molluscs from the Yew Tree Formation at Kennpier have yielded amino-acid ratios of 0.385 and 0.405 (Andrews *et al.*, 1984).

In the pebbly shelly sand pockets of the Kenn Church Member (bed 4), restricted molluscan faunas are dominated by *M. balthica* and *Littorina* spp. Freshwater species are occasionally present.

Fossil material from this bed yielded an amino-acid ratio of 0.2 (Andrews *et al.*, 1984).

Interpretation

The stratigraphy and palaeobiology of the site have been interpreted by Gilbertson (1974), Gilbertson and Hawkins (1978a) and Hunt (1981). Andrews *et al.* (1984) reassessed the stratigraphical significance of the Yew Tree Formation and their model has been accepted by later workers.

The grey silts (bed 6) at the top of the section were laid down during the later Holocene (Gilbertson and Hawkins, 1978a; Butler, 1987). The red silty sands of the Brean Member (bed 5) were interpreted by Gilbertson and Hawkins (1978a) as coversands of aeolian origin. The pockets of pebbly shelly sand of the Kenn Church Member (bed 4) are most likely of estuarine origin. This bed can be correlated aminostratigraphically with Oxygen Isotope Stage 7 (Andrews *et al.*, 1984; Campbell *et al.*, in prep.).

In the Yew Tree Formation (bed 3), several of the mollusc taxa, especially *B. marginata* and *C. fluminalis*, require interglacial conditions. Declining counts of thermophilous taxa in the higher levels of the deposit led Gilbertson and Hawkins (1978a) to suggest that the sediments

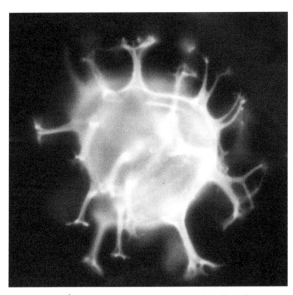

Figure 10.6 The commonest marine dinoflagellate cyst in the Kennpier interglacial deposit (bed 3) – *Achomosphaera andalousiense* Jan du Chene – seen at a magnification of *c.* × 1000 by UV fluorescence microscopy. (Photo: S.A.V. Hall.)

were laid down during an episode of deteriorating climate. The Yew Tree Formation at Kennpier was regarded as of fluvial origin, because most of the Mollusca are freshwater taxa, with the most abundant environmental group being taxa typical of moving water such as *B. tentaculata* and *V. piscinalis*. They suggested that the estuarine taxa in the channel-fill were recycled from nearby estuarine gravels. This interpretation was challenged by Hunt (1981) who described palynological evidence – high counts for Chenopodiaceae – and the presence of marine microfossils as evidence for marine influence: a depositional environment in the backwaters of an estuary was suggested. High counts for Poaceae and herbaceous taxa point to a relatively 'open' coastland environment, but the presence of a variety of broad-leaved tree species is evidence for interglacial conditions (Figure 10.7). Similarly 'open' environments have been noted from other coastal interglacial sites, for instance at Morston (Gale *et al.*, 1988). Marine taxa and broad-leaved tree species decrease in importance up-section, suggesting lessening marine influence and an opening of the landscape through time. The palynological evidence is thus consistent with the molluscan evidence for deteriorating climate and suggests that the Yew Tree Formation at Kennpier was laid down as climate started to deteriorate towards the end of an interglacial. The amino-acid ratios may indicate correlation with Oxygen Isotope Stage 15 (Bowen *et al.*, 1989), though Bowen (pers. comm., 1995), while accepting this correlation, suggested that ratios on *Corbicula* should be treated with some caution.

The Nightingale Member (bed 1) was regarded as glacial outwash and the Kennpier Member (bed 2) as till by Gilbertson and Hawkins (1978a). They correlated these deposits with the high-level drifts in the Bath area, on the Failland Ridge and at Court Hill and argued that together these provided evidence for a glacial incursion into the Avon coastlands. This interpretation was accepted by Campbell *et al.* (in prep.).

Conclusion

Kennpier is a key site for studies of the Pleistocene stratigraphy and palaeobiology of the British Isles. The sequence here contains a temperate-stage channel-fill, estuarine deposits also indicative of temperate conditions, gravels which have been interpreted as glacial outwash and diamictons that have been interpreted as glacial till. In the temperate-stage deposits of the Yew Tree Formation, molluscs, pollen, dinoflagellate cysts and foraminifera suggest interglacial conditions. Two amino-acid ratios have been determined from molluscs in this bed, which ostensibly suggest correlation with the Purfleet interglacial deposits in the Thames Estuary and Oxygen Isotope Stage 15. (This date would seem to be at conflict with Purfleet's position in the post-Anglian course of the Thames, which the river only assumed in Stage 11 (Bridgland, 1994).) The fossil remains from the Yew Tree Formation at Kennpier Footbridge offer considerable potential for future investigation, since the palaeobiology of the temperate stage which they reflect is as yet very poorly known. Further research potential is offered by the 'temperate' estuarine deposits of the Kenn Church Member. The interglacial deposits at Kennpier Footbridge rest on diamictons of the Kenn Formation suggested by Gilbertson and Hawkins (1978a) to be of glacial origin. These represent an ancient glacial episode which is very poorly known elsewhere in the British Isles.

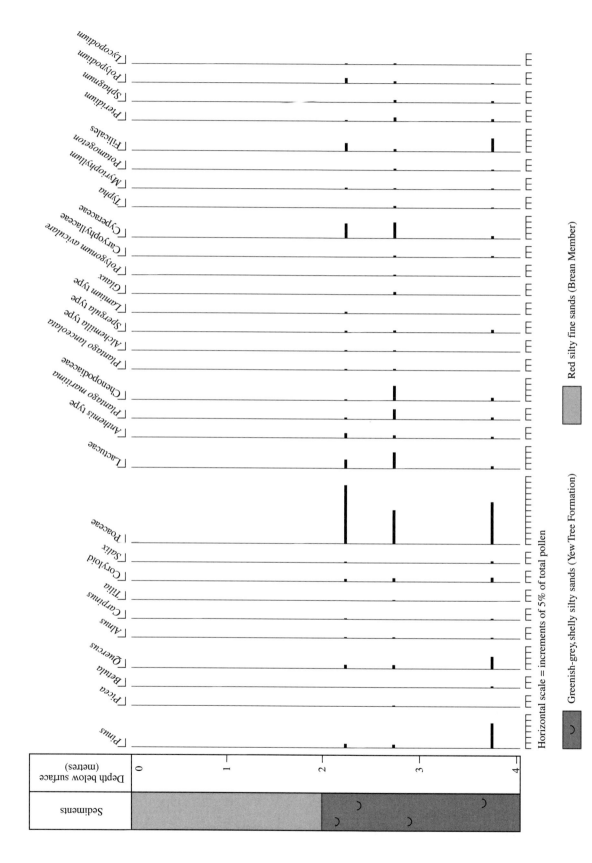

Figure 10.7 Pollen diagram for Kennpier. (Adapted from Hunt, 1981.)

Table 10.1 Fossil molluscs from three profiles through the interglacial channel-fill at Kennpier Footbridge (after Gilbertson and Hawkins, 1978a)

Species	Profiles		
	KPA	**KPB**	**KPD**
Marine			
Littorina littorea (Linné)	3	–	1
Littorina saxatilis (Olivi)	1	–	–
Littorina littoralis (Linné)	5	1	–
Littorina spp.	13	1	2
Retusa sp.	–	–	1
Nucella lapillus (Linné)	3	–	–
Ostrea sp.	1	–	2
Cerastoderma spp.	2	–	frags
Macoma balthica (Linné)	36	1	16
Land and freshwater			
Valvata piscinalis (Müller)	127	72	423
Belgrandia marginata (Michaud)	14	–	13
Bithynia tentaculata (Linné) shells	3	–	4
Bithynia tentaculata (Linné) opercula	1092	3579	5619
Lymnaea peregra (Müller)	12	–	65
Anisus leucostoma Müller	–	–	1
Gyraulus laevis Alder	4	–	1
Armiger crista (Linné)	1	–	1
Planorbis spp.	–	–	1
Ancylus fluviatilis Müller	2	–	–
Trichia hispida (Linné)	–	–	15
Zonitoides nitidus (Müller)	–	–	1
Agrolimax cf. *agrestis* (Linné)	70	70	128
Agrolimax cf. *reticulatus* (Müller)	30	43	35
Agrolimax cf. *laevis* (Müller)	6	–	22
Agrolimax spp.	65	56	162
Corbicula fluminalis (Müller)	–	20	24
Pisidium amnicum (Müller)	4	6	7
Pisidium obtusale (Lamarck)	–	–	1
Pisidium subtruncatum Malm	5	–	1
Pisidium henslowanum (Sheppard)	1	–	–
Pisidium nitidum Jenyns	5	–	1
Pisidium moitessierianum Paladilhe	–	–	1
Pisidium spp.	13	4	13
Total	1560	3854	6564

YEW TREE FARM
C. O. Hunt

Highlights

Yew Tree Farm, Avon, is of considerable significance for its diverse non-marine mollusc faunas dating from Oxygen Isotope Stage 15. The interglacial deposits here overlie glacial outwash deposited during a pre-Anglian glaciation, giving the site considerable stratigraphic significance. It is proposed as the type-site of the Yew Tree Formation.

Introduction

At Yew Tree Farm, estuarine silts of Holocene age overlie fine red silty sands. These in turn overlie gravels and then richly fossiliferous interglacial deposits which lie on coarse cobbly gravels.

Pleistocene gravels have been known in the Kenn area since the work of Woodward (1876) and Greenly (1921). Gilbertson and Hawkins (1978a) described the stratigraphy of the Quaternary deposits at Yew Tree Farm. They also carried out detailed studies on the freshwater and terrestrial molluscs from the site, and described three very small pollen assemblages. They regarded the palaeobiology of the site as indicative of the late stages of an interglacial, probably the Ipswichian. Hunt (1981) described a pollen and organic-walled microplankton assemblage from the site. The pollen was comparable with Ipswichian II–III assemblages. The organic-walled microplankton included both marine and freshwater forms, suggesting estuarine conditions. Andrews *et al.* (1984) and Bowen *et al.* (1989) presented amino-acid ratios derived from *Corbicula* shells from Yew Tree Farm. These are suggestive of an age of 400 to 600 ka BP. The site is proposed as the type-locality of the Yew Tree Formation by Campbell *et al.* (in prep.).

Description

At Yew Tree Farm (ST 42256927), Gilbertson and Hawkins (1978a) recorded the following sequence (maximum bed thicknesses in parentheses).

5. Grey estuarine silts of Holocene age. (0.6 m)
4. Fine red silty sands – Brean Member. (0.5 m)
3. Silty, sandy and cobbly gravels. (1.06 m)
2. Pale-grey, laminated and cross-bedded shelly sandy silt – Yew Tree Formation. (0.3 m)
1. Coarse cobbly gravels containing erratics – Nightingale Member.

A chalk clast from the Kenn gravels (bed 1) yielded foraminifera identified by Dr A. Bahafzallah as *Valvulinaria californica* Cushman, *Gyroidina umbilicata* d'Orbigny and *Atoxoophriagmium subsphaerica* (Marie) (Gilbertson, pers. comm., 1995). These taxa are only present onshore in the British Isles in the highest units of the Chalk of Northern Ireland.

The Yew Tree Formation (bed 2) contains a diverse mollusc assemblage, with over 40 taxa recorded (Gilbertson and Hawkins, 1978a; Table 10.2). The fauna is dominated by *V. piscinalis* and *B. tentaculata* with some *L. peregra*, *G. laevis*, *B. marginata*, *A. crista* and *Pisidium* spp. Other taxa are comparatively rare. Estuarine foraminifera are present in the deposit (Gilbertson and Hawkins, 1978a). The pollen assemblage (Hunt, 1981) contains abundant tree pollen, including *Quercus*, *Alnus*, coryloid, *Pinus*, *Betula*, and *Tilia*, together with pollen of herbs, marsh plants and aquatics, and cryptogam spores. The algal

microfossil assemblage (Hunt, 1981) is species-poor and dominated by the marine dinoflagellate cyst *Achomosphaera andalousiense* Jan du Chene (= *Spiniferites septentrionalis*) with smaller numbers of *Operculodinium centrocarpum* (Deflandre and Cookson). Molluscs from the interglacial deposit have yielded amino-acid ratios of 0.378 (Andrews *et al.*, 1984; Bowen *et al.*, 1989).

Interpretation

The stratigraphy and palaeobiology of the site have been interpreted by Gilbertson and Hawkins (1978a) and Hunt (1981). Andrews *et al.* (1984), Bowen *et al.* (1989) and Campbell *et al.* (in prep.) have reassessed the stratigraphical significance of the site.

The grey silts (bed 5) at the top of the section were laid down in estuarine conditions in the later Holocene (cf. Butler, 1987). The red silty sands (bed 4) underlying the estuarine silts were interpreted by Gilbertson and Hawkins (1978a) as coversands of aeolian origin. The underlying gravels (bed 3) were regarded by them as cold-climate fluvial deposits. They demonstrate the occurrence of cold-stage fluvial activity at the site after the Yew Tree interglacial.

In the Yew Tree Formation (bed 2), several of the mollusc taxa, including *B. marginata*, *C. fluminalis*, *Anisus vorticulus* (Troschel), *Hippentis complanata* (Linné) and *A. lacustris*, require interglacial conditions, with July temperatures perhaps 2°C warmer than today. The interglacial deposit was regarded as of fluvial origin by Gilbertson and Hawkins (1978a), since most of the Mollusca are freshwater taxa, with the most abundant environmental group being taxa typical of moving water such as *B. tentaculata* and *V. piscinalis*. They recorded possible traces of salinity stress in the molluscan assemblages, but no characteristically estuarine or marine species.

From the same bed, Hunt (1981) described palynological evidence for an interglacial climate, with abundant pollen of broad-leaved trees (29%, including 19% *Quercus*), and some marine influence, indicated by the abundant pollen of Chenopodiaceae and *Plantago maritima* and the presence of marine dinoflagellate cysts. The interglacial deposit may have been laid down in a coastal lagoon or in the backwaters of an estuary. The amino-acid ratio is suggestive of Oxygen Isotope Stage 15 and an age of 400 to 600 ka BP

(Andrews *et al.*, 1984; Bowen *et al.*, 1989), although Bowen (pers. comm., 1995) urges caution in interpreting ratios derived from *Corbicula*.

The presence of erratics such as the chalk clast, a non-durable lithology most probably derived from Northern Ireland, is strongly suggestive of a Celtic Sea glacial origin for the Kenn gravels (bed 1). These were regarded as glacial outwash deposits of a sandur plain by Gilbertson and Hawkins (1978a) and this interpretation was accepted by Campbell *et al.* (in prep.).

Conclusion

Yew Tree Farm is an important site for the Pleistocene palaeobiology and stratigraphy of the British Isles. The deposits here preserve a temperate-stage river channel-fill and a complex of cold-stage sediments. In the temperate-stage deposits, molluscs, pollen, dinoflagellate cysts and foraminifera suggest interglacial conditions and a depositional environment at the margins of marine influence. A number of amino-acid ratios have been determined from molluscs in these deposits. These suggest correlation with Oxygen Isotope Stage 15. The fossil remains from the channel-fill at Yew Tree Farm are diverse and abundant and offer great potential for future investigation, since the palaeobiology of the temperate stage which they reflect is as yet very poorly known. The interglacial deposits at Yew Tree Farm and its correlative site at Kennpier, rest on gravels suggested by Gilbertson and Hawkins (1978a) to be of glaciofluvial origin. If, as seems very probable, these gravels are glacigenic, then they must represent an extremely ancient glacial episode, which is very poorly known elsewhere in the British Isles.

KENN CHURCH
C. O. Hunt

Highlights

Kenn Church is an excellent representative of the later transgressive marine deposits in the Kenn area. It contains fossiliferous sands overlying glaciofluvial gravels, and shows a transition from brackish to fully marine conditions with the transgression reaching 14–21 m OD. It is proposed as the type-site of the Kenn Church Member.

Table 10.2 Fossil molluscs from the interglacial deposit at Yew Tree Farm (after Gilbertson and Hawkins, 1978a)

Species	Number
Valvata cristata Müller	12
Valvata piscinalis (Müller)	3324
Belgrandia marginata (Michaud)	970
Bithynia tentaculata (Linné) shells	1039
Bithynia tentaculata (Linné) opercula	1755
Carychium minimum Müller	2
Lymnaea truncatula (Müller)	192
Lymnaea palustris (Müller)	1
Lymnaea peregra (Müller)	886
Planorbis planorbis (Linné)	25
Anisus vorticulus Troschel	147
Anisus leucostoma Müller	141
Gyraulus laevis Alder	1613
Armiger crista (Linné)	615
Planorbis spp.	3
Hippentis complanata (Linné)	21
Acroloxus lacustris (Linné)	8
Oxyloma cf. *pfeifferi* Rossmässler	6
Cochlicopa lubrica (Müller)	2
Pupilla muscorum (Linné)	2
Vallonia costata (Müller)	4
Vallonia pulchella (Müller)	6
Vallonia spp.	1
Cepaea nemoralis (Linné)	1
Trichia hispida (Linné)	3
Punctum pygmaeum Draparnaud	1
Zonitoides nitidus (Müller)	4
Agrolimax cf. *agrestis* (Linné)	6
Agrolimax spp.	42
Sphaerium corneum (Linné)	1
Corbicula fluminalis (Müller)	72
Pisidium amnicum (Müller)	44
Pisidium casertanum (Poli)	7
Pisidium obtusale (Lamarck)	10
Pisidium milium Held	56
Pisidium subtruncatum Malm	138
Pisidium henslowanum (Sheppard)	24
Pisidium nitidum Jenyns	270
Pisidium pulchellum Jenyns	2
Pisidium moitessierianum Paladilhe	3
Pisidium spp.	190
Total	11 649

Introduction

At Kenn Church, interglacial estuarine deposits occupy a channel incised into the glacigenic Kenn gravels. The sequence is overlain by aeolian coversands.

Pleistocene gravels have been known in the Kenn area since the work of Ussher (*in* Woodward,

1876), who described gravels at Kenn and Kennpier and sandy soil over gravels at Yatton. They also noted that ' … small pebbles and large subangular and angular pieces of Carboniferous Limestone, and a few of sandstone, occur in greyish-brown soil … ' near Kenn (Woodward, 1876; p. 154). Greenly (1921) described poorly sorted sediments from Yatton and wrote ' … the formation recalls true boulder clays, but the extreme rarity of striated stones, the feebleness of the striations, and the almost total absence of erratics, forbid us to regard it as such.' (Greenly, 1921; p. 147).

Five feet of sand and gravel with pockets of coarse quartz sand containing *M. balthica* were reported from a degraded pit at St John's Church, Kenn (Welch, 1955). Welch described further gravels with *Macoma* elsewhere in the neighbourhood of Kenn. The gravel lithologies included flint, Greensand chert, quartz and Jurassic rocks. These deposits were equated with the Burtle Beds of King's Sedgemoor, a conclusion endorsed by ApSimon and Donovan (1956) and Kidson (1970). These latter authors also correlated the Kenn gravels with the marine deposits at Weston-in-Gordano and favoured an Ipswichian age for the marine incursion. Tills with striated boulders and coarse gravels overlain by marine and freshwater sands and gravels were briefly described by Hawkins and Kellaway (1971).

Gilbertson (1974) and Gilbertson and Hawkins (1978a) described the stratigraphy of the deposits at Kenn Church in a detailed survey of Pleistocene deposits in the Kenn area. These authors described coversands overlying interglacial deposits which in turn rested on coarse, unfossiliferous cold-stage gravels. Molluscan studies showed an initial brackish-water environment, with marine influence becoming stronger upwards. Amino-acid ratios were determined from a variety of fossil molluscs from the interglacial deposit at Kenn Church by Andrews *et al.* (1984). Most ratios were around 0.2, and were interpreted by these authors as indicating an Ipswichian age. The site was recently proposed as the type-locality of the Kenn Church Member by Campbell *et al.* (in prep.), who suggested assignment of the unit to Oxygen Isotope Stage 7.

Description

The following description is taken from Gilbertson (1974) and Gilbertson and Hawkins (1978a) and the fossil mollusc fauna is listed in Table 10.3. The Pleistocene deposits at Kenn form a low 'island' amidst the Holocene alluvium of the Avon Levels, rising to around 8.2 m OD (Gilbertson, 1974). They overlie Triassic mudrocks of the Mercia Mudstone Formation and are in places over 6 m thick. At Kenn Church (ST 41596890), a channel containing interglacial estuarine deposits is incised into the Kenn gravels. The channel appears to follow a slight rise to the south of the village to ST 412686 where shelly gravel with abundant *M. balthica* was found in 1969 (Gilbertson, 1974). The sequence can be summarized as follows (maximum bed thicknesses in parentheses).

9. Tarmac – made ground. (0.15 m)
8. Pale grey-brown cobbly sand – made ground. (0.38 m)
7. Pale red sand with cobbles – Brean Member. (0.21 m)
6. Dark grey sand – Kenn Church Member. (0.01 m)
5. Reddish-brown shelly sands with occasional well-rounded pebbles – Kenn Church Member. This bed contains a 'raft' of pebbly reddish clayey silt. (0.62 m)
4. Yellow shelly sand – Kenn Church Member. (0.69 m)
3. Yellow fine shelly sand, coarsening upwards – Kenn Church Member. (0.45 m)
2. Pale brown shelly sand – Kenn Church Member. (0.08 m)
1. Coarse, poorly sorted cobbly gravels, base unseen – Nightingale Member. (> 0.15 m)

Amino-acid racemization assays were carried out on a variety of shells from Kenn Church by Andrews *et al.* (1984). Assays on *Macoma* gave ratios of 0.197 ± 0.02? and 0.2 ± 0.02?, on *Corbicula* 0.21 ± 0.03, on *Patella* 0.104 ± 0.005 and on *Littorina* 0.215 ± 0.02.

Interpretation

The stratigraphy and palaeobiology of the site were first interpreted by Gilbertson (1974) and Gilbertson and Hawkins (1978a). Andrews *et al.* (1984) have reassessed the stratigraphical significance of the site in view of their aminostratigraphic results and this is reviewed here, in the light of further research.

The basal Kenn gravels (bed 1) were regarded as sandur deposits by Gilbertson (1974) and

Table 10.3 Molluscs from the interglacial deposit at Kenn Church (after Gilbertson, 1974; Gilbertson and Hawkins, 1978a)

Sample	A	B	C	D	E	G	J	O
Sample depth (m)	2.7	2.2	1.7	1.5	1.2	0.8	unstratified	
Marine/estuarine taxa								
Patella vulgata Linné						2		
Gibbula sp.							1	
Littorina littorea (Linné)	1						2	
Littorina saxatilis (Olivi)	2						1	
Littorina littoralis (Linné)	1				1	1		
Littorina sp.	2f	f	2f	1f		f	f	f
Nucella lapillus (Linné)							1	2
Ocenebra erinacea (Linné)						1	1	
Buccinum undatum (Linné)							1	3
Nassarius reticulatus (Linné)							1	
Cerastoderma spp.	3f	8f	1f	f		f	f	1f
Macoma balthica (Linné)	7f	16f	23f	2f		6f	90f	50f
Brackish-water taxa								
Hydrobia ventrosa Montagu	163	125	16	2		1		
Hydrobia ulvae (Pennant)	103	75	19	4		2	1	
Freshwater taxa								
Valvata piscinalis (Müller)	6	2						
Belgrandia marginata (Michaud)	1							
Bithynia tentaculata (Linné)	3							
Lymnaea peregra (Müller)	20	14	5					
Planorbis planorbis (Linné)	1	2						
Anisus vorticulus Troschel	2							
Gyraulus laevis (Alder)	11	15	1					
Corbicula fluminalis (Müller)						2		
Pisidium subtruncatum Malm	1							
Pisidium nitidum Jenyns	1							
Pisidium moitessierianum Paladilhe	1							
Pisidium spp.	2	2						
Terrestrial taxa								
Vallonia pulchella (Müller)	1							
Vallonia enniensis (Gredler)		1						
Trichia striolata Pfeiffer	1							
Helicella virgata (Da Costa)						1		
Discus rotundatus (Müller)						1		

Gilbertson and Hawkins (1978a). The sands of the interglacial Kenn Church Member (beds 2–6) contain fossil mollusc assemblages which enable detailed palaeoenvironmental reconstruction (Gilbertson, 1974; Gilbertson and Hawkins, 1978a). A fully interglacial but rather continental environment, with July temperatures perhaps 2°C warmer than present, is suggested by the presence of the thermophilous *C. fluminalis, O. erinacea, B. marginata, Vallonia enniensis* (Gredler) and *A. vorticulus* (Gilbertson, 1974). The basal sands of the Kenn Church Member contain abundant brack- ish-water taxa, some marine and some freshwater species, probably reflecting a brackish-water environment with input from a clear freshwater stream. The freshwater taxa decrease rapidly upwards through the deposits, and *H. ulvae* becomes increasingly important at the expense of the less salt-tolerant *H. ventrosa* before both decline rapidly as marine taxa become dominant. Gilbertson (1974) computed a maximum mean sea level of 14–21 m OD for the height of the transgression.

The red cobbly sands of the Brean Member (bed 7), overlying the interglacial deposit, are most likely the result of cold-climate sedimentation, probably having formed as niveo-aeolian coversands with an admixture of cobbles introduced by cryoturbation and solifluction (Gilbertson, 1974; Gilbertson and Hawkins, 1978a).

The aminostratigraphic data, with most ratios of *c.* 0.2, suggest comparison with Oxygen Isotope Stage 7 or older. The initial correlation of these sites with the Ipswichian interglacial is unlikely, as Ipswichian sites are characterized by ratios of about 0.1. (Bowen *et al.*, 1989; Campbell *et al.*, in prep.). Comparison with the Group 4 ratios of Mottershead *et al.* (1987) suggests an age of around 200 ka BP for deposits characterized by amino-acid ratios of *c.* 0.2. The presence of *Corbicula* also points to a pre-Stage 5 age, since Keen (1990) and Bridgland (1994) have argued that this species is not present in Britain after Stage 7.

Conclusion

Kenn Church GCR site is important as a representative of the later interglacial marine transgressive deposits in the Kenn area. Detailed studies of its molluscan fauna have showed the progression of a marine transgression to 14–21 m OD in a warm continental climate, probably around 200 ka BP. The Kenn Church interglacial deposits occupy a channel incised into glaciofluvial gravels and are overlain by niveo-aeolian coversands.

WESTON-IN-GORDANO
C. O. Hunt

Highlights

Weston-in-Gordano is important because it contains a complex of fossiliferous marine and non-marine interglacial deposits which post-date, and thus offer a minimum age for, the glaciation of the valley. The site is of considerable importance for reconstructing ancient sea levels because it contains evidence for as many as three interglacial marine transgressions. The site is the type-locality of the Weston Member.

Introduction

Weston-in-Gordano contains slope deposits overlying a complex of marine and freshwater interglacial deposits, interbedded with stony clays of possible glacial origin and lying against the steep face of the limestone massif of Portishead Down. The site lies almost immediately downslope of the glacigenic sediments of Nightingale Valley (this chapter).

The site was found by ApSimon and Donovan (1956), who recorded marine gravels and sands with *M. balthica*, overlain by stony clays, further marine gravels with *Macoma*, then subtidal sands, all overlain by cryoturbated sandy breccias of sub-aerial origin (Figure 10.8). The highest marine deposit was at 13.6 m OD. The gravels contained fragments of Old Red Sandstone, Carboniferous Limestone, Triassic breccias, flint and Greensand chert.

Spoil from the 1982 excavations for drains at a new police headquarters, yielded fine gravel and grey silts with the interglacial freshwater mollusc *C. fluminalis* (Hunt, in prep.). In 1992, two auger holes were drilled to relocate the freshwater interglacial deposits and establish their relationship with the published stratigraphy (Hunt, in prep.). Colluvial breccias and silts were found to overlie laminated silts, sands and gravel of probable intertidal origin. These in turn overlay laminated sands with freshwater molluscs, most probably the lateral equivalents of the silts with *Corbicula*, and then gravels with occasional marine shells similar to those described by ApSimon and Donovan (1956). The site was proposed as the type-locality of the Weston Member by Campbell *et al.* (in prep.), who assigned its marine and freshwater interglacial deposits to Oxygen Isotope Stage 7 and earlier.

Description

At Woodside, near Weston-in-Gordano (ST 456754), complex Quaternary deposits underlie a gentle slope beneath the steep Carboniferous Limestone of Portishead Down and above the Holocene alluvium and peats of Clapton Moor. The stratigraphy of the site is most complete on the north-west side of the B3124 and is taken here from the work of ApSimon and Donovan (1956) and Hunt (in prep.). Junctions between beds are sharp unless otherwise stated and maximum bed thicknesses are given in parentheses (Figure 10.8).

12. Reddish-brown sandy clayey silt with occasional Carboniferous Limestone fragments. This bed has a transitional junction with bed 11. (0.6 m)
11. Reddish-brown breccia of angular

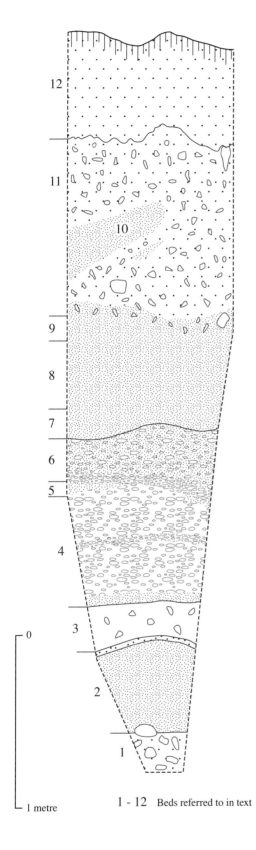

1 - 12   Beds referred to in text

Figure 10.8 The Quaternary sequence at Weston-in-Gordano. (Adapted from ApSimon and Donovan, 1956.)

Carboniferous Limestone, dark red sandstone and green-grey and pink marl fragments in a matrix of sandy clayey silt. (0.25 m)

10. Dark red, slightly silty coarse sand. (0.05 m)

9. Dark reddish-brown, mottled black, laminated clayey silts and occasional sands, passing gradually into bed 8. (0.15 m)

8. Yellow-brown clayey silts with occasional manganese nodules at the top, passing gradually into bed 7. (0.95 m)

7. Yellow-brown laminated coarse silt and very fine sand. (0.4 m)

6. Yellow-brown fine gravel, rich in flint. (0.05 m)

5. Reddish-grey fine silty sand with *L. peregra*, *Trichia* cf. *hispida* and freshwater mollusc fragments. (0.10 m)

4. Strong brown coarse gravel with a matrix of very coarse sand. The gravel is point-contact cemented with calcite. Clasts include flint and Greensand chert, Jurassic micrite, oolitic limestone, ironstone, oysters and belemnites, Carboniferous Limestone, coal, Coal Measures and Old Red sandstones. The bed contains occasional shells of *M. balthica* and *Littorina* sp. (0.95 m)

3. Red clay with a few pebbles. (0.2 m)

2. Grey, mottled red and black coarse sand, patchily cemented at the top. (0.5 m)

1. Sands and gravels with angular fragments of Carboniferous Limestone. Clasts include flint, Old Red Sandstone and Triassic rocks. Occasional shells of *M. balthica* are present. (0.2 m)

Palynological analysis of a coal clast from bed 4 yielded a small, very weathered assemblage including species of *Lycospora*, *Florinites*, *Triquitrites*, *Crassispora*, *Densosporites* and *Dictyotriletes*. These indicate a Westphalian age. A very small mud lens in this bed yielded single specimens of the marine dinoflagellate cysts *A. andalousiense* and *Bitectatodinium tepikense*, a foram cast and grains of Cyperaceae and Poaceae. Analyses of samples from beds 5 and 7 were largely unsuccessful, only two grains of Poaceae and one Filicales spore being found in bed 7.

Interpretation

The basal gravels (bed 1) contain a variety of erratics and thus are likely to post-date a glaciation of the Avon coastlands. The presence of *M. balthica* indicates that this is a marine deposit. ApSimon and Donovan (1956) suggest that this is an 'upper' beach deposit and note that it reaches 11.2 m OD, consistent with a mean sea level of 4.5–6.0 m OD. The overlying sands (bed 2) probably also have a marine origin, having been laid down below High Water of Mean Ordinary Tides (ApSimon and Donovan, 1956). It reaches 11.77 m OD and may indicate a sea level of over 8 m OD.

The red clay (bed 3) was suggested by ApSimon and Donovan (1956) to be derived by weathering from local Triassic rocks, but the presence of flint in this bed is inconsistent with such an origin. It may be a very weathered slope deposit containing erratics from the underlying marine gravel, or it could be till, or partially derived from till, since it is apparently similar to the tills on Portishead Down, though whether it lies *in situ* is not established. Because erratics are present in the underlying (probably marine) deposits, two hypotheses are tenable. First, there may have been two glaciations in the Vale of Gordano, separated by an interglacial marine incursion. Second, and perhaps more likely, there was one glaciation, pre-dating the lower marine deposits, and this bed was soliflucted to its present position after sea level fell following deposition of the basal gravel and sand.

The roundness of the gravels overlying the red clay (bed 4) and the presence of *M. balthica* and *Littorina* sp. and perhaps *B. undatum* (ApSimon and Donovan, 1956) and marine dinoflagellate cysts is consistent with a shallow-marine origin, possibly as upper beach deposits. This gravel reached 12.6 m OD (ApSimon and Donovan, 1956), consistent with a mean sea level of 6.0–7.5 m OD. The presence of a substantial component of Mesozoic rocks in this gravel most probably implies an input of erratic material, possibly from the erratic-rich drifts on nearby Portishead Down (Nightingale Valley; this chapter).

Overlying what is probably the eroded surface of bed 4 are sands and silts with freshwater shells (bed 5). This unit is the probable source of the *Corbicula* shells seen in the temporary excavation. *C. fluminalis* is regarded as a species of clean running water, while *L. peregra* is a generalist species. *T. hispida* is typically found in damp herbaceous vegetation. *Corbicula* has not been found in sites of demonstrable last interglacial age (Bridgland, 1994), and it is therefore likely that this horizon dates from Oxygen Isotope Stage 7 or earlier (Campbell *et al.*, in prep.).

Overlying these sediments are sandy silts (beds 7–9) which were most probably laid down on inter-tidal flats. The deposits are sufficiently weathered and decalcified to be barren of calcareous fossils and virtually barren of palynomorphs. These deposits most probably reflect a third marine transgression. They reach 13.6 m OD and ApSimon and Donovan (1956) suggest that this is evidence for a mean sea level not lower than 14.0 m OD. This is higher than has been suggested for the Oxygen Isotope Stage 5e transgression recorded in the Burtle Formation (Kidson *et al.*, 1978; Chapter 9) and for Stage 7 sites in Somerset (Hunt and Bowen, in prep.; Chapter 9) or elsewhere in South-West England (Mottershead *et al.*, 1987; Chapter 6). It may be more compatible with the sea levels suggested in Stage 9 (reviewed in Jones and Keen, 1993).

Although an interlude of freshwater sedimentation separates marine bed 4 from marine beds 7–9, there is no evidence for climatic deterioration, such as slope deposits or cryoturbation structures between bed 4 and bed 9. It is possible, therefore, that the deposits of beds 4–9 reflect only one, complex temperate period, containing evidence of two marine transgressions. It is also possible that any evidence for an intervening cold stage has been eliminated by erosion.

Finally, these deposits are overlain by terrestrial breccias and silts (beds 11 and 12). These are similar to deposits at Holly Lane, Clevedon and Brean Down which were laid down during periglacial episodes (Gilbertson and Hawkins, 1974; ApSimon *et al.*, 1961; Chapter 9 and this chapter).

Conclusion

This site is of importance since it provides the possibility of establishing a minimum age for the glaciation of the Avon coastlands. The erratic content in the basal gravels provides strong evidence that these post-date a glacial incursion into the area. The red clay overlying these deposits might be till, but further research is needed to confirm this and to establish whether it is *in situ*. Also extremely significant is the presence of sediments relating to what were most probably three interglacial marine transgressions. Further work on this site is clearly necessary.

(B) RIVER TERRACES AND LANDSCAPE DEVELOPMENT

This section describes a series of sites chosen to conserve representative examples of the terrace gravels of the River Avon and of later landscape development in Avon. The Avon terrace gravels were classified as the Avon Formation by Campbell *et al.* (in prep.), and, as has been recognized since the work of Davies and Fry (1929) and Palmer (1931), it can be divided on morphostratigraphic grounds into three distinct aggradations. The Bathampton Member lies approximately 3 m above the modern floodplain, with the Stidham and Ham Green members lying about 15 m and 30 m, respectively, above it. All of these units comprise predominantly trough cross-bedded gravels, and all contain substantial quantities of erratics. Many terrace gravel sites have been known since the last century (Weston, 1850; Dawkins, 1865; Moore, 1870) but few survive to the present day and still fewer have been the subject of modern investigations. Many sites recorded in the literature as 'richly fossiliferous' have been lost under encroaching urban development, for instance the cluster of sites at Twerton.

The stratigraphy and age relationships of the Avon gravels are particularly important because in the valley of the Avon are preserved remains of an extensive pre-Anglian glaciation. The age of this glaciation and its limits have long been subjects for scientific controversy (Kellaway, 1971; Kellaway *et al.*, 1975; Kidson and Bowen, 1976; Gilbertson and Hawkins, 1978a; Andrews *et al.*, 1984; Jones and Keen, 1993), and establishing the chronology of the Avon gravels offers one possible route to determining a timescale for this event.

HAM GREEN
C. O. Hunt

Highlights

Ham Green has importance as a representative of the 100' terrace of the Avon. The terraces of the Avon are of critical importance because of their relationship with the ancient glacial deposits of the Bath and Bristol areas. The site has easily accessible examples of cold-stage fluvial sedimentation and is the type-locality for the Ham Green Member.

Introduction

At Ham Green, a clear terrace surface at 30–31 m OD at the mouth of the Avon Gorge is underlain by fine, massively bedded and imbricated sandy gravels containing a significant proportion of erratics. This site was selected for the GCR as representative of the 'high terrace' of the Bristol Avon.

The site was first described by Davies and Fry (1929) as part of their 100' terrace. They described a 'considerable tract' of gravel, mostly of lithologies derived from the Jurassic, but with the surface gravels rich in Greensand chert and quartzite pebbles. They recorded 'several feet' of gravel in temporary roadside excavations at Bristol Road, Ham Green, and in the nearby railway cutting. Hawkins and Tratman (1977) suggested that the terraces of the Avon west of Bristol may be degraded estuarine terraces or relic glacigenic deposits. The site was re-investigated in 1984 during the compilation of the Geological Conservation Review. Massively bedded gravels were found which provide evidence for cold-stage sedimentation by a precursor of the Bristol Avon. Like other localities in the 'high' terrace of the Avon, a substantial erratic content was noted. The site was proposed as the type-locality of the Ham Green Member by Campbell *et al.* (in prep.). They tentatively attributed the deposits to some part of Oxygen Isotope Stages 10–12, on geomorphological grounds.

Description

At Ham Green, a broad terrace surface lies about 30–31 m above the Avon, to the south and east of the Hospital. Up to 1.2 m of gravels are exposed in the railway cutting near ST 539768, at 31 m OD. The sequence can be summarized as follows (maximum bed thicknesses in parentheses).

4. Dark brown stony loam soil. (0.2 m)
3. Strong brown, massively bedded, stony silty clay, with some large clasts orientated with their A-axes vertical. Clast lithologies include flint, Carboniferous and Greensand chert and quartzites. (0.6 m)
2. Strong brown, massively bedded, clast-supported, imbricated silty gravel. The gravel clasts mostly have B-axes in the size range 10–15 mm. They are predominantly of brown (?Carboniferous) sandstones (24%), Jurassic limestones (oolitic and micritic) and flint (39%), with rarer Carboniferous and Greensand chert (9%), quartzite (5.5%), Carboniferous Limestone (10%), and other lithologies including Triassic mudstone, siltstones, sandstones and vein materials. The quartzite clasts are rounded, the flint and

chert clasts predominantly angular and the other clast lithologies are subrounded to rounded. The clast imbrication direction suggests deposition by a current flowing from the south-east (146°). (0.3 m)

1. Strong brown, imbricated very coarse gravels. The clasts have B-axes up to 80 mm. The gravel rests on a gently undulating erosion surface cut in Triassic mudstones and marly limestones. (0.1 m)

During the compilation of the GCR, a temporary exposure was seen in a sewer trench outside the main entrance to Ham Green Hospital at ST 53307563. The trench exposed 0.4 m of made ground overlying 1.2 m of strong, brown to red-brown, massively bedded, matrix-supported, cobbly sandy gravel. The cobble-sized clasts were mostly of flint and white and grey quartzite, but smaller gravel clasts included Greensand chert, red and yellow sandstone and ironstone.

Interpretation

The Ham Green site lies on a prominent 'flat' in the landscape at the mouth of the Clifton Gorge of the River Avon. Underlying this 'flat' is a sequence of deposits which can be interpreted as follows.

Gilbertson and Hawkins (1978a) have described silty deposits similar to bed 2 from a number of sites in Avon. They regard these deposits as cover-sands of predominantly aeolian origin. The stones may have been incorporated into the bed from below by cryoturbation. The high clay content and the presence of only siliceous clasts in the bed are consistent with prolonged exposure to pedogenic processes.

In bed 3, the size, roundness, sorting and imbrication of the gravels are consistent with a fluvial origin. Palaeocurrent data indicate deposition by an ancient precursor of the modern Avon. The coarseness of the sediment and the lack of Quaternary fossils is perhaps suggestive of aggradation under 'cold-stage' conditions, as suggested for terrace gravels elsewhere in southern England (Briggs and Gilbertson, 1980). Bed 1 is most probably a basal lag deposit.

The geomorphology and sedimentology of the site are thus consistent with the deposits being an early fluvial gravel of the Avon, rather than of glacial or estuarine origin, as suggested by Hawkins and Tratman (1977). The altitude of the deposit points to its considerable antiquity, while the presence of rock types such as flint and Greensand chert, which have no outcrop in the catchment of the Avon, suggests that the site post-dates glaciation of the area.

In particular, the high altitude of the Ham Green Member points to its considerable antiquity. On altitudinal grounds, it must pre-date the Bathampton Member and Stidham Member in the Avon Valley. The Bathampton Member dates from Oxygen Isotope Stage 6 or earlier (this chapter) and the Stidham Member is thus likely to date from Stage 8 or earlier. Considering the significant downcutting separating the Stidham and Ham Green members, the latter is thus likely to date from Stage 10 or 12, or quite possibly from an earlier stage. Its maximum age is constrained by the age of the Kenn Formation glacial deposits, which pre-date Oxygen Isotope Stage 15 (this chapter).

Conclusion

An extensive area of the high (100') terrace of the Avon occurs at Ham Green, downstream from the Clifton Gorge. The site is one of a set which conserves the critical geomorphological and stratigraphical elements of the terraces of the Bath Avon. This terrace sequence is important because of its relationship with the ancient glacial deposits of the Bath and Bristol areas. The site provides easily accessible examples of cold-stage fluvial sedimentation.

NEWTON ST LOE
C. O. Hunt

Highlights

Newton St Loe provides the last surviving example of fossiliferous gravels among the low terraces of the Avon. It is a key site for establishing the stratigraphy and history of the Avon terraces and has wider stratigraphic significance in helping to establish a relative timescale for the ancient glaciation of the Avon coastlands and Bath area.

Introduction

The Avon Valley was famous in the nineteenth century for its fossiliferous Pleistocene gravels. Newton St Loe is the last surviving fossiliferous gravel site; all others downstream from Bath have been built over and lost. At Newton St Loe, thin trough cross-bedded gravels of the Bathampton Member have been exposed in road and railway cuttings.

Owen (1846), Dawkins (1866) and Moore (1870) recorded bones of mammoth and horse from thin gravels here. The site was also mentioned by Woodward (1876) and Winwood (1889), but no modern work was done until the site was revisited during the compilation of the GCR.

Description

A terrace surface approximately 10 m above the Avon can be seen at Newton St Loe. Thin decalcified gravels are exposed in shallow roadside exposures at ST 715664, near the northern edge of the deposit. North of the A4, at ST 71306555, the following section was recorded from a temporary exposure made by road workers (maximum bed thicknesses in parentheses).

3. Dark brown, pebbly loam soil. (0.2 m)
2. Strong brown, matrix-supported, silty clayey fine gravel, with many vertical stones. The clasts are all of insoluble lithologies, including Greensand chert, flint, Carboniferous chert, quartzite and brown sandstone. (0.6 m)
1. Strong brown, clast- and matrix-supported clayey gravel. The gravel is decalcified but shows some traces of trough cross-bedding. Only insoluble lithologies such as flint, Greensand and Carboniferous chert, quartzite, ironstone and brown and yellow sandstone are present. The matrix was probably once sandy, but is now very clay-rich. The bedding appears to be disrupted locally by involutions. The gravels rest on a gently undulating surface cut in Mercia Mudstones. (0.3 m)

Interpretation

The trough cross-bedding seen in bed 1 is consistent with the gravels having been laid down by a braided river, a fluvial style usually associated in the British Isles with cold-stage sedimentation (Briggs and Gilbertson, 1980). The gravels were subsequently weathered and decalcified, probably during episodes of temperate climate, and cryoturbated, most probably under stadial conditions. The altitudinal relationships of the gravels suggest attribution to the Bathampton Member, which

probably accumulated during Oxygen Isotope Stage 6 (this chapter). The fossil remains found at Newton St Loe during the last century must, at face value, reflect a relatively open but not stadial environment, probably one with herbaceous rather than forest vegetation. There is, however, a reasonable probability that the bones are derived and therefore do not reflect conditions during the aggradation of the gravels.

Conclusion

At Newton St Loe, terrace gravels of a low terrace of the river Avon are exposed in road and railway cuttings. The gravels yielded a restricted mammal fauna of elephant and horse in the nineteenth century. Such fossils might be taken as indicative of an open but not necessarily cold environment, perhaps characterized by a herbaceous flora. However, it is quite likely that the bones were derived from older deposits and that they do not reflect environmental conditions at the time of gravel deposition. The exposures show typical features associated with cold-stage fluvial deposition and have been deeply weathered. Newton St Loe forms a key element in a network of sites which can be used to reconstruct the protracted history of the Bath Avon.

STIDHAM FARM
C. O. Hunt

Highlights

Stidham Farm, Saltford, is important as a representative of the middle (50') terrace of the Avon. The site has good examples of cold-stage fluvial sedimentation and is the type-locality of the Stidham Member.

Introduction

At Stidham Farm, the 15 m terrace of the River Avon is underlain by trough cross-bedded gravels of the Stidham Member. The gravels contain abundant clast types not found in the Avon catchment or found only at outcrop downstream. They thus provide important evidence for a glacial advance into the Avon Valley from the west.

The site was first described by Moore (1870),

who noted that mammoth remains had been found there and that the gravels were rich in material derived from the Lias (the local bedrock) and the Coal Measures. It was re-described in some detail (as Steedham Farm) by Woodward (1876), who described gravels to the north-east of Stidham Farm and noted their wide range of clast lithologies. He observed that the gravels he described were one of two patches in the area around Stidham Farm. Winwood (1889) listed the site as containing mammoth remains.

Davies and Fry (1929) briefly described gravels from the other outlier, to the west of the farm. The site was mentioned by Palmer (1931), who assigned it to his 50' terrace, but gave no further details. It was revisited in 1984 during the compilation of the Geological Conservation Review and subsequently designated the type-locality of the Stidham Member by Campbell *et al.* (in prep.). The deposits were attributed, on geomorphological grounds, to Oxygen Isotope Stage 8.

Description

Survey of degraded sections at ST 674684, on the margins of the old gravel excavations described by Davies and Fry (1929), gives a sequence that is essentially similar to those described by Woodward (1876) and Davies and Fry (1929). A similar though thinner sequence was seen nearby in the edge of the old railway cutting (maximum bed thicknesses in parentheses).

3. Dark yellow-brown, stony clayey silt, disturbed in its upper part by ploughing and containing occasional sherds of nineteenth century pottery. The clasts include brown and red sandstone, flint, Greensand chert and quartzite. (0.3 m)
2. Yellow-brown, planar trough cross-bedded and massively bedded, clast-supported sandy gravel, with occasional scour-fills of coarse sand. The clasts are predominantly Jurassic limestones (micritic and oolitic), with some Carboniferous Limestone, brown and red sandstone, flint, Greensand chert and quartzite. Fossils derived from the Carboniferous Limestone, Lower Lias, Upper Lias and Greensand were reported by Davies and Fry (1929). A sample taken from one of the scour-fills yielded no fossil material. (1.2 m)

1. Yellow-brown, imbricated, very coarse, silty sandy gravel, with some cobbles and small boulders of micritic (Lias) limestone. The gravels lie on an undulating erosion surface cut in Lias mudstones. (0.3 m)

Interpretation

Bed 3 is most probably the result of soil formation on, and plough disturbance of, the underlying gravels. Any limestone pebbles present in this bed have probably been removed by weathering. Bed 2 has the hallmarks of braided sandy gravel, typically laid down in the British Pleistocene during stadial phases (Briggs and Gilbertson, 1980). Bed 1 is a 'lag' deposit, consisting of large clasts moved a minimal distance by the river during an erosional phase, when all smaller material was carried farther downstream. This bed is the probable source of the mammoth bones found during the nineteenth century. These may have been *in situ*, but in this context are very likely to have been recycled. The deposits at Stidham Farm, therefore make up a typical cold-stage fluvial aggradation. The virtual absence of fossil material from the site is consistent with this interpretation, although the facies present at the site would generally have been unsuitable for the deposition of molluscs (Briggs *et al.*, 1990), since the sediments are coarse-grained and were most probably laid down near the channel centre. Like most Avon Valley gravel localities, erratic lithologies such as flint, Greensand chert and Carboniferous limestones and sandstones are common. They provide important evidence for a previous glacial advance into the Avon catchment from the west.

Conclusion

Gravels of the Stidham Member, underlying the middle (15 m) terrace of the Avon, are preserved at Stidham Farm, Saltford. The site has importance as one of a set safeguarding representatives of the terrace stratigraphy of the Bath Avon. These terrace deposits have critical importance because of their relationship with the ancient glacial deposits of the Bath and Bristol areas. Deposits at the site provide impressive examples of cold-stage fluvial sedimentation and contain a wide variety of erratics.

HAMPTON ROCKS CUTTING
C. O. Hunt

Highlights

Hampton Rocks Cutting was one of the first Pleistocene sites described from Somerset and Avon. It provides fine exposures through the low terrace of the Bath Avon and is nationally important for its calcretes – soil carbonate induration features. It is the type-site for the Bathampton Member and Bathampton Palaeosol of the Avon Valley Formation.

Introduction

Hampton Rocks railway cutting is one of a series of sites chosen to conserve representative examples of the terrace gravels of the Bath Avon. Hampton Rocks Cutting GCR site (ST 778667) is important for a number of reasons. It was selected as the best-available area of river gravels from the lowest (and therefore probably youngest) river terrace of the Avon. It has historical importance as one of the first sites described from Somerset and Avon, and it demonstrates good examples of cold-stage fluvial and aeolian sedimentation as well as a cold-stage fossil mollusc assemblage. There are fragmentary mammal remains. More significant, however, are the calcretes, which are extremely rare in the British Quaternary record. At Hampton Rocks Cutting, these are especially well developed and well preserved.

The site was initially described by Weston (1850), who recorded horizontal beds of fine gravel and coarse sand, with clast lithologies including flint, Greensand, oolitic limestone, Carboniferous Limestone, Millstone Grit and Old Red Sandstone. The section was re-described by Woodward (1876), who noted 3 feet of reddish-brown clay overlying 'an irregular bed of sand and small gravel' about 6 feet thick. The site was revisited in 1984 in the course of the GCR and again in 1990, and the stratigraphy, sedimentology, clast lithology and palaeontology of the gravels re-described (Hunt, 1990b). The site was nominated as the type-locality of the Bathampton Member (gravels) and the Bathampton Palaeosol (calcreted soil) by Campbell *et al.* (in prep.). The gravels of the Bathampton Member were attributed to Oxygen Isotope Stage 6 and the Bathampton Palaeosol to Stage 5e.

Description

The Bathampton Member is a terrace deposit of the River Avon. The deposits lie at *c.* 30–35 m OD, some 3 m above the floodplain, and are thus part of Davies and Fry's (1929) 'Lower Terrace'. The description given below follows Hunt (1990b), but also includes new data (maximum bed thicknesses in parentheses).

6. Dark brown, silty sandy clay soil with occasional pebbles and extremely rare 'blue and white' (19th–20th century) potsherds and modern sheep bones. (0.2 m)

5. Strong brown, slightly clayey, fine sandy silt with occasional stones, some vertical. This bed has a sharp but somewhat involuted boundary with bed 4. (0.35 m)

4. Pale brown, imbricated, clast-supported, medium trough cross-bedded gravels. The bedding is disturbed at the top, probably by cryoturbation, and irregular masses up to 0.6 m across of calcrete (pedogenic calcium carbonate induration) have formed, often emphasizing and respecting the disturbed bedding and sometimes appearing to have the shape of the roots of substantial trees. The transition to the underlying bed occurs over a distance of 0.05 m. (0.55 m)

3. Pale brown, trough cross-bedded fine gravels and coarse sand, with channel-fills of pale brown, plane-bedded, sandy silty clays with occasional pebbles and extremely rare mollusc remains. (0.30 m)

2. Pale brown, mottled strong brown, trough cross-bedded sandy gravel and gravelly sand, with occasional openwork lenses, generally fining upwards. Trough cross-beds are 0.05–0.10 m deep, 2 to >3 m wide. Maximum clast size in the bed is 0.25 m diameter. (2.0 m)

1. Brown, mottled black, sandy fine gravel, fining upwards, and strongly manganese indurated. Contains occasional very large clasts, mostly Jurassic limestones, up to 0.35 m, at base of horizon and often partially embedded in the bedrock. This bed has a sharp but very irregular junction with Fuller's Earth bedrock. (0.10 m)

The gravels of beds 1 and 2 are relatively well sorted and show a strong imbrication. The clast orientation data are heavily dominated by eastward dips. The clasts comprise two main groups of materials. The most numerically important includes oolitic and micritic limestones, Jurassic fossils, phosphate, flint, beef, Greensand and Greensand chert. All these lithologies have parent outcrops upstream of Bathampton. The other lithologies either have outcrops downstream in the Bristol district (yellow and red sandstones of the Carboniferous and Devonian or Permo-Triassic, Carboniferous Limestone, Bunter Quartzite, and possibly the igneous rocks) or farther afield (metamorphosed siltstones and mudrock probably originating from the Midlands or Wales). Others, such as limonite, are untraceable.

The palaeochannel-fills of bed 3 are laterally very discontinuous. They contain rare *P. muscorum*, *Succinea* cf. *oblonga*, *Trichia* cf. *hispida* and *Pisidium* sp. Fragments of elephant or mammoth tusk ivory and unidentifiable mammal bone were found evenly distributed in the gravels.

The upper parts of the gravel (bed 4) are somewhat involuted, with many vertical stones. Irregular masses of calcrete are present at the top of the gravels. These seem to have formed after the involutions had formed. These deposits are overlain by homogeneous sandy silts (bed 5), which in some places are disturbed by ploughing.

Interpretation

The stratigraphy can be interpreted in the following way. The trough cross-bedded gravels (bed 2) are the product of deposition by a braided (multichannel) stream. Such streams are typical of environments with little vegetation cover and a seasonal precipitation pattern, such as deserts and arctic-alpine areas (Briggs and Gilbertson, 1980). This environmental diagnosis is supported by the molluscan data. The few fossils recovered are typical of cold-stage mollusc assemblages in the British Isles (Kerney, 1976b; Jones and Keen, 1993). The molluscs present have well-known environmental tolerances at the present day and the environment at the time of deposition can therefore be deduced. *P. muscorum* prefers dry, exposed habitats, while *Succinea oblonga* (Draparnaud) inhabits wet muddy places. *T. hispida* is a generalist with a particular liking for wet grassy places. Members of the genus *Pisidium* are all aquatic. A dry landscape, with little vegetation except near watercourses is suggested. The climate was probably arid and rather cold. The

terraces of the Bath Avon are famous for their mammal remains, but the 'Low Terrace' is regarded as virtually devoid of them (Davies and Fry, 1929). There is a high probability that the durable fragments found during the GCR re-survey are recycled and therefore of little palaeoenvironmental significance. Their presence, however, raises the interesting possibility that other, larger fragments of greater significance may be preserved elsewhere in the deposit.

Clast orientations in the gravels are consistent with deposition by a westward-flowing river. The clast lithological data suggest the input of erratic material into the catchment, probably by the ice sheet which laid down the glacial deposits now found at Bath University and elsewhere in Avon (this chapter; Hawkins and Kellaway, 1971; Gilbertson, 1974).

The upper part of the deposit was disrupted by frost action, which led to the formation of involutions and to the stones being turned into an upright position (bed 4). Subsequently, the masses of calcrete formed, most probably around the roots of trees, judging by the morphology of the calcrete bodies. Calcrete does not form at the present day in the British Isles, but is forming in the Mediterranean basin, suggesting a warm, relatively arid climate when this calcrete formed. The last time it was sufficiently warm for calcrete formation in this country was during Oxygen Isotope Stage 5e. Following this episode, the climate deteriorated and wind-blown 'coversands' were laid down. These have been altered significantly by later pedogenic processes to give rise to bed 5.

Conclusion

Hampton Rocks Cutting is important for a number of reasons. It provides an excellent example of the low terrace of the Bath Avon and therefore has stratigraphical significance. It contains a good example of cold-stage river sedimentation, dating from the last part of the Middle Pleistocene. It contains non-local pebbles which can only have been brought into the area by glacial action and thus provides compelling evidence for a former glaciation of the Bath area. Its greatest importance is for its calcretes, extremely unusual soil carbonate induration features probably dating from the last interglacial, Oxygen Isotope Stage 5e.

HOLLY LANE
C. O. Hunt

Highlights

Holly Lane, Clevedon, provides important evidence for landscape evolution in Avon during the Devensian. The site is an important locality for Devensian mammal, bird and mollusc remains and is also notable for its periglacial breccias and aeolian deposits.

Introduction

At Holly Lane, up to 21 m of breccias, sands and silts are banked against an ancient cliff. The sequence consists of a basal talus cone, overlain by sandy loams, breccias, very coarse breccias, sandy loams, breccias, sandy loams and then a final thin breccia. Several beds in this sequence have yielded important mammalian fossils and sparse mollusc assemblages indicative of cold-stage conditions.

The breccias and sands at Holly Lane were first exposed by quarrying in 1905. Initially, interest centred around a small, richly fossiliferous cave, which, with the breccias immediately outside it, yielded over 2000 bones (Greenly, 1922). Davies (1907) demonstrated that the cave was unconnected to a karstic conduit system and unrelated to the structure of the Carboniferous Limestone bedrock. Two sedimentary horizons were identified in the cave: a 'dull red stony loam' about 0.45 m thick, overlain by 1.5 m of the lower breccias (Davies, 1907; Greenly, 1922). Outside the cave, Davies (1907) described a sequence of lower 'gravel' (breccia), clay, middle 'gravel', sand, and upper 'gravel'. The mammal fauna was identified by Hinton (1907a, 1907b, 1910) and Reynolds (1907) and included horse, bear, wolf, fox, ?Arctic fox, ?rabbit, voles and ?lemming. Bird fossils were also present. The fauna included 24 species, notably eagle, buzzard, heron, gull and cormorant. Four fish vertebrae and a mollusc fauna comprising *T. hispida* and *Succinea putris* (Linné) were found in the cave. *T. hispida*, *Hellicella (Cernuella) virgata* (Da Costa), *Helix aspersa* Müller, *P. muscorum* and *S. putris* were found in the breccias outside the cave (Greenly, 1922), though Kerney (1966), Gilbertson (1974) and Gilbertson and Hawkins (1974) have suggested that *Helix* and *Helicella* were contaminants. Gilbertson (1974) quotes a personal communication from W.A.E. Ussher in 1970, who recalled that a local doctor

had unsuccessfully tried to save human remains in the cave when excavation started.

Interest later turned to the breccias and sands at Holly Lane. Greenly (1922) provided a description of these deposits. A basal sandy breccia was overlain by a very coarse breccia. This was sharply overlain by a sandy loam, then a slightly stony loamy sand with abundant *T. hispida* and *P. muscorum* and then an upper sandy breccia. The beds dipped as much as 32° and were banked up against a steep, or even undercut, cliff in the Carboniferous Limestone. The maximum recorded thickness of the sequence was about 21 m. The clasts in the breccias were all derived from the local outcrop of Carboniferous Limestone, but mineralogical analyses of the sandy loam and loamy sand showed the presence of far-travelled minerals. The mineralogy of the sands was comparable with that of sands in South Wales. These were suggested, therefore, to be the parent deposits from which the sands at Holly Lane were derived by the action of wind.

The deposits were discussed again by Hinton (1926), Palmer and Hinton (1929), Palmer (1931, 1934), Kennard and Woodward (1934) and Vink (1949). Hinton (1926) dealt with the vole remains from the site, identifying five taxa. Palmer and Hinton (1929) described mineralogical analyses of the sandy matrix of the breccias and of the sandy loam and loamy sand which showed a common composition and thus probably origin. They described a fauna of voles, horse, wolf, bear and ?Arctic fox, most probably from the lower breccias, and a polished bone point of indeterminate type. Palmer (1931, 1934) integrated the site within a local stratigraphical scheme and suggested that the deposits overlay the '50 foot' shore platform. Palmer (1934) compared mineralogical analyses from Holly Lane with analyses from other sites in the Bristol district. The far-travelled minerals in these deposits were thought to have been derived from the south or south-west. Kennard and Woodward (1934) described mollusc faunas from the cave, lower breccia, 'aeolian sands', 'upper coarse gravel' and 'subsoil'. The faunas from the subsoil and the upper breccia and some components of the fauna from the lower breccia were regarded by Kerney (1966), Gilbertson (1974) and Gilbertson and Hawkins (1974) as intrusive. The probably *in situ* components of the fauna are, in the middle sands, *T. hispida* and *P. muscorum* and in the lower breccias *T. hispida*, *P. muscorum*, *V. costata* and *Lymnaea stagnalis* (Linné). Vink (1949) carried out sedimentological analyses of the sands at Holly Lane. They were comparable with

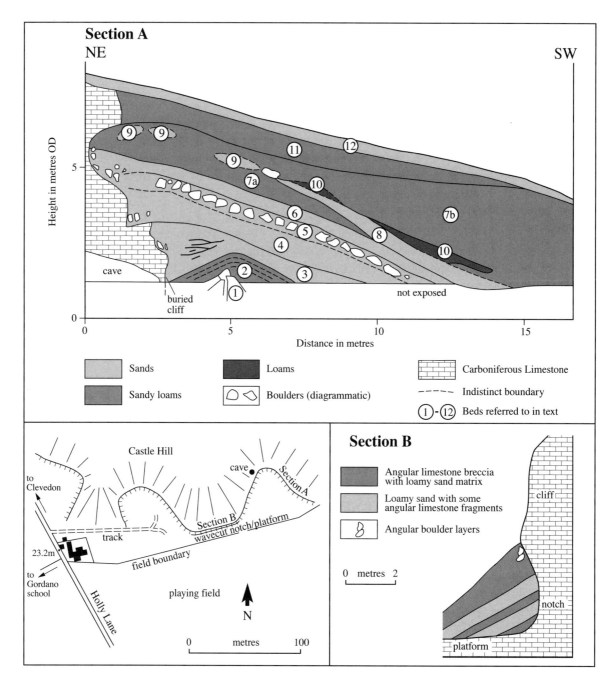

Figure 10.9 The Quaternary sequence at Holly Lane, Clevedon. (Adapted from Gilbertson and Hawkins, 1974.)

the periglacial coversands of the Low Countries.

Gilbertson (1974) and Gilbertson and Hawkins (1974) re-described the sections at Holly Lane and reviewed previous work on the site. They described two sections and recognized a complex stratigraphy, with a basal talus cone, overlain by sandy loams, breccias, very coarse breccias, sandy loams, breccias, sandy loams and then a final thin breccia. The middle sandy loam was equivalent to the middle aeolian sand bed of Greenly (1922), Palmer and Hinton

(1929) and Palmer (1931, 1934). They suggested that the breccias had accumulated during phases of frost-shattering and that the sandy loams were niveo-aeolian coversands, and they compared the sequence with stratified slope-waste deposits elsewhere. These authors reassessed early accounts of the terrestrial molluscs from the site. They obtained occasional specimens of *P. muscorum* from the middle and lower sandy loams and one specimen of *Succinea* sp. from the lower sandy loam.

Holly Lane

Description

At Holly Lane (ST 419727), a thick breccia and sandy loam sequence has been exposed by quarrying (Figure 10.9). The description by Gilbertson (1974) and Gilbertson and Hawkins (1974) is followed here (maximum bed thicknesses in parentheses).

12. Black topsoil developed on breccia, with angular blocks in a sandy matrix. The bed has a sharp boundary with bed 11. (0.5 m)

11. Reddish-brown sandy loam with a few small angular clasts. The bed has a sharp boundary with bed 10. (1.0 m)

10. Breccia containing many angular blocks in a reddish-brown loam matrix, with clasts up to 0.25 m. The deposit is poorly bedded and has an indistinct boundary with bed 9. (4.30 m)

9. Reddish-brown silty sand with occasional angular clasts. It occurs in discontinuous lenses and has an indistinct boundary with bed 8. (0.30 m)

8. Breccia containing angular boulders in a reddish-brown blocky sandy matrix. This occurs in discontinuous pockets and has an indistinct boundary with bed 7. (0.50 m)

7. Breccia containing angular blocks in a reddish-brown sandy loam matrix, and clasts up to 0.3 m. The bed thickens downslope and has a distinct boundary with bed 6. (0.20 m)

6. Breccia containing angular blocks and boulders in a reddish-brown sandy matrix, and clasts up to 0.4 m. Occasional lines of finer clasts are present and the deposit is poorly bedded and coarsens upward. It has a sharp boundary with bed 5. (1.50 m)

5. Reddish-brown loamy sand with occasional tabular clasts up to 0.04 m, lying parallel to the dip of the bed. It contains occasional *P. muscorum* and some foraminifera. It has a sharp boundary with bed 4. (0.40 m)

4. Breccia containing angular boulders and smaller clasts in a reddish-brown sandy matrix. It has an indistinct boundary with bed 3. (0.45 m)

3. Breccia containing angular blocks in a reddish-brown sandy matrix. Clasts average 0.02–0.04 m, and the deposit coarsens upwards. It contains occasional boulders and has a sharp boundary with bed 2. (1.10 m)

2. Reddish-brown, very silty loamy sand with occasional angular clasts up to 0.01 m, becoming more frequent upwards.

Occasional *P. muscorum*, very rare *Succinea* sp. and other molluscs, and some foraminifera are present. This deposit buries the dome of bed 1 and is banked against the limestone cliff at an angle of 30°. It has a sharp junction with bed 1. (> 2.0 m)

1. Breccia containing angular blocks in a red sandy matrix. Clasts average 0.25 m, and occasional boulders are present. It is poorly bedded and forms a low cone-shaped structure. It lies on a fissured rock surface, possibly bedrock. (1.1 m)

Interpretation

No palaeosols or weathering horizons have been found at Holly Lane, so it is probable that the whole sequence dates from the Devensian (Gilbertson and Hawkins, 1974). The alternating breccias and sandy loams at Holly Lane provide evidence for aeolian, perhaps niveo-aeolian, activity and intermittent frost-shattering. Gilbertson and Hawkins (1974) argued that the breccias reflected cold moist phases, while the sandy loams and loamy sands were laid down in cold arid periods.

The biological evidence in the lower beds is for open, relatively exposed landscapes, with molluscs such as *P. muscorum* and *T. hispida* typical of open, perhaps discontinuous herbaceous vegetation. The presence of moisture-loving taxa like *Succinea* in the lower loam is consistent with a locally very damp environment and conflicts with Gilbertson and Hawkins' (1974) suggestion that the sandy loams indicate aridity, unless the shells were brought to Holly Lane from marshy environments nearby. The mammal and bird faunas, which were probably recovered mostly from beds 1–3 and their lateral equivalents are arguably indicative of a cold steppe environment. The decrease upward in the diversity and incidence of faunal remains may perhaps be taken as evidence for a gradual climatic deterioration.

Conclusion

The deposits at Holly Lane, Clevedon, provide evidence for a complex sequence of environmental change and landscape evolution during the Devensian cold stage. They have yielded important mammal, bird and mollusc faunas characteristic of this period, together with sedimentary evidence for aeolian activity and intermittent frost-shattering.

References

Addison, K. (1981) The contribution of discontinuous rock-mass failure to glacier erosion. *Annals of Glaciology*, **2**, 3-10.

Addison, K. and Edge, M.J. (1992) Early Devensian interstadial and glacigenic sediments in Gwynedd, North Wales. *Geological Journal*, **27**, 181-90.

Ager, D.V. and Smith, W.E. (1965) *The coast of South Devon and Dorset between Branscombe and Burton Bradstock*. Geologists' Association Guide No. 23, 23 pp.

Aguirre, E. and Pasini, G. (1985) The Pliocene-Pleistocene boundary. *Episodes*, **8**, 116-20.

Aharon, P. (1983) 140,000-year isotope climatic record from raised coral reefs in New Guinea. *Nature*, **304**, 702-23.

Albers, G. (1930) Notes on the tors and the clitter of Dartmoor. *Report and Transactions of the Devonshire Association for the Advancement of Science, Literature and Art*, **62**, 373-8.

Alexander, J.J. (1933) Section for Archaeology and Geology. Reports of the Committee. *Transactions and Proceedings of the Torquay Natural History Society*, **6**, 259.

Alexander, J.J. (1934) Section for Archaeology and Geology. Reports of the Committee. *Transactions and Proceedings of the Torquay Natural History Society*, **6**, 350.

Alexander, J.J. (1935) Section for Archaeology and Geology. Reports of the Committee. *Transactions and Proceedings of the Torquay Natural History Society*, **7**, 73-4.

Allen, P., Benton, M.J., Black, G.P., Cleal, C.J., Evans, K.M., *et al.* (1989) The future of earth-science site conservation in Great Britain. *Geological Curator*, **5**, 101-9.

Ammann, B. and Lotter, A.F. (1989) Late-Glacial radiocarbon and palynostratigraphy on the Swiss plateau. *Boreas*, **18**, 109-26.

Amstutz, G.C. and Bernard, A.J. (1973) *Ores in sediments*. Springer-Verlag, Berlin.

Anderson, J.G.C. (1968) The concealed rock surface and overlying deposits of the Severn Valley and Estuary from Upton to Neath. *Proceedings of the South Wales Institute of Engineers*, **83**, 27-47.

Andrews, J.T., Bowen, D.Q. and Kidson, C. (1979) Amino acid ratios and the correlation of raised beach deposits in south-west England and Wales. *Nature*, **281**, 556-8.

Andrews, J.T., Gilbertson, D.D. and Hawkins, A.B. (1984) The Pleistocene succession of the Severn Estuary: a revised model based on amino-acid racemization studies. *Journal of the Geological Society of London*, **141**, 967-74.

Anon. (1948) The Joint Mitnor Cave. A preliminary note. *Transactions and Proceedings of the Torquay Natural History Society*, **10**, 1-2.

ApSimon, A.M. (1977) Brean Down sand cliff. In *Quaternary Research Association Field Handbook, Easter Meeting 1977, Bristol* (ed. K. Crabtree), Quaternary Research Association, Bristol, pp. 33-8.

ApSimon, A.M. and Donovan, D.T. (1956) Marine Pleistocene deposits in the Vale of Gordano, Somerset. *Proceedings of the University of Bristol Spelaeological Society*, **7**, 130-6.

ApSimon, A.M., Donovan, D.T. and Taylor, H. (1961) The stratigraphy and archaeology of the Late-Glacial and Post-Glacial deposits at Brean Down, Somerset. *Proceedings of the University of Bristol Spelaeological Society*, **9**, 67-136.

Arber, E.A.N. (1911) *The coastal scenery of north Devon*. J.M. Dent and Sons, London.

References

Arber, M.A. (1940) The coastal landslips of south-east Devon. *Proceedings of the Geologists' Association*, **51**, 257-71.

Arber, M.A. (1946) The valley system of Lyme Regis. *Proceedings of the Geologists' Association*, **57**, 8-15.

Arber, M.A. (1949) Cliff profiles of Devon and Cornwall. *Geographical Journal*, **114**, 191-7.

Arber, M.A. (1960) Pleistocene sea-levels in North Devon. *Proceedings of the Geologists' Association*, **71**, 169-76.

Arber, M.A. (1964) Erratic boulders within the Fremington Clay of North Devon. *Geological Magazine*, **101**, 282-3.

Arber, M.A. (1973) Landslips near Lyme Regis. *Proceedings of the Geologists' Association*, **84**, 121-33.

Arber, M.A. (1974) The cliffs of North Devon. *Proceedings of the Geologists' Association*, **85**, 147-57.

Arber, M.A. (1977) A brickfield yielding elephant remains at Barnstaple, North Devon. *Quaternary Newsletter*, **21**, 19-21.

Arkell, W.J. (1943) The Pleistocene rocks at Trebetherick Point, North Cornwall: their interpretation and correlation. *Proceedings of the Geologists' Association*, **54**, 141-70.

Arkell, W.J. (1945) Three Oxfordshire Palaeoliths and their significance for Pleistocene correlation. *Proceedings of the Prehistoric Society*, **11**, 20-32.

Arkell, W.J. (1947) *The Geology of the Country around Weymouth, Swanage, Corfe and Lulworth.* Memoir of the Geological Survey of Great Britain. H.M.S.O., London.

Atkinson, K. (1975) Observations on the basal hardpan of the St. Agnes Beds, Cornwall (Abstract). *Proceedings of the Ussher Society*, **3**, 288.

Atkinson, K. (1980) The St. Agnes Beds. In *West Cornwall* (ed. P.C. Sims), Quaternary Research Association Field Handbook, West Cornwall Field Meeting, September 1980, pp. 23-6.

Atkinson, K., Boulter, M.C., Freshney, E.C., Walsh, P.T. and Wilson, A.C. (1975) A revision of the geology the the St. Agnes outlier, Cornwall (Abstract). *Proceedings of the Ussher Society*, **3**, 286-7.

Atkinson, K., Eastwood, N.J. and Scott, P.R. (1974) The occurrence and composition of concretionery structures in the Tertiary sands of St. Agnes, Cornwall. *Journal of the Camborne School of Mines*, **74**, 38-45.

Austen, R.A.C. (1835) An account of the raised beach near Hope's Nose in Devonshire, and other recent disturbances in that neighbourhood. *Proceedings of the Geological Society*, **2**, 102-3.

Austen, R.A.C. (1851) On the superficial accumulations of the coasts of the English Channel, and the changes they indicate. *Proceedings of the Geological Society*, **7**, 118-36.

Avery, B.W. (1955) *The soils of the Glastonbury district of Somerset* (Sheet 296). Memoir of the Soil Survey of Great Britain. H.M.S.O., London.

Avery, B.W., Stephen, I., Brown, G. and Yaalon, D.H. (1959) The origin and development of brown earths on clay-with-flints and coombe deposits. *Journal of Soil Science*, **10**, 177-95.

Baden-Powell, D.F.W. (1927) On the present climatic equivalence of British raised beach Mollusca. *Geological Magazine*, **64**, 433-8.

Baden-Powell, D.F.W. (1930) Notes on raised beach Mollusca from the Isle of Portland. *Proceedings of the Malacological Society*, **19**, 67-76.

Baden-Powell, D.F.W. (1955) The correlation of Pliocene and Pleistocene marine beds of Britain and the Mediterranean. *Proceedings of the Geologists' Association*, **66**, 271-92.

Baker, A. (1993) Speleothem growth rate and palaeoclimate. Unpublished PhD thesis, University of Bristol.

Baker, A. and Proctor, C.J. (1996) The caves of Berry Head. In *The Quaternary of Devon and East Cornwall: Field Guide* (eds D.J. Charman, R. M. Newnham and D.G. Croot), Quaternary Research Association, London, pp. 147-62.

Bakker, J.P. (1967) Weathering of granites in different climates, particularly in Europe. In *L'évolution des versants* (ed. P. Macar), Université de Liège and Academie Royale de Belgique, pp. 52-68.

Balaam, N.D., Bell, M.G., David, A.E.U., *et al.* (1987) Prehistoric and Romano–British sites at Westward Ho!, Devon – archaeological and palaeo-environmental surveys 1983 and 1984, Palaeoenvironment and Palaeoeconomy in South-West England.

Balch, H.E. (1937) *Mendip: its Swallet Caves and Rock Shelters*. Clare, Wells.

Balch, H.E. (1947) *Mendip: Cheddar its Gorge and Caves*. (2nd edn) Simpkin Marshall, London.

Balch, H.E. (1948) *Mendip: its Swallet Caves and Rock Shelters*. (2nd edn) Simpkin Marshall, London.

References

Balchin, W.G.V. (1937) The erosion surfaces of north Cornwall. *Geographical Journal*, **90**, 52-63.

Balchin, W.G.V. (1946) The geomorphology of the North Cornish coast. *Transactions of the Royal Geological Society of Cornwall*, **17**, 317-44.

Balchin, W.G.V. (1952) The erosion surfaces of Exmoor and adjacent areas. *Geographical Journal*, **98**, 453-76.

Balchin, W.G.V. (1964) The denudation chronology of South-West England. In *Present Views of some Aspects of the Geology of Cornwall and Devon* (eds K.F.G. Hosking and G.J. Shrimpton) Blackford, Truro, pp. 267-81.

Balchin, W.G.V. (1981) *Concern for Geography*. A selection of the work of Professor W.G.V. Balchin. Department of Geography, University College of Swansea, 268 pp.

Ballantyne, C.K. (1986) Nonsorted patterned ground on the mountains in the Northern Highlands of Scotland. *Biuletyn Peryglacjalny*, **30**, 15-34.

Ballantyne, C.K. and Harris, C. (1994) *The Periglaciation of Great Britain.* Cambridge University Press, Cambridge, 330 pp.

Bard, E., Hamelin, B., Arnold, M., *et al.* (1996) Deglacial sea-level record from Tahiti corals and the timing of global meltwater discharge. *Nature*, **382**, 241-4.

Bard, E., Hamelin, B., Fairbanks, R.G. and Zindler, A. (1990) Calibration of the ^{14}C timescale over the past 30,000 years using mass spectrometric U-Th ages from Barbados. *Nature*, **345**, 405-10.

Barnola, J.M., Raynaud, D., Korotkevich, Y.S. and Lorius, C. (1987) Vostok ice core provides 160,000-year record of atmospheric CO_2. *Nature*, **329**, 405-14.

Barrow, G. (1904) On a striated boulder from the Scilly Isles. *Quarterly Journal of the Geological Society of London*, **60**, 106.

Barrow, G. (1906) *The Geology of the Isles of Scilly*. Memoir of the Geological Survey of Great Britain. H.M.S.O., London, pp. 15-31.

Barrow, G. (1908) The high-level platforms of Bodmin Moor and their relation to the deposits of stream-tin and wolfram. *Quarterly Journal of the Geological Society of London*, **64**, 384-400.

Barton, R.M. (1964) *An Introduction to the Geology of Cornwall*. Truro Bookshop, Truro.

Barton, R.N.E. (1996) The Late Glacial period and Upper Palaeolithic archaeology. In *The Quaternary of Devon and East Cornwall: Field Guide* (eds D.J. Charman, R.M. Newnham, and D.G. Croot), Quaternary Research Association, London, pp. 198-200.

Bate, C.S. (1866) An attempt to approximate the date of the flint flakes of Devon and Cornwall. *Report and Transactions of the Devonshire Association for the Advancement of Science*, **1**, 128-36.

Bate, C.S. (1871) On the clitter of the tors of Dartmoor. *Report and Transactions of the Devonshire Association for the Advancement of Science*, **4**, 517-19.

Battiau-Queney, Y. (1984) The pre-glacial evolution of Wales. *Earth Surface Processes and Landforms*, **9**, 229-52.

Battiau-Queney, Y. (1987) Tertiary inheritance in the present landscape of the British Isles. In *International Geomorphology 1986. Part II.* (ed. V. Gardiner), Proceedings of the First International Conference on Geomorphology, Wiley, Chichester, pp. 979-90.

Bayley, J. (1975) Pollen. In (ed. H. Miles) Excavations at Killibury hillfort, Egloshayle 1975-6. *Cornish Archaeology*, **16**, 89-121.

Baynes, J. and Dearman, W.R. (1978) The microfabric of a chemically weathered granite. *Bulletin of the International Association of Engineering Geology*, **18**, 91-100.

Becker, G.F. (1905) Simultaneous joints. *Proceedings of the Washington Academy of Science*, **7**, 267-75.

Beckett, S.C. (1977) Peat stratigraphy and pollen analysis of the Garvin's tracks. *Somerset Levels Papers*, **3**, 82-4.

Beckett, S.C. (1978a) The environmental setting of the Meare Heath Track. *Somerset Levels Papers*, **4**, 42-6.

Beckett, S.C. (1978b) Palaeobotanical investigations at the Difford's 1 site. *Somerset Levels Papers*, **4**, 101-6.

Beckett, S.C. (1979a) The palaeobotanical background to the Meare Lake Village sites. *Somerset Levels Papers*, **5**, 18-24.

Beckett, S.C. (1979b) The environmental setting of the Sweet track, Drove site. *Somerset Levels Papers*, **5**, 75-83.

Beckett, S.C. (1979c) Pollen influx in peat deposits: values from raised bogs in the Somerset Levels, South-Western England. *New Phytologist*, **83**, 839-47.

Beckett, S.C. and Hibbert, F.A. (1976) An absolute pollen diagram from the Abbot's Way. *Somerset Levels Papers*, **2**, 24-7.

References

Beckett, S.C. and Hibbert, F.A. (1978) The influence of man on the vegetation of the Somerset Levels - a summary. *Somerset Levels Papers*, 4, 86-90.

Beckett, S.C. and Hibbert, F.A. (1979) Vegetational change and the influence of prehistoric man in the Somerset Levels. *New Phytologist*, 83, 577-600.

Bell, F.G., Coope, G.R., Rice, R.J. and Riley, T.H. (1972) Mid-Weichselian fossil-bearing deposits at Syston, Leicestershire. *Proceedings of the Geologists' Association*, 83, 187-211.

Bell, M. (1990) *Brean Down Excavations 1983-1987*. English Heritage Archaeological Report No. 15, London.

Bell, M. (1992a) Brean Down Excavations 1989 and 1991. *Severn Estuary Levels Research Committee Annual Report 1991*. Lampeter.

Bell, M. (1992b) The Prehistory of soil erosion. In *Past and Present Soil Erosion: Archaeological and Geographical Perspectives* (eds M. Bell and J. Boardman), Oxbow Monograph No. 22, Oxford, pp. 21-35.

Bell, M., Caseldine, A., Caseldine, C., Crabtree, K., Maguire, D. and Maltby, E. (1984) Environmental archaeology in South West England. In *Environmental archaeology: a regional review* (ed. H.C.M. Keeley), Directorate of Ancient Monuments and Historic Buildings, Occasional Paper No. 6, DoE, London, 43-133.

Belt, T. (1876) The drift of Devon and Cornwall, its origin, correlation with that of the South-East of England, and place in the Glacial Series. *Quarterly Journal of the Geological Society of London*, 32, pp. 80-90.

Bennett, K.D. (1989) A provisional map of forest types for the British Isles 5,000 years ago. *Journal of Quaternary Science*, 4, 141-4.

Bennett, K.D. and Birks, H.J.B. (1990) Postglacial history of alder (*Alnus glutinosa* (L.) Gaertn.) in the British Isles. *Journal of Quaternary Science*, 5, 123-33.

Bennett, M.R., Mather, A.E. and Glasser, N.F. (1996) Earth hummocks and boulder runs at Merrivale, Dartmoor. In *The Quaternary of Devon and East Cornwall: Field Guide* (eds D.J. Charman, R.M. Newnham and D.G. Croot). Quaternary Research Association, London, pp. 81-96.

Berger, A. (1988) Milankovich and climate. *Reviews of Geophysics*, 26, 624-57.

Berglund, B.E., Birks, H.J.B., Ralska-Jasiewiczowa, M. and Wright, H.E. (1996) *Palaeoecological Events during the last 15,000 Years. Regional Syntheses of Palaeoecological Studies of Lakes and Mires*. Wiley, Chichester, 764pp.

Berridge, P.J. (1996) Later prehistoric and Romano-British evidence from the Torbryan Valley. In *The Quaternary of Devon and East Cornwall: Field Guide* (eds D.J. Charman, R.M. Newnham and D.G. Croot), Quaternary Research Association, London, pp. 202-4.

Beskow, G. (1935) Soil freezing and frost heaving with special application to roads and railroads. *Swedish Geological Society*, **Series C**, 375.

Bevercombe, G.P. (1980) An investigation of the microbial components of timbers from the Baker site. *Somerset Levels Papers*, 6, 43-5.

Beynon, F. (1931) The Cow Cave, Chudleigh. *Transactions and Proceedings of the Torquay Natural History Society*, **1931/2**, 127-32.

Beynon, F., Dowie, H.G. and Ogilvie, A.H. (1929) Report on the excavations at Kent's Cavern. *Transactions and Proceedings of the Torquay Natural History Society*, 5, 237.

Bird, M.I. and Chivas, A.R. (1988) Oxygen isotope dating of the Australian regolith. *Nature*, **331**, 513-16.

Bisdom, E.B.A. (1967) The role of micro-crack systems in the spheroidal weathering of an intrusive granite in Galicia (NW Spain). *Geologie en Mijnbouw*, 46, 333-40.

Bishop, M.J. (1974) A preliminary report on the Middle Pleistocene mammal bearing deposits of Westbury-sub-Mendip, Somerset. *Proceedings of the University of Bristol Spelaeological Society*, 13, 301-18.

Bishop, M.J. (1975) Earliest record of Man's presence in Britain. *Nature*, **253**, 95-7.

Bishop, M.J. (1982) The mammal fauna of the early Middle Pleistocene cavern infill site of Westbury-sub-Mendip, Somerset. *Special Papers in Palaeontology*, pp. 28.

Bishop, W.W. (1967) Discussion. In G.F. Mitchell and A.R. Orme, The Pleistocene deposits of the Isles of Scilly. *Quarterly Journal of the Geological Society of London*, **123**, 90-1.

Blyth, F.G.H. (1962) The structure of the north-eastern tract of the Dartmoor granite. *Quarterly Journal of the Geological Society of London*, **118**, 435-53.

Boase, H.S. (1828) Some observations on alluvial formations of the western part of Cornwall and On the sand-banks of the northern shores of Mount's-Bay. *Transactions of the Royal*

References

Geological Society of Cornwall, **3**, 17-34 and 166-91.

Boase, H.S. (1832) Contributions towards a knowledge of the geology of Cornwall. *Transactions of the Royal Geological Society of Cornwall*, **4**, 166-474.

Bond, G., Broecker, W., Johnsen, S. *et al.* (1993) Correlation between climate records from North Atlantic sediments and Greenland ice. *Nature*, **365**, 143-7.

Bond, G. and Lotti, R. (1995) Iceberg discharges into the North Atlantic on millennial time scales during the last glaciation. *Science*, **267**, 1005-10.

Borlase, W. (1757) An account of some trees discovered under-ground on the shore at Mount's-Bay in Cornwall. *Philosophical Transactions of the Royal Society of London*, **50**, 51-3.

Borlase, W. (1758) *The natural history of Cornwall*. E and W Books, London.

Boswell, P.G.H. (1923) The petrography of the Cretaceous and Tertiary outliers of the west of England. *Quarterly Journal of the Geological Society of London*, **79**, 205-30.

Bott, M.H.P., Day, A.A. and Masson-Smith, D. (1958) The geological interpretation of gravity and magnetic surveys in Devon and Cornwall. *Philosophical Transactions of the Royal Society of London*, **A251**, 161-91.

Bott, M.H.P., Holder, A.P., Long, R.E. and Lucas, A.L. (1970) Crustal structure beneath the granites of South West England. In *Mechanisms of Igneous Intrusion* (eds G. Newell and N. Rast), Geological Journal Special Issue No. 2, Liverpool, pp. 93-102.

Bowen, D.Q. (1969) A new interpretation of the Pleistocene succession in the Bristol Channel area (Abstract). *Proceedings of the Ussher Society*, **2**, 86.

Bowen, D.Q. (1970) South-east and central South Wales. In *The Glaciations of Wales and Adjoining Regions* (ed. C.A. Lewis), Longman, London, pp. 197-227.

Bowen, D.Q. (1971) The Quaternary succession of south Gower. In *Geological excursions in South Wales and the Forest of Dean* (eds D.A. Bassett and M.G. Bassett), Geologists' Association, South Wales Group, Cardiff, pp. 135-42.

Bowen, D.Q. (1973a) The Pleistocene history of Wales and the borderland. *Geological Journal*, **8**, 207-24.

Bowen, D.Q. (1973b) The Pleistocene succession of the Irish Sea. *Proceedings of the Geologists' Association*, **84**, 249-72.

Bowen, D.Q. (1974) The Quaternary of Wales. In *The Upper Palaeozoic and post-Palaeozoic rocks of Wales* (ed. T.R. Owen), University of Wales Press, Cardiff, pp. 373-426.

Bowen, D.Q. (1977) The coast of Wales. In *The Quaternary History of the Irish Sea* (eds C. Kidson and M.J. Tooley), Geological Journal Special Issue No. 7. Seel House Press, Liverpool, pp. 223-56.

Bowen, D.Q. (1978) *Quaternary Geology: a Stratigraphic Framework for Multidisciplinary Work*. Pergamon Press, Oxford, 221 pp.

Bowen, D.Q. (1981) The 'South Wales end-moraine': fifty years after. In *The Quaternary of Britain* (eds J. Neale and J. Flenley), Pergamon Press, Oxford, pp. 60-7.

Bowen, D.Q. (1982) Surface morphology. In *National Atlas of Wales* (eds H. Carter and H.M. Griffiths), University of Wales Press, Cardiff.

Bowen, D.Q. (1991) Time and space in the glacial sediment systems of the British Isles. In *Glacial deposits in Great Britain and Ireland* (eds J. Ehlers, P.L. Gibbard and J. Rose), Balkema, Rotterdam, 3-11.

Bowen, D.Q. (1994a) The Pleistocene of North-West Europe. *Science Progress*, **76**, 1-13.

Bowen, D.Q. (1994b) Late Cenozoic Wales and South-West England. *Proceedings of the Ussher Society*, **8**, 209-13.

Bowen, D.Q. (Ed.) (In Prep.) *Correlation of Quaternary rocks in the British Isles*. Geological Society Special Report.

Bowen, D.Q., Hughes, S., Sykes, G.A. and Miller, G.H. (1989) Land-sea correlations in the Pleistocene based on isoleucine epimerization in non-marine molluscs. *Nature*, **340**, 49-51.

Bowen, D.Q., Phillips, F.M., Maddy, D., Sykes, G.A. and Rose, J. (In Prep.) [36]Cl age determinations in the British Quaternary.

Bowen, D.Q., Rose, J., McCabe, A.M. and Sutherland, D.G. (1986) Correlation of Quaternary glaciations in England, Ireland, Scotland and Wales. *Quaternary Science Reviews*, **5**, 299-340.

Bowen, D.Q. and Sykes, G.A. (1988) Correlation of marine events and glaciations on the north-east Atlantic margin. *Philosophical Transactions of the Royal Society of London*, **B318**, 619-35.

Bowen, D.Q., Sykes, G.A., Reeves, A., *et al.* (1985)

References

Amino acid geochronology of raised beaches in south-west Britain. *Quaternary Science Reviews*, **4**, 279-318.

Bradley, R.S. (1985) *Quaternary Palaeoclimatology: Methods of Palaeoclimatic Reconstruction*. Allen and Unwin, Boston, 472 pp.

Bradley, W.C. (1963) Large-scale exfoliation in massive sandstones of the Colorado Plateau. *Bulletin of the Geological Society of America*, **74**, 519-28.

Bradshaw, M.J. (1961) Aspects of the geomorphology of north-west Devon. MA thesis, University of London.

Brammall, A. (1926a) Gold and silver in the Dartmoor granite. *Mineralogical Magazine*, **21**, 14-20.

Brammall, A. (1926b) The Dartmoor granite. *Proceedings of the Geologists' Association*, **37**, 251-77.

Brammall, A. and Harwood, H.F. (1923) The Dartmoor granite: its mineralogy, structure and petrology. *Mineralogical Magazine*, **20**, 39-53.

Brammall, A. and Harwood, H.F. (1932) The Dartmoor granites: their genetic relationships. *Quarterly Journal of the Geological Society of London*, **88**, 171-237.

Bramwell, D. (1960) Some research into bird distribution in Britain during the Late Glacial and Post Glacial periods. *Bird Report (Merseyside Naturalists' Association)*, 1959-60, pp. 51-8.

Brassell, S.C., Eglington, G., Marlowe, I.T., Pflaumann, U. and Sarnthein, M. (1986) Molecular stratigraphy: a new tool for climatic assessment. *Nature*, **320**, 129-33.

Brayley, E.W. (1830) On the probable connection of rock-basins, in form and situation, with an internal concretionary structure in the rocks on which they occur: introduced by remarks on the alleged artificial origin of those cavities. *Philosophical Magazine*, **8**, 331-42.

Brent, F. (1886) On the occurrence of flint flakes, and small stone implements, in Cornwall. *Journal of the Royal Institution of Cornwall*, **9**, 58-61.

Brent, F. (1901) On the occurrence of flint flakes, and small stone implements in Cornwall. *Journal of the Royal Institution of Cornwall*, **14**, 417-19.

Bridgland, D.R. (1994) *Quaternary of the Thames*. Geological Conservation Review Series No. 7. Chapman and Hall, London, 441 pp.

Briggs, D.J. and Gilbertson, D.D. (1980) Quaternary processes and environments in the upper Thames Basin. *Transactions of the Institute of British Geographers*, New Series **5**, 53-65.

Briggs, D.J., Gilbertson, D.D. and Harris, A.L. (1990) Molluscan taphonomy in a braided river environment and its implications for studies of Quaternary cold-stage river deposits. *Journal of Biogeography*, **17**, 623-37.

Briggs, D.J., Gilbertson, D.D. and Hawkins, A.B. (1991) The raised beach and sub-beach deposits at Swallow Cliff, Middle Hope. *Proceedings of the Bristol Naturalists' Society*, **51**, 63-71.

Bristow, C.M. (1968) The derivation of Tertiary sediments in the Petrockstow Basin, North Devon. *Proceedings of the Ussher Society*, **2**, 29-35.

Bristow, C.M. (1969) Kaolin deposits of the United Kingdom of Great Britain and Northern Ireland. *Proceedings of the 23rd International Geological Congress (1968)*, **15**, 275-88.

Bristow, C.M. (1977) A review of the evidence for the origin of the kaolin deposits in S.W. England. *Proceedings of the 8th International Kaolin Symposium and Meeting on Alunite*. Madrid-Rome, pp. 7-19.

Bristow, C.M. (1988) Ball clays, weathering and climate. *Proceedings of the 24th Forum on the Geology of Industrial Minerals*. South Carolina Geological Survey, Columbia, South Carolina, pp. 25-38.

Bristow, J. (1850) Geological survey map of Portland and Weymouth with notes. Geological Survey of Great Britain. H.M.S.O., London.

Broccoli, A.J. and Manabe, S. (1987) The influence of continental ice, atmospheric CO_2, and land albedo on the climate of the last glacial maximum. *Climate Dynamics*, **1**, 87-99.

Broecker, W.S. and Denton, G.H. (1989) The role of ocean-atmospheric reorganizations in glacial cycles. *Geochimica et Cosmochimica Acta*, **53**, 2465-501.

Broecker, W.S. and Denton, G.H. (1990) What drives glacial cycles? *Scientific American*, **262**, 43-50.

Brown, A.P. (1972) Late-Weichselian and Flandrian vegetation of Bodmin Moor, Cornwall. Unpublished PhD thesis, University of Cambridge.

Brown, A.P. (1977) Late-Devensian and Flandrian

vegetational history of Bodmin Moor, Cornwall. *Philosophical Transactions of the Royal Society of London*, **B276**, 251–320.

Brown, A.P. (1980) Deposits at Hawks Tor, Bodmin Moor. In *West Cornwall* (ed. P.C. Sims), Quaternary Research Association Field Handbook, West Cornwall Field Meeting, September 1980, pp. 37–40.

Bruce Mitford, R.L.S. (1956) A Dark-Age settlement at Morgan Porth, Cornwall. In *Recent Archaeological Excavations in Britain* (ed. R.L.S. Bruce Mitford). Routledge, London.

Brunner, F.K. and Scheidegger, A.E. (1973) Exfoliation. *Rock Mechanics*, **5**, 43–62.

Brunsden, D. (1963) The denudation chronology of the River Dart. *Transactions of the Institute of British Geographers*, **32**, 49–63.

Brunsden, D. (1964) The origin of decomposed granite on Dartmoor. In *Dartmoor essays* (ed. I.G. Simmons), Devonshire Association for the Advancement of Science, Literature and Art. Devonshire Press, Torquay, 97–116.

Brunsden, D. (1968) *Dartmoor. British Landscapes through Maps 12*. The Geographical Association, Sheffield, 55 pp.

Brunsden, D. (1996) Landslides of the Dorset coast: some unresolved questions. *Proceedings of the Ussher Society*, **9**, 1–11.

Brunsden, D., Doornkamp, J.C., Green, C.P. and Jones, D.K.C. (1976) Tertiary and Cretaceous sediments in solution pipes in the Devonian limestone of South Devon. *Geological Magazine*, **113**, 441–7.

Brunsden, D. and Jones, D.K.C. (1976) The evolution of landslide slopes in Dorset. *Philosophical Transactions of the Royal Society of London*, **A283**, 605–31.

Brunsden, D., Kidson, C., Orme, A.R. and Waters, R.S. (1964) Denudation chronology of parts of south-western England. *Field Studies*, **2**, 115–32.

Buckland, W. (1823) *Reliquiae Diluvianae; or, observations on the organic remains contained in caves, fissures, and Diluvial gravel, and on other geological phenomena, attesting the action of an Universal Deluge.* J. Murray, London, 303 pp. (second edition, 1824).

Buckland, W. and Conybeare, W.D. (1824) Observations on the South-western Coal District of England. *Transactions of the Geological Society*, Series 2, **1**, 210–36.

Budel, J. (1957) Die Doppelten Einebnungsfluchen in dem feuchten Tropen. *Zeitschrift für Geomorphologie*, **NF1**, 201–28.

Budge, E. (1841) On the conglomerates and raised beaches of the Lizard district. *Transactions of the Royal Geological Society of Cornwall*, **6**, 1–11.

Budge, E. (1843) On Diluvial action, as exemplified in the Gravel-beds and Sienitic formation of Crousa-Down, in the Parish of St. Keverne. *Transactions of the Royal Geological Society of Cornwall*, **6**, 91–8.

Bull, W.B. (1972) Recognition of alluvial fan deposits in the stratigraphic record. In *Recognition of ancient sedimentary environments* (eds J.K. Rigby and W.K. Hamblin), Society of Economic Paleontologists and Mineralogists Special Publication No. 16, pp. 68–83.

Bulleid, A. and Gray, H. St G. (1911) *The Glastonbury Lake Village I.* Glastonbury Antiquarian Society.

Bulleid, A. and Gray, H. St G. (1917) *The Glastonbury Lake Village II.* Glastonbury Antiquarian Society.

Bulleid, A. and Gray, H. St G. (1948) *The Meare Lake Village.* Privately printed, Taunton.

Bulleid, A. and Jackson, J.W. (1937) The Burtle sand beds of Somerset. *Proceedings of the Somerset Archaeological and Natural History Society*, **83**, 171–95.

Bulleid, A. and Jackson, J.W. (1941) Further notes on the Burtle sand-beds of Somerset. *Proceedings of the Somerset Archaeological and Natural History Society*, **87**, 111–16.

Burek, C.V. and Cubitt, J.M. (1991) Geochemical properties of glacial deposits in the British Isles. In *Glacial Deposits in Great Britain and Ireland* (eds J. Ehlers, P.L. Gibbard and J. Rose), Balkema, Rotterdam, 471–91.

Burrows, R. (1971) *The naturalist in Devon and Cornwall.* David and Charles, Newton Abbot.

Butcher, S.A. (1970) Excavations at Nornour. *Cornish Archaeology*, **9**, 77–91.

Butcher, S.A. (1971) Excavations at Nornour. *Cornish Archaeology*, **10**, 94.

Butcher, S.A. (1972) Excavations at Nornour. *Cornish Archaeology*, **11**, 58–9.

Butcher, S.A. (1974) *Nornour.* Isles of Scilly Museum Publication, No. 7.

Butler, S. (1987) Coastal change since 6000 BP and the presence of man at Kenn Moor. *Proceedings of the Somerset Archaeological and Natural History Society*, **131**, 1–11.

Calkin, J.B. and Green, J.F.N. (1949) Palaeoliths and terraces near Bournemouth. *Proceedings of the Prehistoric Society*, **15**, 21–37.

References

Callow, W.J. and Hassall, G.I. (1969) National Physical Laboratory Radiocarbon Measurements VI. *Radiocarbon*, **11**, 130-6.

Campbell, J.B. (1977) *The Upper Palaeolithic of Britain. A study of Man and Nature in the Late Ice Age.* Clarendon Press, Oxford, 264 pp. and 376 pp. (2 vols).

Campbell, J.B. and Sampson, C.G. (1971) *A new analysis of Kent's Cavern, Devonshire, England.* University of Oregon Occasional Paper No. 3, 40 pp.

Campbell, J.F. (1865) *Frost and Fire. Natural Engines, Tool-marks and Chips with Sketches taken at Home and Abroad by a Traveller* (2 vols). Edmonston and Douglas, Edinburgh.

Campbell, S. (1984) The nature and origin of the Pleistocene deposits around Cross Hands and on west Gower, South Wales. Unpublished PhD thesis, University of Wales (Swansea).

Campbell, S. (1991) Two Bridges Quarry - a classic site for understanding tors. *Earth Science Conservation*, **29**, 24-6.

Campbell, S. and Bowen, D.Q. (1989) *Quaternary of Wales.* Geological Conservation Review Series No. 2. Nature Conservancy Council, Peterborough, 237 pp.

Campbell, S., Hunt, C.O., Scourse, J.D., Keen, D.H. and Croot, D.G. (In Prep.) South-west England. In *Correlation of Quaternary Rocks in the British Isles* (ed. D.Q. Bowen), Geological Society Special Report.

Campbell, S. and Scourse, J.D. (1996) Report on the Annual Field Meeting: Devon and east Cornwall. *Quaternary Newsletter*, **79**, 48-54.

Carlisle, A. and Brown, A.H.F. (1968) Biological flora of the British Isles: *Pinus sylvestris* L. *Journal of Ecology*, **56**, 269-307.

Carne, J. (1827) On the granite of the western part of Cornwall. *Transactions of the Royal Geological Society of Cornwall*, **3**, 208-46.

Carne, J. (1832) An account of the discovery of some varieties of tin-ore in a vein, which have been considered peculiar to streams; with remarks on Diluvial tin in general. *Transactions of the Royal Geological Society of Cornwall*, **4**, 95-112.

Carne, J. (1846) Notice of the remains of a submarine forest in the north-eastern part of Mount's Bay. *Transactions of the Royal Geological Society of Cornwall*, **6**, 230-5.

Carne, J. (1865) Notice of a raised beach lately discovered at Zennor. *Transactions of the Royal Geological Society of Cornwall*, **7**, 176-8.

Carreck, J.N. (1958) A Late Pleistocene rodent and amphibian fauna from Levaton, near Newton Abbot, south Devon. *Proceedings of the Geologists' Association*, **68**, 304-8.

Carreck, J.N. (1960) Whitsun field meeting to Weymouth, Abbotsbury and Dorchester, Dorset. *Proceedings of the Geologists' Association*, **70**, 341-7.

Carreck, J.N. and Davis, A.G. (1955) The Quaternary deposits of Bowleaze Cove, near Weymouth, Dorset. *Proceedings of the Geologists' Association*, **66**, 74-100.

Carruthers, S.M. (1979) Examination of timbers from the Sweet track for evidence of decay and microbial activity. *Somerset Levels Papers*, **5**, 94-7.

Cartwright, C.R. (1996) The wood charcoal assemblages from the Torbryan Caves Project. In *The Quaternary of Devon and East Cornwall: Field Guide* (eds D.J. Charman, R.M. Newnham, and D.G. Croot), Quaternary Research Association, London, pp. 195-7.

Case, D.J. (1977) Horton (loess). In *Wales and the Cheshire-Shropshire Lowland* (ed. D.Q. Bowen), INQUA X Congress. Guidebook for Excursions A8 and C8. Geo Abstracts, Norwich, pp. 26-9.

Case, D.J. (1983) Quaternary airfall deposits in South Wales: loess and coversands. Unpublished PhD thesis, University of Wales.

Case, D.J. (1984) Port Eynon Silt (Loess). In *Wales: Gower, Preseli, Fforest Fawr* (eds D.Q. Bowen and A. Henry), Quaternary Research Association Field Guide, April 1984, pp. 51-4.

Caseldine, A.E. (1980[A]) Palaeoenvironmental reconstruction at the Baker site. *Somerset Levels Papers*, **6**, 29-36.

Caseldine, C.J. (1980) Environmental change in Cornwall during the last 13,000 years. *Cornish Archaeology*, **19**, 3-16.

Caseldine, C.J. (1983) Problems of relating site and environment in pollen-analytical research in South West England. In *Site, environment and economy* (ed. B. Proudfoot), British Archaeological Report No. 173, pp. 61-74.

Caseldine, C.J. and Hatton, J. (1993) The development of high moorland on Dartmoor: fire and the influence of Mesolithic activity on vegetation change. In *Climate Change and Human Impact on the Landscape* (ed. F.M. Chambers), Chapman and Hall, London, pp. 119-31.

Caseldine, C.J. and Hatton, J. (1996) Vegetation history of Dartmoor - Holocene development and the impact of human activity and pollen (Tornewton Caves). In *The Quaternary of*

References

Devon and East Cornwall: Field Guide (eds D.J. Charman, R.M. Newnham and D.G. Croot), Quaternary Research Association, London, pp. 48-61 and pp. 197-8.

Caseldine, C.J. and Maguire, D.J. (1981) A review of the prehistoric and historic environment on Dartmoor. *Proceedings of the Devon Archaeological Society*, **39**, 1-16.

Caseldine, C.J. and Maguire, D.J. (1986) Lateglacial/early Flandrian vegetation change on northern Dartmoor, south-west England. *Journal of Biogeography*, **13**, 255-64.

Catt, J.A. (1981) British pre-Devensian glaciations. In *The Quaternary in Britain* (eds J. Neale and J. Flenley), Pergamon Press, Oxford, pp. 10-19.

Catt, J.A. (1986) Silt mineralogy of loess and 'till' on the Isles of Scilly. In *The Isles of Scilly* (ed. J.D. Scourse), Field Guide. Quaternary Research Association, Coventry, pp. 134-6.

Catt, J.A. and Staines, S.J. (1982) Loess in Cornwall. *Proceedings of the Ussher Society*, **5**, 368-75.

Charlesworth, J.K. (1957) *The Quaternary Era*. Edward Arnold, London, 591 pp. and 1700 pp. (2 vols).

Charman, D.J., Newnham, R.M. and Croot, D.G. (Eds) (1996) *The Quaternary of Devon and East Cornwall: Field Guide*. Quaternary Research Association, London, 224 pp.

Chartres, C.J. and Whalley, W.B. (1975) Evidence for late Quaternary solution of Chalk at Basingstoke, Hampshire. *Proceedings of the Geologists' Association*, **86**, 365-72.

Cheesman, W.A. (1959) Mitnor Cave discovery. Letter dated April 22nd, 1959. *Western Morning News.*

Chisholm, N.W.T. (1996) Morphological changes at Braunton Burrows, north-west Devon. *Proceedings of the Ussher Society*, **9**, 25-30.

Christie, P. (1978) The excavation of an Iron Age souterrain and settlement at Carn Euny, Sancreed, Cornwall. *Proceedings of the Prehistoric Society*, **44**, 309.

Churchill, D.M. (1965) The displacement of deposits formed at sea-level 6,500 years ago in southern Britain. *Quaternaria*, **7**, 239-57.

Churchill, D.M. and Wymer, J.J. (1965) The kitchen midden site at Westward Ho!, Devon, England: ecology, age, and relation to changes in land and sea level. *Proceedings of the Prehistoric Society*, **31**, 74-84.

Clapham, A.R. and Godwin, H. (1948) Studies of the post-glacial history of British vegetation: VIII. Swamping surfaces in peats of the Somerset Levels: IX. Prehistoric trackways in the Somerset Levels. *Philosophical Transactions of the Royal Society of London*, **B233**, 233-73.

Clapperton, C.M. (1970) The evidence for a Cheviot ice cap. *Transactions of the Institute of British Geographers*, **50**, 115-27.

Clark, M.J., Lewin, J. and Small, R.J. (1967) The sarsen stones of the Marlborough Downs and their geomorphological significance. *Southampton Research Series in Geography*, **4**, 3-40.

Clarke, B.B. (1962) The coastal morphology of the Padstow Peninsula. *Transactions of the Royal Geological Society of Cornwall*, **19**, 220-32.

Clarke, B.B. (1965a) The upper and lower surfaces, and some structural features of the frost soils of the Camel Estuary. *Proceedings of the Ussher Society*, **1**, 192-3.

Clarke, B.B. (1965b) The superficial deposits of the Camel Estuary and suggested stages in its Pleistocene history. *Transactions of the Royal Geological Society of Cornwall*, **19**, 257-79.

Clarke, B.B. (1968) Geomorphological features of Mother Ivey's Bay near Padstow, with an account of the under-cliff bank and intertidal reef of cemented limesand at Little Cove. *Transactions of the Royal Geological Society of Cornwall*, **20**, 69-79.

Clarke, B.B. (1969) The problem of the nature, origin and stratigraphical position of the Trebetherick boulder gravel. *Proceedings of the Ussher Society*, **2**, 87-91.

Clarke, B.B. (1971) The Quaternary section at Porth Mear Cove (Abstract). *Proceedings of the Ussher Society*, **2**, 298.

Clarke, B.B. (1973) The Camel Estuary Pleistocene section west of Tregunna House. *Proceedings of the Ussher Society*, **2**, 551-3.

Clarke, B.B. (1976) Quaternary features of Porth Mear Cove, north Cornwall. *Transactions of the Royal Geological Society of Cornwall*, **20**, 275-85.

Clarke, B.B. (1980) Geomorphology of the Camel Valley and Estuary (Abstract). *Proceedings of the Ussher Society*, **5**, 93.

Clarke, R.H. (1970) Quaternary sediments off south-east Devon. *Quarterly Journal of the Geological Society of London*, **125**, 277-318.

Clayden, A.W. (1906) *The history of Devonshire scenery*. J.G. Commin, Exeter, and Chatto and Windus.

Clayden, B. (1964) Soils of Cornwall. In *Present

References

Views of some Aspects of the Geology of Cornwall and Devon (eds K.F.G. Hosking and G.J. Shrimpton), Blackford, Truro, pp. 311-30.

Clayden, B. and Findlay, D.C. (1960) Mendip derived gravels and their relationship to combes. *Abstracts of the Proceedings of the 3rd Conference of Geologists and Geomorphologists Working in the South-West of England. Bristol, 1960.* Royal Geological Society of Cornwall, Penzance, pp. 24-5.

Clayden, B. and Manley, D.J.R. (1964) The soils of the Dartmoor granite. In *Dartmoor essays* (ed. I.G. Simmons), Devonshire Association for the Advancement of Science, Literature and Art. Devonshire Press, Torquay, 117-45.

Codrington, T. (1898) On some submerged rock valleys in South Wales, Devon and Cornwall. *Quarterly Journal of the Geological Society of London*, **54**, 251-76.

Colbourne, G.J., Gilbertson, D.D. and Hawkins, A.B. (1974) Temporary drift exposures on the Failland Ridge. *Proceedings of the Bristol Naturalists' Society*, **33**, 91-7.

Coles, J.M. and Orme, B.J. (1976) The Abbot's Way and The Sweet track: Railway site. *Somerset Levels Papers*, **2**, 7-20 and 34-65.

Coles, J.M. and Orme, B.J. (1978) Multiple track-ways from Tinney's Ground. *Somerset Levels Papers*, **4**, 47-81.

Coles, J.M., Orme, B.J. and Hibbert, F.A. (1975) The Eclipse track. *Somerset Levels Papers*, **1**, 20-4.

Collcutt, S.N. (1984) The analysis of Quaternary cave sediments and its bearing upon Palaeolithic archaeology, with special reference to selected sites from western Britain. Unpublished PhD thesis, University of Oxford.

Collcutt, S.N. (1986) The later Upper Palaeolithic site of Pixie's Hole, Chudleigh, Devon. In *The Palaeolithic of Britain and its Nearest Neighbours: recent trends* (ed. S.N. Collcutt), J.R. Collis Publications, pp. 73-4.

Collier, D. (1961) Mise au point sur les processus de l'altération des granites en pays tempéré. *Anales Agronomicos*, **12**, 273-332.

Collins, J.H. (1878) *The Hensbarrow Granite District: a Geological Description and Trade History*. Lake and Lake, Truro.

Collins, J.H. (1887) On the nature and origin of clays: the composition of kaolinite. *Mineralogical Magazine*, **7**, 205-14.

Collins, J.H. (1909) Geological features visible at the Carpella china clay pit. *Quarterly Journal of the Geological Society of London*, **65**, 155-61.

Conolly, A.P., Godwin, H. and Megaw, E.M. (1950) Studies in the post-glacial history of British vegetation XI: late-glacial deposits in Cornwall. *Philosophical Transactions of the Royal Society of London*, **B234**, 397-469.

Conybeare, W.D. and Phillips, W. (1822) *Outlines of the Geology of England and Wales, with an introductory compendium of the general principles of that science, and comparative views of the structure of foreign countries*. William Phillips, London.

Coombe, D.E., Frost, L.C., Le Bas, M. and Watters, W. (1956) The nature and origin of the soils over the Cornish Serpentine. *Journal of Ecology*, **44**, 605-15.

Coope, G.R. (1977) Coleoptera as clues to the understanding of climatic changes in North Wales towards the end of the last (Devensian) Glaciation. *Cambria*, **4**, 65-72.

Coope, G.R. and Brophy, J.A. (1972) Late-glacial environmental changes indicated by a coleopteran succession from North Wales. *Boreas*, **1**, 97-142.

Cooper, L.H.N. (1948) A submerged ancient cliff near Plymouth. *Nature*, **161**, 280.

Cope, J.C.W. (1994) A latest Cretaceous hotspot and the south-easterly tilt of Britain. *Journal of the Geological Society of London*, **151**, 905-8.

Coque-Delhuille, B. (1982) Importance de l'érosion differentielle et de la tectonique tardi-Hercynienne dans le massif du Dartmoor (Grande-Bretagne). *Hommes et Terres du Nord*, **3**, 9-26.

Coque-Delhuille, B. (1987) Le massif du Sud-Ouest Anglais et sa bordure sédimentaire: études géomorphologiques. Thèse d'état de l'Université de Paris 1. Pantheon-Sorbonne, 2 tomes, 1040 pp.

Coque-Delhuille, B. and Veyret, Y. (1984) La limite de l'englacement Quaternaire dans le Sud-Ouest Anglais (Grande Bretagne). *Revue de Géomorphologique Dynamique*, **33**, 1-24.

Coque-Delhuille, B. and Veyret, Y. (1989) Limite d'englacement et évolution périglaciaire des Iles Scilly: l'intérêt des arènes *in situ* et remaniées. *Zeitschrift für Geomorphologie*, **72**, 79-96.

Corte, A.E. (1963) Relationship between four ground patterns, structure of the active layer and type and distribution of ice in the permafrost. *Biuletyn Peryglacjalny*, **12**, 7-90.

Cosgrove, M.E. (1975) Clay mineral suites in the post-Armorican formations of South-West England. *Proceedings of the Ussher Society*, **3**, 243.

References

Crabtree, K. and Maltby, E. (1975) Soil and land use change on Exmoor. Significance of a buried profile on Exmoor. *Proceedings of the Somerset Archaeology and Natural History Society*, **119**, 38-43.

Crabtree, K. and Round, F.E. (1967) Analysis of a core from Slapton Ley. *New Phytologist*, **66**, 255-70.

Crawford, O.G.S. (1921) The ancient settlements at Harlyn Bay. *Antiquarians Journal*, **1**, 281-99.

Cresswell, D. (1983) Deformation of weathered profiles, below head, at Constantine Bay, North Cornwall (Abstract). *Proceedings of the Ussher Society*, **5**, 487.

Croot, D.G. (1987) Brannam's Clay Pit. Unpublished report of survey and excavations undertaken for the Nature Conservancy Council, Peterborough.

Croot, D.G., Gilbert, A., Griffiths, J. and Van Der Meer, J.J. (1996) The character, age and depositional environments of the Fremington Clay Series, north Devon. In *The Quaternary of Devon and East Cornwall: Field Guide* (eds D.J. Charman, R.M. Newnham and D.G. Croot), Quaternary Research Association, London, pp. 14-34.

Croot, D.G., Gilbert, A., Griffiths, J., Van Der Meer, J.J. and Sims, P.C. (In Prep.) Quaternary deposits at Fremington, north Devon: characteristics, age and origin.

Crosby, W.O. (1893) The origin of parallel and intersecting joints. *American Geologist*, **12**, 368-75.

Crowther, P.R. and Wimbledon, W.A. (Eds) (1988) The use and conservation of palaeontological sites. *Special Papers in Palaeontology*, **40**, 1-200.

Cullingford, C.H.D. (Ed.) (1953) *British Caving: an introduction to speleology*. Routledge and Kegan Paul Limited, London.

Cullingford, C.H.D. (Ed.) (1962) *British Caving: an introduction to speleology*, 2nd edn, Routledge and Kegan Paul, London.

Cullingford, R.A. (1982) The Quaternary. In *The Geology of Devon* (eds E.M. Durrance and D.J.C. Laming), University of Exeter, Exeter, pp. 249-90.

Currant, A.P. (1989) The Quaternary origins of the modern British mammal fauna. *Biological Journal of the Linnean Society*, **38**, 23-30.

Currant, A.P. (1996) Tornewton Cave and the palaeontological succession. In *The Quaternary of Devon and East Cornwall: Field Guide* (eds D.J. Charman, R.M.

Newnham and D.G. Croot), Quaternary Research Association, London, pp. 174-80.

Curry, D., Hamilton, D. and Smith, A.J. (1970) Geological and shallow subsurface investigations of the Western Approaches to the English Channel. Institute of Geological Sciences, London, Report No. 70/3, pp. 1-12.

Czudek, T. and Demek, J. (1970) Pleistocene cryopedimentation in Czechoslovakia. *Acta Geographica Lodziensia*, **24**, 101-8.

Dahl, R. (1966) Blockfields, weathering pits and tor-like forms in the Narvik Mountains, Nordland, Norway. *Geografiska Annaler*, **48**, 55-85.

Dalzell, D. and Durrance, E.M. (1980) The evolution of the Valley of Rocks, North Devon. *Transactions of the Institute of British Geographers*, **5**, 66-79.

Damon, R. (1860) *The Geology of Weymouth and Portland*. Memoir of the Geological Survey of Great Britain. H.M.S.O., London.

Dansgaard, W., Johnsen, S.J., Clausen, H.B. *et al.* (1993) Evidence for general instability of past climate from a 250-kyr ice-core record. *Nature*, **364**, 218-20.

Dansgaard, W., White, J.W.C. and Johnsen, S.J. (1989) The abrupt termination of the Younger Dryas climate event. *Nature*, **339**, 532-4.

Darrah, J. (1993) The Bluestones of Stonehenge. *Current Archaeology*, **134**, 78.

Davies, A.T. and Kitto, B.K. (1878) On some beds of sand and clay in the parish of St. Agnes, Cornwall. *Transactions of the Royal Geological Society of Cornwall*, **9**, 196-204.

Davies, C.M. (1983) Pleistocene rockhead surfaces underlying Barnstaple Bay (SW England). *Marine Geology*, **54**, 9-16.

Davies, D.C. (1988) The amino-stratigraphy of British Pleistocene glacial deposits. Unpublished PhD thesis, University of Wales (Aberystwyth).

Davies, H.N. (1907) Supplementary notes on the Clevedon bone cave and gravels. *Proceedings of the University of Bristol Spelaeological Society*, **7**, 130-6.

Davies, J.A. and Fry, T.R. (1929) Notes on the gravel terraces of the Bristol Avon. *Proceedings of the University of Bristol Spelaeological Society*, **3**, 162-72.

Davies, K.H. (1983) Amino-acid analysis of Pleistocene marine molluscs from the Gower Peninsula. *Nature*, **302**, 137-9.

Davies, K.H. (1985) The aminostratigraphy of British Pleistocene beach deposits.

References

Unpublished PhD thesis, University of Wales (Aberystwyth).

Davies, K.H. and Keen, D.H. (1985) The age of Pleistocene marine deposits at Portland, Dorset. *Proceedings of the Geologists' Association*, **96**, 217-25.

Davis, A.G. and Carreck, J.N. (1958) Further observations on a Quaternary deposit at Bowleaze Cove, near Weymouth, Dorset. *Proceedings of the Geologists' Association*, **69**, 120-2.

Davis, W.M. (1909) The systematic description of land forms. *Geographical Journal*, **34**, 300-26.

Davison, E.H. (1930) *Handbook of Cornish Geology*, (2nd edn). Blackford, Truro.

Dawkins, W.B. (1862) On a hyaena-den at Wookey-Hole, near Wells. *Proceedings of the Geological Society*, **18**, 115-25.

Dawkins, W.B. (1863a) On a hyaena-den at Wookey Hole, near Wells. No. II. *Proceedings of the Geological Society*, **19**, 260-74.

Dawkins, W.B. (1863b) Wookey Hole hyaena den. *Proceedings of the Somerset Archaeological and Natural History Society*, **11**, 197-219.

Dawkins, W.B. (1865) On the Mammalia of the Newer Pliocene Age in the caverns and river-deposits of Somersetshire. *Geological Magazine*, **2**, 43-4.

Dawkins, W.B. (1866) Bone caverns and river deposits. *Geological Magazine*, **3**, 183-4.

Dawson, A.G. (1990) Shore erosion by frost: an example from the Scottish Lateglacial. In *Studies in the Lateglacial of North-west Europe* (eds J.J. Lowe, J.M. Gray and J.E. Robinson), pp. 45-53.

Dawson, A.G. (1992) *Ice Age Earth. Late Quaternary Geology and Climate*. Routledge, London, 293 pp.

Day, E.C.H. (1866) On a raised beach and other recent formations near Weston-super-Mare. *Geological Magazine*, **3**, 115.

Dearman, W.R. (1963) Wrench-faulting in Cornwall and south Devon. *Proceedings of the Geologists' Association*, **74**, 265-87.

Dearman, W.R. (1964) Dartmoor: its geological setting. In *Dartmoor essays* (ed. I.G. Simmons), Devonshire Association for the Advancement of Science, Literature and Art. Devonshire Press, Torquay, pp. 1-29.

Dearman, W.R. and Baynes, F.J. (1978) A field study of the basic controls of weathering patterns in the Dartmoor granite. *Proceedings of the Ussher Society*, **4**, 192-203.

Dearman, W.R., Baynes, F.J. and Irfan, T.Y. (1976) Practical aspects of periglacial effects on weathered granite. *Proceedings of the Ussher Society*, **3**, 373-81.

Dearman, W.R., Baynes, F.J. and Irfan, T.Y. (1978) Engineering grading of weathered granite. *Engineering Geology*, **12**, 345-74.

Dearman, W.R. and Butcher, N.E. (1959) Geology of the Devonian and Carboniferous rocks of the N.W. Border of the Dartmoor granite. *Proceedings of the Geologists' Association*, **70**, 51-92.

Dearman, W.R. and Fattohi, Z.R. (1974) The variation of rock properties with geological setting: a preliminary study of chert from S.W. England. *Proceedings of the 2nd International Congress of the International Association of Engineering Geology*, **1**, IV-26, 1-10.

Dearman, W.R. and Fookes, P.G. (1972) The influence of weathering on the layout of quarries in South-west England. *Proceedings of the Ussher Society*, **2**, 372-87.

Debenham, N. (1996) Thermoluminescence dating of calcite and sediment from the Torbryan Caves. In *The Quaternary of Devon and East Cornwall: Field Guide* (eds D.J. Charman, R.M. Newnham and D.G. Croot), Quaternary Research Association, London, 186-7.

De Dekker, P. and Forester, R.M. (1988) The use of ostracods to reconstruct continental palaeoenvironmental records. In *Ostracoda in the Earth Sciences* (eds P. De Dekker, J.P. Colin and J.-P. Peypouquet), Elsevier, Amsterdam, 175-99.

De Freitas, M.H. (1972) Some examples of cliff failure in S.W. England. *Proceedings of the Ussher Society*, **2**, 388-97.

Dejou, J., Guyot, J., Pedro, G., Chaumont, C. and Huguette, A. (1968) Nouvelles données concernant la présence de gibbsite dans les formations d'altération superficielle des massif granitique (cas du Cantal et du Limousin). *Comptes Rendus. Academie des Sciences*, **266D**, 1825-7.

De La Beche, H.T. (1826) On the Chalk and sands beneath it (usually termed Green-Sand) in the vicinity of Lyme Regis, Dorset, and Beer, Devon. *Transactions of the Geological Society of London*, **2**, 109-18.

De La Beche, H.T. (1835) Note on the Trappean rocks associated with the (New) Red Sandstone of Devonshire. *Proceedings of the Geological Society of London*, **2**, 196-8.

References

De La Beche, H.T. (1839) *Report on the geology of Cornwall, Devon and west Somerset.* Memoir of the Geological Survey of Great Britain. Longman, Orme, Brown, Green, and Longmans, London, pp. 395-434.

De La Beche, H.T. (1853) *The geological observer* (2nd edn). Longman, Brown, Green, and Longmans, London.

Demek, J. (1968) Cryoplanation terraces in Yakutia. *Biuletyn Peryglacjalny*, **17**, 91-116.

Dewar, H.S.L. and Godwin, H. (1963) Archaeological discoveries in the raised bogs of the Somerset Levels, England. *Proceedings of the Prehistoric Society*, **29**, 17-49.

Dewey, H. (1910) Notes on some igneous rocks from North Devon. *Proceedings of the Geologists' Association*, **21**, 429-34.

Dewey, H. (1913) The raised beach of North Devon: its relation to others and to Palaeolithic Man. *Geological Magazine*, **10**, 154-63.

Dewey, H. (1916) On the origin of some river gorges in Cornwall and Devon. *Quarterly Journal of the Geological Society of London*, **72**, 63-76.

Dewey, H. (1935) *South-West England.* British Regional Geology. H.M.S.O., London.

Dewey, H. (1949) The geology of South Devon. *Transactions and Proceedings of the Torquay Natural History Society*, **10**, 59-69.

Dewolf, Y. (1970) Les argiles à silex: paléosols ou pédolithes. *Bulletin Association Française pour l'étude du Quaternaire*, **2/3**, 117-19.

Dimbleby, G.W. (1962) *The Development of the British Heathlands and their Soils.* Oxford Forestry Memoirs, Clarendon Press, Oxford.

Dimbleby, G.W. (1963) Pollen analysis at a Mesolithic site at Addington, Kent. *Grana Palynologica*, **4**, 140-8.

Dimbleby, G.W. (1977) A buried soil at Innisidgen, St. Mary's, Isles of Scilly. *Cornish Studies*, **4/5**, 5-10.

Dimbleby, G.W., Greig, J.R.A. and Scaife, R.G. (1981) Vegetational history of the Isles of Scilly. In *Environmental Aspects of Coasts and Islands* (eds D. Brothwell and G.W. Dimbleby). Symposia of the Association for Environmental Archaeology No. 1. *British Archaeological Reports International Series*, **94**, 127-44.

Dineley, D.L. (1963) Contortions in the Bovey Beds (Oligocene), SW England. *Biuletyn Peryglacjalny*, **12**, 151-60.

Dines, H.G. (1956) *The Metalliferous Mining Region of South West England.* Memoir of the Geological Survey of Great Britain. H.M.S.O., London.

Dines, H.G., Hollingworth, S.E., Edwards, W., Buchan, S. and Welch, F.B.A. (1940) The mapping of head deposits. *Geological Magazine*, **76**, 198-226.

Doeglas, D.J. (1962) The structure of sedimentary deposits of braided rivers. *Sedimentology*, **1**, 167-90.

Dollar, A.T.J. (1957) Excursion to the Scilly Isles. *Circular of the Geologists' Association*, p. 597.

Donovan, D.T. (1954) A bibliography of the Palaeolithic and Pleistocene sites of the Mendip, Bath and Bristol area. *Proceedings of the University of Bristol Spelaeological Society*, **7**, 23-34.

Donovan, D.T. (1955) The Pleistocene deposits at Gough's Cave, Cheddar, including an account of recent excavations. *Proceedings of the University of Bristol Spelaeological Society*, **7**, 76-104.

Donovan, D.T. (1962) Sea levels of the last glaciation. *Bulletin of the Geological Society of America*, **73**, 1297-8.

Donovan, D.T. (1964) A bibliography of the Palaeolithic and Pleistocene sites of the Mendip, Bath and Bristol area. First supplement. *University of Bristol Spelaeological Society*, **10**, 89-97.

Donovan, D.T. and Stride, A.H. (1975) Three drowned coastlines of probable Late Tertiary age around Devon and Cornwall. *Marine Geology*, **19**, 35-40.

Doornkamp, J.C. (1974) Tropical weathering and the ultra-microscopic characteristics of regolith quartz on Dartmoor. *Geografiska Annaler*, **56A**, 73-82.

Doré, A.G. (1976) Preliminary geological interpretation of the Bristol Channel approaches. *Journal of the Geological Society of London*, **132**, 453-9.

Dowdeswell, J.A. and Sharp, M.J. (1986) Characterization of pebble fabrics in modern terrestrial glacigenic sediments. *Sedimentology*, **33**, 699-710.

Dowie, H.G. (1925) The excavation of a cave at Torbryan. *Transactions and Proceedings of the Torquay Natural History Society*, **4**, 261-8.

Dowie, H.G. (1928) Note on recent excavations in Kent's Cavern. *Proceedings of the Prehistorical Society of East Anglia*, **5**, 306-7.

References

Drew, D.P. (1975) The caves of Mendip. In *Limestone Caves of the Mendip Hills* (eds D.I. Smith and D.P. Drew), David and Charles, Newton Abbot, pp. 214-312.

D'urban, W.S.M. (1878) Palaeolithic implements from the valley of the Axe. *Geological Magazine*, **5**, 37-8.

Durrance, E.M. (1969) The buried channels of the Exe. *Geological Magazine*, **106**, 174-89.

Durrance, E.M. (1971) The buried channel of the Teign Estuary. *Proceedings of the Ussher Society*, **2**, 299-306.

Durrance, E.M. (1974) Gradients of buried channels in Devon. *Proceedings of the Ussher Society*, **3**, 111-19.

Durrance, E.M., Bromley, A.V., Bristow, C.M., Heath, M.J. and Penman, J.M. (1982) Hydrothermal circulation and post-magmatic changes in granites of south-west England. *Proceedings of the Ussher Society*, **5**, 304-20.

Durrance, E.M. and Laming, D.J.C. (Eds) (1982) *The geology of Devon*. University of Exeter Press, Exeter, 346 pp.

Dury, G.H. (1971) Relict deep weathering and duricrusting in relationship to the palaeoenvironments of middle latitudes. *Geographical Journal*, **137**, 511-22.

Eakin, H.M. (1916) The Yukon-Koyukuk region, Alaska. *Bulletin of the United States Geological Survey*, **63**, 1-88.

Eden, M.J. and Green, C.P. (1971) Some aspects of granite weathering and tor formation on Dartmoor, England. *Geografiska Annaler*, **53A**, 92-9.

Edmonds, E.A. (1972) The Pleistocene history of the Barnstaple area. Report of the Institute of Geological Sciences, 72/2. H.M.S.O., London, 12 pp.

Edmonds, E.A., McKeown, M.C. and Williams, M. (1975) *South-west England. British Regional Geology* (4th edn). H.M.S.O., London.

Edmonds, E.A., Whittaker, A. and Williams, B.J. (1985) *Geology of the country around Ilfracombe and Barnstaple*. Memoir for 1:50,000 geological sheets 277 and 293, New Series. British Geological Survey England and Wales. H.M.S.O., London.

Edmonds, E.A. and Williams, B.J. (1985) *Geology of the country around Taunton and the Quantock Hills*. Memoir for 1:50,000 geological sheet 295, New Series. H.M.S.O., London, pp. 49-83.

Edmonds, E.A., Williams, B.J. and Taylor, R.T. (1979) *Geology of Bideford and Lundy Island*. Memoir for 1:50,000 geological sheet 292, New Series, with sheets 275, 276, 291 and part of sheet 308. Institute of Geological Sciences. H.M.S.O., London, 143 pp.

Edmonds, E.A., Wright, J.E., Beer, K.E., Hawkes, J.R., Williams, M., Freshney, E.C. and Fenning, P.J. (1968) *Geology of the country around Okehampton*. Memoir of the Geological Survey of Great Britain, Sheet No. 324. H.M.S.O., London.

Edmonds, R. (1848) Notice of land shells found beneath the surface of sand-hillocks on the coasts of Cornwall. *Transactions of the Royal Geological Society of Cornwall*, **7**, 70-1.

Edwards, R.A. (1973) The Aller Gravels: Lower Tertiary braided river deposits in South Devon. *Proceedings of the Ussher Society*, **2**, 608-16.

Edwards, R.A. (1976) Tertiary sediments and structures of the Bovey Basin, south Devon. *Proceedings of the Geologists' Association*, **87**, 1-26.

Edwards, R.A. and Freshney, E.C. (1982) The Tertiary sedimentary rocks. In *The Geology of Devon* (eds E.M. Durrance and D.J.C. Laming), University of Exeter Press, Exeter, pp. 204-37.

Ehlen, J. (1989) Geomorphic, petrographic and structural relations in the Dartmoor granite, Southwest England. Unpublished PhD thesis, University of Birmingham.

Ehlen, J. (1991) Significant geomorphic and petrographic relations with joint spacing in the Dartmoor granite, Southwest England. *Zeitschrift für Geomorphologie*, **NF35**, 425-38.

Ehlen, J. (1992) Analysis of spatial relationships among geomorphic, petrographic and structural characteristics of the Dartmoor tors. *Earth Surface Processes and Landforms*, **17**, 53-67.

Ehlen, J. (1994) Classification of Dartmoor tors. In *Rock Weathering and Landform Evolution*. (eds D.A. Robinson and R.B.G. Williams), John Wiley, Chichester, pp. 393-412.

Ehlers, J., Gibbard, P.L. and Rose, J. (1991) Glacial deposits of Britain and Europe: general overview. In *Glacial Deposits in Great Britain and Ireland* (eds J. Ehlers, P.L. Gibbard and J. Rose), Balkema, Rotterdam, pp. 493-501.

Einarsson, T. and Albertson, K.J. (1988) The glacial history of Iceland during the past three million years. *Philosophical Transactions of the Royal Society of London*, **B318**, 637-44.

Ellis, H.S. (1866) On a flint-find in a submerged

forest of Barnstaple Bay, near Westward Ho! *Report and Transactions of the Devonshire Association for the Advancement of Science, Literature and Art*, **1**, 80-1.

Ellis, H.S. (1867) On some mammalian bones and teeth, found in the submerged forest at Northam. *Report and Transactions of the Devonshire Association for the Advancement of Science, Literature and Art*, **2**, 162-3.

Ellis, N.V. (Ed.), Bowen, D.Q., Campbell, S. *et al.* (1996) *An introduction to the Geological Conservation Review.* GCR Series No. 1, Joint Nature Conservation Committee, Peterborough, 131 pp.

Emiliani, C. (1955) Pleistocene temperatures. *Journal of Geology*, **63**, 538-78.

Emiliani, C. (1957) Temperature and age analysis of deep-sea cores. *Science, New York*, **125**, 383-7.

Esteoule-Choux, J. (1983) Kaolinitic weathering profiles in Brittany: genesis and economic importance. In *Residual Deposits: surface-related weathering processes and materials* (ed. R.C.L. Wilson), Blackwell, Oxford, 33-8.

Evans, A.L., Fitch, F.J. and Miller, J.A. (1973) Potassium-argon age determinations on some British Tertiary igneous rocks. *Journal of the Geological Society of London*, **129**, 419-43.

Evans, C.D.R. and Hughes, M.R. (1984) The Neogene succession of the South-Western Approaches, Great Britain. *Journal of the Geological Society of London*, **141**, 315-26.

Evans, H.M. (1912) Sand formation against the Saunton Down cliffs, North Devon. *Report and Transactions of the Devonshire Association for the Advancement of Science, Literature and Art*, **44**, 692-702.

Evans, J. (1872) *The ancient stone implements, weapons and ornaments of Great Britain* (1st edn). Longmans, Green, London.

Evans, J. (1897) *The ancient stone implements, weapons and ornaments of Great Britain* (2nd edn). Longmans, Green, London, 747 pp.

Evans, J.G. (1979) The palaeoenvironment of coastal blown-sand deposits in western and northern Britain. *Scottish Archaeological Forum*, **9**, 16-26.

Evans, J.G. (1984) Excavations at Bar Point, St. Mary's, Isles of Scilly, 1979-1980. *Cornish Studies*, **11**, 7-32.

Evans, J.W. (1922) The geological structure of the country round Combe Martin, North Devon. *Proceedings of the Geologists' Association*, **33**, 201-34.

Eve, R.M. (1970) The Pleistocene deposits between Croyde and Woolacombe in north Devon. Unpublished BA dissertation, Queen's University, Belfast.

Everard, C.E. (1960a) Mining and shoreline evolution near St. Austell, Cornwall. *Transactions of the Royal Geological Society of Cornwall*, **19**, 199-219.

Everard, C.E. (1960b) Valley and structural trends in West Cornwall. *Abstracts of the Proceedings of the Third Conference of Geologists and Geomorphologists Working in the South-West of England. Bristol, 1960.* Royal Geological Society of Cornwall, Penzance, 19-20.

Everard, C.E. (1977) *Valley direction and geomorphological evolution in west Cornwall, England.* Occasional Paper No. 10, Department of Geography, Queen Mary College, University of London, 72 pp.

Everard, C.E., Lawrence, R.H., Witherick, M.E. and Wright, L.W. (1964) Raised beaches and marine geomorphology. In *Present views of some aspects of the geology of Cornwall and Devon* (eds K.F.G. Hosking and G.J. Shrimpton). Blackford, Truro, 283-310.

Exley, C.S. (1959) Magmatic differentiation and alteration in the St. Austell granite. *Quarterly Journal of the Geological Society of London*, **114**, 197-230.

Exley, C.S. (1964) Some factors bearing on the natural synthesis of clays in the granites of south-west England. *Clay Minerals Bulletin*, **5**, 411-26.

Exley, C.S. (1965) Some structural features of the Bodmin Moor granite mass. *Proceedings of the Ussher Society*, **1**, 157-9.

Exley, C.S. (1976) Observations on the formation of kaolinite in the St. Austell granite, Cornwall. *Clay Minerals*, **11**, 51-63.

Exley, C.S. and Stone, M. (1964) The granitic rocks of South-West England. In *Present views of some aspects of the geology of Cornwall and Devon* (eds K.F.G. Hosking and G.J. Shrimpton), 150th Anniversary Volume, Royal Geological Society of Cornwall. Blackford, Truro, pp. 131-84.

Eyles, N. and Mccabe, A.M. (1989) The Late Devensian (<22,000 BP) Irish Sea basin: the sedimentary record of a collapsed ice-sheet margin. *Quaternary Science Reviews*, **8**, 307-51.

Eyles, N. and McCabe, A.M. (1991) Glaciomarine deposits of the Irish Sea Basin: the role of glacio-isostatic disequilibrium. In *Glacial deposits in Great Britain and Ireland* (eds J. Ehlers, P.L. Gibbard and J. Rose), Balkema, Rotterdam, pp. 311-31.

Eyles, V.A. (1952) *The composition and origin of the Antrim laterites and bauxites.* Memoir of the Geological Survey of Northern Ireland.

Fairbanks, R. (1989) A 17,000 year glacio-eustatic sea level record: influence of glacial melting rates on the Younger Dryas event and deep ocean circulation. *Nature*, **342**, 637-42.

Fairbridge, R.W. (1961) Eustatic changes of sea level. *Physics and Chemistry of the Earth*, **5**, 99-185.

Findlay, D.C. (1965) *The Soils of the Mendip District of Somerset.* Memoir of the Soil Survey of Great Britain, Harpenden.

Findlay, D.C. (1977) Bourne (Avon): a temporary section in gravelly head. In *Field Meeting Handbook, Easter Meeting 1977, Bristol* (ed. K. Crabtree). Quaternary Research Association, Cambridge, pp. 21-5.

Findlay, D.C., Hawkins, A.B. and Lloyd, C.R. (1972) A gravel deposit on Bleadon Hill, Mendip. *Proceedings of the University of Bristol Spelaeological Society*, **13**, 83-7.

Fitzpatrick, E.A. (1963) Deeply weathered rock in Scotland, its occurrence, age and contribution to the soils. *Journal of Soil Science*, **14**, 33-43.

Fleming, A. (1988) *The Dartmoor Reaves: Investigating Prehistoric Land Divisions.* Batsford, London.

Flett, J.S. and Hill, J.B. (1912) *Geology of the Lizard and Meneage (explanation of sheet 359* 1st edn), Memoirs of the Geological Survey of Great Britain. H.M.S.O., London.

Flett, J.S. and Hill, J.B. (1946) *Geology of the Lizard and Meneage,* (2nd edn), Memoirs of the Geological Survey of Great Britain. H.M.S.O., London.

Floyd, P.A., Exley, C.S. and Styles, M.T. (1993) *Igneous rocks of South-West England.* Geological Conservation Review Series No. 5. Chapman and Hall, London, 256 pp.

Fookes, P.G., Dearman, W.R. and Franklin, J.A. (1971) Some engineering aspects of rock weathering with field examples from Dartmoor and elsewhere. *Quarterly Journal of Engineering Geology*, **4**, 139-85.

Forbes, J.D. (1846) On the connexion between the distribution of the existing fauna and flora of the British Isles, and the geological changes which have affected their area, especially dur-ing the epoch of the Northern Drift. *Geological Survey of Great Britain Memoir*, **1**, 336-432.

Ford, D.C. (1968) Features of cavern development in central Mendip. *Transactions of the British Cave Research Association*, **10**, 11-25.

Fowler, A. and Robbie, J.A. (1961) *The Geology of the Country around Dungannon.* Memoir of the Geological Survey of Northern Ireland.

Fowler, P.J. and Thomas, C. (1979) Lyonesse revisited: the early walls of Scilly. *Antiquity*, **53**, 175-89.

Fox, H. (1891) The Cavouga boulder. *Transactions of the Royal Geological Society of Cornwall*, **11**, 334-5.

French, H.M. (1976) *The periglacial environment.* Longman, London.

Freshney, E.C., Beer, K.E. and Wright, J.E. (1979a) *Geology of the Country around Chulmleigh.* Memoir of the Geological Survey of Great Britain, Sheet No. 309. H.M.S.O., London.

Freshney, E.C., Edmonds, E.A., Taylor, R.T. and Williams, B.J. (1979b) *Geology of the Country around Bude and Bradworthy.* Memoir of the Geological Survey of Great Britain, Sheet Nos 307 and 308. H.M.S.O., London.

Freshney, E.C., Edwards, R.A., Isaac, K.P. *et al.* (1982) A Tertiary basin at Dutson, near Launceston, Cornwall, England. *Proceedings of the Geologists' Association*, **93**, 395-402.

Freshney, E.C., McKeown, M.C. and Williams, M. (1972) *Geology of the Coast between Tintagel and Bude.* Memoir of the British Geological Survey, Sheet No. 322 (part). H.M.S.O., London.

Fryer, G. (1960) Evolution of the land forms of Kerrier. *Transactions of the Royal Geological Society of Cornwall*, **19**, 122-53.

Funnell, B.M. (1987) Late Pliocene and early Pleistocene stages of East Anglia and the adjacent North Sea. *Quaternary Newsletter*, **52**, 1-11.

Funnell, B.M. (1988) Foraminifera in the late Tertiary and early Quaternary Crags of East Anglia. In *The Pliocene-Middle Pleistocene of East Anglia: Field Guide* (eds P.L. Gibbard and J.A. Zalasiewicz), Quaternary Research Association, Cambridge, pp. 50-2.

Gale, S.J., Hoare, P.G., Hunt, C.O. and Pye, K. (1988) The Middle and Upper Quaternary deposits at Morston, north Norfolk, UK. *Geological Magazine*, **125**, 521-33.

Garrard, R.A. (1977) The sediments of the south Irish Sea and Nymphe Bank area of the Celtic

Sea. In *The Quaternary History of the Irish Sea* (eds C. Kidson. and M.J. Tooley), Geological Journal Special Issue No. 7, Seel House Press, Liverpool, pp. 69-92.

Garrard, R.A. and Dobson, M.R. (1974) The nature and maximum extent of glacial sediments off the west coast of Wales. *Marine Geology*, **16**, 31-44.

Gascoyne, M., Currant, A.P. and Lord, T. (1981) Ipswichian fauna of Victoria Cave and the marine palaeoclimatic record. *Nature*, **294**, 652-4.

Gearey, B. and Charman, D.J. (1996) Rough Tor, Bodmin Moor: testing some archaeological hypotheses with landscape palaeoecology. In *The Quaternary of Devon and East Cornwall: Field Guide* (eds D.J. Charman, R.M. Newnham and D.G. Croot), Quaternary Research Association, London, pp. 101-19.

Gerrard, A.J.W. (1974) The geomorphological importance of jointing in the Dartmoor granite. In *Progress in geomorphology* (eds E.H. Brown, and R.S. Waters), Institute of British Geographers Special Publication No. 7, pp. 39-51.

Gerrard, A.J.W. (1978) Tors and granite landforms of Dartmoor and eastern Bodmin Moor. *Proceedings of the Ussher Society*, 4, 204-10.

Gerrard, A.J.W. (1982) Granite structures and landforms. In *Papers in Earth Studies* (eds B.H. Adlam, C.R. Fenn and L. Morris), Geo Books, Norwich, pp. 69-105.

Gerrard, A.J.W. (1983) *Periglacial landforms of the Cox Tor - Staple Tors area of western Dartmoor*. Department of Geography, University of Birmingham, Working Paper Series, No. 13, 36 pp.

Gerrard, A.J.W (1984) Multiple working hypotheses and equifinality in geomorphology: comments on the recent article by Haines-Young and Petch. *Transactions of the Institute of British Geographers*, **NS9**, 364-6.

Gerrard, A.J.W. (1988) Periglacial modification of the Cox Tor - Staples Tors area of western Dartmoor, England. *Physical Geography*, **9**, 280-300.

Gerrard, A.J.W. (1989a) The nature of slope materials on the Dartmoor granite, England. *Zeitschrift für Geomorphologie*, **33**, 179-88.

Gerrard, A.J.W. (1989b) Drainage basin analysis of the granitic upland of Dartmoor, England. *Geographical Review of India*, **51**, 1-17.

Gerrard, A.J.W. (1990) Variations within and between weathered granite and head on Dartmoor. *Proceedings of the Ussher Society*, 7, 285-8.

Gerrard, A.J.W. (1991) The status of temperate hillslopes in the Holocene. *The Holocene*, **1**, 86-90.

Gerrard, A.J.W. (1993) Landscape sensitivity and change on Dartmoor. In *Landscape Sensitivity* (eds D.S.G. Thomas and R.J. Allison), John Wiley, Chichester, pp. 49-63.

Gerrard, A.J.W. (1994a) Fractal dimensions of rock fractures: an analysis of the Dartmoor granite, Southwest England. Report to USARDSG-UK, Contract number DAJA45-92-C-0043.

Gerrard, A.J.W. (1994b) Classics in Physical Geography revisited: Linton, D.L. 1955: The Problem of Tors. The *Geographical Journal*, **121**, 470-87. *Progress in Physical Geography*, **18**, 559-63.

Gerrard, A.J.W. (1994c) A specific example of the nature and relationship of weathered granite and head on Dartmoor, South West England. *Geograficky Casopis*, **46**, 3-15.

Gerrard, A.J.W. (1994d) Weathering of granitic rocks: environment and clay mineral formation. In *Rock Weathering and Landform Evolution* (eds D.A. Robinson, and R.B.G. Williams), John Wiley and Sons, Chichester, pp. 3-20.

Gibbard, P.L. (1988) The history of the great northwest European rivers during the past three million years. *Philosophical Transactions of the Royal Society of London*, **B318**, 559-602.

Gibbard, P.L., West, R.G., Zagwijn, W.H. *et al.* (1991) Early and early Middle Pleistocene correlations in the southern North Sea Basin. *Quaternary Science Reviews*, **10**, 23-52.

Gilbert, A. (1996) The raised shoreline sequence at Saunton in North Devon. In *The Quaternary of Devon and East Cornwall: Field Guide* (eds D.J. Charman, R.M. Newnham and D.G. Croot), Quaternary Research Association, London, pp. 40-7.

Gilbert, A. (In Prep.) Quaternary history of Barnstaple Bay. Unpublished PhD thesis, University of Plymouth.

Gilbert, G.K. (1904) Domes and dome structures of the High Sierra. *Bulletin of the Geological Society of America*, **15**, 29-36.

Gilbertson, D.D. (1974) The Pleistocene succession in the coastal lowlands of Somerset. Unpublished PhD thesis, University of Bristol.

Gilbertson, D.D. (1979) The Burtle Sand Beds of

References

Somerset: the significance of freshwater interglacial molluscan faunas. *Proceedings of the Somerset Archaeological and Natural History Society*, **123**, 115-17.

Gilbertson, D.D. (1980) The palaeoecology of Middle Pleistocene Mollusca from Sugworth, Oxfordshire. *Philosophical Transactions of the Royal Society of London*, **B289**, 107-18.

Gilbertson, D.D. and Beck, R.B. (1975) A molluscan interglacial fauna from terrace gravels of the River Cary, Somerset (Abstract). *Proceedings of the Ussher Society*, **3**, 293.

Gilbertson, D.D. and Hawkins, A.B. (1974) Upper Pleistocene deposits and landforms at Holly Lane, Clevedon, Somerset (ST 419727). *Proceedings of the University of Bristol Spelaeological Society*, **13**, 349-60.

Gilbertson, D.D. and Hawkins, A.B. (1977) The Quaternary deposits at Swallow Cliff, Middlehope, County of Avon. *Proceedings of the Geologists' Association*, **88**, 255-66.

Gilbertson, D.D. and Hawkins, A.B. (1978a) *The Pleistocene Succession at Kenn, Somerset*. Bulletin of the Geological Survey of Great Britain, No. 66. Institute of Geological Sciences, H.M.S.O., London, 44 pp.

Gilbertson, D.D. and Hawkins, A.B. (1978b) The col-gully and glacial deposits at Court Hill, Clevedon, near Bristol, England. *Journal of Glaciology*, **20**, 173-88.

Gilbertson, D.D. and Hawkins, A.B. (1983) Periglacial slope deposits and frost structures along the southern margins of the Severn Estuary. *Proceedings of the University of Bristol Spelaeological Society*, **16**, 175-84.

Gilbertson, D.D. and Mottershead, D.N. (1975) The Quaternary deposits at Doniford, west Somerset. *Field Studies*, **4**, 117-29.

Gilbertson, D.D. and Sims, P.C. (1974) Some Pleistocene deposits and landforms at Ivybridge, Devon. *Proceedings of the Geologists' Association*, **85**, 65-77.

Girling, M.A. (1976) Fossil Coleoptera from the Somerset Levels: the Abbot's Way. *Somerset Levels Papers*, **2**, 28-33.

Girling, M.A. (1977a) Fossil insect assemblages from Rowland's track. *Somerset Levels Papers*, **3**, 51-60.

Girling, M.A. (1977b) Bird pellets from a Somerset Levels Neolithic trackway. *Naturalist*, **102**, 49-52.

Girling, M.A. (1978) Fossil insect assemblages from Difford's 1 site. *Somerset Levels Papers*, **4**, 107-13.

Girling, M.A. (1979a) The fossil insect assemblages from the Meare Lake Village. *Somerset Levels Papers*, **5**, 25-32.

Girling, M.A. (1979b) Fossil insects from the Sweet track. *Somerset Levels Papers*, **5**, 84-93.

Girling, M.A. (1980) The fossil insect assemblage from the Baker site. *Somerset Levels Papers*, **6**, 36-42.

Gleed-Owen, C. (1996) Amphibians and reptile remains. In *The Quaternary of Devon and East Cornwall: Field Guide* (eds D.J. Charman, R.M. Newnham and D.G. Croot), Quaternary Research Association, London, pp. 191-2.

Gleed-Owen, C. (1997) The Devensian late-glacial arrival of Natterjack Toad, *Bufo calamita*, in Britain and its implications for colonisation routes and land-bridges. *Quaternary Newsletter*, **81**, 18-24.

Goddard, A. and Coque-Delhuille, B. (1982) L'ile de Lundy (Bristol Channel – GB): bilan d'une reconnaissance géomorphologique. *Hommes et Terres du Nord*, **3**, 27-38.

Godwin, H. (1941) Studies in the post-glacial history of British vegetation. IV. Correlations in the Somerset Levels. *New Phytologist*, **40**, 108-32.

Godwin, H. (1943) Coastal peat-beds of the British Isles and North Sea. *Journal of Ecology*, **31**, 199-247.

Godwin, H. (1947) The Late-Glacial period. *Science Progress*, **138**, 185.

Godwin, H. (1948) Studies of the post-glacial history of British vegetation. X. Correlation between climate, forest composition, prehistoric agriculture and peat stratigraphy in Sub-Boreal and Sub-Atlantic peats in the Somerset Levels. *Philosophical Transactions of the Royal Society of London*, **B233**, 275-86.

Godwin, H. (1955a) Botanical and geological history of the Somerset Levels. *British Association for the Advancement of Science*, **12**, 319-22.

Godwin, H. (1955b) Studies of the post-glacial history of British vegetation. XIII. The Meare Pool region of the Somerset Levels. *Philosophical Transactions of the Royal Society of London*, **B239**, 161-90.

Godwin, H. (1956) *The History of the British Flora*. Cambridge University Press, Cambridge, 384 pp.

Godwin, H. (1960) Prehistoric wooden trackways of the Somerset Levels: their construction, age and relation to climatic change. *Proceedings of the Prehistoric Society*, **26**, 1-36.

Godwin, H. (1967) Discoveries in the peat near Shapwick Station, Somerset. *Somerset Archaeological and Natural History Society*, **111**, 20-3.

Godwin, H., Suggate, R.P. and Willis, E.H. (1958) Radiocarbon dating and the eustatic rise in ocean level. *Nature*, **181**, 1518-19.

Godwin, H. and Willis, E.H. (1959) Radiocarbon dating of prehistoric wooden trackways. *Nature*, **184**, 490-1.

Goode, A.J.J. and Taylor, R.T. (1988) *Geology of the Country around Penzance*. Memoir for 1:50,000 geological sheets 351 and 358 (England and Wales). British Geological Survey, H.M.S.O., London.

Goode, A.J.J. and Wilson, A.C. (1976) The geomorphological development of the Penzance area. *Proceedings of the Ussher Society*, **3**, 367-72.

Gordon, J.E. (1994) Conservation of geomorphology and Quaternary sites in Great Britain: an overview of site assessment. In *Conserving our Landscape: evolving landforms and Ice-Age heritage* (eds C.P. Green, J.E. Gordon, M.G. Macklin and C. Stevens), English Nature, Peterborough, pp. 11-21.

Gordon, J.E. (1997) *Reflections on the Ice Age in Scotland. An update on Quaternary studies*. Scottish Natural Heritage and Scottish Association of Geography Teachers, Edinburgh.

Gordon, J.E. and Campbell, S. (1992) Conservation of glacial deposits in Great Britain: a framework for assessment and protection of Sites of Special Scientific Interest. *Geomorphology*, **6**, 89-97.

Gordon, J.E. and Sutherland, D.G. (1993) *Quaternary of Scotland*. Geological Conservation Review Series No. 6. Chapman and Hall, London, 695 pp.

Gouldstone, T.M. (1975) Some aspects of the Quaternary history of the Bovey Basin. Unpublished MSc thesis, University of Exeter.

Grainger, P. and Kalaugher, P.G. (1996) Renewed landslide activity at Pinhay, Lyme Regis. *Proceedings of the Ussher Society*, **8**, 421-5.

Grainger, P., Kalaugher, P.G. and Kirk, S. (1996) The relation between coastal landslide activity at Pinhay, east Devon and rainfall and groundwater levels. *Proceedings of the Ussher Society*, **9**, 12-16.

Grainger, P., Tubb, C.D.N. and Nielson, A.P.M. (1985) Landslide activity at the Pinhay water source. *Proceedings of the Ussher Society*, **6**, 246-52.

Gray, H. St G. (1908) On the stone circles of east Cornwall. *Archaeologia*, **61**, 10-11.

Gray, H. St G. (1927) Palaeolith from the Axe gravels. *Antiquarians Journal*, **7**, 71-2.

Gray, H. St G. (1966) *The Meare Lake Village* Vol. III. Privately printed, Taunton.

Gray, H. St G. and Bulleid, A. (1953) *The Meare Lake Village* Vol. II. Privately printed, Taunton.

Gray, J.M. and Coxon, P. (1991) The Loch Lomond Stadial glaciation in Britain and Ireland. In *Glacial Deposits in Great Britain and Ireland* (eds J. Ehlers, P.L. Gibbard and J. Rose), Balkema, Rotterdam, pp. 89-105.

Green, C.P. (1973) Pleistocene river gravels and the Stonehenge problem. *Nature*, **243**, 214-16.

Green, C.P. (1974a) The Haldon gravels of South Devon. *Proceedings of the Geologists' Association*, **85**, 293-4.

Green, C.P. (1974b) Pleistocene gravels of the River Axe in south-western England, and their bearing on the southern limit of glaciation in Britain. *Geological Magazine*, **111**, 213-20.

Green, C.P. (1985) Pre-Quaternary weathering residues, sediments and landform development: examples from southern Britain. In *Geomorphology and soils* (eds K.S. Richards, R.R. Arnett, and S. Ellis), George Allen and Unwin, London, Boston and Sydney, pp. 58-77.

Green, C.P. (1988) The Palaeolithic site at Broom, Dorset, 1932-41: from the record of C.E. Bean, Esq. *Proceedings of the Geologists' Association*, **99**, 173-80.

Green, C.P. (1993) The Stonehenge Bluestones – Ice Age or Bronze Age? *Geology Today*, **9**, 177-8.

Green, C.P. (1997) Stonehenge: geology and prehistory. *Proceedings of the Geologists' Association*, **108**, 1-10.

Green, C.P. and Eden, M.J. (1971) Gibbsite in the weathered Dartmoor granite. *Geoderma*, **6**, 315-17.

Green, C.P. and Eden, M.J. (1973) Slope deposits on the weathered Dartmoor granite, England. *Zeitschrift für Geomorphologie*, **18**, 26-37.

Green, C.P. and Gerrard, A.J.W. (1977) Dartmoor. In *South West England* (ed. D.N. Mottershead), INQUA X Congress. Guidebook for Excursions A6 and C6. Geo Abstracts, Norwich, pp. 29-33.

References

Green, C.P., Stephens, R.A. and Shakesby, R.A. (In Prep.) Quaternary deposits at Broom Gravel Pits in the Axe Valley, South-West England.

Green, G.W. and Welch, F.B.A. (1965) *Geology of the country around Wells and Cheddar*. Memoir of the Geological Survey of Great Britain, Sheet No. 280. H.M.S.O., London.

Green, J.F.N. (1936) The terraces of southernmost England. *Quarterly Journal of the Geological Society of London*, **92**, 58-88.

Green, J.F.N. (1941) The high platforms of east Devon. *Proceedings of the Geologists' Association*, **52**, 36-52.

Green, J.F.N. (1943) The age of the raised beaches of south Britain. *Proceedings of the Geologists' Association*, **54**, 129-40.

Green, J.F.N. (1947) Some gravels and gravel-pits in Hampshire and Dorset. *Proceedings of the Geologists' Association*, **58**, 128-43.

Green, J.F.N. (1949) The history of the river Dart, Devon. *Proceedings of the Geologists' Association*, **60**, 105-24.

Greenly, E. (1921) The Pleistocene formations of Claverham and Yatton. *Proceedings of the Bristol Naturalists' Society*, Series 4, **5**, 145-7.

Greenly, E. (1922) An aeolian deposit at Clevedon. *Geological Magazine*, **59**, 365-76 and 414-21.

Greenwood, B. (1972) Modern analogues and the evaluation of a Pleistocene sedimentary sequence. *Transactions of the Institute of British Geographers*, **56**, 145-69.

Gregory, K.J. (1969) Geomorphology. In *Exeter and its Region* (ed. F. Barlow), University of Exeter Press, Exeter, pp. 27-42.

Gregory, K.J. (1971) Drainage density changes in South-West England. In *Exeter Essays in Geography* (eds K.J. Gregory and W.L.D. Ravenhill), University of Exeter Press, Exeter, pp. 33-53.

Groves, A.W. (1931) The unroofing of the Dartmoor granite and the distribution of its detritus in the sediments of southern England. *Quarterly Journal of the Geological Society of London*, **87**, 62-96.

Guilcher, A. (1949) Aspects et problèmes morphologiques du massif de Devon–Cornwall comparés à ceux d'Armorique. *Revue de Géographie Alpine*, **37**, 689-717.

Guilcher, A. (1950) Nivation, cryoplanation et solifluxion Quaternaires dans les collines de Bretagne occidentale et du nord du Devonshire. *Revue de Géomorphologie Dynamique*, **1**, 53-77.

Guilcher, A. (1957) *On Pleistocene coastal deposits at Trebetherick, Cornwall, and in Gower Peninsula, Glamorgan, Wales*. INQUA 5th Congress, Madrid-Barcelona, Résumés des Communications, pp. 71-2.

Guilcher, A. (1969) Pleistocene and Holocene sea level changes. *Earth-Science Reviews*, **5**, 69-97.

Guilcher, A. (1974) L'âge des plages anciennes de bas niveau dans le Nord-Ouest de l'Europe dans son interet morphologique. *Memoir of the Societa Geografica Italiano*, pp. 283-95.

Gullick, C.F.W.R. (1936) A physiographical survey of West Cornwall. *Transactions of the Royal Geological Society of Cornwall*, **16**, 380-99.

Hails, J.R. (1975a) Submarine geology, sediment distribution and Quaternary history of Start Bay, Devon. *Journal of the Geological Society of London*, **131**, 1-5.

Hails, J.R. (1975b) Sediment distribution and Quaternary history. *Journal of the Geological Society of London*, **131**, 19-35.

Hall, A. (1974) *West Cornwall*. Geologists' Association Guide No. 19. Benham and Company, Colchester, pp. 1-39.

Hall, A.M. (1983) Deep weathering and landform evolution in north-east Scotland. Unpublished PhD thesis, University of St Andrews.

Hall, A.M., Mellor, A.M. and Wilson, M.J. (1989) The clay mineralogy and age of deeply weathered rocks in north-east Scotland. *Zeitschrift für Geomorphologie*, NF, Supplementband, **72**, 97-108.

Hall, T.M. (1870) The raised beaches and submerged forests of Barnstaple Bay. *The Student and Intellectual Observer (of Science Literature and Art)*, **4**, 338-49.

Hall, T.M. (1879a) The submerged forest of Barnstaple Bay. *Quarterly Journal of the Geological Society of London*, **35**, 106.

Hall, T.M. (1879b) Note on the occurrence of granite boulders near Barnstaple, and a vein of granitoid rock at Portledge. *Report and Transactions of the Devonshire Association for the Advancement of Science, Literature and Art*, **11**, 429-32.

Hallégouët, B. and Van Vliet-Lanoë, B. (1989) Héritage glaciels sur les côtes du massif Armoricain, France. *Géographie Physique et Quaternaire*, **43**, 223-32.

Hamblin, R.J.O. (1968) Results of mapping the Haldon Hills. *Proceedings of the Ussher Society*, **2**, 21-2.

Hamblin, R.J.O. (1973a) The clay mineralogy of

the Haldon Gravels. *Clay Mineralogy*, **10**, 87-97.

Hamblin, R.J.O. (1973b) The Haldon Gravels of South Devon. *Proceedings of the Geologists' Association*, **84**, 459-76.

Hamblin, R.J.O. (1974a) On the correlations of the Haldon and Aller gravels, South Devon. *Proceedings of the Ussher Society*, **3**, 103-10.

Hamblin, R.J.O. (1974b) The Haldon Gravels of South Devon - a reply to C.P. Green. *Proceedings of the Geologists' Association*, **85**, 294-7.

Hamling, J.G. and Rogers, I. (1910) Excursion to North Devon, Easter 1910. *Proceedings of the Geologists' Association*, **21**, 457-72.

Hancock, J.M. (1969) Transgression of the Cretaceous sea in south west England. *Proceedings of the Ussher Society*, **2**, 61-83.

Hancock, J.M. (1975) The petrology of the Chalk. *Proceedings of the Geologists' Association*, **86**, 499-536.

Harding, J.R. (1950) Prehistoric sites on the north Cornish coast between Newquay and Perranporth. *Antiquarians' Journal*, **30**, 156-69.

Harland, W.B., Armstrong, R.L., Cox, A.V., Craig, L.E., Smith, A.G. and Smith, D.G. (1982) *A Geologic Time Scale 1989.* Cambridge University Press, Cambridge, 263 pp.

Harmer, F.W. (1907) On the origin of certain canon-like valleys associated with lake-like areas of depression. *Quarterly Journal of the Geological Society of London*, **63**, 470-514.

Harris, C. (1973) The Ice Age in Gower, as illustrated by coastal landforms and deposits, Heatherslade to Hunts Bay. *Journal of the Gower Society*, **24**, 74-9.

Harrison, R.A. (1977) The Uphill Quarry Caves, Weston-super-Mare, a reappraisal. *Proceedings of the University of Bristol Spelaeological Society*, **14**, 233-54.

Harrison, S., Anderson, E. and Winchester, V. (1996) Large boulder accumulations and evidence for permafrost creep, Great Mis Tor, Dartmoor. In *The Quaternary of Devon and East Cornwall: Field Guide* (eds D.J. Charman, R.M. Newnham and D.G. Croot), Quaternary Research Association, London, pp. 97-100.

Harrod, T.R., Catt, J.A. and Weir, A.H. (1973) Loess in Devon. *Proceedings of the Ussher Society*, **2**, 554-64.

Hart, M.B. (1982) The marine rocks of the Mesozoic. In *The Geology of Devon* (eds E.M.

Durrance and D.J.C. Laming), University of Exeter Press, Exeter, pp. 179-203.

Hatton, J. (1991) Environmental change on northern Dartmoor during the Mesolithic Period. Unpublished PhD thesis, University of Exeter.

Hawkes, C.F.C. (1943) Two palaeoliths from Broom, Dorset. *Proceedings of the Prehistoric Society*, **9**, 48-52.

Hawkes, J.R. (1982) The Dartmoor granite and later volcanic rocks. In *The Geology of Devon* (eds E.M. Durrance and D.J.C. Laming), University of Exeter Press, Exeter, pp. 85-116.

Hawkes, J.R. (1991) Appendix 2. Identifications of erratics from the Hell Bay Gravel. In J.D. Scourse, Late Pleistocene stratigraphy and palaeobotany of the Isles of Scilly. *Philosophical Transactions of the Royal Society of London*, **B334**, 405-48.

Hawkins, A.B. (1962) The buried channel of the Bristol Avon. *Geological Magazine*, **99**, 369-74.

Hawkins, A.B. (1969) Post glacial sea level changes in the Bristol Channel (Abstract). *Proceedings of the Ussher Society*, **2**, 86-7.

Hawkins, A.B. (1971) The late Weichselian and Flandrian transgressions of South-West Britain. *Quaternaria*, **14**, 115-30.

Hawkins, A.B. (1972) Some gorges of the Bristol district. *Proceedings of the Bristol Naturalists' Society*, **32**, 167-85.

Hawkins, A.B. (1973) Sea level changes around South-West England. In *Marine Archaeology* (ed. D.J. Blackman), Butterworths, London, pp. 67-87.

Hawkins, A.B. (1977) South-west Avon and Court Hill Channel. In *Field Guide to the Bristol Meeting* (ed. K. Crabtree), Quaternary Research Association, Cambridge, pp. 1-20 and 7-11.

Hawkins, A.B. and Kellaway, G.A. (1971) Field meeting at Bristol and Bath with special reference to new evidence of glaciation. *Proceedings of the Geologists' Association*, **82**, 267-91.

Hawkins, A.B. and Kellaway, G.A. (1973) 'Burtle Clay' of Somerset. *Nature*, **243**, 216-17.

Hawkins, A.B. and Tratman, E.K. (1977) The Quaternary deposits of the Mendip, Bath and Bristol areas; including a reprinting of Donovan's 1954 and 1964 bibliographies. *Proceedings of the University of Bristol Spelaeological Society*, **14**, 197-232.

Hawkins, J. (1832) On a very singular deposit of alluvial matter on St. Agnes Beacon, and on the granitical rock which occurs in the same situation. *Transactions of the Royal Geological Society of Cornwall*, **4**, 135-44.

Hays, J.D., Imbrie, J. and Shackleton, N.J. (1976) Variations in the earth's orbit, pacemaker of the Ice Ages. *Science, New York*, **194**, 1121-32.

Heal, G.J. (1970) A new Pleistocene mammal site, Mendip Hills, Somerset. *Proceedings of the University of Bristol Spelaeological Society*, **12**, 135-6.

Hedges, R.E.M., Housley, R.A., Law, I.A. and Bronk, C.R. (1989) Radiocarbon dates from the Oxford AMS system: *Archaeometry* datelist 9. *Archaeometry*, **31**, 207-34.

Hendriks, E.M.L. (1923) The physiography of South-West Cornwall, the distribution of Chalk flints, and the origin of the gravels of Crousa Common. *Geological Magazine*, **60**, 21-31.

Henry, A. (1984a) The lithostratigraphy, biostratigraphy and chronostratigraphy of coastal Pleistocene deposits in Gower, South Wales. Unpublished PhD thesis, University of Wales (Aberystwyth).

Henry, A. (1984b) Horton. In *Wales: Gower, Preseli, Fforest Fawr* (eds D.Q. Bowen and A. Henry), Quaternary Research Association Field Guide, April 1984, pp. 48-51

Henson, M.R. (1972) The form of the Permo-Triassic basin in south-east Devon. *Proceedings of the Ussher Society*, **2**, 447-57.

Henwood, W.J. (1843) On the metalliferous deposits of Cornwall and Devon. *Transactions of the Royal Geological Society of Cornwall*, **5**, 1-512.

Henwood, W.J. (1858) *Notice of the submarine forest near Padstow*. 40th Annual Report of the Royal Institution of Cornwall, Appendix 1, pp. 17-19.

Henwood, W.J. (1873) On the detrital tin-ore of Cornwall. *Journal of the Royal Institution of Cornwall*, **4**, 191-254.

Heyworth, A. and Kidson, C. (1982) Sea level changes in south-west England and Wales. *Proceedings of the Geologists' Association*, **93**, 91-111.

Heyworth, A., Kidson, C. and Wilks, P. (1985) Late-glacial and Holocene sediments at Clarach Bay, near Aberystwyth. *Journal of Ecology*, **73**, 459-80.

Hickling, G.H. (1908) China clay: its nature and origin. *Transactions of the Institution of Mining Engineers*, **36**, 10-33.

Hill, J.B. and MacAlister, D.A. (1906) *The geology of Falmouth and Truro and of the mining district of Camborne and Redruth*. Memoirs of the Geological Survey (explanation of sheet 352), H.M.S.O., London, 335 pp.

Hinton, M.A.C. (1907a) Notes on the occurrence of the Alpine Vole (*Microtus nivalis*) in the Clevedon cave deposit. *Proceedings of the Bristol Naturalists' Society*, **1**, 190-1.

Hinton, M.A.C. (1907b) On the existence of the Alpine Vole (*Microtus nivalis*) in Britain during Pleistocene times. *Proceedings of the Geologists' Association*, **20**, 39-58.

Hinton, M.A.C. (1910) A preliminary account of the British fossil voles and lemmings; with some remarks on Pleistocene climate and geography. *Proceedings of the Geologists' Association*, **20**, 489-507.

Hinton, M.A.C. (1926) *Monograph of the Voles and Lemmings (Microtinae), Living and Extinct*. Volume 1. British Museum (Natural History), London.

Hodgson, J.M., Catt, J.A. and Weir, A.H. (1967) The origin and development of Clay-with-flints and associated soil horizons of the South Downs. *Journal of Soil Science*, **18**, 85-102.

Hodgson, R.A. (1961) Classification of structures on joint surfaces. *American Journal of Science*, **259**, 493-502.

Hoegbom, B. (1913/1914) Uber die geologische bedeutung des frostes. *Bulletin of the Geological Institute of Uppsala*, **12**, 1913-14.

Holder, M.T. and Leveridge, B.E. (1986) A model for the tectonic evolution of south Cornwall. *Journal of the Geological Society of London*, **143**, 125-34.

Hollin, J.T. (1977) Thames interglacial sites, Ipswichian sea levels and Antarctic ice surges. *Boreas*, **6**, 32-52.

Hollingworth, S.E. (1939) The recognition and correlation of high level erosion surfaces in Britain. *Quarterly Journal of the Geological Society of London*, **94**, 55-79.

Holman, J.A. (1988) Herpetofauna of the late Devensian/early Flandrian Cow Cave site, Chudleigh, Devon. *Herpetological Journal*, **1**, 214-18.

Hooper, J.H.D. (1950) Reed's Cave, Buckfastleigh. *Report and Transactions of the Devonshire Association for the Advancement of Science, Literature and Art*, **82**, 291-4.

Horner, L. (1816) Sketch of the geology of the south-western part of Somersetshire. *Transactions of the Geological Society of London*, **1**, 338-84.

Hosking, K.F.G. and Camm, G.S. (1980)

References

Occurrences of pyrite framboids and polyframboids in West Cornwall. *Journal of the Camborne School of Mines*, **80**, 33–42.

Hosking, K.F.G. and Pisarski, J.B. (1964) Chemical characteristics of the cement of the Godrevy '10ft' Raised Beach and of certain deposition pipes in the St. Agnes Pliocene deposits, Cornwall. *Transactions of the Royal Geological Society of Cornwall*, **19**, 328–48.

Hosking, K.F.G. and Shrimpton, G.J. (Eds) (1964) *Present views of some aspects of the Geology of Cornwall and Devon*. Royal Geological Society of Cornwall 150th Anniversary Paper.

Hughes, C.E. (1980) Interglacial marine deposits and strandlines of the Somerset Levels. Unpublished PhD thesis, University of Wales (Aberystwyth).

Hughes, T.J. (1987) Deluge II and the continent of doom: rising sea level and collapsing Antarctic ice. *Boreas*, **16**, 89–100.

Hughes, T. Mck. (1887) On the ancient beach and boulders near Braunton and Croyde, in N. Devon. *Quarterly Journal of the Geological Society of London*, **43**, 657–70.

Hunt, A.R. (1888) The raised beach on the Thatcher Rock; its shells and their teaching. *Report and Transactions of the Devonshire Association for the Advancement of Science, Literature and Art*, **20**, 225–53.

Hunt, A.R. (1894) Four theories of the age and origin of the Dartmoor granites. *Geological Magazine*, **1**, 97–108.

Hunt, A.R. (1903) Notes and comments on the raised beaches of Torbay and Sharkham Point. *Report and Transactions of the Devonshire Association for the Advancement of Science, Literature and Art*, **35**, 318–37.

Hunt, A.R. (1913a) The age of the Torbay raised beaches. *Report and Transactions of the Devonshire Association for the Advancement of Science, Literature and Art*, **50**, 106–8.

Hunt, A.R. (1913b) Torbay and its raised beaches. *Report and Transactions of the Devonshire Association for the Advancement of Science, Literature and Art*, **64**, 377–93.

Hunt, C.O. (1981) Pollen and organic-walled microfossils in interglacial deposits from Kenn, Avon. *Proceedings of the Somerset Archaeological and Natural History Society*, **125**, 73–6.

Hunt, C.O. (1984) Erratic palynomorphs from some British tills. *Journal of Micropalaeontology*, **3**, 71–4.

Hunt, C.O. (1987) The Pleistocene history of the Langport-Chard area of Somerset. Unpublished PhD thesis, University of Wales (Aberystwyth).

Hunt, C.O. (1989) Molluscs from A.L. Armstrong's excavations in Pin Hole Cave, Creswell Crags. *Cave Science*, **16**, 97–100.

Hunt, C.O. (1990a) An interglacial pollen assemblage from the Chadbrick Gravel at Hurcott Farm, Somerset. *Proceedings of the Somerset Archaeological and Natural History Society*, **134**, 267–70.

Hunt, C.O. (1990b) A4/A36/A46 Swainswick-Batheaston bypass: geological survey of Hampton Rocks Railway Cutting. Unpublished report, Departments of Environment and Transport, Bristol.

Hunt, C.O. (In Press) Pleistocene gravels at Langport, Somerset. *Proceedings of the Somerset Archaeological and Natural History Society*.

Hunt, C.O. (In Prep.) Glacial deposits on Portishead Down, Avon. *Proceedings of the Somerset Archaeological and Natural History Society*.

Hunt, C.O. and Bowen, D.Q. (In Prep.) Stage 7 deposits at Portfield, Somerset, UK. stratigraphy, molluscs and aminostratigraphy. *Journal of Quaternary Science*.

Hunt, C.O., Bowen, D.Q. and Whatley, R. (In Prep.) A late Stage 5 high sea-level stand in Somerset, UK. *Journal of Quaternary Science*.

Hunt, C.O. and Clark, G. (1983) The palaeontology of the Burtle Beds at Middlezoy, Somerset. *Proceedings of the Somerset Archaeological and Natural History Society*, **127**, 129–30.

Hunt, C.O., Gilbertson, D.D. and Thew, N.M. (1984) The Pleistocene Chadbrick river gravels of the Cary Valley, Somerset: amino-acid racemisation and molluscan studies. *Proceedings of the Ussher Society*, **6**, 129–33.

Huntley, B. and Birks, H.J.B. (1983) *An atlas of past and present pollen maps for Europe 0–13000 years ago*. Cambridge University Press, Cambridge.

Imbrie, J. and Imbrie, K.P. (1979) *Ice Ages. Solving the mystery*. Macmillan, London, 224pp.

Imbrie, J., Hayes, J.D., Martinson, D.G. *et al.* (1984) The orbital theory of Pleistocene climate: support from revised chronology of the marine $\delta^{18}O$ record. In *Milankovich and climate. Understanding the Response to Astronomical Forcing* (eds A. Berger, J. Imbrie, J. Hays, G. Kukla and B. Saltzman), D. Reidel, Dordrecht, pp. 269–305.

References

Irfan, T.Y. and Dearman, W.R. (1978) The engineering petrography of weathered granite in Cornwall, England. *Journal of Engineering Geologists*, **11**, 233-44.

Irfan, T.Y. and Dearman, W.R. (1979a) Characterisation of weathering grades in granite using standard tests on aggregates. *Annales de la Société Géologique de Belgique*, **101**, 67-9.

Irfan, T.Y. and Dearman, W.R. (1979b) Micropetrographic and engineering characterization of a weathered granite. *Annales de la Société Géologique de Belgique*, **101**, 71-7.

Irving, B. (1996) Icthyofaunal remains from Three Holes Cave and Broken Shelter. In *The Quaternary of Devon and East Cornwall: Field Guide* (eds D.J. Charman, R.M. Newnham and D.G. Croot), Quaternary Research Association, London, pp. 193.

Irwin, D.J. and Knibbs, A.J. (1987) *Mendip Underground*. Mendip Publishing, Somerset, pp. 190-7.

Isaac, K.P. (1979) Tertiary silcretes of the Sidmouth area, East Devon. *Proceedings of the Ussher Society*, **4**, 341-54.

Isaac, K.P. (1981) Tertiary weathering profiles in the plateau deposits of East Devon. *Proceedings of the Geologists' Association*, **92**, 159-68.

Isaac, K.P. (1983a) Tertiary lateritic weathering in Devon, England, and the Palaeogene continental environment of South-West England. *Proceedings of the Geologists' Association*, **94**, 105-14.

Isaac, K.P. (1983b) Silica diagenesis of Palaeogene residual deposits in Devon, England. *Proceedings of the Geologists' Association*, **94**, 181-6.

Iversen, J. (1954) The late-glacial flora of Denmark and its relation to climate and soil. *Danmarks Geologiske Undersoegelse*, Series II, **80**, 87-119.

Jacobi, R.M. (1975) Aspects of the Postglacial archaeology of England and Wales. Unpublished PhD thesis, University of Cambridge.

Jacobi, R.M. (1979) Early Flandrian hunters in the South-West. *Proceedings of the Devon Archaeological Society*, **37**, 48-93.

Jacobi, R.M., Tallis, J.H. and Mellars, P.A. (1976) The southern Pennine Mesolithic and the ecological record. *Journal of Archaeological Science*, **3**, 307-20.

Jahn, A. (1962) Geneza skalek granitowych. *Czasopismo Geograficzne*, **33**, 19-44.

Jahn, A. (1969) Some problems concerning slope developments in the Sudetes. *Biuletyn Peryglacjalny*, **18**, 331-48.

Jahns, R.H. (1943) Sheet structure in granites: its origin and use as a measure of glacial erosion in New England. *Journal of Geology*, **51**, 71-98.

James, H.C.L. (1968) Aspects of the raised beaches of South Cornwall. *Proceedings of the Ussher Society*, **2**, 55-6.

James, H.C.L. (1974) Problems of dating raised beaches in South Cornwall. *Transactions of the Royal Geological Society of Cornwall*, **20**, 260-74.

James, H.C.L. (1975a) A Pleistocene section at Gunwalloe Fishing Cove, Lizard Peninsula. *Proceedings of the Ussher Society*, **3**, 294-8.

James, H.C.L. (1975b) An examination of recently exposed Pleistocene sections at Godrevy. *Proceedings of the Ussher Society*, **3**, 299-301.

James, H.C.L. (1981a) Evidence for Late Pleistocene environmental changes at Towan Beach, south Cornwall. *Proceedings of the Ussher Society*, **5**, 238.

James, H.C.L. (1981b) Pleistocene sections at Gerrans Bay, south Cornwall. *Proceedings of the Ussher Society*, **5**, 239-40.

James, H.C.L. (1994) Late Quaternary coastal landforms and associated sediments of west Cornwall. Unpublished PhD thesis, University of Reading.

James, H.C.L. (1995) Raised beaches of west Cornwall and their evolving geochronology. *Proceedings of the Ussher Society*, **8**, 437-40.

Jardine, W.G. (1979) The western (United Kingdom) shore of the North Sea in Late Pleistocene and Holocene times. *Acta Universitatis Upsaliensis. Annum Quingentesimum Celebrantis*, **2**, 159-74.

Jelgersma, S. (1966) Sea-level changes during the last 10,000 years. Proceedings of the Symposium of World Climate, 8000, o.b.c. 54-71.

Jelgersma, S. (1979) Sea level changes in the North Sea basin. In *The Quaternary history of the North Sea* (eds E. Oele, R.T.E. Schüttenhelm and A.J. Wiggers), Almqvist and Wiksell, Stockholm, pp. 233-48.

Jenkins, C.A. and Vincent, A. (1981) Periglacial features in the Bovey Basin, south Devon. *Proceedings of the Ussher Society*, **5**, 200-5.

Jenkins, D.A. (1986) Personal communication to

References

J.D. Scourse. In J.D. Scourse (1991) Late Pleistocene stratigraphy and palaeobotany of the Isles of Scilly. *Philosophical Transactions of the Royal Society of London*, **B334**, 430.

Jenkins, D.G., Whittaker, J.E. and Carlton, R. (1986) On the age and correlation of the St Erth Beds, South-West England, based on planktonic foraminifera. *Journal of Micropalaeontology*, **5**, 93–105.

Jessen, K. (1949) Studies on late Quaternary deposits and flora history of Ireland. *Proceedings of the Royal Irish Academy*, **B52**, 85–290.

John, B.S. (1968) Directions of ice movement in the southern Irish Sea basin during the last major glaciation: an hypothesis. *Journal of Glaciology*, **7**, 507–10.

John, B.S. (1971) Pembrokeshire. In *The Glaciations of Wales and Adjoining Regions* (ed. C.A. Lewis), Longman, London, pp. 229–65.

Johnsen, S.J., Dansgaard, W., Clausen, H.B. and Langway, C.C. Jr (1972) Oxygen isotope profiles through the Antarctic and Greenland ice sheets. *Nature*, **235**, 429–34.

Jones, A. and Keigwin, L.D. (1988) Evidence from Fram Straight (78°N) for early deglaciation. *Nature*, **336**, 56–9.

Jones, D.K.C. (1980) The Tertiary evolution of south-east England with particular reference to the Weald. In *The shaping of Southern England* (ed. D.K.C. Jones), Academic Press, London.

Jones, E.W. (1959) Biological flora of the British Isles. *Quercus* L. *Journal of Ecology*, **47**, 169–222.

Jones, O.T. (1951) The drainage systems of Wales and the adjacent areas. *Quarterly Journal of the Geological Society of London*, **107**, 201–25.

Jones, R.L. and Keen, D.H. (1993) *Pleistocene Environments in the British Isles*. Chapman and Hall, London, 346 pp.

Jones, R.L., Keen, D.H., Birnie, J.F. and Waton, P.V. (1990) *Past Landscapes of Jersey – Vegetation Development and Sea-level Change during the Flandrian Stage*. Société Jersiaise, St Helier, 145 pp.

Jones, T.R. (1859) Sketches from note-books. No. 1. Note on some granite-tors. *The Geologist*, **2**, 301–12.

Jouzel, J., Lorius, C., Petit, J.R., *et al.* (1987) Vostok ice core: a continuous isotope temperature record over the last climatic cycle (160,000 years). *Nature*, **329**, 403–8.

Jouzel, J., Barkov, N.I., Barnola, J.M. *et al.* (1993) Extending the Vostok ice-core record of palaeoclimate to the penultimate glacial period. *Nature*, **364**, 407–12.

Jouzel, J., Petit, J.R. and Raynaud, D. (1990) Palaeoclimatic information from ice cores: the Vostok records. *Transactions of the Royal Society of Edinburgh: Earth Sciences*, **81**, 349–55.

Jowsey, N.L., Parker, D.L., Slipper, I.J., Smith, A.P.C. and Walsh, P.T. (1992) The geology and geomorphology of the Beacon Cottage Farm Outlier, St Agnes, Cornwall. *Geological Magazine*, **129**, 101–21.

Jukes-Browne, A.J. (1903) *The Cretaceous rocks of Britain 2. The Lower and Middle Chalk of England*. Memoir of the Geological Survey of Great Britain. H.M.S.O., London.

Jukes-Browne, A.J. (1904a) The geology of the country around Chard. *Proceedings of the Somerset Archaeological and Natural History Society*, **49**, 12–22.

Jukes-Browne, A.J. (1904b) The valley of the Teign. *Quarterly Journal of the Geological Society of London*, **60**, 319–34.

Jukes-Browne, A.J. (1905) *The Geology of the Country South and East of Devizes*. Memoir of the Geological Survey of Great Britain.

Jukes-Browne, A.J. (1906) The clay-with-flints: its origin and distribution. *Quarterly Journal of the Geological Society of London*, **62**, 132–61.

Jukes-Browne, A.J. (1907) The age and origin of the plateau around Torquay. *Quarterly Journal of the Geological Society of London*, **63**, 106–23.

Jukes-Browne, A.J. (1911) The making of Torbay. *Report and Transactions of the Devonshire Association for the Advancement of Science, Literature and Art*, **39**, 103–36.

Kalaugher, P.G. and Grainger, P. (1981) A coastal landslide at West Dawn Beacon, Budleigh Salterton, Devon. *Proceedings of the Ussher Society*, **5**, 217–21.

Kalaugher, P.G., Grainger, P. and Hodgson, R.L.P. (1996) Tidal influence on the intermittent surging movements of a coastal mudslide. *Proceedings of the Ussher Society*, **8**, 416–20.

Keen, D.H. (1978) *The Pleistocene deposits of the Channel Islands*. Report of the Institute of Geological Sciences, 78/26, 15 pp.

Keen, D.H. (1985) The Pleistocene deposits and Mollusca from Portland, Dorset. *Geological Magazine*, **122**, 181–6.

Keen, D.H. (1990) Significance of the record provided by Pleistocene fluvial deposits and their included molluscan faunas for palaeoenvironmental reconstruction and stratigraphy. *Palaeogeography, Palaeoclimatology, Palaeoecology*, **80**, 25-34.

Keen, D.H. (Ed.) (1993) *Quaternary of Jersey: Field Guide*. Quaternary Research Association, London, 162 pp.

Keen, D.H., Harmon, R.S. and Andrews, J.T. (1981) U series and amino acid dates from Jersey. *Nature*, **289**, 162-4.

Keen, D.H., Van Vliet-Lanoë, B. and Lautridou, J.P. (1996) Two long sedimentary records from Jersey, Channel Islands: stratigraphic and pedologic evidence for environmental change during the last 200 kyr. *Quaternaire*, **7**, 3-13.

Keene, P. and Cornford, C. (1995) *The Cliffs of Saunton*. Thematic Trails, Oxford, 44 pp.

Kelland, N.C. (1975) Submarine geology of Start Bay determined by continuous seismic profiling and core sampling. *Journal of the Geological Society of London*, **131**, 7-17.

Kellaway, G.A. (1967) Geomorphology and hydrology of the central Mendips. *Proceedings of the University of Bristol Spelaeological Society*, **12**, 63-74.

Kellaway, G.A. (1971) Glaciation and the stones of Stonehenge. *Nature*, **233**, 30-5.

Kellaway, G.A., Redding, J.H., Shephard-Thorn, E.R. and Destombes, J.P. (1975) The Quaternary history of the English Channel. *Philosophical Transactions of the Royal Society of London*, **A279**, 189-218.

Kellaway, G.A. and Welch, F.B.A. (1948) *British Regional Geology: Bristol and Gloucester District* (2nd edn). H.M.S.O., London, 91 pp.

Kellaway, G.A. and Welch, F.B.A. (1955) The Upper Old Red Sandstone and Lower Carboniferous rocks of Bristol and the Mendips compared with those of Chepstow and the Forest of Dean. *Bulletin of the Geological Survey Great Britain*, **9**, 1-21.

Kellaway, G.A. and Welch, F.B.A. (1993) *Geology of the Bristol District*. Memoir of the Geological Survey of Great Britain. H.M.S.O., London.

Keller, G. and Barron, J.A. (1982) Palaeoceanographic implications of Miocene deep-sea hiatuses. *Bulletin of the Geological Society of America*, **94**, 590-613.

Keller, W.D. (1976) Scan electron micrographs of kaolins collected from diverse environments of origin, 1 and 2. *Clay and Clay Minerals*, **24**, 107-13, 114-17.

Kendall, H.G.O. (1906) The flint supplies of the Ancient Cornish. *Man*, **97**, 150-1.

Kennard, A.S. (1945) The early digs in Kent's Hole, Torquay, and Mrs. Cazalet. *Proceedings of the Geologists' Association*, **56**, 156-213.

Kennard, A.S. and Woodward, B.B. (1934) Non-marine Mollusca from the Clevedon Breccias. *Proceedings of the Geologists' Association*, **45**, 158-60.

Kerney, M.P. (1963) Late-glacial deposits on the Chalk of South-East England. *Philosophical Transactions of the Royal Society of London*, **B246**, 203-54.

Kerney, M.P. (1966) Snails and Man in Britain. *Journal of Conchology*, **26**, 3-14.

Kerney, M.P. (1976a) A list of the fresh and brackish-water Mollusca of the British Isles. *Journal of Conchology*, **29**, 26-8.

Kerney, M.P. (1976b) Two Postglacial molluscan faunas from South-West England. *Journal of Conchology*, **29**, 71-3.

Kerney, M.P., Brown, E.H. and Chandler, T.J. (1964) The Late-Glacial and Post-Glacial history of the Chalk escarpment near Brook, Kent. *Philosophical Transactions of the Royal Society of London*, **B248**, 135-204.

Kerney, M.P. and Cameron, R.A.D. (1979) *A Field Guide to the Land Snails of Britain and North-West Europe*. Collins, London, 288 pp.

Kerr, M.H. (1955) On the occurrence of silcretes in southern England. *Proceedings of the Leeds Philosophical and Literary Society*, **6**, 328-37.

Kidson, C. (1950) Dawlish Warren. *Transactions of the Institute of British Geographers*, **16**, 67-80.

Kidson, C. (1962) The denudation chronology of the River Exe. *Transactions and Papers of the Institute of British Geographers*, **31**, 43-66.

Kidson, C. (1964) The coasts of south and south-west England. In *Field Studies in the British Isles* (ed. J.A. Steers), Nelson, London, pp. 26-42.

Kidson, C. (1970) The Burtle Beds of Somerset. *Proceedings of the Ussher Society*, **2**, 189-91.

Kidson, C. (1971) The Quaternary history of the coasts of South-West England, with special reference to the Bristol Channel coast. In *Exeter Essays in Geography* (eds K.J. Gregory and W.L.D. Ravenhill), University of Exeter Press, Exeter, pp. 1-22.

Kidson, C. (1974) The Quaternary of South-West England; Westward Ho!; Saunton Down; Burtle Beds of Somerset. In *Exeter Field*

References

Meeting, Easter 1974 (ed. A. Straw), Quaternary Research Association Field Handbook, Exeter, pp. 2-4; 23-4; and 52-3.

Kidson, C. (1977) Some problems of the Quaternary of the Irish Sea and The coast of South-West England. In *The Quaternary History of the Irish Sea* (eds C. Kidson and M.J. Tooley), Geological Journal Special Issue No. 7, Seel House Press, Liverpool, pp. 1-12 and 257-98.

Kidson, C., Beck, R.B. and Gilbertson, D.D. (1981) The Burtle Beds of Somerset: temporary sections at Penzoy Farm, near Bridgwater. *Proceedings of the Geologists' Association*, **92**, 39-45.

Kidson, C. and Bowen, D.Q. (1976) Some comments on the history of the English Channel. *Quaternary Newsletter*, **18**, 8-10.

Kidson, C., Collin, R.L. and Chisholm, N.W.T. (1989) Surveying a major dune system – Braunton Burrows, north-west Devon. *Geographical Journal*, **55**, 94-105.

Kidson, C., Gilbertson, D.D., Haynes, J.R. *et al.* (1978) Interglacial marine deposits in the Somerset Levels, South West England. *Boreas*, **7**, 215-28.

Kidson, C. and Haynes, J.R. (1972) Glaciation in the Somerset Levels: the evidence of the Burtle Beds. *Nature*, **239**, 390-2.

Kidson, C., Haynes, J.R. and Heyworth, A. (1974) The Burtle Beds of Somerset – glacial or marine? *Nature*, **251**, 211-13.

Kidson, C. and Heyworth, A. (1973) The Flandrian sea-level rise in the Bristol Channel. *Proceedings of the Ussher Society*, **2**, 565-84.

Kidson, C. and Heyworth, A. (1976) The Quaternary deposits of the Somerset Levels. *Quarterly Journal of Engineering Geology*, **9**, 217-35.

Kidson, C. and Heyworth, A. (1977) Barnstaple Bay. In *South West England* (ed. D.N. Mottershead), INQUA X Congress. Guidebook for Excursions A6 and C6. Geo Abstracts, Norwich, pp. 38-44.

Kidson, C. and Tooley, M.J. (Eds) (1977) *The Quaternary History of the Irish Sea*. Geological Journal Special Issue No. 7. Seel House Press, Liverpool, 345 pp.

Kidson, C. and Wood, R. (1974) The Pleistocene stratigraphy of Barnstaple Bay. *Proceedings of the Geologists' Association*, **85**, 223-37.

Kieslinger, A. (1958) Restspannung und entspannung im gestein. *Geologie und Bauaresen*, **24**, 95-112.

King, L. (1958) Correspondence: The problem of tors. *Geographical Journal*, **124**, 289-91.

King, W.B.R. (1954) The geological history of the English Channel. *Quarterly Journal of the Geological Society of London*, **110**, 77-101.

Kirby, R.P. (1967) The fabric of head deposits in South Devon. *Proceedings of the Ussher Society*, **1**, 288-90.

Kirkaldy, J.F. (1950) Solution of the Chalk in the Mimms Valley, Hertfordshire. *Proceedings of the Geologists' Association*, **61**, 219-24.

Knight, F.A. (1902) *The sea-board of Mendip*. Dent, London.

Knill, J.L. (1972) Engineering geology in reservoir construction in South-West England. *Proceedings of the Ussher Society*, **2**, 359-71.

Konta, J. (1969) Comparison of the proofs of hydrothermal and supergene kaolinisation in two areas of Europe. *Proceedings of the International Clay Conference, Tokyo*, **1**, 281-90.

Kowalski, K. (1967) *Lagurus lagurus* (Palleo, 1773) and *Cricetus cricetus* (Linnaeus, 1758) (Rodentia Mammalia) in the Pleistocene of England. *Acta Zoologica Cracoviensia*, **12**, 111-22.

Kukla, G.J. (1977) Pleistocene land-sea correlations. I. Europe. *Earth Science Reviews*, **13**, 307-74.

Kurten, B. (1968) *Pleistocene Mammals of Europe*. Weidenfeld and Nicolson, London.

Lambeck, K. (1995) Late Devensian and Holocene shorelines of the British Isles and North Sea from models of glacio-hydro-isostatic rebound. *Journal of the Geological Society of London*, **152**, 437-48.

Laming, D.J.C. (1982) In *The geology of Devon* (eds E.M. Durrance and D.J.C. Laming), University of Exeter Press, Exeter, pp. 148-78.

Lautridou, J.P. (Ed.) (1982) *The Quaternary of Normandy*. Quaternary Research Association, Cambridge.

Lee, J.E. (1880) Letter to A.W. Franks, dated 3rd January, 1880. *Proceedings of the Society of Antiquaries*, **8**, 247-9.

Leveridge, B.E., Holder, M.T. and Goode, A.J.J. (1990) *Geology of the country around Falmouth*. Memoir for 1:50,000 geological sheet 352. British Geological Survey. H.M.S.O., London.

Lewis, C.A. (Ed.) (1970) *The Glaciations of Wales and Adjoining Regions*. Longman, London, 378 pp.

Lewis, W.V. (1939) Snow-patch erosion in

Iceland. *Geographical Journal*, **94**, 153-61.

Lewis, W.V. (1955) Discussion: The problem of tors. *Geographical Journal*, **121**, 483-4.

Libby, W.F. (1952) *Radiocarbon dating*. University of Chicago, Chicago, 75 pp.

Lidmar-Bergström, K. (1982) Pre-Quaternary geomorphological evolution in southern Scandinavia. *Sveriges Geologiska Undersökning*, Series C, No. 785.

Linton, D.L. (1951) Problems of Scottish scenery. *Scottish Geographical Journal*, **67**, 65-85.

Linton, D.L. (1955) The problem of tors. *Geographical Journal*, **121**, 470-87.

Linton, D.L. (1971) The low raised beach of southern Ireland, South Wales, Cornwall, Devon and Brittany, and its relations to earlier weathering and later gelifluxtion. *Quaternaria*, **15**, 91-8.

Linton, D.L. and Waters, R.S. (1966) The Exeter symposium: discussion. *Biuletyn Peryglacjalny*, **15**, 133-49.

Lister, A.M. (1987) Giant deer and the giant red deer from Kent's Cavern, and the status of *Strongyloceros spelaeus* Owen. *Transactions and Proceedings of the Torquay Natural History Society*, **19**, 189-98.

Lloyd, W. (1933) *The geology of the country around Torquay* (explanation of sheet 350) (second edition). Memoirs of the Geological Survey of Great Britain, H.M.S.O., London.

Lousley, J.E. (1971) *The Flora of the Isles of Scilly*. David and Charles, Newton Abbot.

Loveday, J. (1962) Plateau deposits of the southern Chiltern Hills. *Proceedings of the Geologists' Association*, **73**, 83-102.

Lowe, H.J. (1918) The caves of Tor Bryan, their excavator, excavation, products, and significance. *Transactions and Proceedings of the Torquay Natural History Society*, **2**, 198-213.

Lowe, J.J., Ammann, B., Birks, H.H. *et al.* (1994) Climate changes in areas adjacent to the North Atlantic during the Last Glacial-Interglacial Transition (14-9 ka BP): a contribution to IGCP-253. *Journal of Quaternary Science*, **9**, 185-98.

Lowe, J.J., Coope, G.R., Harkness, D.D., Sheldrick, C. and Walker, M.J.C. (1995) Direct comparison of UK temperatures and Greenland snow accumulation rates, 15-12,000 years ago. *Journal of Quaternary Science*, **10**, 175-80.

Lowe, J.J. and Walker, M.J.C. (1984) *Reconstructing Quaternary Environments*. Longman, London and New York, 389 pp.

MacAlister, D.A. (1906) *The Geology of Falmouth and Truro and the Mining District of Camborne and Redruth*. Memoir of the Geological Survey of Great Britain. H.M.S.O., London.

Macar, P. (1936) Quelques remarques sur la géomorphologie des Cornouailles et du sud de Devonshire. *Bulletin of the Geological Society of Belgium*, **62**, 152-68.

MacCulloch, J. (1848) On the granite tors of Cornwall. *Transactions of the Geological Society*, **2**, 66-78.

MacEnery, J. (1859) *Cavern Researches* (ed. E. Vivian). 8 Vols, ed. 78, 4-25, Torquay.

Macfadyen, W.A. (1970) *Geological Highlights of the West Country. A Nature Conservancy Handbook*. Butterworths, London, 296 pp.

Mackintosh, D. (1867) Railway geology, No I. – from Exeter to Newton-Bushell and Moretonhampstead. *Geological Magazine*, **4**, 390-401.

Mackintosh, D. (1868a) On a striking instance of apparent oblique lamination in granite (Abstract). *Proceedings of the Geological Society*, **24**, 278-9.

Mackintosh, D. (1868b) On the mode and extent of encroachment of the sea on some parts of the shores of the Bristol Channel. *Proceedings of the Geological Society*, **24**, 279-83.

Macklin, M.G. (1985) Floodplain sedimentation in the upper Axe Valley, Mendip, England. *Transactions of the Institute of British Geographers*, **NS10**, 235-44.

Macklin, M.G. (1986) Quaternary sedimentary environments in north Somerset and south Avon. Unpublished PhD thesis, University of Wales (Aberystwyth).

Macklin, M.G., Bradley, S.B. and Hunt, C.O. (1985) Early mining in Britain: the stratigraphic implications of heavy metals in alluvial sediments. In *Palaeoenvironmental Investigations research design methods and data analysis* (eds N.R.J. Fieller, D.D. Gilbertson and N.G.A. Ralph), British Archaeological Reports International Series No. 258, pp. 45-59.

Macklin, M.G. and Hunt, C.O. (1988) Late Quaternary alluviation and valley floor development in the upper Axe Valley, Mendip, southwest England. *Proceedings of the Geologists' Association*, **99**, 49-60.

Madgett, P.A. and Inglis, E.A. (1987) A re-appraisal of the erratic suite of the Saunton and Croyde areas, North Devon. *Report and Transactions of the Devonshire Association for the Advancement of Science*, **119**, 135-44.

References

Madgett, P.A. and Madgett, R.A. (1974) A giant erratic on Baggy Point, North Devon. *Quaternary Newsletter*, **14**, 1-2.

Maguire, D.J. (1983) The inception and growth of blanket peat: a study of northern Dartmoor. Unpublished PhD thesis, University of Bristol.

Maguire, D.J. and Caseldine, C.J. (1985) The former distribution of forest and moorland on northern Dartmoor. *Area*, **17**, 193-203.

Manley, G. (1951) The range of variation of the British climate. *Geographical Journal*, **117**, 43-65.

Mansel-Pleydell, J.C. (1857) On the Pleistocene tufaceous deposit at Blashenwell. In *A Guide to the geology of the Isle of Purbeck and South-west Coast of Hampshire* (ed. J.H. Austen), Purbeck Society, pp. 120-3.

Mansel-Pleydell, J.C. (1886) On a tufaceous deposit at Blashenwell, Isle of Purbeck. *Proceedings of the Dorset Natural Historians and Antiquarians Field Club*, **7**, 109-13.

Mansfield, R.W. and Donovan, D.T. (1989) Palaeolithic and Pleistocene sites of the Mendip, Bath and Bristol areas. Recent bibliography. *Proceedings of the University of Bristol Spelaeological Society*, **18**, 367-89.

Martin, J.M. (1872) Exmouth Warren and its threatened destruction. *Report and Transactions of the Devonshire Association for the Advancement of Science, Literature and Art*, **5**, 84-9.

Martin, J.M. (1876) Exmouth Warren and its threatened destruction. *Report and Transactions of the Devonshire Association for the Advancement of Science, Literature and Art*, **8**, 453-60.

Martin, J.M. (1893) Exmouth Warren and its threatened destruction. *Report and Transactions of the Devonshire Association for the Advancement of Science, Literature and Art*, **25**, 406-15.

Martinson, D.G., Pisias, N.G., Hays, J.D., Imbrie, J., Moore, T.C. and Shackleton, N.J. (1987) Age dating and the orbital theory of the ice ages: development of a high resolution 0 to 300,000-year chronostratigraphy. *Quaternary Research*, **27**, 1-27.

Masson-Phillips, E.N. (1958) Bunter quartzite artefacts from coastal sites in south Devon. *Report and Transactions of the Devonshire Association for the Advancement of Science, Literature and Art*, **90**, 129-45.

Mathieu, C. (1971) Contribution à l'étude des formations argileuses à silex de Thierache (France). *Pédologie*, **21**, 5-94.

Maurel, P. (1968) Sur la présence de gibbsite dans les arenes du massif du Sidobre (Tarn) et de la Montagne Noir. *Comptes Rendus. Academie des Sciences*, **266D**, 652-3.

Maw, G. (1864) On a supposed deposit of boulder-clay in North Devon. *Quarterly Journal of the Geological Society of London*, **20**, 445-51.

McCabe, A.M. (1996) Dating and rhythmicity from the last deglacial cycle in the British Isles. *Journal of the Geological Society of London*, **153**, 499-502.

McCann, S.B., Howarth, P.J. and Cogley, J.G. (1972) Fluvial processes in a periglacial environment: Queen Elizabeth Islands, North-west Territories. *Transactions of the Institute of British Geographers*, **55**, 69-82.

McFarlane, P.B. (1955) Survey of two drowned river valleys in Devon. *Geological Magazine*, **92**, 419-29.

McKeown, M.C., Edmonds, E.A., Williams, M., Freshney, E.C. and Masson-Smith, D.J. (1973) *Geology of the Country around Boscastle and Holsworthy*. Memoir of the Geological Survey of Great Britain, Sheet Nos 322 and 323. H.M.S.O., London.

McKirdy, A.P.M. (1990) Engineering for conservation: the Brean Down foreshore berm. *Earth Science Conservation*, **27**, 19.

McMahon, C.A. (1893) Notes on Dartmoor. *Quarterly Journal of the Geological Society of London*, **49**, 385-97.

Megaw, J.V.S. (1976) Gwithian, Cornwall: some notes on the evidence for Neolithic and Bronze Age settlement. In *Settlement and economy in the third and second millennium BC* (eds C. Burgess and R. Miket), British Archaeological Association Report No. 33, p. 51.

Megaw, J.V.S., Thomas, A.C. and Wailes, B. (1961) The Bronze Age settlement at Gwithian, Cornwall. *Proceedings of the West Cornwall Field Club*, **2**, 200-15.

Mellars, P.A. (1976) Fire, ecology, animal populations and man: a study of some ecological relationships in prehistory. *Proceedings of the Prehistoric Society*, **42**, 15-45.

Mellor, A. and Wilson, M.J. (1989) Origin and significance of gibbsitic montane soils in Scotland, UK. *Arctic and Alpine Research*, **21**, 417-24.

Mellors, T.W. (1977) Geological and engineering characteristics of some Kent brickearths. Unpublished PhD thesis, University of London (Imperial College).

References

Merryfield, D.L. and Moore, P.D. (1974) Prehistoric human activity and blanket peat initiation on Exmoor. *Nature*, **250**, 439-41.

Meunier, A.R. (1961) Contribution à l'étude géomorphologique du nord-ouest du Brésil. *Bulletin de la Société Géologique de France*, 7, 492-500.

Meyer, C.J.A. (1874) On the Cretaceous rocks of Beer Head and the adjacent cliff-sections, and on the relative horizons therein of the Warminster and Blackdown fossiliferous deposits. *Quarterly Journal of the Geological Society of London*, **30**, 369-93.

Miall, D. (1977) A review of the braided river depositional environment. *Earth Science Reviews*, **13**, 1-62.

Millett, F.W. (1887-1902) Notes on the fossil foraminifera of St Erth Clay Pits. *Transactions of the Royal Geological Society of Cornwall*, **10**, 213-16, 222-6.

Millett, F.W. (1887-1902) Notes on the fossil foraminifera of St Erth Clay Pits. *Transactions of the Royal Geological Society of Cornwall*, **11**, 655-61.

Millett, F.W. (1887-1902) Notes on the fossil foraminifera of St Erth Clay Pits. *Transactions of the Royal Geological Society of Cornwall*, **12**, 43-6, 174-6, 719-20.

Millot, G. (1970) *Geology of Clays.* Springer-Verlag, New York.

Milner, H.B. (1922) The nature and origin of the Pliocene deposits of the county of Cornwall and their bearing on the Pliocene geography of the south-west of England. *Quarterly Journal of the Geological Society of London*, **78**, 348-77.

Mitchell, G.F. (1948) Late-glacial deposits in Berwickshire. *New Phytologist*, **47**, 262.

Mitchell, G.F. (1951) Studies in Irish Quaternary deposits, 7. *Proceedings of the Royal Irish Academy*, **B53**, 111-206.

Mitchell, G.F. (1960) The Pleistocene history of the Irish Sea. *British Association for the Advancement of Science*, **17**, 313-25.

Mitchell, G.F. (1965) The St. Erth Beds - an alternative explanation. *Proceedings of the Geologists' Association*, **76**, 345-66.

Mitchell, G.F. (1968) Glacial gravel on Lundy Island. *Transactions of the Royal Geological Society of Cornwall*, **20**, 65-8.

Mitchell, G.F. (1972) The Pleistocene history of the Irish Sea: second approximation. *Scientific Proceedings of the Royal Dublin Society*, Series **A4**, 181-99.

Mitchell, G.F. (1986) The Scilly Isles in the Quaternary. In *The Isles of Scilly* (ed. J.D. Scourse), Field Guide. Quaternary Research Association, Coventry, pp. 33-4.

Mitchell, G.F., Catt, J.A., Weir, A.H., McMillan, N.F., Margarel, J.P. and Whatley, R.C. (1973a) The Late Pliocene marine formation at St Erth, Cornwall. *Philosophical Transactions of the Royal Society of London*, **266B**, 1-37.

Mitchell, G.F. and Orme, A.R. (1965) The Pleistocene deposits of the Scilly Isles. *Proceedings of the Ussher Society*, **1**, 190-2.

Mitchell, G.F. and Orme, A.R. (1967) The Pleistocene deposits of the Isles of Scilly. *Quarterly Journal of the Geological Society of London*, **123**, 59-92.

Mitchell, G.F., Penny, L.F., Shotton, F.W. and West, R.G. (1973b) *A correlation of Quaternary deposits in the British Isles.* Special Report of the Geological Society of London No. 4, 99 pp.

Mizzen, V.J. (1984) Ostracods from the Quaternary deposits at Low Ham, Somerset. Unpublished MSc thesis, University of Wales (Aberystwyth).

Moir, J.R. (1936) Ancient Man in Devon. *Proceedings of the Devonshire Archaeological Exploration Society*, **2**, 264-81.

Moore, C. (1870) The mammalia and other remains from drift deposits in the Bath Basin. *Proceedings of the Bath Natural History Field Club*, **2**, 37-55.

Moore, P.D. (1968) Human influence upon vegetational history in north Cardiganshire. *Nature*, **217**, 1006-9.

Moore, P.D. (1973) The influence of prehistoric cultures upon the initiation and spread of blanket bog in upland Wales. *Nature*, **241**, 350-3.

Moore, P.D., Merryfield, D.L. and Price, M.D.R. (1984) The vegetation and development of blanket mires. In *European Mires* (ed. P.D. Moore), Academic Press, London, pp. 203-35.

Morawiecka, I. (1993) Palaeokarst phenomena in the Pleistocene raised beach formations of the South West Peninsula of England. Preliminary report. *Kras i Speleologii*, **7**, 79-92.

Morawiecka, I. (1994) The nature and origin of palaeokarst phenomena in late Quaternary calcareous sandrock with special reference to the coasts of SW England and S Wales. Unpublished PhD thesis, University of Silesia.

Morey, C.R. (1983) The evolution of a barrier-lagoon system - a case study from Start Bay.

Proceedings of the Ussher Society, **5**, 454-9.

Morgan, A. (1973) Late Pleistocene environmental changes indicated by insect faunas of the English Midlands. *Boreas*, **2**, 109-29.

Morgan, C.L. (1888) The Mendips: a geological reverie. *Proceedings of the Bristol Naturalists' Society*, **5**, 236-60.

Morgan, R.A. (1976a) Dendrochronological analysis of the Abbot's Way timbers. *Somerset Levels Papers*, **2**, 21-4.

Morgan, R.A. (1976b) Tree-ring studies in the Somerset Levels: the Sweet track. *Somerset Levels Papers*, **2**, 66-77.

Morgan, R.A. (1977) Tree-ring studies in the Somerset Levels: the hurdle tracks on Ashcott Heath (Rowland's) and Walton Heath. *Somerset Levels Papers*, **3**, 61-5.

Morgan, R.A. (1978) Tree-ring studies in the Somerset Levels: the Meare Heath track. *Somerset Levels Papers*, **4**, 40-1.

Morgan, R.A. (1980) Tree-ring studies in the Somerset Levels: the Baker site. *Somerset Levels Papers*, **6**, 24-8.

Morgan, R.A. (1982) Tree-ring studies in the Somerset Levels: the Meare Heath track, 1974-1980. *Somerset Levels Papers*, **8**, 39-45.

Morton, A.C. (1982) The provenance and diagenesis of Palaeogene sandstones of southeast England as indicated by heavy mineral analysis. *Proceedings of the Geologists' Association*, **93**, 262-74.

Moss, A.J. (1966) Origin, shaping and significance of quartz sand grains. *Journal of the Geological Society of Australia*, **13**, 97-136.

Mottershead, D.N. (1964) *The Evolution of the River Lyn*. London, pp. 11-15.

Mottershead, D.N. (1967) The evolution of the Valley of Rocks and its landforms. *Exmoor Review*, **8**, 69-72.

Mottershead, D.N. (1971) Coastal head deposits between Start Point and Hope Cove, Devon. *Field Studies*, **3**, 433-53.

Mottershead, D.N. (1972) Some quantitative aspects of periglacial slope deposits in southwest England. In *International Geography* (eds W.P. Adams and F.M. Helleiner), University of Toronto Press, pp. 43-5.

Mottershead, D.N. (1976) Quantitative aspects of periglacial slope deposits in southwest England. *Biuletyn Peryglacjalny*, **25**, 35-57.

Mottershead, D.N. (Ed.) (1977a) *South West England*. INQUA X Congress. Guidebook for Excursions A6 and C6. Geo Abstracts, Norwich, 60 pp.

Mottershead, D.N. (1977b) The Quaternary evolution of the south coast of England. In *The Quaternary History of the Irish Sea* (eds C. Kidson. and M.J. Tooley), Geological Journal Special Issue No. 7. Seel House Press, Liverpool, pp. 299-320.

Mottershead, D.N. (1977c) Devon valley of rugged rocks. *Geographical Magazine*, **49**, 711-14.

Mottershead, D.N. (1981) The persistence of oil pollution on a rocky shore. *Applied Geography*, **15**, 16-29.

Mottershead, D.N. (1982a) Coastal spray weathering of bedrock in the supratidal zone at East Prawle, South Devon. *Field Studies*, **5**, 663-84.

Mottershead, D.N. (1982b) Some sources of systematic variation in the main head deposits of southwest England. *Biuletyn Peryglacjalny*, **29**, 117-28.

Mottershead, D.N. (1982c) Rapid weathering of greenschist by coastal salt spray, east Prawle, south Devon: a preliminary report. *Proceedings of the Ussher Society*, **5**, 347-53.

Mottershead, D.N., Gilbertson, D.D. and Keen, D.H. (1987) The raised beaches and shore platforms of Torbay: a re-evaluation. *Proceedings of the Geologists' Association*, **98**, 241-57.

Mourant, A.E. (1933) The raised beaches and other terraces of the Channel Islands. *Geological Magazine*, **70**, 58-66.

Mourant, A.E. (1935) The Pleistocene deposits of Jersey. *Bulletin Annuel de la Société Jersiaise*, **12**, 489-96.

Murchison, R.I. (1839) *The Silurian System*. John Murray, London, 768 pp.

Murray, J.W. and Wright, C.A. (1974) Palaeogene Foraminiferida and palaeoecology, Hampshire and Paris basins and the English Channel. *Special Papers in Palaeontology*, **14**.

Nathorst, A.A. (1873) Om den arktiska vegetationens utbredning ofver Europa norr om Alpernaunder istiden. *Ofversight af Kongl, Vetenskaps-Akademiens Forhandlingar*, **6**, Stockholm, 11-20.

Norman, C. (1975) Four Mesolithic assemblages from West Somerset. *Proceedings of the Somerset Archaeological and Natural History Society*, **119**, 26-37.

Norman, C. (1977) Archaeological finds from the Doniford gravels. In *South West England* (ed. D.N. Mottershead), INQUA X Congress. Guidebook for Excursions A6 and C6. Geo Abstracts, Norwich, pp. 46-7.

Northmore, T. (1868) Letter to the editor. *Report*

and Transactions of the Devonshire Association for the Advancement of Science, Literature and Art, **2**, 479-91.

Oeschger, H. and Langway, C.C. Jr (1989) *The Environmental record in glaciers and Ice Sheets*. John Wiley, Chichester, 401 pp.

Ogilvie, A.H. (1939-1941) *Diary of the Joint Mitnor Excavation*. Torquay Museum.

Ogilvie, A.H. (1941) Kent's Cavern. School Nature Study, 120-5.

Ollier, C.D. (1977) Applications of weathering studies. In *Applied Geomorphology* (ed. J.R. Hails), Elsevier, pp. 9-50.

Ollier, C.D. (1983) Weathering or hydrothermal alteration? *Catena*, **10**, 57-9.

Orme, A.R. (1960a) *Morphological Mapping and Geomorphic Analysis in the South Hams*. Abstracts of the Proceedings of the Third Conference of Geologists and Geomorphologists Working in the South-West of England. Bristol, 1960. Royal Geological Society of Cornwall, Penzance, pp. 20-2.

Orme, A.R. (1960b) The raised beaches and strandlines of South Devon. *Field Studies*, **1**, 109-30.

Orme, A.R. (1960c) Abandoned and composite sea cliffs in Britain and Ireland. *Irish Geography*, **4**, 279-91.

Orme, A.R. (1961) The geomorphology of the South Hams. Unpublished PhD thesis, University of Birmingham.

Orme, A.R. (1964) The geomorphology of southern Dartmoor and the adjacent area. In *Dartmoor Essays* (ed. I.G. Simmons), Devonshire Association for the Advancement of Science, Literature and Art. Devonshire Press, Torquay, 31-72.

Orme, B.J., Coles, J.M., Caseldine, A.E. and Bailey, G.N. (1981) Meare Village West, 1979. *Somerset Levels Papers*, **7**, 12-69.

Ormerod, G.W. (1858) On the rock-basins in the granite of the Dartmoor district, Devonshire. *Proceedings of the Geological Society of London*, **15**, 16-29.

Ormerod, G.W. (1868) Notes on the geology of the valleys of the upper part of the River Teign and its feeders. *Quarterly Journal of the Geological Society of London*, **23**, 418-29.

Ormerod, G.W (1869) On some of the results arising from the bedding, joints, and spheroidal structure of the granite on the eastern side of Dartmoor, Devonshire. *Quarterly Journal of the Geological Society of London*, **25**, 273-80.

Osborne-White, H.J. (1903) *The Geology of the Country around Basingstoke*. Memoir of the Geological Survey of Great Britain.

O'Sullivan, P.E., Coard, M.A. and Pickering, D.A. (1982) The use of laminated lake sediments in the estimation and calibration of erosion rates. In *Recent developments in the explanation and prediction of erosion and sediment yield* (ed. D.E. Walling), Proceedings of the Exeter Symposium, July 1982. I.A.H.S Publication No. 137, pp. 385-96.

Owen, R. (1846) *A history of British fossil Mammals and Birds*. London.

Oxaal, J. (1916) Norsk Granite. *Norges Geologiske Undersøkelse*, **76**.

Paddon (1797) A letter to Rev'd R. Polwhele. In R. Polwhele, *The history of Devonshire* (3 Vols). Caddell, Johnson and Dilly, London.

Page, N.R. (1972) On the age of the Hoxnian Interglacial. *Geological Journal*, **8**, 129-42.

Palmer, J. and Nielson, R.A. (1960) *The Origin of Tors on Dartmoor*. Abstracts of the Proceedings of the Third Conference of Geologists and Geomorphologists Working in the South-West of England. Bristol, 1960. Royal Geological Society of Cornwall, Penzance, pp. 25-6.

Palmer, J. and Nielson, R.A. (1962) The origin of granite tors on Dartmoor, Devonshire. *Proceedings of the Yorkshire Geological Society*, **33**, 315-40.

Palmer, J. and Radley, J. (1961) Gritstone tors of the English Pennines. *Zeitschrift für Geomorphologie*, **5**, 37-52.

Palmer, L.S. (1930) The Pleistocene deposits of the Bristol district. In *The Geology of the Bristol District*. British Association, Bristol.

Palmer, L.S. (1931) On the Pleistocene succession of the Bristol district. *Proceedings of the Geologists' Association*, **42**, 345-61.

Palmer, L.S. (1934) Some Pleistocene breccias near the Severn Estuary. *Proceedings of the Geologists' Association*, **45**, 145-61.

Palmer, L.S. and Hinton, M.A.C. (1929) Some gravel deposits at Walton, near Clevedon. *Proceedings of the University of Bristol Spelaeological Society*, **3**, 154-61.

Palmer, S. (1976) The Mesolithic habitation site Culver Well, Portland, Interim Note. *Proceedings of the Prehistoric Society*, **42**, 324-7.

Palmer, S. (1977) *The Mesolithic cultures of England*. Dolphin, Poole.

Pantin, H.M. and Evans, C.D.R. (1984) The Quaternary history of the central and south-western Celtic Sea. *Marine Geology*, **57**, 259-93.

Parkinson, J. (1903) The geology of the Tintagel and Davidstow district. *Quarterly Journal of the Geological Society of London*, **59**, 408-28.

Pascoe, C. (1970) Peat remains of Portmellon and Maenporth. *Journal of the Camborne and Redruth Natural History Society*, **2**, 18-21.

Pattison, S.R. (1852) On the geology of the south coast of Cornwall. *Transactions of the Royal Geological Society of Cornwall*, **7**, 208-10.

Pattison, S.R. (1865) On the geology of the Tintagel district. *Transactions of the Royal Geological Society of Cornwall*, **7**, 3-12.

Payne, A.J. and Sugden, D.E. (1990) Topography and ice sheet growth. *Earth Surface Processes and Landforms*, **15**, 625-39.

Peach, C.W. (1841) An account of the fossil organic remains found on the south-east coast of Cornwall, and in other parts of that county. *Transactions of the Royal Geological Society of Cornwall*, **6**, 12-23.

Peacock, J.D. and Harkness, D.D. (1990) Radiocarbon ages and the full-glacial to Holocene transition in seas adjacent to Scotland and southern Scandinavia: a review. *Transactions of the Royal Society of Edinburgh: Earth Sciences*, **81**, 385-96.

Pearce, F.B. (1972) The Pleistocene contribution to the development of the Valley of Rocks, Lynton, Devonshire. Unpublished manuscript, University of Ulster.

Pearce, F.B. (1982) Some theories on the evolution of the Valley of Rocks. Unpublished Manuscript, pp. 1-5.

Pemberton, M. (1980) Earth hummocks at low elevation in the Vale of Eden, Cumbria. *Transactions of the Institute of British Geographers*, **N55**, 487-501.

Penck, W. (1953) *Morphological Analysis of Land Forms: a contribution to physical geology*. Translated by Hella Czech and Katherine Cumming Boswell. Macmillan, London.

Pengelly, W. (1855) Observations on the geology of the south-western coast of Devonshire. *Transactions of the Royal Geological Society of Cornwall*, **7**, 291-7.

Pengelly, W. (1867) The raised beaches in Barnstaple Bay, North Devon. *Report and Transactions of the Devonshire Association for the Advancement of Science, Literature and Art*, **1**, 43-56.

Pengelly, W. (1868a) The submerged forest and the pebble ridge of Barnstaple Bay. *Report and Transactions of the Devonshire Association for the Advancement of Science, Literature and Art*, **2**, 415-22.

Pengelly, W. (1868b) The literature of Kent's Cavern, Torquay, prior to 1859. *Report and Transactions of the Devonshire Association for the Advancement of Science, Literature and Art*, **1**, 469-522.

Pengelly, W. (1869) The literature of Kent's Cavern Part II. *Report and Transactions of the Devonshire Association for the Advancement of Science, Literature and Art*, **3**, 191-202.

Pengelly, W. (1871) The literature of Kent's Cavern Part III. *Report and Transactions of the Devonshire Association for the Advancement of Science, Literature and Art*, **4**, 467-90.

Pengelly, W. (1873a) The granite boulder on the shore of Barnstaple Bay, North Devon. *Report and Transactions of the Devonshire Association for the Advancement of Science, Literature and Art*, **6**, 211-22.

Pengelly, W. (1873b) The ossiferous caverns and fissures in the neighbourhood of Chudleigh, Devonshire. *Report and Transactions of the Devonshire Association for the Advancement of Science, Literature and Art*, **6**, 46-60.

Pengelly, W. (1873c) The literature of the caverns at Buckfastleigh, Devonshire. *Report and Transactions of the Devonshire Association for the Advancement of Science, Literature and Art*, **6**, 70-2.

Pengelly, W. (1878) The literature of Kent's Cavern Part IV. *Report and Transactions of the Devonshire Association for the Advancement of Science, Literature and Art*, **10**, 141-81.

Pengelly, W. (1884) The literature of Kent's Cavern Part V. *Report and Transactions of the Devonshire Association for the Advancement of Science, Literature and Art*, **16**, 189-488.

Pennington, W. (1947) Studies of the Post-Glacial history of British vegetation. VII. Lake sediments: pollen diagrams from the bottom deposits of the North Basin of Windermere. *Philosophical Transactions of the Royal Society of London*, **B233**, 137-75.

Pennington, W. (1974) *The History of British vegetation*, (2nd edn). The English Universities Press, London, 152 pp.

Pennington, W. (1977) The Late Devensian flora and vegetation of Britain. *Philosophical*

Transactions of the Royal Society of London, **B280**, 247-71.

Pepper, D.M. (1973) A comparison of the 'Argile à silex' of northern France with the 'Clay with flints' of southern England. *Proceedings of the Geologists' Association*, **84**, 331-52.

Perkins, J.W. (1971) *Geology Explained in South and East Devon*. David and Charles, Newton Abbot, 192 pp.

Perkins, J.W. (1972) *Geology Explained: Dartmoor and the Tamar Valley*. David and Charles, Newton Abbot, 195 pp.

Perrin, R.M.S., Rose, J. and Davies, H. (1979) The distribution, variation and origins of pre-Devensian tills in eastern England. *Philosophical Transactions of the Royal Society of London*, **B287**, 535-70.

Phillips, J.A. (1876) On the so-called Greenstones of western Cornwall. *Quarterly Journal of the Geological Society of London*, **32**, 115.

Phillips, J.A. (1878) On the so-called Greenstones of central and eastern Cornwall. *Quarterly Journal of the Geological Society of London*, **34**, 471.

Pickard, R. (1943) Glaciation on Dartmoor. *Report and Transactions of the Devonshire Association for the Advancement of Science, Literature and Art*, **75**, 25-52.

Pickard, R. (1946) The high level gravels in East and North Devon. *Report and Transactions of the Devonshire Association for the Advancement of Science, Literature and Art*, **78**, 207-28.

Pilcher, J. (1991) Radiocarbon dating for the Quaternary scientist. In (ed.) J.J. Lowe, Radiocarbon Dating: Recent Applications and Future Potential. *Quaternary Proceedings*, **1**, 27-33.

Pillar, J.E. (1917) Evidences of glaciation in the West. *Transactions of the Plymouth Institution*, **16**, 179-87.

Pinchemel, P. (1954) *Les plaines de Craie du nord-ouest du Bassin Parisien et du sud-est du Bassin de Londres et leur bordures*. Armand Colin, Paris.

Pitts, M.W. (1981) Stones, pits and Stonehenge. *Nature*, **290**, 46-47.

Planchais, N. (1967) Analyse pollinique de la tourbière de Gizeux (Indre-et-Loire) et étude du chêne vert a l'optimum climatique. *Pollen et Spores*, **9**, 505-20.

Polwhele, R. (1797) *The History of Devonshire* (3 Vols). Caddell, Johnson and Dilly, London.

Poole, G.S. (1864) On the recent geological changes in Somerset and their date relative to the existence of Man and of certain extinct Mammalia. *Quarterly Journal of the Geological Society of London*, **20**, 118-20.

Pounder, E.J. and Macklin, M.G. (1985) The alluvial fan at Burrington Coombe, Mendip: a study of its morphology and development. *Proceedings of the Bristol Naturalists' Society*, **45**, 29-38.

Pounds, N.J.G. (1939) The Helford Depression and the 200 ft platform. *Transactions of the Royal Cornish Polytechnical Society*, **9**, 33-7.

Preece, R.C. (1990) The molluscan fauna of the Middle Pleistocene interglacial deposits at Little Oakley, Essex, and its environmental and stratigraphic implications. *Philosophical Transactions of the Royal Society of London*, **B328**, 387-407.

Prell, W.L., Imbrie, J., Martinson, D.G. *et al.* (1986) Graphic correlation of oxygen isotope stratigraphy application to the late Quaternary. *Paleoceanography*, **1**, 137-62.

Prentice, J.E. (1960) Dinantian, Namurian and Westphalian rocks of the district southwest of Barnstaple. *Quarterly Journal of the Geological Society of London*, **115**, 261-90.

Prestwich, J. (1875) Notes on the phenomena of the Quaternary Period in the Isle of Portland and around Weymouth. *Quarterly Journal of the Geological Society of London*, **31**, 29-54.

Prestwich, J. (1890) On the relation of the Westleton Beds, or Pebbly Sands of Suffolk, and on their extension inland; with some observations on the period of the final elevation and denudation of the Weald and of the Thames Valley. *Quarterly Journal of the Geological Society of London*, **46**, 120-54.

Prestwich, J. (1892) The raised beaches and 'head' or rubble drift of the south of England: and their relation to the valley drifts and to the glacial period; and on a late post-glacial submergence. *Quarterly Journal of the Geological Society of London*, **48**, 263-343.

Price, C. (1996) Evidence from Holocene and Late Pleistocene small mammal remains. In *The Quaternary of Devon and East Cornwall: Field Guide* (eds D.J. Charman, R.M. Newnham and D.G. Croot), Quaternary Research Association, London, pp. 188-9.

Proctor, C.J. (1994) A British Pleistocene chronology based on uranium series and electron spin resonance dating of speleothems. Unpublished PhD thesis, University of Bristol.

Proctor, C.J. (1996) Kent's Cavern. In *The*

References

Quaternary of Devon and East Cornwall: Field Guide (eds D.J. Charman, R.M. Newnham and D.G. Croot), Quaternary Research Association, London, pp. 163-7.

Proctor, C.J. and Smart, P.L. (1989) A new survey of Kent's Cavern, Devon. *Proceedings of the University of Bristol Spelaeological Society*, **18**, 173-86.

Proctor, C.J. and Smart, P.L. (1991) A dated cave sediment record of Pleistocene transgressions on Berry Head, Southwest England. *Journal of Quaternary Science*, **6**, 233-44.

Proctor, C.J. and Smart, P.L. (1996) Uranium series and ESR dating of speleothems from Three Holes and Tornewton caves. In *The Quaternary of Devon and East Cornwall: Field Guide* (eds D.J. Charman, R.M. Newnham and D.G. Croot), Quaternary Research Association, London, pp. 181-6.

Prokopovich, N.P. (1965) Pleistocene periglacial weathering in the Sierra Nevada, California. *Geological Society of America Special Paper*, **82**, 271.

Pryce, W. (1778) *Mineralogica Cornubiensis.* London.

Pugh, M.E. and Shearman, D.J. (1967) Cryoturbation structures at the south end of the Isle of Portland. *Proceedings of the Geologists' Association*, **78**, 463-71.

Rankine, W.F. (1956) The Mesolithic of southern England. *Research Papers of the Surrey Archaeological Society*, **4**, 63 pp.

Ravis, C.F. (1869) Supplementary notes on some late movements of the Somersetshire coast. *Proceedings of the Bristol Naturalists' Society*, **2**, 89-94.

Reid, C. (1890) *The Pliocene Deposits of Britain.* Memoirs of the Geological Survey, UK, 326pp.

Reid, C. (1896) *The Pliocene Deposits of Britain.* Memoir of the Geological Survey of the United Kingdom.

Reid, C. (1898) The Eocene deposits of east Devon. *Quarterly Journal of the Geological Society of London*, **54**, 234-8.

Reid, C. (1904) On a probable Palaeolithic floor at Prah Sands, Cornwall, and On the probable occurrence of an Eocene outlier off the Cornish coast. *Quarterly Journal of the Geological Society of London*, **60**, 106-112 and 113-19.

Reid, C. (1907) *The geology of the country around Mevagissey (explanation of sheet 353).* Memoirs of the Geological Survey, England and Wales. H.M.S.O., London, pp. 58-63.

Reid, C., Barrow, G. and Dewey, H. (1910) *The geology of the country around Padstow and Camelford (explanation of sheets 335 and 336).* Memoirs of the Geological Survey, England and Wales. H.M.S.O., London, pp. 51-7 and 79-89.

Reid, C., Barrow, G., Sherlock, R.L., MacAlister, D.A. and Dewey, H. (1911) *The geology of the country around Tavistock and Launceston (explanation of sheet 337).* Memoirs of the Geological Survey, England and Wales. H.M.S.O., London, pp. 84-8.

Reid, C., Barrow, G., Sherlock, R.L. *et al.* (1912) *The Geology of Dartmoor.* Memoirs of the Geological Survey, England and Wales. H.M.S.O., London.

Reid, C. and Flett, J.S. (1907) *The geology of the Land's End district (explanation of sheets 351 and 358).* Memoirs of the Geological Survey, England and Wales. H.M.S.O., London, pp. 68-84.

Reid, C. and Scrivenor, J.B. (1906) *The geology of the country near Newquay (explanation of sheet 346).* Memoirs of the Geological Survey, England and Wales. H.M.S.O., London, pp. 65-71.

Reid Moir, J. (1936) Ancient Man in Devon. *Proceedings of the Devon Archaeological Exploration Society*, **2**, 264-81.

Reynolds, S.H. (1902-1922) A Monograph of the British Pleistocene Mammalia: Vol. 2(1), The cave hyaena, 1902. Vol. **2**(2), The bears, 1906. Vol. **2**(3), The Canidae, 1909. Vol. 3(1), Hippopotamus, 1922. *Palaeontographical Society. Monographs (London).*

Reynolds, S.H. (1907) A bone cave at Walton, near Clevedon. *Proceedings of the Bristol Naturalists' Society*, **1**, 183-7.

Richardson, C.A. (1944) Sensitivity changes due to bleaching of the Infra-Red Stimulated Luminescence of potassium-rich feldspars. Effects of the dating of sands from the Mojave Desert, California. Unpublished PhD thesis, University of Wales (Aberystwyth).

Ridson, T. (1811) *Survey of the county of Devon.* Rees and Curtis, London. A reprint (with additions) of Ridson's original survey carried out between 1605 and 1630 – *The Chorographical Description or Survey of the County of Devon.*

Roberts, A. (1996) Evidence for Late Pleistocene and early Holocene human activity and environmental change from the Torbryan Valley, south Devon. In *The Quaternary of Devon* and East Cornwall: Field Guide (eds D.J. Charman, R.M. Newnham and D.G. Croot), Quaternary Research Association, London, pp. 168-74 and 201.

Roberts, A., Currant, A.P. *et al.* (*In Prep.*) British Museum Torbryan Caves Research Project.

Roberts, J.C. (1961) Feather-fracture and the mechanics of rock jointing. *American Journal of Science*, **259**, 481-92.

Roberts, M.B. (1986) Excavations of the Lower Pleistocene site at Amey's Eartham Pit, Boxgrove, West Sussex; a preliminary report. *Proceedings of the Prehistoric Society*, **52**, 215-45.

Roberts, M.C. (1985) The geomorphology and stratigraphy of the Lizard Loess in south Cornwall, England. *Boreas*, **14**, 75-82.

Robin, G. De Q. (1983) *The Climatic Record in Polar Ice Sheets*. Cambridge University Press, Cambridge, 212 pp.

Robinson, A.H.W. (1955) The harbour entrances of Poole, Christchurch and Pagham. *Geographical Journal*, **121**, 33-50.

Robinson, A.H.W. (1961) The hydrography of Start Bay. *Geographical Journal*, **127**, 63-77.

Robson, J. (1944) The recent geology of Cornwall: a review. *Transactions of the Royal Geological Society of Cornwall*, **17**, 132-63.

Robson, J. (1950) Coastline development in Cornwall. *Transactions of the Royal Geological Society of Cornwall*, **18**, 215-28.

Robson, J. and Simpson, S. (1960) Abstracts of the Proceedings of the Third Conference of Geologists and Geomorphologists Working in the south-west of England, Bristol 1960. *Royal Geological Society of Cornwall*. 30 pp.

Roe, D.A. (1968a) *A gazetteer of British Lower and Middle Palaeolithic sites.* Council for British Archaeology Research Report No. 8, London.

Roe, D.A. (1968b) British Lower and Middle Palaeolithic hand-axe groups. *Proceedings of the Prehistoric Society*, **34**, 1-82.

Roe, D.A. (1975) Some Hampshire and Dorset handaxes and the question of 'Early Acheulian' in Britain. *Proceedings of the Prehistoric Society*, **41**, 1-9.

Roe, D.A. (1981) *The Lower and Middle Palaeolithic Periods in Britain*. Routledge and Kegan Paul, London, Boston and Henley.

Rogers, E.H. (1946) The raised beach, submerged forest and kitchen midden of Westward Ho! and the submerged stone row of Yelland. *Proceedings of the Devon Archaeological*

Exploration Society, **3**, 109-35.

Rogers, I. (1908) On the submerged forest at Westward Ho! Bideford Bay. *Report and Transactions of the Devonshire Association for the Advancement of Science, Literature and Art*, **40**, 249-59.

Rogers, I. and Simpson, B. (1937) The flint gravel deposit of Orleigh Court, Buckland Brewer, North Devon. *Geological Magazine*, **74**, 309-16.

Rogers, J.J. (1865) Strata of the Cober Valley, Loe-Pool near Helston. *Transactions of the Royal Geological Society of Cornwall*, **7**, 352-4.

Rogers, W. (1909) Notes on the raised beaches and head of rubble in the neighbourhood of Falmouth. 76th Annual Report of the Royal Polytechnic Society of Cornwall, **1**, pp. 91-5.

Rogers, W. (1910) The raised beaches and head of the Cornish coast. *Transactions of the Royal Geological Society of Cornwall*, **13**, 351-84.

Rose, J. (1974) Small-scale spatial variability of some sedimentary properties of lodgement till and slumped till. *Proceedings of the Geologists' Association*, **85**, 239-58.

Rose, J. (1987) Status of the Wolstonian glaciation in the British Quaternary. *Quaternary Newsletter*, **53**, 1-9.

Rose, J. (1988) Stratigraphic nomenclature for the British Middle Pleistocene - procedural dogma or stratigraphic common sense. *Quaternary Newsletter*, **54**, 15-20.

Rose, J. (1989) Tracing the Baginton-Lillington Sands and Gravels from the West Midlands to East Anglia. In *West Midlands* (ed. D.H. Keen), Quaternary Research Association Field Guide, Cambridge, pp. 102-10.

Rose, J. (1991) Stratigraphic basis of the 'Wolstonian glaciation', and retention of the term 'Wolstonian' as a chronostratigraphic stage name - a discussion. In *Central East Anglia and the Fen Basin* (eds S.G. Lewis, C.A. Whiteman and D.R. Bridgland), Quaternary Research Association Field Guide, London, pp. 15-20.

Rosenfeld, A. (1964) Excavations in the Torbryan Caves, Devonshire. II. Three Holes Cave. *Proceedings of the Devon Archaeological Exploration Society*, **22**, 3-26.

Rosenfeld, A. (1969) The Palaeolithic and Mesolithic. In *Exeter and its region* (ed. F. Barlow), University of Exeter Press, Exeter, pp. 129-36.

Round, E. (1944) Raised beaches and platforms of the Marazion area. *Transactions of the Royal*

Geological Society of Cornwall, **17**, 97–108.

Ruddiman, W.F. and Kutzbach, J.E. (1990) Late Cenozoic plateau uplift and climatic change. *Transactions of the Royal Society of Edinburgh: Earth Sciences*, **81**, 301–14.

Ruddiman, W.F. and McIntyre, A. (1981) The North Atlantic during the last deglaciation. *Palaeogeography, Palaeoclimatology, Palaeoecology*, **35**, 145–214.

Ruddiman, W.F. and Raymo, M.E. (1988) Northern Hemisphere climate régimes during the past 3 Ma: possible tectonic connections. *Philosophical Transactions of the Royal Society of London*, **B318**, 411–30.

Ruddiman, W.F., Raymo, M.E. and McIntyre, A. (1986) Matuyama 41,000-year cycles: North Atlantic Ocean and Northern Hemisphere ice sheets. *Earth and Planetary Science Letters*, **80**, 117–29.

Ruddiman, W.F., Raymo, M.E., Martinson, D.G., Clement, B.M. and Backman, J. (1989) Pleistocene evolution: northern hemisphere ice sheets and North Atlantic ocean. *Paleoceanography*, **4**, 353–412.

Ruddiman, W.F. and Wright, H.E.J. (Eds) (1987) *North America and adjacent oceans during the last deglaciation*. The geology of North America, Vol. K-3. Geological Society of America, pp. 463–78.

Rzebik, B. (1968) *Crocidura* Wagler and other Insectivora (Mammalia) from the Quaternary deposits of Tornewton Cave in England. *Acta Zoologica Cracoviensia*, **13**, 251–63.

Salter, A.E. (1899) Pebbly and other gravels in southern England. *Proceedings of the Geologists' Association*, **15**, 264–86.

Saltzman, B. and Maasch, K.A. (1990) A first-order global model of late Cenozoic climatic change. *Transactions of the Royal Society of Edinburgh: Earth Sciences*, **81**, 315–25.

Sandeman, E. (1901) The Burrator works for the water-supply of Plymouth. *Minutes of Proceedings of the Institution of Civil Engineers*, **146**, 2–42.

Sanders, W. (1841) Account of a raised beach at Woodspring Hill, near Bristol. *Report of the British Association for the Advancement of Science, Transactions Section*, 102–3.

Savigear, R.A.G. (1960) *The seaward and valley slopes and cliffs at Porth Nanven, West Penwith*. Abstracts of the Proceedings of the Third Conference of Geologists and Geomorphologists Working in the South-West of England. Bristol, 1960. Royal Geological

Society of Cornwall, Penzance, pp. 22–3.

Savigear, R.A.G. (1962) Some observations on slope development in north Devon and north Cornwall. *Transactions of the Institute of British Geographers*, **31**, 23–42.

Savin, S.M. and Epstein, S. (1970) The oxygen and hydrogen isotope geochemistry of clay minerals. *Geochimica et Cosmochimica Acta*, **34**, 25–42.

Sawicki, L. (1912) Die einebungsflächen in Wales and Devon. *Comptes Rendus. Société des Sciences et des Lettres de Varsovie*, **5**, 123–34.

Scaife, R.G. (1980) Vegetational history of the Isles of Scilly. In Pollen analysis of Higher Moors, St. Mary's. *Department of the Environment Ancient Monuments Laboratory Report*, 3,047.

Scaife, R.G. (1981) Vegetational history of the Isles of Scilly, II: radiocarbon dating and interpretation of Higher Moors, St. Mary's. *Department of the Environment Ancient Monuments Laboratory Report*, 3,500.

Scaife, R.G. (1984) A history of Flandrian vegetation in the Isles of Scilly: palynological investigations of Higher Moors and Lower Moors peat mires, St Mary's. *Cornish Studies*, **11**, 33–47.

Scaife, R.G. (1986) Flandrian palaeobotany. In *The Isles of Scilly. Field Guide* (ed. J.D. Scourse), Quaternary Research Association, Coventry, pp. 28–30.

Scourse, J.D. (1984) Appendix. In R.A. Shakesby and N. Stephens, The Pleistocene gravels of the Axe Valley, Devon. *Report and Transactions of the Devonshire Association for the Advancement of Science*, **116**, 86–7.

Scourse, J.D. (1985a) Late Pleistocene stratigraphy of the Isles of Scilly and adjoining regions. Unpublished PhD thesis, University of Cambridge.

Scourse, J.D. (1985b) The Trewornan 'Lake Flat': a reinterpretation. *Quaternary Newsletter*, **46**, 11–18.

Scourse, J.D. (1986) *The Isles of Scilly. Field Guide*. Quaternary Research Association, Coventry, 151 pp.

Scourse, J.D. (1987) Periglacial sediments and landforms in the Isles of Scilly and West Cornwall. In *Periglacial Processes and Landforms in Britain and Ireland* (ed. J. Boardman), Cambridge University Press, London, pp. 225–36.

Scourse, J.D. (1991) Late Pleistocene stratigraphy and palaeobotany of the Isles of Scilly.

References

Philosophical Transactions of the Royal Society of London, **B334**, 405-48.

Scourse, J.D. (1996a) Trace fossils of talitrid sand-hoppers in interglacial littoral calcareous sandstones, Cornwall, U.K. *Quaternary Science Reviews*, **15**, 607-15.

Scourse, J.D. (1996b) Unpublished manuscript.

Scourse, J.D. (1996c) Late Pleistocene stratigraphy of north and west Cornwall. *Transactions of the Royal Geological Society of Cornwall*, **22**, 2-56.

Scourse, J.D. (1997) Transport of the Stonehenge bluestones: testing the glacial hypothesis. *Proceedings of the British Academy*, **92**, 271-314.

Scourse, J.D. and Austin, W.E.N. (1994) A Devensian Late-glacial and Holocene sea-level and water depth record from the central Celtic Sea. *Quaternary Newsletter*, **74**, 22-9.

Scourse, J.D., Austin, W.E.N., Bateman, R.M. *et al.* (1990) Sedimentology and micropalaeontology of glacimarine sediments from the Central and Southwestern Celtic Sea. In *Glacimarine Environments: processes and sediments* (eds J.A. Dowdeswell and J.D. Scourse), Geological Society Special Publication No. 53, pp. 329-47.

Scourse, J.D., Robinson, E. and Evans, C. (1991) Glaciation of the central and south-western Celtic Sea. In *Glacial Deposits in Great Britain and Ireland* (eds J. Ehlers, P.L. Gibbard and J. Rose), Balkema, Rotterdam, pp. 301-10.

Seagrief, S.C. (1959) Pollen diagrams from southern England: Wareham, Dorset and Nursling, Hampshire. *New Phytologist*, **58**, 316-25.

Seagrief, S.C. (1960) Pollen diagrams from southern England: Cranes Moor, Hampshire. *New Phytologist*, **59**, 73-83.

Seagrief, S.C. and Godwin, H. (1960) Pollen diagrams from southern England: Elstead, Surrey. *New Phytologist*, **59**, 84-91.

Seaward, D.R. (Ed.) (1982) *Sea Area Atlas of the Marine Molluscs of Britain and Ireland*. Nature Conservancy Council, Peterborough, 242 pp.

Seddon, M. (1996) Land Mollusca. In *The Quaternary of Devon and East Cornwall: Field Guide* (eds D.J. Charman, R.M. Newnham and D.G. Croot), Quaternary Research Association, London, pp. 193-5.

Sedgwick, A. and Murchison, R.I. (1840) Description of a raised beach in Barnstaple or Bideford Bay, on the north-west coast of Devonshire. *Transactions of the Geological Society of London*, **5**, 279-88.

Selby, M.J. (1977) Bornhardts of the Namib Desert. *Zeitschrift für Geomorphologie*, **NF21**, 1-13.

Selwood, E.B., Edwards, R.A., Simpson, S. *et al.* (1984) *Geology of the Country around Newton Abbot*. Memoir for 1:50,000 geological sheet 339, New Series. British Geological Survey. H.M.S.O., London.

Selwood, E.B., Thomas, J.M., Williams, B.J. *et al.* (1997) *Geology of the Country around Trevose Head and Camelford*. Memoir for 1:50,000 geological sheets 335 and 336. British Geological Survey. H.M.S.O., London.

Shackleton, N.J. (1969) The last interglacial in the marine and terrestrial records. *Proceedings of the Royal Society of London*, **B174**, 135-54.

Shackleton, N.J., Backman, J., Zimmerman, H. *et al.* (1984) Oxygen isotope calibration of the onset of ice-rafting and history of glaciation in the North Atlantic region. *Nature*, **307**, 620-3.

Shackleton, N.J., Berger, A. and Peltier, W.A. (1990) An alternative astronomical calibration of the Lower Pleistocene timescale based on ODP site 677. *Transactions of the Royal Society of Edinburgh: Earth Sciences*, **81**, 251-61.

Shackleton, N.J., Hall, M.A., Line, J. and Cang Shuxi (1983) Carbon isotope data in Core V19-30 confirm reduced carbon dioxide concentration in the ice age atmosphere. *Nature*, **306**, 319-22.

Shackleton, N.J. and Opdyke, N.D. (1973) Oxygen isotope and palaeomagnetic stratigraphy of Equatorial Pacific Core V28-238, oxygen isotope temperatures and ice volumes on a 10^5 year – 10^6 year scale. *Quaternary Research*, **3**, 39-55.

Shackleton, N.J. and Pisias, N.G. (1985) Atmospheric carbon dioxide, orbital forcing and climate. In *The carbon cycle and atmospheric CO_2: natural variations Archean to present* (eds E.T. Sundquist and W.S. Broecker), American Geophysical Union, Geophysical Monograph No. 32, pp. 303-17.

Shakesby, R.A. and Stephens, N. (1984) The Pleistocene gravels of the Axe Valley, Devon. *Report and Transactions of the Devonshire Association for the Advancement of Science*, **116**, 77-88.

Shannon, W.G. (1927) The petrography and correlation of the surface deposits in south-east Devon. *Geological Magazine*, **64**, 145-53.

Shannon, W.G. (1928) The geology of the Torquay district. *Proceedings of the Geologists' Association*, **39**, 103-36.

Sharp, R.P. (1942) Soil structures in the St Elias Range. *Journal of Geomorphology*, **5**, 274-301.

Shaw, T.R. (1949a) Pixie's Hole: Chudleigh. *British Caver*, **19**, 70-7.

Shaw, T.R. (1949b) The caves at Chudleigh. *Report and Transactions of the Devonshire Association for the Advancement of Science, Literature and Art*, **81**, 341-5.

Shaw, T.R. (1953) Lamb Leer in the 17th Century. *Proceedings of the Spelaeological Society*, **9**, 183-7.

Shearman, D.J. (1967) On Tertiary fault movements in north Devonshire. *Proceedings of the Geologists' Association*, **78**, 555-66.

Sheppard, S.M.F. (1977) The Cornubian batholith, SW England: D/H and $^{18}O/^{16}O$ studies of kaolinite and other alteration minerals. *Journal of the Geological Society of London*, **133**, 573-91.

Sheppard, S.M.F., Nielson, R.L. and Taylor, H.P. Jr (1969) Oxygen and hydrogen isotope ratios of clay minerals from porphyry copper deposits. *Economic Geology*, **64**, 757-77.

Sherman, C.E., Glenn, G.R., Jones, A.T., Burnett, W.C. and Schwarcz, H.P. (1993) New evidence for two highstands of the sea during the last interglacial, oxygen isotope substage 5e. *Geology*, **21**, 1079-82.

Shorter, A.H., Ravenhill, W.L.D. and Davies, M.S. (1964) Geography in National Parks. In *Field Studies in the British Isles* (ed. J.A. Skers), Nelson, London, pp. 43-58.

Shorter, A.H., Ravenhill, W.L.D. and Gregory, K.J. (1969) *Southwest England*. Nelson, London.

Shotton, F.W. (1973) A reply to 'On the age of the Hoxnian Interglacial' by N.R. Page. *Geological Journal*, **9**, 387-94.

Shotton, F.W. (Ed.) (1977) *British Quaternary Studies: Recent Advances*. Clarendon Press, Oxford, 298 pp.

Shotton, F.W., Keen, D.H., Coope, G.R. *et al.* (1993) The Middle Pleistocene deposits of Waverley Wood Pits, Warwickshire, England. *Journal of Quaternary Science*, **8**, 293-325.

Sibrava, V., Bowen, D.Q. and Richmond, G.M. (Eds) (1986) Quaternary glaciations in the Northern Hemisphere. *Quaternary Science Reviews*, **5**, 510 pp.

Simmons, I.G. (1962) An outline of the vegetation history of Dartmoor. *Report and Transactions of the Devonshire Association for the Advancement of Science, Literature and Art*, **92**, 555-74.

Simmons, I.G. (1964a) Pollen diagrams from Dartmoor. *New Phytologist*, **63**, 165-80.

Simmons, I.G. (1964b) An ecological history of Dartmoor. In *Dartmoor essays* (ed. I.G. Simmons), Devonshire Association for the Advancement of Science, Literature and Art. Devonshire Press, Torquay, pp. 191-215.

Simmons, I.G. (1969) Environment and early Man on Dartmoor, Devon, England. *Proceedings of the Prehistoric Society*, **35**, 203-19.

Simmons, I.G. (1975) The ecological setting of Mesolithic Man in the Highland Zone. In *The effect of Man on the landscape: the Highland Zone* (eds J.G. Evans, S. Limbrey and H. Cleere), Council for British Archaeology Report No. 11, 57-63.

Simmons, I.G. (1979) Late Mesolithic societies and the environment of the uplands of England and Wales. *Bulletin of the Institute of Archaeology of the University of London*, **16**, 111-29.

Simmons, I.G. and Innes, J.B. (1981) Tree remains in a North York Moors peat profile. *Nature*, **294**, 76-8.

Simmons, I.G., Rand, J.I. and Crabtree, K. (1983) A further pollen analytical study of the Blacklane peat section on Dartmoor, England. *New Phytologist*, **94**, 655-67.

Simmons, I.G., Rand, J.I. and Crabtree, K. (1987) Dozmary Pool, Bodmin Moor, Cornwall: a new radiocarbon dated pollen profile. In *Periglacial Landforms in Britian and Ireland* (ed. J. Boardman), Cambridge University Press, Cambridge, pp. 125-33.

Simons, J.W. (1963) *Cow Cave, Chudleigh, Devon.* Manuscript held in Torquay Museum, 6 pp.

Simpson, S. (1953) The development of the Lyn drainage system and its relation to the origin of the coast between Combe Martin and Porlock. *Proceedings of the Geologists' Association*, **64**, 14-23.

Simpson, S. (1964) The supposed 690 ft marine platform in Devon. *Proceedings of the Ussher Society*, **1**, 89-91.

Simpson, S. and Kidson, C. (1954) Field meeting Whitsun 1953, at Lynton, north Devon. *Proceedings of the Geologists' Association*, **65**, 178-81.

Simpson, S. and Robson, J. (Eds) (1958) Abstracts of the Proceedings of the (Second) Conference of Geologists and

Geomorphologists in the South-West of England. Exeter, 1958. Royal Geological Society of Cornwall, Penzance, 26 pp.

Sims, P.C. (1980) Godrevy. In *West Cornwall* (ed. P.C. Sims), Quaternary Research Association Field Handbook, West Cornwall Field Meeting, September 1980, pp. 13-14.

Sims, P.C. (1996) The erratics of Barnstaple Bay. In *The Quaternary of Devon and East Cornwall: Field Guide* (eds D.J. Charman, R.M. Newnham and D.G. Croot), Quaternary Research Association, London, 35-9.

Smith, A. (1858) On the chalk flint and greensand fragments found on the Castle Down of Tresco, one of the islands of Scilly. *Transactions of the Royal Geological Society of Cornwall*, **7**, 343-4.

Smith, A.G. (1970) The influence of Mesolithic and Neolithic Man on British vegetation: a discussion. In *Studies in the Vegetational History of the British Isles. Essays in Honour of Harry Godwin* (eds D. Walker and R.G. West), Cambridge University Press, Cambridge, pp. 81-96.

Smith, A.G. and Pilcher, J.R. (1973) Radiocarbon dates and vegetational history of the British Isles. *New Phytologist*, **72**, 903-14.

Smith, B.W., Rhodes, E.J., Stokes, S., Spooner, N.A. and Aitken, M.J. (1990) Optical dating of sediments: initial quartz results from Oxford. *Archaeometry*, **32**, 19-31.

Smith, R. (1931) *The Sturge Collection*. British Museum. 109 pp.

Smith, R.A. (1940) Some recent finds in Kent's Cavern. *Proceedings of the Torquay Natural History Society*, **8**, 59-73.

Smith, W. (1815) *A memoir to the map and delineation of the strata of England and Wales, with part of Scotland*. London.

Somervail, A. (1897) On the absence of small lakes, or tarns, from the area of Dartmoor. *Report and Transactions of the Devonshire Association for the Advancement of Science, Literature and Art*, **29**, 386-9.

Southgate, G.A. (1985) Thermoluminescence dating of beach and dune sands: potential of single-grain measurements. *Nuclear Tracks*, **10**, 743-7.

Spencer, P.J. (1974) Environmental change in the coastal sand dune belt of the British Isles. Unpublished BSc dissertation, Department of Archaeology, University of Wales, Cardiff.

Spencer, P.J. (1975) Habitat change in coastal sand-dune areas: the molluscan evidence. In *The effect of Man on the landscape: the Highland Zone* (eds J.G. Evans and S. Limbrey), Council for British Archaeology Research Report No. 11, pp. 96.

Sperling, C.H.B., Goudie, A.S., Stoddart, D.R. and Poole, G.G. (1977) Dolines of the Dorset Chalklands and other areas in southern Britain. *Transactions of the Institute of British Geographers*, **2**, 205-23.

Spratt, P. (1856) *An investigation of the movements of Teignmouth Bar*. Privately published, London.

Staines, S. (1979) Environmental change on Dartmoor. *Proceedings of the Devon Archaeological Exploration Society*, **37**, 21-47.

Steers, J.A. (1946) *The coastline of England and Wales*. Cambridge University Press, London, 644 pp.

Stephens, N. (1952) Erosion cycles in south-west Devon. Unpublished MSc thesis, University of Bristol.

Stephens, N. (1961a) *Re-examination of some Pleistocene sections in Cornwall and Devon*. Abstracts of the Proceedings of the Fourth Conference of Geologists and Geomorphologists Working in the South-West of England. Camborne, 1961. Royal Geological Society of Cornwall, Penzance, pp. 21-3.

Stephens, N. (1961b) Pleistocene sea-levels in North Devon. A reply to M.A. Arber. *Proceedings of the Geologists' Association*, **72**, 469-71.

Stephens, N. (1966a) Geomorphological studies in Ireland and Western Britain with special reference to the Pleistocene Period. Unpublished PhD thesis (2 Vols), The Queen's University, Belfast.

Stephens, N. (1966b) Some Pleistocene deposits in North Devon. *Biuletyn Peryglacjalny*, **15**, 103-14.

Stephens, N. (1970a) The West Country and Southern Ireland. In *The Glaciations of Wales and Adjoining Regions* (ed. C.A. Lewis), Longman, London, pp. 267-314.

Stephens, N. (1970b) The Lower Severn Valley. In *The Glaciations of Wales and Adjoining Regions* (ed. C.A. Lewis), Longman, London, pp. 107-24.

Stephens, N. (1973) South-west England. In G.F. Mitchell *et al.* (1973b) *A correlation of Quaternary Deposits in the British Isles*. Geological Society of London Special Report No. 4, pp. 36-45.

References

Stephens, N. (1974) Some aspects of the Quaternary of South-West England; Westward Ho!; The Fremington area; North Devon; Hartland Quay and Damhole Point; Chard area and the Axe Valley section. In *Exeter Field Meeting, Easter 1974* (ed. A. Straw), Quaternary Research Association Field Handbook, Exeter, pp. 5-7; 25-7; 28-9; 35-42; 45; and 46-51.

Stephens, N. (1977) The Axe Valley. In *South West England* (ed. D.N. Mottershead), INQUA X Congress. Guidebook for Excursions A6 and C6. Geo Abstracts, Norwich, pp. 24-9.

Stephens, N. (1980) Introduction and Porthleven. In *West Cornwall* (ed. P.C. Sims), Quaternary Research Association Field Handbook, West Cornwall Field Meeting, September 1980, pp. 1-7 and 17.

Stephens, N. (1990) Coastal landforms. In *Natural Landscapes of Britain from the Air* (ed. N. Stephens), Cambridge University Press, Cambridge, pp. 191-231.

Stephens, N. and Green, C.P. (1978) Quaternary deposits at Broom, Devon. *Quaternary Newsletter*, **26**, 14.

Stephens, N. and Sims, P.C. (1980) Pendower. In *West Cornwall* (ed. P.C. Sims), Quaternary Research Association Field Handbook, West Cornwall Field Meeting, September 1980, pp. 31-2.

Stephens, N. and Synge, F.M. (1966) Pleistocene shorelines. In *Essays in Geomorphology* (ed. G. Dury), Heinemann, London, pp. 1-66.

Stewart, J.R. (1996) Quaternary birds from the Torbryan Valley. In *The Quaternary of Devon and East Cornwall: Field Guide* (eds D.J. Charman, R.M. Newnham and D.G. Croot), Quaternary Research Association, London, pp. 190-1.

Stone, M. and Austin, W.G.C. (1961) The metasomatic origin of the potash feldspar megacrysts in the granites of south-west England. *Journal of Geology*, **69**, 464-72.

Strahan, A. (1898) *The Geology of the Isle of Purbeck and Weymouth*. Memoir of the Geological Survey of Great Britain. H.M.S.O., London.

Straw, A. (Ed.) (1974) *Exeter Field Meeting, Easter 1974.* Quaternary Research Association Field Handbook, Exeter, 61 pp.

Straw, A. (1983) Kent's Cavern. *Devon Archaeology*, **1**, 14-21.

Straw, A. (1995) Kent's Cavern: whence and whither. *Transactions and Papers, Torquay Natural History Society*, **21**, 198-211.

Straw, A. (1996) The Quaternary record of Kent's Cavern - a brief reminder and update. *Quaternary Newsletter*, **80**, 17-25.

Stride, A.H. (1962) Low Quaternary sea levels. *Proceedings of the Ussher Society*, **1**, 6-7.

Stringer, C.B., Andrews, P. and Currant, A.P. (1996) Palaeoclimatic significance of mammalian faunas from Westbury Cave, Somerset, England. In *The early Middle Pleistocene in Europe* (ed. C. Turner), Balkema, Rotterdam, pp. 135-43.

Stringer, C.B., Currant, A.P., Schwarcz, H.P. and Collcutt, S.N. (1986) Age of Pleistocene faunas from Bacon Hole, Wales. *Nature*, **320**, 59-62.

Stuart, A. and Hookway, R.J.S. (1954) *Coastal erosion at Westward Ho!, Devon.* Report to the Coast Protection Committee (Special) of the Devon County Council.

Stuart, A. and Simpson, B. (1950) The shore sands of Cornwall and Devon from Land's End to the Taw-Torridge Estuary. *Transactions of the Royal Geological Society of Cornwall*, **17**, 13-40.

Stuart, A.J. (1982a) *Pleistocene Vertebrates in the British Isles*. Longman, London and New York, 212 pp.

Stuart, A.J. (1982b) Unpublished manuscript produced for the Nature Conservancy Council.

Stuart, A.J. (1983) Pleistocene bone caves in Britain and Ireland: a short review. *Studies in Speleology*, **4**, 9-36.

Stuart, A.J. (1995) Insularity and Quaternary vertebrate faunas in Britain and Ireland. In *Island Britain: a Quaternary Perspective* (ed. R.C. Preece), Geological Society Special Publication No. 96, pp. 111-25.

Sugden, D.E. (1968) The selectivity of glacial erosion in the Cairngorm Mountains, Scotland. *Transactions of the Institute of British Geographers*, **45**, 79-92.

Sumbler, M.G. (1983a) A new look at the type Wolstonian glacial deposits of Central England. *Proceedings of the Geologists' Association*, **94**, 23-31.

Sumbler, M.G. (1983b) The type Wolstonian sequence - some further comments. *Quaternary Newsletter*, **40**, 36-9.

Summerfield, M.A. and Goudie, A.S. (1980) The sarsens of southern England: their palaeoenvironmental interpretation with reference to other silcretes. In *The Shaping of Southern England* (ed. D.K.C. Jones), Academic Press, London, pp. 71-100.

Sutcliffe, A.J. (1957) Cave fauna and cave sediments. Unpublished PhD thesis, University of London.

Sutcliffe, A.J. (1960) Joint Mitnor Cave, Buckfastleigh. *Proceedings of the Torquay Natural History Society*, **13**, 1-26.

Sutcliffe, A.J. (1966) Joint Mitnor Cave: notes on the sequence of deposits so far exposed and some suggestions on further lines of work likely to add to this sequence. Unpublished manuscript produced for the Association of the William Pengelly Cave Research Centre. 4 pp.

Sutcliffe, A.J. (1969) Pleistocene faunas of Devon. In *Exeter and its Region* (ed. F. Barlow), University of Exeter Press, Exeter, pp. 66-70.

Sutcliffe, A.J. (1974) The caves of south Devon, The William Pengelly Cave Studies Centre and Joint Mitnor Cave and The Torbryan Caves, including Tornewton Cave. In *Exeter Field Meeting, Easter 1974* (ed. A. Straw), Quaternary Research Association Field Handbook, Exeter, pp. 8-10; 16-19; and 20-2.

Sutcliffe, A.J. (1975) A hazard in the interpretation of glacial-interglacial sequences. *Quaternary Newsletter*, **17**, 1-5.

Sutcliffe, A.J. (1976) The British glacial-interglacial sequence - a reply. *Quaternary Newsletter*, **18**, 1-7.

Sutcliffe, A.J. (1977) The caves of south Devon. In *South West England* (ed. D.N. Mottershead), INQUA X Congress. Guidebook for Excursions A6 and C6. Geo Abstracts, Norwich, pp. 35-7.

Sutcliffe, A.J. (1981) Progress report on excavations in Minchin Hole, Gower. *Quaternary Newsletter*, **33**, 1-17.

Sutcliffe, A.J. (1995) Insularity of the British Isles 250,000-30,000 years ago: the mammalian including human evidence. In *Island Britain: a Quaternary Perspective* (ed. R.C. Preece), Geological Society Special Publication No. 96, pp. 127-40.

Sutcliffe, A.J., Currant, A.P. and Stringer, C.B. (1987) Evidence of sea-level changes from coastal caves with raised beach deposits, terrestrial faunas and dated stalagmites. *Progress in Oceanography*, **18**, 243-71.

Sutcliffe, A.J. and Kowalski, K. (1976) Pleistocene rodents of the British Isles. *Bulletin of the British Museum (Natural History). Geology*, **27**, 33-147.

Sutcliffe, A.J. and Zeuner, F.E. (1962) Excavations in the Torbryan Caves, Devonshire I. Tornewton Cave. *Proceedings of the Devon Archaeological Exploration Society*, **5**, 127-45.

Symons, R. (1877) Alluvium in the Par Valley. *Journal of the Royal Institution of Cornwall*, **5**, 383-4.

Synge, F.M. (1977) The coasts of Leinster (Ireland). In *The Quaternary history of the Irish Sea* (eds C. Kidson and M.J. Tooley), Geological Journal Special Issue No. 7, Seel House Press, Liverpool, pp. 199-222.

Synge, F.M. (1979) Quaternary glaciation in Ireland. *Quaternary Newsletter*, **28**, 1-18.

Synge, F.M. (1981) Quaternary glaciation and changes of sea level in the south of Ireland. *Geologie en Mijnbouw*, **60**, 305-15.

Synge, F.M. (1985) Coastal evolution. In *The Quaternary History of Ireland* (eds K.J. Edwards and W.P. Warren), Academic Press, London, pp. 115-31.

Synge, F.M. and Stephens, N. (1960) The Quaternary Period in Ireland. *Irish Geographer*, **4**, 121-30.

Tallis, J.H. and Switsur, V. (1973) Studies on southern Pennine peats. VI. A radiocarbon-dated pollen diagram from Featherbed Moss, Derbyshire. *Journal of Ecology*, **61**, 743-51.

Taylor, C.W. (1956) Erratics of the Saunton and Fremington areas. *Report and Transactions of the Devonshire Associaton for the Advancement of Science, Literature and Art*, **88**, 52-64.

Taylor, C.W. (1958) Some supplementary notes on Saunton erratics. *Report and Transactions of the Devonshire Association for the Advancement of Science, Literature and Art*, **90**, 187-91.

Taylor, H. and Taylor, E.E. (1949) An Early Beaker burial (?) at Brean Down near Weston-super-Mare. *Proceedings of the University of Bristol Spelaeological Society*, **6**, 88-92.

Taylor, J.A. (1980) Environmental changes in Wales during the Holocene period. In *Culture and Environment in Prehistoric Wales* (ed. J.A. Taylor), British Archaeological Association Report No. 76, Oxford, pp. 101-30.

Taylor, K.C., Lamorey, G.W., Doyle, G.A. *et al.* (1993) The 'flickering switch' of Late Pleistocene climate change. *Nature*, **361**, 432-6.

Taylor, R.T. and Beer, K.E. (1981) Raised beach and mined fluvial deposits near Marazion, Cornwall. *Proceedings of the Ussher Society*, **5**, 247-50.

Te Punga, M.T. (1956) Altiplanation terraces in southern England. *Biuletyn Peryglacjalny*, **4**, 331-8.

References

Te Punga, M.T. (1957) Periglaciation in southern England. *Tijdschrift Koninklijk Nederlandsch Aardrijkskundig Genootschap*, **74**, 400–12.

Ternan, J.L. and Williams, A.G. (1979) Hydrological pathways and granite weathering on Dartmoor. In *Geographical approaches to fluvial processes* (ed. A. Pitty), Geo Abstracts, Norwich, pp. 5–30.

Thomas, A.N. (1940) The Triassic rocks of North-West Somerset. *Proceedings of the Geologists' Association*, **51**, 1–43.

Thomas, C. (1985) *Exploration of a Drowned Landscape. Archaeology and History of the Isles of Scilly*. Batsford, London, 64 pp.

Thomas, M.F. (1978) The study of inselbergs. *Zeitschrift für Geomorphologie Supplementband*, **31**, 1–41.

Thorez, J., Bullock, P., Catt, J.A. and Weir, A.H. (1971) The petrography and origin of deposits filling solution pipes in the Chalk near South Mimms, Hertfordshire. *Geological Magazine*, **108**, 413–23.

Thorpe, R.S., Williams-Thorpe, O., Jenkins, D.G. and Watson, J.S. (1991) The geological source and transport of the bluestones of Stonehenge, Wiltshire, UK. *Proceedings of the Prehistoric Society*, **57**, 103–57.

Thurston, E. (1930) *British and Foreign Trees and Shrubs in Cornwall*. Cambridge University Press, Cambridge.

Todd, M. (1987) *The South-West to AD 1000*. Longman, London and New York, pp. 1–54.

Toy, H.S. (1934) The Loe Bar near Helston. *Geographical Journal*, **83**, 40–7.

Tratman, E.K., Donovan, D.T. and Campbell, J.B. (1971) The Hyaena Den (Wookey Hole), Mendip Hills, Somerset. *Proceedings of the University of Bristol Spelaeological Society*, **12**, 245–79.

Tricart, J. (1951) *Le modélé périglaciaire*. Tournier et Constans, Paris.

Tricart, J. (1956) *Cartes des phenomenes périglaciares Quaternaires en France*. Memoires Carte Géologique Detaillee, Ministere de l'Industrie et du Commerce, pp. 1–40.

Trimmer, J. (1853) On the southern termination of the erratic Tertiaries; and on the remains of a bed of gravel on the summit of Clevedon Down, Somersetshire. *Quarterly Journal of the Geological Society of London*, **9**, 282–6.

Trueman, A.E. (1938) Erosion levels in the Bristol district and their relation to the development of the scenery. *Proceedings of the Bristol Naturalists' Society*, **4**, 402–28.

Tucker, M.E. (1969) The sedimentological history of the Padstow area, North Cornwall (Abstract). *Proceedings of the Ussher Society*, **2**, 111.

Tufnell, L. (1972) Ploughing blocks with special reference to north-west England. *Biuletyn Peryglacjalny*, **21**, 237–70.

Tufnell, L. (1975) Hummocky microrelief in the Moor House area of the northern Pennines, England. *Biuletyn Peryglacjalny*, **24**, 353–68.

Turner, C. (1975) Der Einfluss grosser Mammalier auf die interglaziale vegetation. *Quartär-paleontologie*, **1**, 13–19.

Turner, C. and West, R.G. (1968) The subdivision and zonation of interglacial periods. *Eiszeitalter und Gegenwart*, **19**, 93–101.

Turner, J. (1970) Post-Neolithic disturbance of British vegetation. In *Studies in the Vegetational History of the British Isles. Essays in honour of Harry Godwin* (ed. D. Walker and R.G. West), Cambridge University Press, Cambridge, pp. 97–116.

Twidale, C.R. (1982) *Granite Landforms*. Elsevier Scientific, Amsterdam and Oxford, 372 pp.

Tyack, W. (1875) On a deposit of quartz gravel at Blue Pool, in Crowan. *Transactions of the Royal Geological Society of Cornwall*, **9**, 177–81.

Ussher, W.A.E. (1876) On the Triassic rocks of Somerset and Devon. *Quarterly Journal of the Geological Society of London*, **32**, 274–7.

Ussher, W.A.E. (1878) The chronological value of the Pleistocene deposits of Devon. *Quarterly Journal of the Geological Society of London*, **34**, 449–58.

Ussher, W.A.E. (1879a) *The post-Tertiary geology of Cornwall*. Stephen Austin, Hertford.

Ussher, W.A.E. (1879b) On the deposits of Petrockstow in Devon. *Report and Transactions of the Devonshire Association for the Advancement of Science, Literature and Art*, **11**, 422–8.

Ussher, W.A.E. (1879c) Pleistocene geology of Cornwall. Part IV. – Submerged forests and stream tin gravels. *Geological Magazine*, **6**, 251–63.

Ussher, W.A.E. (1888) The granite of Dartmoor. Parts 1 and 2. *Report and Transactions of the Devonshire Association for the Advancement of Science, Literature and Art*, **20**, 141–57.

Ussher, W.A.E. (1890) The Devonian rocks of south Devon. *Quarterly Journal of the Geological Society of London*, **46**, 487.

References

Ussher, W.A.E. (1903) *The Geology of the Country around Torquay* (1st edn). Memoir of the Geological Survey of Great Britain. H.M.S.O., London.

Ussher, W.A.E. (1904) *The Geology of the Country around Kingsbridge and Salcombe.* Memoirs of the Geological Survey, England and Wales. H.M.S.O., London.

Ussher, W.A.E. (1906) *Geology of the Country between Wellington and Chard*. Memoir of the Geological Survey of England and Wales, New Series. H.M.S.O., London, 311.

Ussher, W.A.E. (1908) *Geology of the Quantock Hills and of Taunton and Bridgwater.* Memoir of the Geological Survey of England and Wales. H.M.S.O., London.

Ussher, W.A.E. (1910) Cornwall, Devon, and West Somerset. In (eds) H.W. Monchton, and R. Herries *Geology in the field. The Jubilee volume of the Geologists' Association*, **3**, 859-96.

Ussher, W.A.E. (1912) *Geology of the Country around Ivybridge and Modbury* (sheet 349). Memoir of the Geological Survey of Great Britain. H.M.S.O., London.

Ussher, W.A.E. (1913) *The Geology of the Country around Newton Abbot. Explanation of sheet 339.* Memoirs of the Geological Survey, England and Wales. H.M.S.O., London.

Ussher, W.A.E. (1914) A geological sketch of Brean Down and its environs, with special reference to the marsh deposits of Somerset. *Proceedings of the Somerset Archaeological and Natural History Society*, **40**, 17-40.

Ussher, W.A.E., Barrow, G. and MacAlister, D.A. (1909) *Geology of the Country around Bodmin and St Austell*. Memoirs of the Geological Survey of Great Britain. H.M.S.O., London.

Vachell, E.T. (1963) Fifth report on geology. *Report and Transactions of the Devonshire Association for the Advancement of Science, Literature and Art*, **95**, 100-7.

Vancouver, C. (1808) *General view of the agriculture of the County of Devon.* London, 479 pp.

Van Der Veen, C.J. (1987) The West Antarctic ice sheet: the need to understand its dynamics. In *Dynamics of the West Antarctic ice sheet* (eds C.J. Van der Veen and J. Oerlemans), D. Reidel, Dortrecht, pp. 1-16.

Van Zeist, W. (1963) Recherches palynologiques en Bretagne occidentale. *Norois*, **10**, 5-19.

Varney, W.D. (1921) The geological history of the Pewsey Vale. *Proceedings of the Geologists' Association*, **32**, 189-205.

Vink, A.P.A. (1949) *Bejtrage tot des kennis van Loess en Dekzanden bes bijzonder van Zuidoostelijke Veluwe*. Waginen, The Netherlands.

Vivian, E. (1856) Researches in Kent's Cavern, Torquay with the original MS memoir of its first opening by the late Rev. J. MacEnery (long supposed to have been lost) and the report of the Sub-Committee of the Torquay Natural History Society. *Report of the British Association*, Cheltenham, 1856, pp. 78-80.

Wahrhaftig, C. (1965) Stepped topography of the southern Sierra Nevada, California. *Bulletin of the Geological Society of America*, **76**, 1165-90.

Wainwright, G.J. (1960) Three microlithic industries from south-west England and their affinities. *Proceedings of the Prehistoric Society*, **26**, 193-201.

Walker, H.H. and Sutcliffe, A.J. (1968) James Lyon Widger 1823-1892 and the Torbryan Caves. *Report and Transactions of the Devonshire Association for the Advancement of Science, Literature and Art*, **99**, 49-110.

Walling, D.E. (1971) Streamflow from instrumented catchments in south-east Devon. In *Exeter Essays in Geography* (eds K.J. Gregory and W.L.D. Ravenhill), University of Exeter Press, Exeter, pp. 55-81.

Walsh, P.T., Atkinson, K., Boulter, M.C. and Shakesby, R.A. (1987) The Oligocene and Miocene outliers of west Cornwall and their bearing on the geomorphological evolution of Oldland Britain. *Philosophical Transactions of the Royal Society of London*, **A323**, 211-45.

Walsh, P.T., Boulter, M.C., Ijtaba, M. and Urbani, D.M. (1972) The preservation of the Neogene Brassington Formation of the southern Pennines and its bearing on the evolution of Upland Britain. *Journal of the Geological Society of London*, **128**, 519-59.

Walsh, P.T., Edler, G.A., Edwards, B.R. *et al.* (1973) Large-scale surveys of solution subsidence deposits in the Carboniferous and Cretaceous limestones of Great Britain and Belgium and their contribution to an understanding of the mechanisms of karstic subsidence. International Association of Engineering Geology. *Proceedings of the Symposium: sink-holes and subsidence, Hanover 1973*. Deutsche Gesellschaft für Erd-und Grundbau and Dr R. Waters.

References

Walsh, P.T., Morawiecka, I. and Skawińska-Wieser, K. (1996) A Miocene palynoflora preserved by karstic subsidence in Anglesey and the origin of the Menaian Surface. *Geological Magazine*, **133**, 713-19.

Waltham, A.C., Simms, M.J., Farrant, A.R. and Goldie, H.S. (1996) *Karst and Caves of Great Britain*. Geological Conservation Review Series No. 12. Chapman and Hall, London, 358 pp.

Waters, R.S. (1951) Some aspects of the denudation chronology of south-west Devon. Unpublished MA thesis, University of Reading.

Waters, R.S. (1953) Aits and breaks of slope on Dartmoor streams. *Geography*, **38**, 67.

Waters, R.S. (1954) Pseudo-bedding in the Dartmoor granite. *Transactions of the Royal Geological Society of Cornwall*, **18**, 456-62.

Waters, R.S. (1955) Discussion. In D.L. Linton (1955) The problem of tors. *Geographical Journal*, **121**, 484-5.

Waters, R.S. (1957) Differential weathering and erosion on Oldlands. *Geographical Journal*, **123**, 503-9.

Waters, R.S. (1960a) Pre-Würm periglacial phenomena in Britain. *Les Congrès et Colloques de l'Université de Liège*, **17**, 163-76.

Waters, R.S. (1960b) Pre-Würm periglacial phenomena in Britain. *Biuletyn Peryglacjalny*, **9**, 163-76.

Waters, R.S. (1960c) *Erosion surfaces on Dartmoor and adjacent areas*. Abstracts of the Proceedings of the Third Conference of Geologists and Geomorphologists Working in the South-West of England. Bristol, 1960. Royal Geological Society of Cornwall, Penzance, pp. 28-9.

Waters, R.S. (1960d) *The bearing of superficial deposits on the age and origin of the upland plain of East Devon*. Abstracts of the Proceedings of the Third Conference of Geologists and Geomorphologists Working in the South-West of England. Bristol, 1960. Royal Geological Society of Cornwall, Penzance, pp. 26-8.

Waters, R.S. (1960e) The bearing of superficial deposits on the age and origin of the upland plain of east Devon, west Dorset and south Somerset. *Transactions of the Institute of British Geographers*, **28**, 89-97.

Waters, R.S. (1961) Involutions and ice-wedges in Devon. *Nature*, **189**, 389-90.

Waters, R.S. (1962) Altiplanation terraces and slope development in west Spitsbergen and South-West England. *Biuletyn Peryglacjalny*, **11**, 89-101.

Waters, R.S. (1964) The Pleistocene legacy to the geomorphology of Dartmoor. In *Dartmoor essays* (ed. I.G. Simmons), Devonshire Association for the Advancement of Science, Literature and Art. Devonshire Press, Torquay, pp. 73-96.

Waters, R.S. (1965) The geomorphological significance of Pleistocene frost action in south-west England. In *Essays in geography for Austin Miller* (eds J.B. Whittow and P.D. Wood), University of Reading, pp. 39-57.

Waters, R.S. (1966a) Dartmoor excursion. *Biuletyn Peryglacjalny*, **15**, 123-8.

Waters, R.S. (1966b) North Devon excursion. *Biuletyn Peryglacjalny*, **15**, 129-32.

Waters, R.S. (1971) The significance of Quaternary events for the landform of South-West England. In *Exeter essays in geography* (eds K.J. Gregory and W.L.D. Ravenhill), University of Exeter Press, Exeter, pp. 23-31.

Waters, R.S. (1974) The Haldon Hills and Dartmoor tors. In *Exeter Field Meeting, Easter 1974* (ed. A. Straw), Quaternary Research Association Field Handbook, Exeter, pp. 11 and 14-15.

Watkins, J.R. (1967) The relationship between climate and the development of landforms in the Cainozoic rocks of Queensland. *Journal of the Geological Society of Australia*, **14**, 153-68.

Watts, S. (1978) A petrographic study of silcrete from inland Australia. *Journal of Sedimentary Petrology*, **48**, 987-94.

Watts, W.A. (1980) Regional variation in the response of vegetation to Lateglacial climatic events in Europe. In *Studies in the Lateglacial of North West Europe* (eds J.J. Lowe, J.M. Gray and J.E. Robinson), Pergamon Press, Oxford, pp. 1-22.

Wedlake, A.L. (1950) Mammoth remains and Pleistocene implements found on the West Somerset coast. *Proceedings of the Somerset Archaeological and Natural History Society*, **95**, 167-8.

Wedlake, A.L. and Wedlake, D.J. (1963) Some Palaeoliths from the Doniford gravels on the coast of West Somerset. *Proceedings of the Somerset Archaeological and Natural History Society*, **107**, 93-100.

Welch, F.B.A. (1955) Note on gravels at Kenn, Somerset. *Proceedings of the University of Bristol Spelaeological Society*, **7**, 137.

Welin, E., Engstrand, L. and Vaczy, S. (1971)

Institute of Geological Sciences radiocarbon dates III. *Radiocarbon*, **14**, 331-5.

Welin, E., Engstrand, L. and Vaczy, S. (1973) Institute of Geological Sciences radiocarbon dates IV. *Radiocarbon*, **15**, 299-302.

Weller, M.R. (1959) A contribution to the geomorphology of Bodmin Moor and adjacent districts. Unpublished MA thesis, University of London.

Weller, M.R. (1960) The erosion surfaces of Bodmin Moor. *Transactions of the Royal Geological Society of Cornwall*, **19**, 233-42.

Weller, M.R. (1961) *The palaeogeography of the 430-ft shoreline stage in East Cornwall*. Abstracts of the Proceedings of the Fourth Conference of Geologists and Geomorphologists Working in the South-West of England. Camborne, 1961. Royal Geological Society of Cornwall, Penzance, pp. 23-4.

West, I.M. (1973) Carbonate cementation of some Pleistocene temperate marine sediments. *Sedimentology*, **20**, 229-49.

West, R.G. (1963) Problems of the British Quaternary. *Proceedings of the Geologists' Association*, **74**, 147-86.

West, R.G. (1968) *Pleistocene geology and biology* (1st edn). Longman, London, 379 pp.

West, R.G. (1977a) *Pleistocene geology and biology* (2nd edn). Longman, London, 440 pp.

West, R.G. (1977b) Early and Middle Devensian flora and vegetation. *Philosophical Transactions of the Royal Society of London*, **B280**, 229-46.

West, R.G. and Sparks. B.W. (1960) Coastal interglacial deposits of the English Channel. *Philosophical Transactions of the Royal Society of London*, **243**, 95-133.

West, S., Charman, D.J. and Grattan, J. (1996) Palaeoenvironmental investigations at Tor Royal, central Dartmoor. In *The Quaternary of Devon and East Cornwall: Field Guide* (eds D.J. Charman, R.M. Newnham and D.G. Croot), Quaternary Research Association, London, pp. 62-80.

Weston, C.H. (1850) On the Diluvia and valleys in the vicinity of Bath. *Quarterly Journal of the Geological Society of London*, **6**, 449-51.

Weston, C.H. (1852) On the sub-escarpments of the Ridgeway Range and their contemporary deposits in the Isle of Portland. *Quarterly Journal of the Geological Society of London*, **8**, 110-20.

Whimster, R. (1977) Harlyn Bay reconsidered: the excavations of 1900-1905 in the light of recent work. *Cornish Archaeology*, **16**, 61-88.

Whitaker, W. (1869) On the raised beach at Portland Bill, Dorset. *Geological Magazine*, **6**, 438-40.

White, W.A. (1945) Origin of granite domes in the southeastern Piedmont. *Journal of Geology*, **53**, 276-82.

Whitley, D.G. (1907) The head of rubble on the Cornish coast. *Journal of the Royal Institution of Cornwall*, **17**, 63-81.

Whitley, D.G. (1908) On the occurrence of trees and vegetable remains in the stream tin in Cornwall. *Transactions of the Royal Geological Society of Cornwall*, **13**, 237-56.

Whitley, D.G. (1910) The angular quartz drift of Cornwall. *Transactions of the Royal Geological Society of Cornwall*, **13**, 329-50.

Whitley, N. (1866) On recent flint finds in the south-west of England. *Journal of the Royal Insitution of Cornwall*, **2**, 121-4.

Whitley, N. (1868) On supposed glacial markings in the valley of the Exe, north Devon. *Quarterly Journal of the Geological Society of London*, **24**, 3-4.

Whitley, N. (1882) The evidence of glacial action in Cornwall and Devon. *Transactions of the Royal Geological Society of Cornwall*, **10**, 132-41.

Whitley, N. (1885) Traces of a great post-glacial flood in Cornwall. *Journal of the Royal Institution of Cornwall*, **8**, 240-2.

Whittaker, A. and Green, G.W. (1983) *Geology of the country around Weston-super-Mare*. Memoir for 1:50 000 geological sheet 279, New Series, with parts of sheets 263 and 295. Institute of Geological Sciences. H.M.S.O., London, 147 pp.

Whittington, G., Fallick, A.E. and Edwards, K.J. (1996) Stable oxygen isotope and pollen records from eastern Scotland and a consideration of Late-glacial and early Holocene climate change for Europe. *Journal of Quaternary Science*, **11**, 327-40.

Widger, J.L. (1892) The Torbryan Caves. *Torquay Directory* [Newspaper], 15th June.

Widger, J.L. (Ms.) Unpublished manuscript held in the Torquay Natural History Archives.

Wilkinson, G.C. and Boulter, M.C. (1980) Oligocene pollen and spores from the western part of the British Isles. *Palaeontographica Abteilung*, **B175**, 27-83.

Willemsen, G.F. (1992) A revision of the Pliocene and Quaternary Lutrinae from Europe. *Scripta Geologica*, **101**, 1-115.

References

Williams, D. (1837) On the raised beaches of Saunton Downend and Baggy Point (Abstract). *Proceedings of the Geological Society*, **2**, 535.

Williams, P.F. and Rust, B.R. (1969) The sedimentology of a braided river. *Journal of Sedimentary Petrology*, **39**, 649-79.

Williams, P.J. (1957) Some investigations into solifluction features in Norway. *Geographical Journal*, **123**, 42-58.

Williams, R.B.G. (1969) Permafrost and temperature conditions in England during the last glacial period. In *The Periglacial Environment* (ed. T.L. Péwé), McGill-Queen's University Press, Montreal, pp. 399-410.

Williams, R.B.G. (1975) The British climate during the last glaciation: an interpretation based on periglacial phenomena. In *Ice Ages ancient and modern* (eds A.E. Wright and F. Moseley), Seel House Press, Liverpool, pp. 95-120.

Willing, M.J. (1985) The biostratigraphy of Flandrian tufa deposits in the Cotswold and Mendip districts. Unpublished PhD thesis, University of Sussex.

Wilson, A.C. (1974) Some aspects of Quaternary fluvial activity in south-west Cornwall, England. *Proceedings of the Ussher Society*, **3**, 120-2.

Wilson, A.C. (1975) A late-Pliocene marine transgression at St. Erth, Cornwall, and its possible geomorphic significance. *Proceedings of the Ussher Society*, **3**, 289-92.

Wilson, V., Welch, F.B.A., Robbie, J.A. and Green, G.W. (1958) *Geology of the Country around Bridport and Yeovil*. Memoir of the Geological Survey of Great Britain. H.M.S.O., London.

Wimbledon, W.A., Benton, M.J., Bevins, R.E. *et al.* (1995) The development of a methodology for the selection of British geological sites for conservation: Part 1. *Modern Geology*, **20**, 159-202.

Wintle, A.G. (1981) Thermoluminescence dating of late Devensian loesses in southern England. *Nature*, **289**, 479-80.

Winwood, H.H. (1889) List of fossil Mammalia found near Bath. *Proceedings of the Bath Natural History and Antiquarian Field Club*, **6**, 95.

Witherick, M.E. (1963) Stages in the growth of urban settlement in central Cornwall. Unpublished PhD thesis, University of Birmingham.

Wood, T.R. (1970) Some aspects of the Quaternary deposits of South-West England. Unpublished PhD thesis, University of Wales (Aberystwyth).

Wood, T.R. (1974) Quaternary deposits around Fremington. In *Exeter Field Meeting, Easter 1974* (ed. A. Straw), Quaternary Research Association Handbook, Exeter, pp. 30-4.

Woodward, H.B. (1876) *Geology of East Somerset and the Bristol Coal-Fields*. Memoir of the Geological Survey of England and Wales. H.M.S.O., London.

Woodward, H.B. (1905) Geology of the railway cuttings between Langport and Castle Cary. *Summary of the Progress of the Geological Survey for 1904*. H.M.S.O., London.

Woodward, H.B. and Ussher, W.A.E. (1911) *The geology of the country near Sidmouth and Lyme Regis. Explanation of sheets 326 and 340*, (2nd edn). Memoirs of the Geological Survey, England and Wales. H.M.S.O., London.

Wooldridge, S.W. (1950) The upland plains of Britain: their origin and geographical significance. *Advancement of Science*, **7**, 162-75.

Wooldridge, S.W. (1954) The physique of the South-West. *Geography*, **39**, 231-42.

Wooldridge, S.W. and Kirkaldy, J.F. (1937) The geology of the Mimms Valley. *Proceedings of the Geologists' Association*, **48**, 307-15.

Wooldridge, S.W. and Linton, D.L. (1939) Structure, surface and drainage in south-east England. *Transactions of the Institute of British Geographers*, **10**, 1-124.

Wooldridge, S.W. and Linton, D.L. (1955) *Structure, surface and drainage in south-east England*. Philip, London.

Worssam, B.C. (1963) *Geology of the Country around Maidstone*. Memoir of the Geological Survey of Great Britain. H.M.S.O., London.

Worssam, B.C. (1981) Pleistocene deposits and superficial structures, Allington Quarry, Maidstone, Kent. In *The Quaternary in Britain. Essays, reviews and original work on the Quaternary published in honour of Lewis Penny on his retirement* (eds J. Neale and J. Flenley), Pergamon Press, Oxford, pp. 20-31.

Worth, R.H. (1898) Evidences of glaciation in Devonshire. *Report and Transactions of the Devonshire Association for the Advancement of Science, Literature and Art*, **30**, 378-90.

Worth, R.H. (1904) Hallsands and Start Bay. *Report and Transactions of the Devonshire Association for the Advancement of Science, Literature and Art*, **36**, 302-46.

Worth, R.H. (1909) Hallsands and Start Bay.

Report and Transactions of the Devonshire Association for the Advancement of Science, Literature and Art, **41**, 301-8.

Worth, R.H. (1923) Hallsands and Start Bay. *Report and Transactions of the Devonshire Association for the Advancement of Science, Literature and Art*, **55**, 131-47.

Worth, R.H. (1930) The physical geography of Dartmoor. *Report and Transactions of the Devonshire Association for the Advancement of Science, Literature and Art*, **62**, 49-115.

Worth, R.H. (1934) An antler from the submerged forest at Westward Ho! *Report and Transactions of the Devonshire Association for the Advancement of Science, Literature and Art*, **46**, 127.

Worth, R.H. (1967) *Worth's Dartmoor* (eds G.M. Spoones and F.S. Russell). David and Charles, Newton Abbot.

Wright, W.B. (1924) Age and origin of the Lough Neagh Clays. *Quarterly Journal of the Geological Society of London*, **80**, 468-88.

Wunsch, E.A. (1895) On raised beaches. *Transactions of the Royal Geological Society of Cornwall*, **11**, 605-10.

Wymer, J.J. (1968) *Lower Palaeolithic Archaeology in Britain as Represented by the Thames Valley*. Baker, London.

Wymer, J.J. (1970) Clactonian and Acheulian industries in Britain. *Proceedings of the Geologists' Association*, **85**, 391-421.

Wymer, J.J. (1977) A chert hand-axe from Chard, Somerset. *Proceedings of the Somerset Archaeology and Natural History Society*, **120**, 101-3.

Wymer, J.J. (1985) *The Palaeolithic sites of East Anglia*. Geobooks, Norwich.

Zagwijn, W.H. (1989) The Netherlands during the Tertiary and Quaternary: a case history of coastal lowland evolution. *Geologie en Mijnbouw*, **68**, 107-20.

Zbinden, H., Andrée, M., Oeschger, H. *et al.* (1989) Atmospheric radiocarbon at the end of the last glacial; an estimate based on AMS radiocarbon dates on terrestrial macrofossils from lake sediments. *Radiocarbon*, **31**, 795-804.

Zeuner, F.E. (1945) *The Pleistocene Period: its climate, chronology and faunal successions* (1st edn). Hutchinson, London.

Zeuner, F.E. (1959) *The Pleistocene Period* (2nd edn). Hutchinson, London.

Zeuner, F.E. (1960) Excavations at the site called 'The Old Grotto', Torbryan. *Report and Transactions of the Devonshire Association for the Advancement of Science, Literature and Art*, **92**, 311-30.

Index

Note: Page numbers in **bold** and *italic* type refer to **figures** and *tables* respectively

Index

Index